Tree Physiology
and
Yield Improvement

A compendium of papers given at a
meeting held near Edinburgh in July 1975

Tree Physiology and Yield Improvement

Edited by

M. G. R. CANNELL and F. T. LAST

*Institute of Terrestrial Ecology, Bush Estate,
Penicuik, Midlothian EH26 0QB, Scotland*

1976
ACADEMIC PRESS
London   New York   San Francisco
A Subsidiary of Harcourt Brace Jovanovich, Publishers

ACADEMIC PRESS INC. (LONDON) LTD.
24/28 Oval Road,
London NW1

United States Edition published by
ACADEMIC PRESS INC.
111 Fifth Avenue
New York, New York 10003

Copyright © 1976 by
ACADEMIC PRESS INC. (LONDON) LTD.

All Rights Reserved
No part of this book may be reproduced in any form by photostat, microfilm, or any other means, without written permission from the publishers

Library of Congress Catalog Card Number: 76 1073
ISBN: 0 12 158750 9

TYPESET IN GREAT BRITAIN BY SANTYPE INTERNATIONAL LTD.
(COLDTYPE DIVISION), SALISBURY, WILTS
PRINTED IN GREAT BRITAIN BY WHITSTABLE LITHO

Participants and Contributors

ARMSTRONG, W. *Department of Plant Biology, University of Hull, HU6 7RX, England.*

BARADAT, P. *Laboratoire d'amélioration des Conifers, Domaine de l'Hermitage, Pierroton, 33610 Cestas Principal, France.*

BECK, R. C. *School of Forestry and Wildlife Management, Louisiana State University, Baton Rouge, Louisiana 70803, U.S.A.*

BILAN, M. V. *Stephen F. Austin State University, Nocogdoches, Texas, U.S.A.*

BRUNIG, E. F. *Institute of World Forestry, University of Hamburg, Leuschnerstrasse 91, 2050 Hamburg 80.*

BUIJTENEN, J. P. van *Forest Genetics Laboratory, Texas A & M University College Station, Texas 77843, U.S.A.*

BUNTING, A. H. *Agricultural Development Overseas, Plant Science Laboratories, University of Reading, Whiteknights, Reading RG6 2AS, England.*

BURDON, R. D. *Forest Research Institute, Private Bag, Rotorua, New Zealand.*

BURLEY, J. *Commonwealth Forestry Institute, Oxford University, South Parks Road, Oxford OX1 3RB, England.*

CANNELL, M. G. R. *Institute of Terrestrial Ecology, Bush Estate, Penicuik, Midlothian EH26 0QB, Scotland.*

COOPER, J. P. *Welsh Plant Breeding Station, Plas Gogerddan, Aberystwyth WY23 3EB, Wales.*

COUTTS, M. P. *Forestry Commission Northern Research Station, Roslin, Midlothian EH25 9SY, Scotland.*

CRAWFORD, R. M. M. *Department of Botany, The University, St. Andrews, KY16 9AL Scotland.*

DAVIES, W. J. *Department of Botany, Duke University, Durham, North Carolina 27706, U.S.A.*

DENNE, M. P. *Department of Forestry and Wood Science, University College of North Wales, Bangor LL57 2UW, Wales.*

DORMLING, I. *Department of Forest Ecophysiology, The Phytotron, Royal College of Forestry, S-104 05, Stockholm 50, Sweden.*

EKBERG, I. *Department of Forest Genetics, Royal College of Forestry, S-104 05, Stockholm 50, Sweden.*

ERIKSSON, G. *Department of Forest Genetics, Royal College of Forestry, S-104 05, Stockholm 50, Sweden.*

FARMER, R. E., Jnr. *Division of Forestry, Fisheries and Wildlife Development, Tennessee Valley Authority, Norris, Tennessee 37828, U.S.A.*

FAULKNER, R. *Forestry Commission Northern Research Station, Roslin, Midlothian EH25 9SY, Scotland.*

FLETCHER, A. M. *Forestry Commission Northern Research Station, Roslin, Midlothian EH25 9SY, Scotland.*

FORD, E. D. *Institute of Terrestrial Ecology, Bush Estate, Penicuik, Midlothian EH26 0QB, Scotland.*

GIERTYCH, M. *Instytut Dendrologii, Polskiej Akademii Nauk, 63—120 Kórnik, ul. Parkowa 5, Poland.*

GLERUM, C. *Ontario Ministry of Natural Resources, Division of Forests, Forest Research Branch, Maple, Ontario LO5 1EO, Canada.*

GODDARD, R. E. *School of Forest Resources and Conservation, University of Florida, 305 Rolfs Hall, Gainesville, Florida 32611, U.S.A.*

GOODMAN, P. J. *Welsh Plant Breeding Station, Plas Gogerddan, Aberystwyth SY23 3EB, Wales.*

GORDON, J. C. *Department of Forestry, University of Iowa, 251 Bessey Hall, Ames, Iowa 50010, U.S.A.*

HAGNER, M. *Department of Forestry, University of Umea, 90235, Umea, Sweden.*

HALL, S. M. *Institute of Terrestrial Ecology, Bush Estate, Penicuik, Midlothian, EH26 0QB, Scotland.*

HELMS, J. A. *College of Natural Resources, University of California, Berkeley, California 94720, U.S.A.*

HOLLIS, C. A. *School of Forest Resources and Conservation, University of Florida, 305 Rolfs Hall, Gainesville, Florida 32611, U.S.A.*

JANKIEWICZ, L. S.   *Vegetable Crops Research Institute, 22 Lipca 1/3, 96–100 Skierniewice, Poland.*

JANSEN, E. C.   *Dorschkamp Research Institute for Forestry and Landscape Planning, Bosrandweg 20, Postbox 23, Wageningen, The Netherlands.*

JARVIS, P. G.   *Department of Forestry and Natural Resources, University of Edinburgh, King's Buildings, Mayfield Road, Edinburgh EH9 3JU, Scotland.*

JOHNSTONE, R. C. B.   *Forestry Commission Northern Research Station, Roslin, Midlothian EH25 9SY, Scotland.*

JUNTILLA, O.   *Institute of Biology and Geology, University of Tromsö, P.O. Box 790, N-9001 Tromsö, Norway.*

KLEINSCHMIT, J.   *Niedersächsische Forstliche Versuchsanstalt, Abteilung Forstpflanzenzüchtung, 3511 Escherode, F.R. Germany.*

KOZLOWSKI, T. T.   *Department of Forestry, 126 Russell Laboratories, University of Wisconsin, Madison, Wisconsin 53706, U.S.A.*

LANNER, R. M.   *Department of Forestry and Outdoor Recreation, College of Natural Resources, Utah State University, Logan, Utah 84322, U.S.A.*

LARSON, P. R.   *Institute of Forest Genetics, P.O. Box 898, Rhinelander, Wisconsin 54501, U.S.A.*

LAST, F. T.   *Institute of Terrestrial Ecology, Bush Estate, Penicuik, Midlothian EH26 0QB, Scotland.*

LEDIG, F. T.   *School of Forestry and Environmental Studies, Greeley Memorial Laboratory, Yale University, New Haven, Connecticut 06511, U.S.A.*

LINES, R.   *Forestry Commission Northern Research Station, Roslin, Midlothian EH25 9SY, Scotland.*

LITTLE, C. H. A.   *Canadian Forest Service, Maritime Forest Research Centre, P.O. Box 4000, Fredericton, New Brunswick E3B 544, Canada.*

LOGAN, K. T.   *Petawawa Forest Experiment Station, Canadian Forest Service, Chalk River, Ontario K0J 1J0, Canada.*

LONGMAN, K. A. L.   *Institute of Terrestrial Ecology, Bush Estate, Penicuik, Midlothian EH26 0QB, Scotland.*

LUUKKANEN, O.   *Department of Silviculture, University of Helsinki, Unioninkatu 40B, SF-00170, Helsinki 17, Finland.*

MacGREGOR, M.   *Department of Botany, University of Toronto, Toronto M5S 1A1, Canada.*
MALCOLM, D. C.   *Department of Forestry and Natural Resources, University of Edinburgh, King's Buildings, Mayfield Road, Edinburgh EH9 3JU, Scotland.*
MASON, P. A.   *Institute of Terrestrial Ecology, Bush Estate, Penicuik, Midlothian EH26 0QB, Scotland.*
MATTHEWS, J. D.   *Department of Forestry, University of Aberdeen, St. Machar Drive, Aberdeen AB9 2UU, Scotland.*
PELHAM, J.   *Institute of Terrestrial Ecology, Bush Estate, Penicuik, Midlothian EH26 0QB, Scotland.*
PERRY, T. O.   *School of Forest Resources, Department of Forestry, North Carolina State University, Box 5488 Raleigh, North Carolina 27607, U.S.A.*
PHARIS, R. P.   *Department of Biology, University of Calgary, 2920 24 Ave. N.W. Calgary, Alberta T2N 1N4, Canada.*
POLLARD, D. F. W.   *Petawawa Forest Experiment Station, Canadian Forest Service, Chalk River, Ontario K0J 1J0, Canada.*
PROMNITZ, L. C.   *Department of Forestry, University of Iowa, 251 Bessey Hall, Ames, Iowa 50010, U.S.A*
SAUER, A.   *Niedersächsische Forstliche Versuchsanstalt, Abteilung Forstpflanzenzüchtung, 3511 Escherode, F.R. Germany.*
SCHOLZ, F.   *Bundesforschungsanstalt für Forst-und Holzwirtschaft, Institut für Forstgenetik und Forstpflanzenzüchtung, D.207 Schmalenbeck, F.R. Germany.*
STECKI, Z. J.   *Polska Akademia Nauk, Zaktad Dendrologii i Arboretum Kórnickie, 63–120 Kórnik, Poland.*
SWEET, G. B.   *Forest Research Institute, Private Bag, Rotorua, New Zealand.*
THIELGES, B. A.   *(New address) U.S. Forest Service, Southern Forest Experiment Station, P.O. Box 906, Starkville, Mississippi 39759, U.S.A.*
THOMPSON, S.   *Department of Forestry, University of Aberdeen, St. Machar Drive, Aberdeen AB9 2UU, Scotland.*
TIMMIS, R.   *Forest Research Center, Weyerhauser Company, 505 North Pearl Street, Centralia, Washington 98531 U.S.A.*
TUBBS, F. R.   *John Innes Institute, Colney Lane, Norwich NR4 7UH, England.*

TYREE, M. T.   Department of Botany, University of Toronto, Toronto M5S 1A1, Canada.
WAREING, P. F.   Department of Botany and Microbiology, University College of Wales, Penglais, Aberystwyth SY23 3DA, Wales.
WELLENDORF, H.   Arboretum, Royal Veterinary and Agricultural University, Copenhagen, 2970 Hørsholm, Denmark.
WETTSTEIN D. von,   Institute of Genetics, University of Copenhagen, Denmark.
WILLETT, S. C.   Institute of Terrestrial Ecology, Bush Estate, Penicuik, Midlothian EH26 0QB, Scotland.
WORRALL, J.   Faculty of Forestry, University of British Columbia, Vancouver 8, Canada.
ŻELAWSKI, W.   Institute of Silviculture, Warsaw Agricultural University, 02-528-Warsawa, uli Rakowlecka 26/30, Poland
ZIMMERMAN, R. H.   Plant Genetics and Germplasm Institute, Beltsville, Maryland, U.S.A.
ZOBEL, B. J.   School of Forest Resources, North Carolina State University, Box 5488 Raleigh, North Carolina 27607, U.S.A.

Preface

These are the proceedings of a conference on "Physiological Genetics of Forest Tree Yield" held at Middleton Hall Conference Centre, Gorebridge, Scotland, from 13th to 21st July 1975, under the auspices of the International Union of Forest Research Organisations (working party S2.01.4, chairman Dr M. Giertych, Poland). The objectives were, first, to examine physiological and morphological characteristics which limit wood yield and underlie inherent differences in forest tree yield, and secondly, to consider, where possible, the heritability of those characters and the ways in which they might be exploited by breeding.

To limit the field of interest, and allow discussion in depth, the conference was restricted to wood yield as the breeding objective. It was not concerned with the physiology of flowering, seed germination, vegetative propagation, wood quality or factors conferring resistance to pests and pathogens. These are all relatively well reviewed areas of research in which silviculturists and tree improvers are already engaged together, and it was doubtful whether a conference was needed to stimulate further exchanges of information. On the other hand, the physiological basis of wood yield has tended to be the preserve of physiologists with, perhaps, too little opportunity for feedback to people directly concerned with wood production.

Agronomists have realized for several decades that genetic gains might be made more rapidly or predictably if it were possible to identify desirable physiological attributes among parents and their progenies. Rice and wheat provide examples where breeders have consciously selected for particular critical photoperiodic responses and canopy architectures and achieved large gains in yield. There are, of course, numerous other examples where crop breeders have selected for desirable attributes unconsciously by selecting for final yield itself. However, tree breeders cannot do this. Normally they have selected big elite trees within heterogeneous stands and evaluated their progenies by measuring height and diameter growth well before the economic yield can be evaluated. The "ideotype" for wood yield has been the big fast-growing individual. This may prove satisfactory while large gains are being made in tree form, pest and pathogen resistance and silvicultural characteristics, but it may not be the best method of increasing wood yield within advanced, well adapted breeding populations. Experience gained on field crops suggests that some crosses can usefully be planned on the basis of potential physiological complementation — selecting parents with superior phenotypes for different desirable components of yield — and, furthermore, that progenies should be selected with some understanding of the physiology associated with rapid rates of dry matter production per hectare in stands. With this in mind, we called this conference of tree physiologists and tree breeders.

Three features became evident when planning the conference. First, that it would have to be multi-disciplinary and draw on the expertise of specialist physiologists in many fields as well as on the observations and viewpoints of practising tree breeders. Secondly, so few heritability studies had yet been done on physiological attributes of trees that the main subjects for discussion would be comparative studies on species, provenances or clones. The emphasis of the meeting would be therefore on physiology rather than genetics. Thirdly, because all components of yield vary genetically, we could focus on only those major aspects which seemed most relevant to tree yield improvement. Clearly, it was beyond our expectation to construct selection indices or ideotypes. We shall be rewarded if, as a result of the conference, more tree physiologists interest themselves in the large, poorly exploited pools of genetic variation that exist within forest tree species, and, conversely, if more breeders begin to consider the component capabilities of their parent trees and progenies.

When considering the subjects for discussion we began with the assumption that any major increase in wood yield implies an increase in annual dry-matter production per hectare. It is impossible to separate assimilate "sources" and "sinks" in forest trees as clearly as in many field crops; the cambium is an integral part of the structure and not a storage organ. Thus, there are several introductory chapters dealing with photosynthesis and processes affecting the efficiency with which trees accumulate dry matter. Following this, it was logical to review observations on the performance of trees with differing canopy characteristics, and consider factors underlying widely reported variation in the activity and development of the shoot and cambial sinks. This, in turn, led one to consider in more detail the adaptive physiological attributes that are so important for forest crops to survive continuous or occasional inhospitable conditions such as drought, waterlogging, frost and nutrient deficiencies. Evidently, many problems and pitfalls would be encountered by breeders and physiologists when seeking to interpret and apply physiological information obtained on individual parent trees, or young progenies, and some of the major ones are reviewed in the final chapters. At the end there is a brief account of the major comments and suggestions recorded by discussion groups during the conference.

We are grateful to the Royal Society and the British Council for providing funds for overseas participants. It is also a pleasure to thank Mr R. Faulkner, staff of the Forestry Commission Northern Research Station, Professor J. D. Matthews, Dr M. Giertych, Professor A. H. Bunting, Professor J. P. Cooper and staff of the Welsh Plant Breeding Station, for contributing to the success of the conference. We also wish to acknowledge the help of our fellow organizers, Mrs Yvonne Foubister, Mr Stephen Willett, Dr Shelagh Hall and Mrs Riet Cannell.

July, 1976
Melvin G. R. Cannell
Fred T. Last

Contents

List of Participants	v
Preface	xi
1. Inaugural Lecture: Maximizing the product, or how to have it both ways. A. H. BUNTING	1

Carbon Fixation Efficiency

2. Physiological genetics, photosynthesis and growth models. F. T. LEDIG	21
3. Factors influencing net photosynthesis in trees: an ecological viewpoint. J. A. HELMS	55
4. Photosynthetic and enzymatic criteria for the early selection of fast-growing *Populus* clones. J. C. GORDON and L. C. PROMNITZ	79
5. Variation in the photosynthetic capacity of *Pinus sylvestris*. W. ZELAWSKI	99
6. Relationship between the CO_2 compensation point and carbon fixation efficiency in trees. O. LUUKKANEN	111

Shoot and Cambial Growth

7. Relationships between genetic differences in yield of deciduous tree species and variation in canopy size, structure and duration. R. E. FARMER . . .	119
8. Tree forms in relation to environmental conditions: an ecological viewpoint. E. F. BRUNIG	139
9. Some mechanisms responsible for differences in tree form. L. S. JANKIEWICZ and Z. J. STECKI	157
10. An analysis of inherent differences in shoot growth within some north temperate conifers. M. G. R. CANNELL, S. THOMPSON and R. LINES	173

11. Inheritance of the photoperiodic response in forest trees. I. EKBERG, I. DORMLING, G. ERIKSSON and D. VON WETTSTEIN 207
12. Patterns of shoot development in *Pinus* and their relationship to growth potential. R. M. LANNER . . . 223
13. Inherent variation in "free" growth in relation to numbers of needles produced by provenances of *Picea mariana*. D. F. W. POLLARD and K. T. LOGAN 245
14. Control of bud-break and its inheritance in *Populus deltoides*. B. A. THIELGES and R. C. BECK . . 253
15. The leaf-cambium relation and some prospects for genetic improvement. P. R. LARSON 261
16. Predicting differences in potential wood production from tracheid diameters and leaf cell dimensions of conifer seedlings. M. P. DENNE 283
17. Probable roles of plant hormones in regulating shoot elongation, diameter growth and crown form of coniferous trees. R. P. PHARIS 291

Water Stress and Waterlogging

18. Water relations and tree improvement. T. T. KOZLOWSKI 307
19. Physical parameters of the soil–plant–atmosphere system: breeding for drought resistance characteristics that might improve wood yield. M. T. TYREE . . 329
20. Morpho-physiological characteristics related to drought resistance in *Pinus taeda*. J. P. VAN BUIJTENEN, M. V. BILAN and R. H. ZIMMERMAN 349
21. Role of oxygen transport in the tolerance of trees to waterlogging. M. P. COUTTS and W. ARMSTRONG . 361
22. Tolerance of anoxia and the regulation of glycolysis in tree roots. R. M. M. CRAWFORD 387

Frost Hardiness

23. Frost hardiness of forest trees. C. GLERUM . . 403
24. Methods of screening tree seedlings for cold hardiness. R. TIMMIS 421

Mineral Nutrition

25. Genetic factors affecting the response of trees to mineral nutrients.   P. A. MASON and J. PELHAM . . . 437
26. Response of *Pinus taeda* and *Pinus elliottii* to varied nutrition.   R. E. GODDARD, B. J. ZOBEL and C. A. HOLLIS 449

Problems Concerning the use of Physiological Selection Criteria

27. Competition, genetic systems and improvement of forest yield.   E. D. FORD 463
28. Maternal effects on the early performance of tree progenies.   T. O. PERRY 473
29. Problems of interpreting inherent differences in tree growth shortly after planting.   R. D. BURDON and G. B. SWEET 483
30. Variation in morphology, phenology and nutrient content among *Picea abies* clones and provenances, and its implications for tree improvement.   J. KLEINSCHMIT and A. SAUER 503

Discussion 519

Species Index 533

Subject Index 543

Inaugural Lecture

Maximizing the Product, or How to Have it Both Ways

A. H. BUNTING
Professor of Agricultural Development Overseas, University of Reading, England

I.	Objects and Scope	2
II.	Adaptation	3
	A. Adaptation to Seasonal Régimes	4
	B. Adaptation to Soil Conditions	5
	C. Adaptation to Extreme Aerial Conditions and Unstable Soils	6
	D. Adaptation to Economic Requirements	6
III.	Growth	7
	A. Closing the Canopy and the Nutrient Cycles	7
	B. Accumulation of Yield: the Analysis of Growth	9
	C. Carbon Dioxide Uptake	10
	D. Leaf Area and Canopy Structure	11
IV.	Distribution of the Products of Growth Within Individual Plants	12
	A. Sinks	12
	B. Reserves	13
	C. Distribution Ratios	13
V.	Distribution Between and Within Individuals of the Products of Growth of the Forest as a Whole	14
	A. Variation in Forest Communities	15
	B. Effects of Density on the Physiology of Individuals	16
	C. Maximizing the Products	16
VI.	Quality and Resistance	17
VII.	The International Board for Plant Genetic Resources	17
References		18

I. OBJECTS AND SCOPE

During the past 30 years, physiologists working on short-term crop plants, on pasture grasses, on a number of tree species grown for their fruits, and on *Hevea* rubber, have learnt a good deal about the more detailed heritable morphological and physiological attributes which produce that blunderbuss character, final yield (Blackman, 1965; Bunting, 1971; Ferwerda and Wit, 1969; Bingham, 1971; Wallace *et al.*, 1972; Evans, L. T., 1975). The object of this address is to draw from these experiences some ideas and questions which may be relevant to studies of growth and yield in managed, even-aged forests of single botanical species. It does not consider the physiology of yield in wild, self-propagated mixed forest of many species and diverse age, from which yield is obtained by clear felling or selective culling.

Most of conventional plant physiology is concerned with the study of states and processes in cells, tissues, organs and whole plants; but crop physiology deals with the even more complex level of organization represented by assemblages of plants, in which individuals affect one another within a finite environment in which important natural resources, particularly energy and water, are provided per unit of land area, regardless of the number of individual plants it carries. It seeks to study economic vegetation as a physiological system.

Perhaps physiology's most important contribution to tree improvement is its practice of analysing complex characters like yield into less complex components and subcomponents. The morphology and physiology of these components can then be studied separately, sometimes in young seedlings or in cultures of organs, tissues or cells. How far it will actually be possible to study the inheritance of such components in forest tree species and to use the results in breeding and propagation is, I think, another matter. At the least, to study inheritance we must make a cross and grow the progeny to a sufficient size to assess the characters that interest us. The breeder has to add recombination and selection to the sequence.

Now in breeding maize populations for the humid tropics at Ibadan, we can go through a complete cycle of recombination, selecting and testing every 12 months. In forest trees crosses are becoming easier to make as we learn how to make more and more species flower when they are very young. We can produce some hybrids by fusing haploid cells in culture, and we may before long be able to induce meiosis in cultures of such hybrid cells. It might then be possible to analyse genetically any sub-components that can be identified in cultures or plantlets. But even for species whose economic characteristics can be reliably assessed in as little as 6 or 7 years, selection and testing, at least for complex characters, seems bound to be a tedious business. As to propagation, Green's experiences in Kenya (1973) suggest that we cannot rely unquestioningly on vegetative propagation (see also Burdon and Shelbourne, 1974); and seed production is never straightforward in outbreeding species.

The physiological attributes with which we are specifically concerned in this Conference may be grouped into four categories — concerned with adaptation in general, growth, the distribution of the proceeds of growth and growth in communities. Evidently all these topics overlap but they will, I hope, provide a convenient framework for discussion.

II. ADAPTATION

We are well accustomed to the idea that particular provenances are adapted to particular environmental situations. It is perhaps worth pointing out that the situation in which a provenance grows in the wild is not necessarily particularly well suited for the growth of that provenance: it may mean only that the provenance is better able to endure the limiting conditions, or can use them more effectively, than any other potential occupant. In culture, where management modifies limiting conditions and discourages alternative occupants, the provenance may grow and yield best in some quite different situation. Perhaps the stock examples are coffee, cocoa, and tea, all of which occur wild as small understory trees of tropical woodlands or forests. Given effective protection against insects and moderate dressings of fertilizer they can yield many times more in full light than under the shade which used conventionally to be provided for "ecological" reasons (Cunningham and Burridge, 1960; Murray and Nichols, 1966; Hadfield, 1974). Sugar beet began its career a couple of centuries ago as a wild plant of the seacoasts of Europe; it succeeds as a crop on some of the most fertile soils of East Anglia, though admittedly it retains the marks of its ancestry by being able to use sodium as a nutrient in partial replacement of potassium (Crowther, 1947; Draycott et al., 1970).

Adaptation, then, depends at least in part on attributes which enable a provenance to deal with or endure the limiting or adverse features of a particular environment; and it will be within the pool of variation and variability of an adapted provenance that we may hope to find those more positive attributes which make for the maximum yield that environment can produce.

A. Adaptation to Seasonal Régimes

To survive, let alone to succeed, in a given environment, a provenance must be adapted to the seasonal régimes of temperature and water relations. Much tree breeding seeks to improve adaptations of these kinds, as we shall hear in the later sessions of this Conference. To succeed in a climate where the season favourable to plant growth is limited by extremes of temperature, a provenance must either have an inbuilt timetable of appropriate length, or be able to endure the extremes. Among crop plants, there is a surprisingly wide range of adaptations to extreme temperatures. Many varieties of maize become chlorotic and may die at temperatures below 18°C; others remain green at freezing temperatures. Andean types of maize have been used in the breeding of hybrids for the cool highlands of Kenya. Seedlings of some African varieties of cowpeas studied at the International Institute of Tropical Agriculture, Ibadan, are little affected by soil surface temperatures approaching 45°C, which severely check some American varieties of soya bean. Such differences may reflect differences between the effects of temperature on the rates of assimilation and respiration, of the synthesis and photo-bleaching of chlorophyll (Millerd and McWilliam, 1968; Millerd et al., 1969), of the rates of division and expansion of cells in meristems, internodes and leaves, of the differentiation of reproductive structures, and no doubt many more.

The seasonal régime of water relations is most usefully described in terms of the time course of the difference between the rates of precipitation and evaporation (Bunting, 1961) and of the amounts of water stored in the profile. Between them these features determine the length of the favourable season for growth in seasonally arid regions. The most valuable adaptation to dry conditions is an inbuilt timetable of appropriate length, supported by a pattern of root growth and distribution which allows maximum possible access to the water in the profile. Special structural and biophysical properties of the leaves or other organs may be useful for survival, but they seldom seem to have more than marginal significance for growth and yield in crop plants (see also the papers by Kozlowski and Tyree in this volume, Chapters 18 and 19).

Nevertheless at least two species of trees, the neem tree *Azadirachta indica*, and *Acacia albida*, seem to flourish in conditions so dry that some special adaptation may be involved. *Acacia albida*, which grows in the seasonally arid Sudan and Guinea savannah zones of West Africa, is particularly curious: it passes the rains in a bare and leafless condition, and comes into leaf only when the rains are over and the desert wind has ushered in the dry season (Radwanski and Wickens, 1967). The ecophysiology of the drought-adapted species of *Eucalyptus* also seems to me to deserve special investigation.

The seasonal timetables required for adaptations to temperature and water régime are regulated in plants by limiting temperatures and by changes in photoperiod. Such effects are well enough known, though how they work is usually far less clear. In Curtis' investigations (1968) of indigenous Nigerian cultivars of sorghum, which are very precisely adapted in respect of flowering time to the season-lengths and latitudes in which they are grown, flowering does not appear to be induced by a specific threshold daylength; the simplest hypothesis suggests that these plants can count the number of successively longer nights. In cowpea and soya, there are complex and sometimes compensatory interactions between night temperature and photoperiod. Some varieties are sensitive to both, some to one or the other, some to neither; and these characteristics appear to be heritable (Huxley and Summerfield, 1974).

B. Adaptation to Soil Conditions

Many wild plants appear to be specifically adapted to adverse soil conditions — shortages of particular nutrients or minor elements, seasonal waterlogging, acidity, alkalinity, salinity, presence of toxic elements (e.g. Garcia-Novo and Crawford, 1973). Important variation of this kind is known in rice. Forest vegetation, particularly in the tropics, seems to be particularly efficient at handling nutrient deficiencies, apparently by establishing a closed system within which nutrients are cycled as rapidly as possible (Milne, 1937; Nye and Greenland, 1960). But it may well be that in addition to the mycorrhizal and other relations with more or less endogenous micro-organisms which appear both to increase the rate of cycling of phosphate and to add nitrogen to the ecosystem, some trees provide a rhizosphere environment particularly favourable for free-living nitrogen-fixing organisms (Rambelli, 1973). It seems clear that certain tropical grasses do this, though whether this is peculiarly linked with the Krantz syndrome and the C4 pathway in photosynthesis, as has been suggested, it is too early to say. In relation to salt adaptation, I have never lost my youthful wonder at the *Avicennia* of the Gambian mangrove swamps, which can actually take up salt water and transport it to the surfaces of its leaves, regardless of what people think the textbooks have to say about osmosis.

C. Adaptation to Extreme Aerial Conditions and Unstable Soils

We should include in this catalogue of physiological/morphological adaptations the attributes which enable many trees to withstand high wind and airborne salt, and (like *Prosopis juliflora*, mesquite) to colonize and stabilize unstable, particularly sandy, habitats.

D. Adaptation to Economic Requirements

The last adaptation in my list is required by the human rather than the natural environment — adaptation to the economic time frame. A successful forest crop species must deliver the economic goods in an appropriate time, and to do this it will require not only good management but also inherent physiological and structural attributes, which bring it to a maximum rate of increase of stem volume, or to a sufficient level of yield of the required quality, at an appropriate time.

III. GROWTH

The growth of the sort of forest crop I am considering is largely vegetative. The trees may flower and fruit, but the forester's main purpose is to produce as much dry matter as he can per hectare in the trunks and perhaps the larger branches. The nearest example the crop physiologist can offer to this is sugar cane, with sugar beet and some other root crops as runners up.

A. Closing the Canopy and the Nutrient Cycles

The conventional man-made forest, however, differs from them in one very important respect: the period before it closes canopy may be several years rather than several weeks. During this extended period heat and rain can affect and perhaps damage the surface of the soil, and the rain can leach nutrients and bases from the profile. Many other plants are present which (except in so far as management can afford to hold them back) will also draw on the resources of the environment, particularly plant nutrients, and in some circumstances water.

In wild forests after clearing, the growth of weeds — including rapidly extending trees like *Musanga*, the aptly named *parasolier* (parasol tree) of the forests of Zaire and other humid parts of West Africa (Coombe and Hadfield, 1962) — soon establishes a sufficiently continuous leaf cover, and a large enough quantity of root surface in the profile, to protect the soil from damage, restore the evaporative component of the hydrological cycle, and close the nutrient cycles. In this way the essential features of forest are re-established. Over time the taller-growing species begin to shade the shorter ones and a developing forest structure emerges. Whether it will resemble the "primary" forest, which the inward eye of faith so clearly sees but the ecologist so seldom finds, is another matter. In the English Breckland, an area of prehistoric industrial dereliction, the environmental disaster of heavy podsolization, associated with unimproved traditional agricultural technology, has prevented to this day the return of the primeval birch forest. Similarly, it seems to me, the original forest, which is now represented only by the "emergents", may never again be re-established naturally in the central basin of

Zaire, because the leaching and other processes associated with clearing and agriculture have removed phosphate, bases, and other metallic elements from the deep profile of windblown Kalahari sand, which weathering cannot replace because there is no accessible parent rock to be weathered.

The forester, however, cannot look to plant succession to establish his crop. His budget seldom permits much control of weeds, and indeed, he must often consciously or unconsciously use them, to some extent at least, to conserve nutrients and perhaps add some nitrogen to the developing ecosystem.

Traditional farmers in tropical forest country have resolved a comparable problem by mixed cropping of various kinds. In the Bamliki country of Western Cameroon, a principal crop such as cassava, cocoyam or coffee is preceded or accompanied by one or more early-planted short-season crops, of varying stature and including a legume, which rapidly cover the ground, tend to offset leaching, and produce a crop or crops before the principal crop has produced so much leaf that it needs the whole area of ground to itself. Some of the outstanding performers are cucurbits, which make more leaf area per unit of dry matter accumulated than any other plants known to me. Professor Roche, at Ibadan, has proposed such a system of agri-silviculture, using *Gmelina*, from the foresters' side. In current surveys in Eastern Nigeria scientists of the International Institute of Tropical Agriculture, also at Ibadan, have found that the trees which farmers who already employ such a system plant or conserve, in increasing numbers per hectare as their farming becomes more intensive, are all sources of food, or of textile and other raw materials, rather than of timber. A system of this sort has been devised for poplar plantations on fertile agricultural land in Britain, and cattle are grazed in pine plantations in New Zealand, but on much of the land devoted to forestry in temperate regions the investment in equipment and other facilities for producing crops is not often likely to be economically justifiable.

Consequently many foresters have to depend on the trees themselves to cover the ground, grow tall and shade unwanted species as quickly as possible. For this, they may seek vigorous provenances which branch profusely and extend rapidly in height, to shade other plants and establish the framework of stem on which yield will later be accumulated. In this, I believe, may lie the roots of a morphological/physiological dilemma. The characteristics required for rapid establishment and dimensional growth, and consequent control of unwanted species, are *par excellence* those of secondary pioneers: many of these, when fully grown, lack the mechanical strength, the energy density or the chemical or aesthetic characteristics necessary if they are to be as valuable, area for area or volume for volume, as many species which grow more slowly (in the sense of producing fewer nodes or leaf bud units per unit of time, or having characteristically shorter internodes), branch less freely, and live longer. However, species like *Chlorophora excelsa*, which I know as a plant of secondary successions in southern Tanzania, some of the less soft softwoods, and seemingly some acacias and eucalypts, appear to be able to have it both ways. Maybe by grafting, we can produce compound trees which can also have it both ways — as the pomologists have done with top fruits for many years and as *Hevea* agronomists have done experimentally (Blackman, 1965).

B. Accumulation of Yield: the Analysis of Growth

Once the forest canopy has closed and the geometry of the stems and branches is established, the forester is interested, as the agronomist is, in the largest possible rate of increase in dry weight per unit of land area and of time. Up to 1947, the model used to analyse this process represented the relative growth rate of a plant, $1/w \cdot dw/dt$, as the product of the ratio of leaf area to plant weight, l/w, a size factor, and that blunderbuss intensity factor, the unit leaf rate or net assimilation rate $1/l \cdot dw/dt$. Since, in many plants, the proportion of respiring tissue tends to increase as the plant ages, while the leaf area tends to become constant as older leaves die, net assimilation rate tends to fall with age in the so-called ontogenetic drift. Moreover, as measured in the field by many experimenters, net assimilation rate inevitably measures losses by decay as well as losses by internal respiration. So in an old pasture, in early summer, net assimilation rate may seem very small even though the pasture is growing rapidly, because the dry matter of old leaves and stems is being lost by decay. Measurements of growth rate in stable forests indicate average values of crude net assimilation rate (including decomposition) approximating to zero (see discussion in Evans, G. C., 1972, Chapter 31).

The model of growth analysis which we now use to study growth in the crop as a whole was introduced by Watson in 1947. It relates the crop growth rate per unit area of land, $dw/dt \cdot 1/a$, to the net assimilation rate $1/l \cdot dw/dt$ and the leaf area index l/a. Once again, this is no more than a simple set of crude definitions, a sort of axiomatic identity, but it has proved of substantial value in analysing differences in yield between varieties, and the effects of weather and climate on crop growth. Like our colleagues in forestry, we find, on the whole, that leaf area index and its duration are generally more important than net assimilation rate as determinants of growth rate and economic yield respectively (see Farmer, Chapter 7).

We are not, however, generally dogged by the very large endogenous respiration losses with which foresters have to contend in older plantations. Maybe a crop like sugar cane has some way of switching off respiration as its internodes mature.

C. Carbon Dioxide Uptake

Faced with these difficulties, both crop physiologists and foresters have turned to direct measurements of net rates of carbon dioxide uptake by leaves. These rates are still net, but they are net over a much smaller bulk of tissue, whose rate of respiration in the dark can be measured with little difficulty, and in the light with rather more. In my view, surprisingly little of value in crop improvement has so far come from this work. For over 10 years we have been impressed by the differences in assimilatory arrangements between the so-called C3 and C4 plants. Though some C4 plants, which include sugar cane, maize and bulrush millet, have set up world records for rate of accumulation of biomass, the advantage in initial CO_2 uptake rates becomes more and more diluted as the carbon makes its way along the tortuous road which ends up in the sack in the market place (Gifford, 1974). Nevertheless, it would be good to know whether in the formally somewhat simpler situation of the forest, any of the outstanding accumulators of biomass are able to use the C4-dicarboxylic acid entry as the first step in assimilation.

D. Leaf Area and Canopy Structure

Since leaf area is so important for yield, we should, I suggest, be concerned with comparative studies of the morphology of expansion of the leaf surface in forests — how many apical meristems are produced where and when, how frequently do they produce leaves and how rapidly do the leaves expand? What are the effects on these numbers and processes of variations in climate and weather — particularly of variations in temperature and water supply? Coupled with this there will no doubt be interest in the effects of these variations on the behaviour of the stomata and therefore on water, gas and energy exchanges.

So far I have spoken as if all is leaf that is green. One of the difficulties of growth analysis in field crops is that it has to average net assimilation rate, the intensity factor, over all green surfaces. Now it was discovered 20 years ago that there is a reciprocal relation between leaf area index L and net assimilation rate E, apparently because if there are more leaves more of them are shaded, so that the average rate of assimilation per unit area is lessened. It was also surmised, and has since been verified, that the size of the product of E and L depends on posture and the general structure of the foliage system or canopy: with inclined or curved leaves in an open structure in bright light, a larger proportion of leaf area can be productively illuminated than in a structure in which leaves are horizontally extended and arrayed more closely together, so that only the topmost leaves are productively illuminated. In other words, though E and L are reciprocally related, plants and plant communities have ways of increasing the product — and indeed this is really an example of the art of our business, from which the title of this address is derived. Nonetheless, though we have bred varieties of maize, rice and other crops which have what appear to be more effective sorts of radiation-capturing geometries, we have always improved other attributes at the same time; and I doubt whether geometry by itself has often added more than 15 or 20% to the yield. Moreover many of our most important plants have it all ways round by having curved leaves — maybe *Pinus patula* is the outstanding example.

Where an open canopy allows more light to reach the ground it may perhaps diminish the heat load on its leaves and the amount of water it has to find to balance its energy income; but the price of this will be the growth of other plants. Leaves, in addition to all their other functions, are organs of aggression, as Snoad and Davies (1972) found when they grew a leafless green pea, which assimilated and yielded at least as well as a normal one, but ran up large bills for weed control. We have encountered the same problem with the open-canopy cottons of the "okra-leaf" and "super-okra" types.

There is however another element in the $E \times L$ complex. More L may mean that individual leaves are surviving longer. In several species which have been examined, the leaves not only assimilate less rapidly as they become more and more deeply shaded, they also respond less to brighter light. Greater L may therefore include a larger proportion of older leaves which are not only shaded but would be poor users of bright light anyway. Perhaps in selecting for apparently more effective canopy structures, we should also select for leaves which can take advantage of a better light climate for a longer time and so gain on both roundabouts and swings.

IV. DISTRIBUTION OF THE PRODUCTS OF GROWTH WITHIN INDIVIDUAL PLANTS

The different ecological or economic roles to which plant species, forest provenances, tree crops or annual crop plants are adapted, or for which they are managed, are performed by inbuilt programmes of growth and development, supported by associated programmes for the distribution of the products of assimilation. Apple trees, rubber trees, tanbark wattles, coffee trees, pine trees, baobabs and chestnut coppice are all members of tree species, yet the programmes they exhibit include very different patterns of distribution of the products of assimilation.

A. Sinks

In annual crop plants, the important and substantial increases in yield of the past 30 or 40 years in many temperate and some tropical crops have of course rested on a large number of physiological and morphological bases. Modern varieties make more effective use of time, and they resist or tolerate pests and diseases better than their predecessors. They respond more readily to fertilizer, irrigation and other tools of improved management, which increase the number of shoots that survive, the rate at which leaf area expands, and the total leaf area duration. But perhaps more important than all these are the large increases in the potential size of the yield-forming sinks — particularly fruits, storage organs and stems, in dense field populations of plants and shoots. These improvements have increased the potential sizes of both the sources and the sinks, but this would not have been enough without changes in the patterns of distribution of dry matter. The result has been to increase the product of the numbers and sizes of sinks, which seemed, 50 years ago, to set a limit, long since exceeded by the breeders, to the potential contribution of physiology to cereal breeding.

B. Reserves

Over the same period crop physiologists have moved away from the insistence on the role of reserves in perennial plants that was so prominent in the twenties and thirties. Valuable as it would be to crop plants if they could accumulate surplus carbohydrate early in life and then use it later on to fill yield organs, it is now clear that even in grasses virtually all yield is produced on current account. There are of course some important exceptions. Plants are not very good at storing nitrogen, and in most plants nitrogen taken up early in life, or in the season, is transferred later to younger parts, including new leaves and growing fruits. This nitrogen can move only on carbon skeletons, and so some dry matter has to be transferred along with the nitrogen. Reserve carbohydrate and nitrogen evidently help to start up the new season's growth in many perennial plants; and this is perhaps a special case of the transfer of previously accumulated material to new growth after what it is nowadays fashionable to call periods of stress.

C. Distribution Ratios

These ideas are contained in the concept of distribution ratios, such as the leaf to total growth ratio (LTGR) developed by Jackson

(1963, see also Rees, 1963) to describe the seasonal course of the distribution of newly accumulated dry matter in cotton. It represents the proportion of total dry matter, accumulated in a particular growth interval, which is laid down in new leaves (sources): it is in effect a re-investment ratio in new productive equipment. *Musanga cecropoides* (Coombe and Hadfield, 1962), *Trema guineensis* (Coombe, 1960) and *Coffea arabica* (Cannell, 1971) grow faster than most forest trees when they are young because their reinvestment ratios are large. In a strict annual, LTGR rapidly falls to zero; in a longer lived form it falls more gradually. This sort of idea could perhaps be used in forestry to compare provenances in respect of the seasonal course of distribution of dry matter between extension growth of the main stem and growth of branches and leaves, on the one hand, and growth of older parts of the stem on the other, at different ages and stand densities. If this were feasible, distribution ratios of suitable kinds might be useful indices for selection. They might help particularly in the search for forms that are better than others at turning, at an appropriate time, from the linear expansionist phase which is important for the establishment and early life of a forest, to the later phase of consolidation or "cambial growth" (when we would like the main axis internodes to expand more slowly and most of the new dry matter to go to the older parts of the trunk).

Differences between sorghum and maize in the distribution of new dry matter between root and shoot seem to be evident in seedlings, and we have wondered whether this approach might not be useful in selecting maize for dry areas. Though this seems a long shot, it may well be possible to find some reflection of the courses of mature distribution patterns in the behaviour of young trees grown at various densities.

V. DISTRIBUTION BETWEEN AND WITHIN INDIVIDUALS OF THE PRODUCTS OF GROWTH OF THE FOREST AS A WHOLE

We come now to what I think is the nub of the matter, and at which I have hinted several times already. It is evidently not difficult to identify a considerable number of desirable physiological attributes in individual plants: the important thing is that these attributes should form part of a productive pattern of growth not only within the individuals but also in a community of many trees, which have to share the resources of the environment amongst them, and consequently affect each other's growth and form. How they do this cannot readily be predicted from observations on isolated trees, let alone individual plant organs, tissues or cells.

A. Variation in Forest Communities

It is evident that in a single-species stand, important variation develops between individuals as the forest grows. Much of this may well be due to random variation in the specific environment, and perhaps particularly in the soil, to which individuals have access. But some of it may have a partly heritable physiological basis, in differences in seed size, in the number of leaf and branch initials present in the seed, in the rate of expansion of the main axis, in the rate of initiation and expansion of new leaf and bud primordia and thus of the branch and leaf systems, in the earlier stages. All of this may enable different individuals to make different effective claims on environmental resources, including light, nutrients and water. Consequently it leads to the death of some plants and, among those which survive, to a range of plant sizes which, as Ford (1975) has shown, may be bimodal in character, so that the survivors in the population consist of two nations, of dominant and subordinate individuals. It is particularly interesting that the larger individuals are not necessarily clumped, nor randomly distributed: they tend to be more evenly distributed — perhaps because light, the most critical environmental resource of all for individual success in a community of plants, is evenly distributed (Ford, Chapter 27), and plants obey a sort of Parkinson's law by expanding to fill the space available.

It would seem important to minimize, as far as possible, by selection and breeding as well as by management (for example by planting in wider rows), the identifiable sources of variability. It is presumably an important objective of forest thinning operations to avoid waste by anticipating the death of suppressed individuals.

B. Effects of Density on the Physiology of Individuals

In many field crops, the effects of individual plants on one another reinforce the patterns of distribution of dry matter and other materials within individuals. As density increases in improved forms of cereals and sugar cane, fewer tillers (branches) are formed, but most of those which are formed come to maturity, so that the product of shoot number and yield per shoot is at least maintained: in the most successful varieties it is increased. But in an amply fertilized cereal crop it is actually possible to decrease the product by increasing density, because tillers die late in the life of the crop, so that the radiation and nutrients used to form them are largely wasted. In somewhat the same way, so many pods may be lost in broad bean crops, as density increases, that yield declines.

C. Maximizing the Products

To maximize the product of numbers of individuals or shoots and yield per individual or shoot in an assemblage of plants requires an appropriate balance of many potentially conflicting components. Except perhaps in extreme environments, there are no single-factor solutions, nor does it seem possible, in the present stage of our knowledge, to conceive, by rational process, the most productive combination of form and function, which is known in agronomy as the plant type or ideotype. At this stage it seems most useful to proceed by analysing the reasons for the success of those tree cultures which produce large amounts of timber per hectare in a relatively short time — perhaps poplars, eucalypts, the Australian acacias, some of the *Julbernardia–Isoberlinia* formations, *Chlorophora excelsa* and maybe other species of secondary or disturbed situations. The grotesque baobab (*Adansonia digitata*) may provoke some ideas: in it the apical meristem of the main shoots appears to abort, leaving the leaf system to be maintained by a "whorl" of seemingly long-lived branches. After this it seems that all subsequent growth is confined to the thickening of the trunk, with no further increase in height.

Though we may become able to define at least some sorts of forest ideotypes, the genetics of the factors which interact to produce larger yield will not be at all easy to study. It has proved difficult enough to do this in annual crops, and will evidently be even more difficult in forest trees. In so far as it is possible to advance by selection, perhaps the most useful idea I can offer is that selection should be aimed not so much at "competitiveness", whatever that may mean, as at its opposite, "commensalism" (see Ford, Chapter 27). So important do I believe this distinction to be that I have, up to this point, completely avoided the use of the word "competition", or any derivative of it, in this address.

VI. QUALITY AND RESISTANCE

So far, I have spoken as if the whole purpose of forestry is to produce dry matter in the right place. But we want this dry matter for a purpose — as an energy-dense fuel, as a structurally strong building or engineering material, as cellulose for pulp and paper, as an aesthetically pleasing and workable material for joinery and ornamental woodwork, or as plain board for containers, or for some special chemical property. Moreover, we would like our trees to resist pests and diseases of all sorts. All these properties have physiological bases, and they are evidently heritable. So, though we do not plan to discuss them this week, we shall not forget, at the end of the day, that these attributes too must form part of the physiological assemblages we hope to construct for the man-made forests of the future.

VII. THE INTERNATIONAL BOARD FOR PLANT GENETIC RESOURCES

For many years new and improved crop varieties have displaced their older predecessors. The world-wide advance of agricultural technology has now proceeded so far and extended so widely that it has become essential to collect and conserve the older varieties in order to ensure that their heritable characteristics remain available for use by plant breeders and students of the evolution of crop plants. The Food and Agriculture Organisation of the United Nations has been concerned with these questions for a number of years. It established a Crop Ecology and Genetic Resources Unit, advised by two Panels of Experts, on crop plant resources and forest resources, to develop international action on them. In 1974, in the wake of the Stockholm Conference, the Consultative Group for International Agricultural Research set up the International Board for Plant Genetic Resources, to carry the work forward in collaboration with FAO (which provides the secretariat of the Board) on the one hand and the world community of crop and forest scientists on the other.

Of the 15 members of the Board one, M. P. Bouvarel, of the French Institut National de Recherche Agronomique, is a forest geneticist; and the Board will allocate its first grants in the field of forest genetic resources later this year. These grants are in part for certain *in situ* conservation measures and in part for exploration work in some tropical hardwood species.

As valuable and heritable physiological attributes of the kind we are to discuss are identified, international means now exist by which action can be planned and financed to collect and conserve them. Much of the Board's action will be directed to strengthening the work of appropriate national, regional and international agencies, like IUFRO, and to ensuring as far as possible that the potential users, world-wide, are involved in what is done. On behalf of the Board, I wish to invite your Union and its members to interest yourselves in our work and to help us to develop it so that it becomes as useful as possible to you.

REFERENCES

Bingham, J. (1971). Physiological objectives in breeding for grain yield in wheat. *Proceedings of 6th Eucarpia Congress, Cambridge*, 15–29.

Blackman, G. E. (1965). Factors affecting the production of latex. In "Proceedings of the Natural Rubber Producers' Research Association Jubilee Conference, Cambridge, 1964", (L. Mullins, ed.). Maclaren and Sons Ltd., London.

Bunting, A. H. (1961). Some problems of agricultural climatology in tropical Africa. *Geography* **46**, 283–294.

Bunting, A. H. (1971). Productivity and profit, or Is your vegetative phase really necessary? *Ann. appl. Biol.* **67**, 265–272.

Burdon, R. D. and Shelbourne, C. J. A. (1974). The use of vegetative propagation for obtaining genetic information, *N.Z. Jl For. Sci.* **4**, 418–425.

Cannell, M. G. R. (1971). Production and distribution of dry matter in trees of *Coffea arabica* L. in Kenya as affected by seasonal climatic differences and the presence of fruit. *Ann. appl. Biol.* 67, 99—120.

Coombe, D. E. (1960). An analysis of the growth of *Trema guineensis*. *J. Ecol.* 48, 219—231.

Coombe, D. E. and Hadfield, W. (1962). An analysis of the growth of *Musanga cecropoides*. *J. Ecol.* 50, 221—234.

Crowther, E. M. (1947). The use of salt for sugar beet. *British Sugar-Beet Rev.* 16, 19—22.

Cunningham, R. K. and Burridge, J. C. (1960). The growth of cacao (*Theobroma cacao*) with and without shade. *Ann. Bot.* 24, 458—62.

Curtis, D. L. (1968). The relation between the date of heading of Nigerian sorghums and the duration of the growing season. *J. appl. Ecol.* 5, 215—226.

Draycott, A. P., Marsh, J. A. P. and Tinker, P. B. H. (1970). Sodium and potassium relationships in sugar beet. *J. agric. Sci., Camb.* 74, 568—573.

Evans, G. C. (1972). "The Quantitative Analysis of Leaf Growth". Blackwell Scientific Publications, Oxford.

Evans, L. T. (1975). "Crop physiology". Cambridge University Press, London.

Ferwerda, F. P. and Wit, F. (1969). "Outlines of Perennial Crop Breeding in the Tropics". Misc. Pap. 4, Landbouwhogesch. Wageningen.

Ford, E. D. (1975). Competition and stand structure in some even-aged plant monocultures. *J. Ecol.* 63, 311—333.

Garcia-Novo, F. and Crawford, R. M. M. (1973). Soil aeration, nitrate reduction and flooding tolerance in higher plants. *New Phytol.* 72, 1031—1039.

Gifford, R. M. (1974). A comparison of potential photosynthesis, productivity and yield of plant species with differing photosynthetic metabolism. *Aust. J. Plant Physiol.* 1, 107—17.

Green, M. J. (1973). "Studies on Criteria used to Predict Yield in Clonal Tea". Ph.D. thesis, University of Reading.

Hadfield, W. (1974). Shade in north east Indian tea plantations. I. The shade pattern. II. Foliar illumination and canopy characteristics. *J. appl. Ecol.* 11, 151—178, 179—200.

Huxley, P. A. and Summerfield, R. J. (1974). Effects of night temperature and photoperiod on the reproductive ontogeny of cultivars of cowpea and of soyabean selected for the wet tropics. *Plant Science Letters* 3, 11—17.

Jackson, J. E. (1963). The relationship of relative leaf growth rate to net assimilation rate and its relevance to the physiological analysis of plant growth. *Nature, Lond.* 200, 909.

Millerd, A., Goodchild, D. J. and Spencer, D. (1969). Studies on a maize mutant sensitive to low temperature. II. Chloroplast structure, development, and physiology. *Plant Physiol.* 44, 569—583.

Millerd, A. and McWilliam, J. R. (1968). Studies on a maize mutant sensitive to low temperature. I. Influence of temperature and light on the production of chloroplast pigments. *Plant Physiol.* 43, 1967–1972.

Milne, G. (1937). Essays in applied pedology. I. Soil type and soil management in relation to plantation agriculture in East Usambara. *E Afr. agric. J.* 3, 7–20.

Murray, D. B. and Nichols, R. (1966). Light, shade and the growth of some tropical plants. *In* "Light as an Ecological Factor", (R. Bainbridge, G. C. Evans and O. Rackham, eds.). Blackwell Scientific Publications, Oxford.

Nye, P. H. and Greenland, D. J. (1960). "The Soil under Shifting Cultivation". Technical Communication 51, Commonwealth Agricultural Bureaux, Farnham Royal.

Radwanski, S. A. and Wickens, G. E. (1967). The ecology of *Acacia albida* on mantle soils in Zalingei, Jebel Marra, Sudan. *J. appl. Ecol.* 4, 569–579.

Rambelli, A. (1973). The rhizosphere of mycorrhizae. *In* "Ectomycorrhizae, their Ecology and Physiology" (G. C. Marks and T. T. Kozlowski, eds). Academic Press, New York and London.

Rees, A. R. (1963). Relationship between crop growth rate and leaf area index in the oil palm. *Nature, Lond.* 197, 63–64.

Snoad, B. and Davies, D. R. (1972). Breeding peas without leaves. *Span* 15, 87–89.

Wallace, D. H., Ozbun, J. L. and Munger, H. M. (1972). Physiological genetics of crop yield. *Adv. Agron.* 24, 97–146.

Watson, D. J. (1947). Comparative physiological studies on the growth of field crops. I. Variation in net assimilation rate and leaf area between species and varieties, and within and between years. *Ann. Bot.* 11, 41–76.

CARBON FIXATION EFFICIENCY

2
Physiological Genetics, Photosynthesis and Growth Models

F. THOMAS LEDIG
School of Forestry and Environmental Studies, Yale University, Connecticut, U.S.A.

I.	Introduction	21
II.	Photosynthesis and Growth — A Question of Causation . . .	22
III.	CO_2 Exchange	23
	A. Genetic Variation in the Rate of CO_2 Exchange . . .	23
	B. The Rate of CO_2 Uptake and Genetic Variation in Growth	26
	C. Seasonal Patterns of Photosynthesis	28
	D. The Carbon Balance and Models of Growth	31
IV.	Developing Models of Growth	33
	A. Objectives and Philosophy of Modelling	33
	B. A Basic Framework	34
	C. Partitioning CO_2 Exchange	35
	D. The Distribution of Photosynthate	39
	E. An Expanded Model	42
V.	Summary	43
Acknowledgements		45
References		46

I. INTRODUCTION

An understanding of growth and its genetic control is a fundamental objective of biological science. It also has an immediate practical value in designing agricultural and silvicultural systems, including the selection and breeding of more productive types. Genetic testing of trees requires a much larger area and longer time than testing of short-term crop plants, and this results in large environmental variances and low precision. It would be a major benefit to tree breeders to predict growth to harvest age from parameters measured at the seedling stage, eliminating or reducing long-term field trials. Early evaluation would increase the rate of genetic gains, and provide social and economic benefits on a global scale.

The rate of CO_2 exchange, regulated by photosynthetic CO_2 fixation and respiration, is one of the most obvious factors determining the rate of dry matter growth, but not the only one. Plant morphogenesis interacts with the process of CO_2 exchange in many ways, both simple and complex. Morphogenetic patterns have been treated empirically but have defied *a priori* explanation, constituting a major obstacle in growth prediction. Only simulation models will be able to integrate all the diverse factors that contribute to genetic variation in growth.

This contribution reviews what is known of genetic variation in CO_2 exchange, particularly in woody plants, and reports progress in relating CO_2 exchange to growth. The second part of the review introduces a classic growth model and discusses how it might be expanded to include environmental responses, canopy and solar geometry, leaf acclimatization, maintenance and constructive respiration, diffusion pathways, enzyme kinetics, and feedback mechanisms for integrating leaf, stem and root development. Whether such a model can or cannot predict tree growth based on seedling parameters, it will increase appreciation of the components of growth, both providing a set of guidelines for plant breeding and revealing those areas in which knowledge is deficient.

II. PHOTOSYNTHESIS AND GROWTH – A QUESTION OF CAUSATION

The bulk of tree growth is an accumulation of carbon compounds, derived ultimately from fixation of CO_2 by ATP and $TPNH_2$ generated in photosynthesis. Only a small fraction of total dry weight is represented by mineral elements. Over three decades ago, Smith (1943, 1944) showed that the carbohydrate fraction of a leaf increased in amount equivalent to the carbon assimilated as CO_2 during a 45 min period of photosynthesis. Therefore, if total CO_2 assimilation could be measured during longer periods, it should be directly related to accumulation of dry weight. In fact, only recently has this premise been convincingly demonstrated for whole plants. Cumulative CO_2 uptake of orchard-grass (*Dactylis glomerata*) over a 24 h period was closely related to accumulated dry weight (Eagles, 1974). In *Populus tremuloides* carbon accumulation over a 56 day period was within 7–14% of that estimated by continuously recording CO_2 exchange (Bate and Canvin, 1971).

Many foresters and agronomists have suggested the rate of CO_2 uptake as a screen for genetic differences in growth (e.g. Bourdeau, 1958; Moss and Musgrave, 1971; Delaney and Dobrenz, 1974; Eagles, 1974). Inherent in this proposal is the assumption that growth is regulated by the availability of carbohydrate. An alternative hypothesis is that the rate of growth regulates the rate of CO_2 uptake (Sweet and Wareing, 1966). According to one theory, in the absence of sinks (i.e. actively growing regions), carbohydrates accumulate in the leaf and limit photosynthetic carbon fixation by feedback inhibition. Another theory is that hormones produced in the meristems, or "sinks", are translocated to the leaves and stimulate photosynthesis (Bidwell et al., 1968; Bidwell, 1973). It seems likely that hormones and not feedback inhibition provide the regulatory mechanism, because (a) concentrations of soluble carbohydrates at twice the level observed in the field fail to inhibit photosynthesis (Austin, 1972), while (b) exogenous applications of indoleacetic acid, kinetin, or gibberellic acid stimulate photosynthesis (Bidwell *et al.*, 1968; Wareing *et al.*, 1968). Whether "sinks" regulate photosynthesis or not, it is obvious that the rate of CO_2 uptake is limited by genetic restraints on morphological and physiological characteristics.

III. CO_2 EXCHANGE

A. Genetic Variation in the Rate of CO_2 Exchange

Several studies have demonstrated genetic variation in the rate of photosynthetic and respiratory CO_2 exchange among species, provenances, families, or clones of trees (e.g. Huber and Polster, 1955; Bourdeau, 1958, 1963; Rüsch, 1959; Huber and Rüsch, 1961; McGregor *et al.*, 1961; Polster and Weiss, 1962; Reines, 1963;

Gatherum, 1965; Krueger and Ferrell, 1965; Robertson and Reines, 1965; Campbell and Rediske, 1966; Burkhalter et al., 1967; Gatherum et al., 1967; Ledig and Perry, 1967; Zelawski, 1967; Gordon and Gatherum, 1968; Schultz and Gatherum, 1971; Zelawski et al., 1971; Fryer and Ledig, 1972; Luukkanen and Kozlowski, 1972; Corley et al., 1973; Sorensen and Ferrell, 1973). There was a 2-fold difference in rate of CO_2 uptake among 26 full-sib families of *Pinus taeda* (Ledig and Perry, 1967) and the difference was not the result of more active "sink" production by the rapidly photosynthesizing families.

Frequently, differences in CO_2 exchange are obvious only at certain temperatures, radiation flux densities, or ages. For example, southern provenances of *Pinus strobus* only differed from northern provenances at low radiation flux densities and low temperatures (Bourdeau, 1963; see also Townsend et al., 1972), and differences among *Pinus sylvestris* provenances were observed at 1 year of age but not at 2 years (Gordon and Gatherum, 1968). An excellent review of the effect of environmental variables and growth stage on CO_2 exchange was written by Larcher (1969b).

The pattern of photosynthetic response to environmental variables is itself under genetic control. The photosynthetic response to temperature varied among populations of *Abies balsamea* from different elevations (Fig. 1; Fryer and Ledig, 1972). The optimum temperature for CO_2 uptake decreased from c. 23°C for populations from 732 m altitude to 17°C for populations from 1463 m, demonstrating genetic adaptation to summer temperatures, which decrease with increasing elevation (Fig. 2).

In addition to short-term responses to environmental fluctuation, there is genetic variation in the ability to acclimatize or adapt to new conditions over periods of days or months. When previously *grown* at 26°C and 100 ft-c, simulating conditions under a forest canopy, *Picea rubens* had a higher rate of CO_2 uptake than *Picea mariana* no matter what the light and temperature conditions were when CO_2 exchange was *measured*. When grown at 12°C and 100 ft-c, the situation was reversed, and *P. mariana* had higher rates of CO_2 uptake than *P. rubens*, even when measured at 26°C, 100 ft-c (Fig. 3). The photosynthetic responses determine the habitat of *P. rubens* and *P. mariana* in the climax forest, where the former occurs on the warmer uplands and the latter is found in bogs and drainages which are pockets for colder air.

2. Physiological Genetics, Photosynthesis and Growth Models

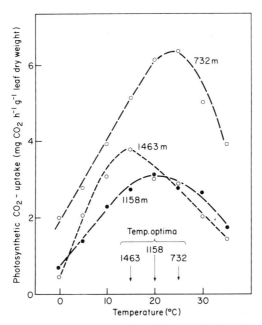

FIG. 1 Temperature response curves for CO_2 uptake in *Abies balsamea* from populations originating at 1463, 1158 and 732 m altitude but grown under uniform conditions. (Fryer and Ledig, 1972)

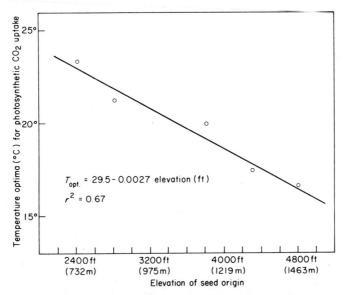

FIG. 2 Relationship of the photosynthetic temperature optimum to elevational origin in *Abies balsamea*. (Fryer and Ledig, 1972)

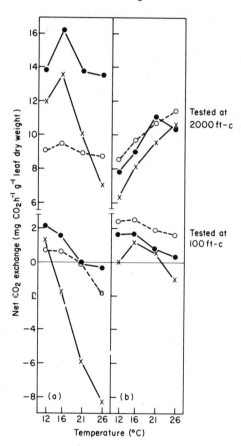

FIG. 3 Effects of long-term acclimatization temperatures of either $12°C$ (a) or $26°C$ (b) on rates of CO_2 exchange in *Picea mariana* (●), *Picea rubens* (○) and their hybrid (x) when measured at a range of temperatures. Upper curves measured at 2000 ft-c, lower curves at 100 ft-c. Acclimatization light intensity was 100 ft-c. (Manley and Ledig, *in litt.*)

B. The Rate of CO_2 Uptake and Genetic Variation in Growth

Frequently, the rate of CO_2 uptake bears no relation to, or even is negatively correlated with, seedling "growth" as measured by dry or fresh weight of the top (e.g. Reines, 1963; Gatherum, 1965; Krueger and Ferrell, 1965; Borsdorf, 1967; Burkhalter *et al.*, 1967; Ledig and Perry, 1967; Neuwirth, 1967; Gordon and Gatherum, 1968).

2. Physiological Genetics, Photosynthesis and Growth Models

Negative correlations could result from greater mutual shading among leaves of large seedlings compared to small seedlings. Pine needles placed so that they did not overlap reached their highest rate of CO_2 uptake at 3000 ft-c, while entire seedlings did not reach their maximum rate even at 9000 ft-c because of the shading of lower needles by upper needles (Kramer and Clark, 1947). The rate of CO_2 uptake in pine species decreased as the number of needles per fascicle increased (Uhl, 1937), which might be explained as a result of greater shading in denser clusters. Further evidence for the importance of mutual shading was found by Zelawski et al., (1973) working with *Pinus sylvestris* seedlings enclosed in an integrating sphere. Shading was less in the diffuse, multi-directional light of the integrating sphere than in unidirectional lighting, and as expected, highest rates of CO_2 uptake occurred in the integrating sphere (see Zelawski, Chapter 5).

Positive correlations between CO_2 exchange and genetic variation in dry matter accumulation are rarer. However, CO_2 exchange did account for differences between two varieties of maize grown for a period of 30 days (Heichel, 1971), and 1½ month old rooted cuttings of four clones of *Populus grandidentata* x *P. alba* had similar ranks for net CO_2 uptake and total dry weight (Gatherum et al., 1967). In the classic work of Huber and Polster (1955), CO_2 uptake accounted for some variation in growth, but relative leaf area explained more of the difference among *Populus* clones.

Positive relationships between the rate of CO_2 uptake and growth are most commonly observed in controlled environments, such as glasshouses or growth chambers. When grown at a mean temperature of 27°C, seedling dry weight of five populations of *Abies balsamea* from different elevational zones was correlated with rate of CO_2 uptake at the same temperature (Fig. 4; Fryer and Ledig, 1972). In *Picea rubens* and *P. mariana*, their hybrids, backcrosses and recurrent backcrosses, growth in a series of controlled environments was related to rate of CO_2 uptake under the same conditions (Fig. 5; Manley and Ledig, *in litt.*). Within environments, (2000 or 100 ft-c in factorial combination with temperatures of 26° or 12°C), correlations of CO_2-uptake and growth ranged from 0.45 to 0.87. Over all environments and crosses, the correlation coefficient was 0.87.

In the field, differences in height among 10 provenances of *Pinus banksiana* at 1, 2, 6, and 7 years of age were positively related to

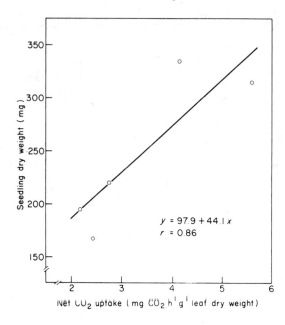

FIG. 4 Correlation between dry weight and rate of CO_2 exchange in *Abies balsamea* from different elevational origins. (Fryer and Ledig, 1972)

rates of CO_2 uptake in October, but not in other months (Logan, 1971, 1974). The rate of CO_2 uptake in temperate tree species varies from near zero in winter to maximum rates during the summer. These changes are partly a result of internal controls and partly a result of acclimatization to changing temperatures and photoperiod. Because adaptability (i.e. the ability to acclimatize) is itself under genetic control (e.g. Manley and Ledig, *in litt.*), genetic variation in CO_2 exchange at some times of year but not at others could be very common (e.g. Ledig and Perry, 1969; Ledig and Botkin, 1974).

C. Seasonal Patterns of Photosynthesis

Under conditions of standard temperature such as 20° or 25°C, many north temperate zone evergreens, both gymnosperm and angiosperm, have seasonal patterns of photosynthesis characterized by a low winter capacity for CO_2 uptake, rapid spring increase to a maximum around June, and high rates maintained until September

2. Physiological Genetics, Photosynthesis and Growth Models

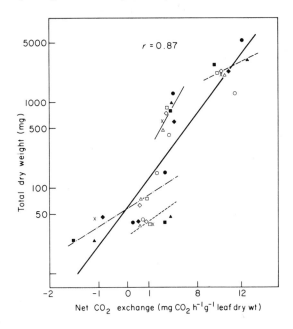

FIG. 5 Correlation between dry weight and rate of CO_2 uptake in *Picea mariana* (●), *Picea rubens* (○), and hybrid types (▲, ■, ◆, x, ◇, □, △) grading from *Picea mariana* through the F_1 hybrid (x) to *Picea rubens*. Seedlings grown and measured at 100 ft-c, 12°C -----; 100 ft-c, 26°C —·—; 2000 ft-c, 12°C ⎯⎯; 2000 ft-c, 26°C ———. (Manley and Ledig, *in litt.*)

or October (Saeki and Nomoto, 1958; Bourdeau, 1959; McGregor and Kramer, 1963; Tranquillini, 1963a; Negisi, 1966; Shiroya *et al.*, 1966; Lister *et al.*, 1967; Steinhubel and Halas, 1969; Larcher, 1969b; Clark and Bonga, 1970; Halas, 1971; Albrecht, 1972). In contrast, deciduous gymnosperms and angiosperms apparently reach maximum capacity in July or early August followed by a steady decline to near zero rates in mid-October (Saeki and Nomoto, 1958; Logan and Krotkov, 1968; Schulze, 1970; Ledig and Botkin, 1974). When CO_2 uptake is measured in the field at ambient temperatures, even gymnosperms may show an early decline in net photosynthesis (Parker, 1961; Rastorgueva, 1966; Helms, 1965; Mooney *et al.*, 1966). Exceptions to these generalizations have been reported (Clark and Bonga, 1970; Zelawski *et al.*, 1971). The seasonal pattern of CO_2 exchange is affected by edaphic as well as climatic factors. For example, the rate of decline in photosynthetic CO_2 uptake was more rapid when tung (*Aleurites fordii*) seedlings were grown under low levels of nitrogen than when they were grown with high nitrogen (Loustalot et al., 1950). Also, shade-grown plants maintain high rates of CO_2 uptake later in the fall than sun-grown plants (Schulze, 1970; Halas, 1971).

Interspecific variation in the seasonal pattern has been well documented. The period of near maximum photosynthetic capacity was much longer for *Pinus densiflora* than for either *Cryptomeria japonica* or *Chamaecyparis obtusa* (Negisi, 1966), and for *Larix decidua* and *L. leptolepis* than for *L. russica* (Ledig and Botkin, 1974; see Fig. 6). Conifers have been reported capable of net CO_2 uptake during winter even at temperatures below zero (e.g. Parker, 1953; but cf. Schulze et al., 1967). Substantial overwinter increases in seedling dry weight were observed in the deciduous *Liquidambar styraciflua* and the evergreen *Pinus taeda* in North Carolina (Perry, 1971) and *Pinus sylvestris* in Great Britain (Rutter, 1957), but not in

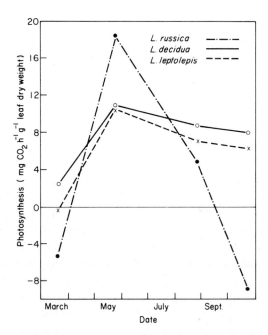

FIG. 6 Seasonal patterns of net CO_2 uptake for second-year seedlings of *Larix decidua* (○), *L. leptolepis* (x), and *L. russica* (●) grown in Connecticut, measured at 4000 ft-c and temperatures corresponding to the mean daily maximum. (Ledig and Botkin, 1974)

Larix sp. or *Platanus occidentalis* in Connecticut (Ledig and Botkin, 1974). These differences may be largely climatic rather than species specific.

Intraspecific genetic variation in the yearly pattern of assimilation was shown for *P. sylvestris* provenances (Zelawski, 1967) and *P. taeda* families (Ledig and Perry, 1969). Some provenances and families with superior rates of CO_2 uptake in spring had relatively low rates in the autumn. Seasonal change in rank is one reason for the frequent failure to relate photosynthesis to growth.

One method of comparison when there are seasonal changes in ranking is to calculate the integral of assimilation rate over an entire period. In 26 half-sib families of *P. taeda*, this summation was closely related to observed differences in dry weight growth (Ledig and Perry, 1969). The integral of photosynthetic rate was suggested as a selection index for wheat in the semi-arid provinces of Canada (Kaul and Crowle, 1974). Larcher (1969a) stressed that final yield depends on the "production period" as well as the maximum attained rates of CO_2 uptake and therefore must be related to the integral of seasonal assimilation rates. The seasonal aspect of CO_2 uptake is much more important in trees than in agricultural crops because most agricultural plants are annuals with a relatively short growing season, while trees must cope with a variety of environmental conditions experienced throughout the entire year.

D. The Carbon Balance and Models of Growth

The carbon balance is determined by the product of leaf area times the rate of CO_2 uptake, minus respiring mass times the rate of CO_2 efflux. Biomass increment for single trees or whole forests were paralleled by estimates of the carbon balance in several studies (Tranquillini, 1963a, b; Botkin *et al.*, 1970; Schulze, 1970; Connor *et al.*, 1971; Cartledge and Connor, 1972). Carbon balance techniques also accounted for dry matter production with some precision in short-term growth chamber experiments. For example, over weekly periods it was possible to estimate the dry weight growth of clover (*Trifolium repens*: McCree and Troughton, 1966) and fodder cabbage (*Brassica oleracaeae*: Čatsky *et al.*, 1967). In longer trials (30 days), estimates of carbon accumulation were 99% of harvest value for sunflower (*Helianthus annuus*) and 85—93% of harvest value for aspen, although such precision required continuous monitoring of CO_2 exchange of the whole plant (Bate and Canvin, 1971). When CO_2-exchange rates were measured for only brief periods, total diurnal CO_2 uptake was estimated within 25% (Bate and Canvin, 1972).

A mathematical formalization of the carbon balance method was presented by Ledig (1969) as an integral equation. In an attempt to make the method predictive, the amount of tissue in leaf, stem and root was incremented according to the allometric equation (see also Chang and Huang, 1973). Using a computer, the expression was evaluated iteratively for 26 full-sib families of *Pinus taeda*. The correlation between predicted and actual dry weights at 21 weeks was $r = 0.97$ (Fig. 7; Ledig, 1969). In two other tests, dry weights of three *Larix* species and three half-sib families of *Platanus occidentalis* were simulated over two growing seasons. Ranks for simulated and observed dry weight were identical (Fig. 8; Ledig and Botkin, 1974). However, terms for the distribution of growth among leaf, stem and root, and the seasonal patterns of CO_2 exchange were empirical, and it was not determined that the model could predict growth outside

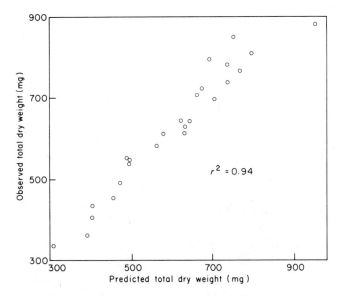

FIG. 7 Relationship between observed dry weight growth of 26 full-sib families of *Pinus taeda* at 21 weeks of age and dry weight calculated by a model employing NAR and allometric constants (Ledig, 1969).

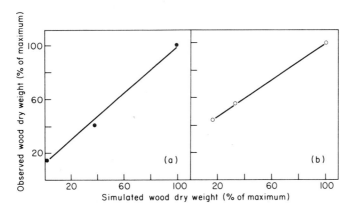

FIG. 8 Relationship between observed dry weight (% of maximum) after two growing seasons and dry weight (% of maximum) calculated from rates of CO_2 exchange and the distribution of photosynthate. (a) *Larix decidua, L. leptolepis* and *L. russica*. (b) Three half-sib families of *Platanus occidentalis*. (Ledig and Botkin, 1974)

the range of observation. Nevertheless, there are no other examples where models have been used to account for genetic variation in growth, so the results encourage further development, perhaps by constructing combinations of several more realistic models.

Negisi's (1966) use of carbon balance concepts was more complex and more realistic than Ledig's (1969) because carbon accumulation was calculated from light and temperature response curves plus monthly frequencies of hourly mean temperatures and radiation flux densities. The models of Monsi and Murata (1970) and Ross (1966) are philosophically superior because they explicitly include maintenance and constructive respiration of each organ and account for redistribution of assimilate from organ to organ.

IV. DEVELOPING MODELS OF GROWTH

A. Objectives and Philosophy of Modelling

There are at present several models available for photosynthesis and for plant growth. But, objectives vary and determine the nature of input and output. In genetics the desire is to predict which of several provenances, families or individuals will grow most rapidly to harvestable size — a "pick the winner" approach. Inputs to the model should be heritable plant characteristics. Physical variables such as light, temperature, humidity and soil moisture will be important only if plants differ in their response to these factors and genetic response surfaces are included in the model. On the other hand, many crop physiologists are more concerned with defining climatic and physical conditions that influence growth in order to guide cultural decisions on spacing, soil management and irrigation regimes, and their statistical models (e.g. Gopal, 1972), relating growth to climatic variables, are not relevant for the geneticist.

B. A Basic Framework

We attacked the practical problem of defining an algorithm that would explain genetic differences in growth based on seedling input (Ledig, 1969, 1974). The method we adopted was to start with the most obvious, gross relationships that would describe dry weight growth and then to seek more basic causes and effects to substitute for the original, empirical rules. The major objection to empirical or statistical rules is that if a relationship must be measured first, a model is not required for prediction. Our first approximation was based on the classic model:

$$\frac{dW}{dt} = P \cdot W - R \cdot W, \qquad (1)$$

where dW/dt is the change in total dry weight, P is a row matrix for the rate of CO_2 uptake in leaf, stem, and root, W is a vector of leaf, stem and root dry weight, and R is a row matrix of the rate of respiration in leaf, stem and root. While photosynthesis in the root is zero, the stem may contribute appreciably to the carbon balance and cannot be neglected. Net CO_2 uptake occurs in the twigs of *Quercus robur, Q. phellos,* and *Platanus occidentalis* (Perry, 1971; Longman and Coutts, 1974), and in other species, bark contributes to the carbon economy even though rates do not reach the compensation point (e.g. Keller, 1973). In order to evaluate the effect of variation in proportion of leaves, stems and roots, the respiration of each

2. Physiological Genetics, Photosynthesis and Growth Models

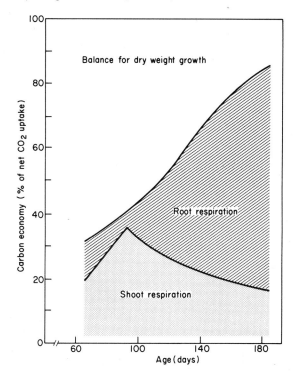

FIG. 9 Ontogenetic changes in the carbon balance and the proportions of diurnal CO_2 fixation accounted for by nocturnal shoot respiration and daily root respiration in *Pinus rigida* seedlings grown in a constant environment. (Ledig *et al.*, 1976)

component should be considered separately. Rates of respiration are generally much higher in leaves than in stems or roots. Root respiration is infrequently measured, but is important. In pine seedlings it can dissipate 69% of diurnal CO_2 assimilation (Ledig *et al.*, 1976; Fig. 9).

C. Partitioning CO_2 Exchange

1. Genetic Response to Environmental Variables

As stated, the model (1) is too naive. CO_2 exchange responds to environmental parameters, the responses are not linear, and there is genetic variation in response. Therefore, P should be substituted by a response surface for each genotype, expressing the rate of CO_2 uptake in relation to ambient light, temperature, soil moisture, wind speed and vapour pressure deficit. Moreover, an individual's response depends on its developmental reaction to the conditions under which it grew (Fig. 3), therefore the response surface must include the effects of acclimatization to previous environments. In previous applications of simulation models (Ledig, 1969; Ledig and Botkin, 1974), the empirical relationship of CO_2 exchange to time accounted for such effects, but this may not be adequate for prediction beyond the observed range.

2. Forest Canopies and Solar Geometry

To predict growth in forest stands, models must eventually account for canopy and solar geometry. A model conceptually similar to ours (Ledig, 1969) was extended to stands by considering the influence of spacing on leaf area index and the extinction of light within the canopy (Promitz and Rose, 1974). It includes as a subprogram the model of Duncan et al. (1967) for canopy photosynthesis. Allen et al. (1974) expanded the Duncan model and applied it to CO_2—uptake of tropical trees. Van Bavel (1974) modified the iterative Duncan-Stewart type model to reduce costs.

The biological factors influencing the distribution of radiation flux density within the canopy are the numbers of layers of leaves, leaf dispersion and leaf angle. These are determined in part by branch angle and length, characteristics that can be modified by the breeder (Dorman, 1952). At high leaf area indexes, the model suggests that leaf angles vertical or near vertical are optimum for most of the crown (Duncan, 1971). Success in the breeding of IR-8 rice is in part the result of selection for more erect leaves. For crown shape, tall conical forms will intercept more light diurnally than plants with their foliage oriented in a horizontal plane, at least in the absence of shading by adjacent trees (Jahnke and Lawrence, 1965). In forestry, selection against steeply ascending branches and wide crowns has been common and may have unanticipated effects on photosynthesis. To our knowledge, the effect of genetic variation in leaf angle and dispersion has not been adequately quantified in trees (see Farmer, Chapter 7). When such information becomes available, its effect on photosynthesis can be tested by the Duncan-Stewart light model to determine the optimum crown geometry, enabling breeders to select accordingly.

One failure of the Duncan-Stewart model is the assumption that light response of leaves in the shaded lower crown is the same as that in the fully illuminated upper crown. In fact, leaves may adapt to sun or shade as a result of acclimatization (Björkman and Holmgren, 1963; Zelawski, Chapter 5). Shade compensation may be complete for leaves whose entire development occurs in the lower crown, but partial compensation is possible even for leaves that mature in high illumination and subsequently are "transferred" to lower light regimes as they are overtopped by new shoots and leaves (Gauhl, 1969). Acclimatization to changing temperatures must also be considered. Models of photosynthesis that included adaptation of leaves to changing conditions were proposed by Moldau (1971) and Tooming (1967). The empirical seasonal changes in photosynthesis incorporated in our model (Ledig, 1969) reflected, in part, acclimatization to seasonal temperatures and radiation as well as direct effects of climate.

3. Maintenance and Constructive Respiration

In the basic equation of growth (1), respiration is apparently a loss to the system. This is unrealistic because respiration provides the energy for growth, therefore higher rates of respiration mean higher rates of growth (Beevers, 1970). A reduction in respiration will increase growth only if the efficiency of converting substrate to structural components can be improved. Respiration (R) can be expressed in terms of two components, maintenance respiration (R_m in g CO_2 respired g^{-1} dry weight) and constructive respiration (R_c in g CO_2 respired g^{-1} dry weight growth). Maintenance respiration, the respiration necessary to maintain tissue in a functional state, should be proportional to the amount of tissue, or mass (W). Constructive respiration, required to produce a unit of new growth, may be proportional to the change in mass, or growth rate. The change in mass (dW/dt) is proportional to the net CO_2 uptake (P_N) in carbohydrate equivalents (McCree, 1970, 1974):

$$R = R_m W + R_c P_N. \qquad (2)$$

The rates of respiration and photosynthesis vary during the day and their ratio is not constant; Hesketh *et al.* (1971), Baker *et al.* (1972), and Thornley and Hesketh (1972) prefer to use:

$$R = R_m W + R_c dW/dt. \qquad (3)$$

In fact, the ratio of photosynthesis to respiration does fluctuate because of transport lags, but over periods of a day or longer, may be very stable. Ledig *et al.* (1976) found no relationship between R and current dW/dt in *Pinus rigida* because high-energy-requiring syntheses, that result in little change in dry weight, precede secondary cell wall construction that requires less energy but is responsible for large increases in dry weight. But, R_c in *P. rigida* was closely related to the rate of CO_2 fixation in the light, and equation (2) may therefore be more useful than (3) in trees. Leaves, stem and roots differ in R_m and R_c; e.g. R_m in shoots is four times greater than R_m in roots and R_c is three times greater for shoots than for roots of *P. rigida* (Ledig *et al.*, 1976). Therefore, R must be partitioned for each individual component.

Breeding for lower rates of respiration is not desirable because it would result in slower rates of growth, but selection for lower R_c would be selection for more efficient conversion of substrate to dry matter. Lower R_m may or may not be desirable. On the one hand, a low R_m could indicate less labile cell constituents and more efficient maintenance, but on the other hand, low R_m might result in earlier senescence and a shorter growing season. R_m could be decreased only to the point where it does not impair the rate of CO_2 uptake in the leaves, water and mineral uptake in the roots, or transport in the stem.

4. Leaf Models

The gross characteristic, rate of CO_2 uptake, can be modelled using more basic principles. Several models of leaf photosynthesis are available, based on the electrical analogue to CO_2-diffusion resistance, on elementary enzyme kinetics, or a combination of both (Brown, 1969; Waggoner, 1969; Laisk, 1970; Chartier, 1970; Hall, 1971; Charles-Edwards, 1971a, b; Lommen *et al.*, 1971). Many of the models include stomatal resistance to diffusion, which can itself be modelled using other plant characteristics, including thickness of the guard cell walls, chloroplast density, mesophyll cell frequency, and internal exposed area (Sharpe and DeMichele, 1974). Some leaf models depend on characteristics not readily measurable. For example, the resistance between the cell wall and the site of photosynthesis is an important component of Waggoner's (1969) model, but is not measurable at present and would have to be assumed constant. Nevertheless, many characteristics such as stomatal resistance, activity of carboxylating enzymes, rate of light respiration, and the Q_{10} of these activities can be estimated. In fact, they may be explained in terms of leaf size and shape, stomatal size and frequency, thickness of the guard cell wall, specific leaf weight, mesophyll cell size, activation energy, thermal stability of enzymes, and other measurable characteristics (e.g. Wilson and Cooper, 1967; Sharpe and DeMichele, 1974).

For practical purposes of predicting the winner, models which include compound characteristics such as the rate of photosynthetic CO_2 uptake are sufficient. In fact, statistical models may suffice as well as those that work toward first principles. But in the long run, the greatest value of modelling may be to illuminate the separate characteristics that contribute to increased productivity. Selection for a compound characteristic, in reality, could be selection for one or several major components. Genes with small effects in low frequency might be lost. For example, modern wheats have higher yields than primitive wheats because they have a larger leaf area, but primitive wheats have higher photosynthetic rates (Evans and Dunstone, 1970). Variation in relative leaf growth rate has a greater effect on yield than photosynthetic rate dm^{-2} leaf, so loss of genes for higher photosynthetic rate might have occurred by chance during selection for higher yield. Ultimately, breeding for the physiological and morphological components of CO_2 uptake may result in greater gain than blind selection for CO_2 uptake alone.

D. The Distribution of Photosynthate

1. Dynamic Models for Shoot: Root Balance

Any model for growth must contain components that account for the distribution of photosynthate among organs. To date, this has been the most difficult aspect in building predictive models (de Wit *et al.*, 1970). The model used by us relied on allometric relationships among organs (Ledig, 1969). Though allometric relationships may be constant for brief periods or even reflect a general pattern throughout a tree's life, they will not be sufficiently constant in a number of situations to permit precise simulations of growth. An alternative algorithm for growth distribution is the proportion of total growth distributed to leaf, stem or root during a period (Monsi and Murata, 1970). For deciduous trees, we (Ledig and Botkin, 1974) followed this course. But, there is little predictive value in such a simulation. It requires that the distribution of growth be empirically determined for each period. To be useful, genetic differences in the seasonal pattern would have to be repeated year after year, so that measurement in one year would suffice for an entire rotation. This may be too much to hope for. What is required is a subprogram that distributes growth according to some algorithm whose parameters can be determined in juvenile phases of growth and remain constant thereafter. A basic, conceptual model with feedback control would be more valuable than an empirical, statistical relationship.

One possibility is a model such as Thornley's (1972a, b) in which shoot : root ratio was modelled as the balance between nitrogen uptake of the root and gross photosynthesis of the shoot. The balance depends on (a) utilization of substrate (nitrate or photosynthate) which is dependent on substrate concentrations and Michaelis-Menten relationships and (b) translocation which is dependent on substrate gradients and transport resistances. To apply models similar to Thornley's, availability of photosynthate could be simulated using leaf models referred to above, combined with climatic and biologic inputs. Michaelis-Menten constants could be determined. Methods of estimating transport resistance do constitute a problem (Kramer and Fiscus, 1972), and opinions differ on whether transport resistance actually limits CO_2 exchange (contrast Liu *et al.*, 1973 with Servaites and Geiger, 1974). However, it is neither impossible nor inconceivable that transport resistance could exhibit genetic variation. The weakest link in our knowledge is the lack of data on genetic variation in nitrogen uptake activity by the root.

In addition to nitrogen uptake, the root has other functions important to growth but too little understood to be included in present simulators. Besides supplying mineral nutrients, the root

2. Physiological Genetics, Photosynthesis and Growth Models

functions in water uptake. Borchert (1973) proposed a model explaining the episodic extension growth of tropical trees as a feedback resulting from water stress. In this model, shoot growth continues until limited by water stress. Once shoot growth has stopped, more photosynthate is available for root growth, a favourable water balance is reestablished, and shoot growth recommences. Unfortunately, growth of *Quercus rubra* does not support the model (Farmer, 1974).

2. The Dependence of CO_2 Uptake on Root and Stem Growth

In all plant growth models, an increase in leaf area increases total photosynthesis, provided leaves are dispersed adequately in space. The impossible outcome is that total growth would be maximized if the distribution of growth to roots and stem was zero. No models, except perhaps those of Moldau (1971), Thornley (1972a, b) and Borchert (1973), give explicit value to the roots. Of course, roots are necessary for the supply of water and nutrients to the top, and water stress or nutrient deficiency will reduce the rate of photosynthesis (e.g. Brix, 1962, 1971). In addition, roots probably produce hormones or precursors that affect the level of carboxylating enzymes in the leaf. Partial defoliation reduces the "sink" for such products, increasing their concentration and resulting in higher rates of CO_2 uptake in the remaining leaves (Wareing *et al.*, 1968). In rootless cuttings, CO_2 uptake declines to 58% of its initial value in 3 days even though water stress is no problem (Brix and Barker, 1973). Therefore, photosynthetic rate must be a function of the leaf : root ratio, independent of the influence of the root on water stress and nutrient uptake. There will be an optimum allocation of photosynthate between leaves and roots for a given moisture and nutrient regime, and it would be possible to solve for the optimum. Empirical data quantifying the influence of leaf : root ratio are lacking, but it is highly probable that there is genetic variation in this relationship as there is in more familiar aspects of physiology.

Total photosynthesis is also dependent on the leaf : stem ratio because it affects the pattern of leaf dispersal in space, and therefore the effectiveness with which solar radiation is intercepted. Thus, there will be a distribution of growth between leaves and stem that will optimize photosynthesis for a given phyllotaxy, branch strength, and leaf angle; i.e. for a given canopy geometry. In all crops, economic yield is measured in terms of particular products such as fruits, roots or leaves, and forestry is no exception. Foresters market tree boles, and therefore maximum stem growth rather than total growth is desired. To determine the optimum allocation of growth among organs is certainly one of the most important tasks of physiologists.

3. Seasonality

There are aspects of tree growth that would not obviously be predicted by any of the factors discussed above. Two of these are the seasonal shed of leaves and the spring flush of growth. Genetic response to photoperiod, chilling and heat sum could be incorporated in models to shift from a leafy to leafless condition or vice versa. The death of roots is a more difficult problem. Perhaps a root mortality rate might be employed using life tables, but methods of measuring such losses are not clear.

Spring regrowth is dependent on redistribution of stored products from the stem or root to new shoots (Tepper, 1967). The problems of redistribution and storage pools must eventually be solved, but there are at present no good basic models to link these topics to rates of CO_2 uptake, respiration, or water and mineral absorption.

E. An Expanded Model

All of the factors discussed above are linked in a proposed growth model in Fig. 10. The detail with which components are described is a reflection of current knowledge, not true complexity. The entire model is iterative. Leaf, stem and root increments occur (positively or negatively) as a result of daily net C-assimilation. These values are then fed back to calculate assimilation the following day, and so on. Routines for leaf photosynthesis (P) based on Waggoner's (1969) model and for the distribution of meteorological fluxes (DMF/C) based on the Duncan-Stewart models (Duncan et al., 1967; Allen et al., 1974) are iterative and costly in computer time. A model like Fig. 10, that computed growth over a several year period, would

2. Physiological Genetics, Photosynthesis and Growth Models

probably be impractical unless the Waggoner and Duncan-Stewart models were modified or substituted with simpler functions. Improper scale and complexity may be as much a problem as over-simplification.

V. SUMMARY

A priori, growth should be dependent on carbon fixation. There is genetic variation in the rate of CO_2 exchange and in a few instances it has been the major component of variation in growth. Often, the relationship is obscured by seasonal changes in CO_2 exchange and the rate at which photosynthate is allocated to leaf growth. Carbon balance models have been proposed to handle these complex relationships. In some cases, models have successfully simulated growth using rates of CO_2 exchange and parameters describing the distribution of growth among leaves, stems and roots. Unfortunately, the carbon balance models are empirical and it may not be possible to predict future growth by extrapolation.

Starting with the gross generalizations used in simple plant growth simulators, it is possible to build more realistic models. Environmental responses, both long-term response associated with acclimatization phenomena and short-term response to rapid fluctuations, are under genetic control and should be included in growth models. Since plants function in stands, canopy and solar geometry must eventually be considered. Further development includes the partition of respiration into maintenance and constructive components and the incorporation of leaf models with input such as leaf size and shape, stomatal frequency, pore size and mesophyll resistances. The most difficult aspect of modelling is to understand and simulate feedback mechanisms regulating the relative growth of leaves, stems and roots. Feedback may be incorporated by modelling the capacity of the roots to supply nitrate, water or hormone requirements of the leaves, and the capacity of stems to disperse the leaves and transport materials. Finally, simulation of phenological events such as leaf shed and leaf flush is indispensable for long-lived perennials and requires measurement of heat sum, photoperiod, and chilling requirements. Physiological genetics produces information that will permit a start in the formulation of growth models, and modelling in turn directs attention to areas where research is needed.

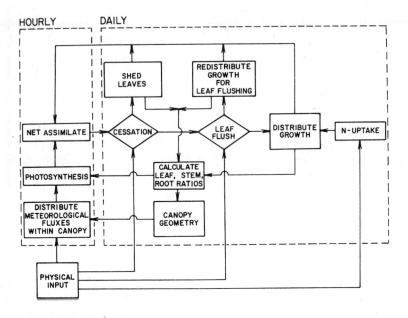

FIG. 10 Flow chart for a proposed plant growth model. Subprograms are: PHYSICAL INPUT (PI). Supplies geographic coordinates, date, hour, radiation flux density, temperature, vapour pressure deficit, soil moisture.
DISTRIBUTE METEOROLOGICAL FLUXES WITHIN CANOPY (DMF/C) after Duncan et al. (1967) and Allen et al. (1974) models as modified by Van Bavel (1974). Computes light and temperature climate for each canopy layer. Input: leaf area index, leaf angle, and leaf dispersion as function of leaf: stem ratio from CG; soil moisture, light, temperature, and vapour pressure deficit at canopy top from PI.
PHOTOSYNTHESIS (P) after Waggoner (1969). Computes rate of CO_2 uptake for each canopy layer, j. Biological parameters: maximum gross photosynthesis (P_x), concentration of CO_2 at half P_x, radiation flux density for half P_x, concentration of CO_2 for half maximum stomatal resistance resulting from high CO_2 concentration, radiation flux density for half stomatal resistance in dark, maximum light respiration (R_{LX}), radiation flux density for half R_{LX}, Q_{10} for photosynthesis, Q_{10} for light respiration, Q_{10} for dark respiration, maximum dark respiration, boundary layer resistance as a function of leaf size and shape, resistance between respiratory and photosynthetic CO_2 streams, resistance between substomatal cavity and respiratory stream, minimum stomatal resistance, carboxylation resistance and stomatal resistance as functions of the leaf : root ratio and climatic variables from DMF/C. The leaf : root ratio accounts for the dependence of the top on hormones and water supplied by the root. Light and temperature acclimatization and ageing change the biological responses among canopy layers and are accounted for by computing the parameters as functions of (a) the integral over time of the leaf area index about

2. Physiological Genetics, Photosynthesis and Growth Models 45

the jth level and (b) heat sum. Input: leaf:root ratio from CR; radiation flux density, temperature, and vapour pressure deficit for jth leaf layer from DMF/C.
NET ASSIMILATE (NA). Computes difference between diurnal CO_2 uptake and daily maintenance respiration for plants. Input: CO_2 exchange rates from P; leaf, stem, and root weights from SL, RG, or DISTG.
CESSATION (C). Tests to determine whether water stress, photoperiod, and temperature are low enough to signal growth cessation. Biological parameters: threshold requirements for growth cessation. Input: soil moisture, photoperiod, and negative heat sum from PI.
SHED LEAVES (SL). Distributes photosynthate (+ or —) between stem and root and computes leaf loss as a function of PI. Biological parameters: constants of regression equation for leaf loss on climatic data. Input: photoperiod, negative heat sum from PI.
LEAF FLUSH (LF). Tests to determine whether chilling requirement and subsequent heat sum are satisfied for growth to resume. Biological parameters: threshold requirements for growth flush. Input: negative and positive heat sums from PI.
REDISTRIBUTE GROWTH FOR LEAF FLUSHING (RG). Distributes dormant season photosynthate (+ or —) between stems and roots and computes spring leaf growth from reserves in stem and root as functions of PI. Biological parameters: empirical constants of regression equations. Input: climatic data from PI (through LF).
N UPTAKE (NU). Computes nitrogen uptake of root. Biological parameters: rate of nitrogen uptake as a function of temperature. Input: temperature from PI.
DISTRIBUTE GROWTH (DISTG) — after Thornley (1972a, b). Computes distribution of growth among leaves, stems, and roots — the final output. Biological parameters: transport resistances, rate constants for substrate utilization, and partitioning coefficients (for each tissue for both photosynthate and nitrogen), and growth or constructive respiration. May require stochastic process to account for root mortality. Input: photosynthate from NA and nitrogen from NU.
CALCULATE LEAF, STEM, ROOT RATIOS (CR). Calculates new ratios. Input: leaf, stem, root weights from SL, RG, or DISTG.
CANOPY GEOMETRY (CG). Disperses leaf growth among and within canopy layers. Biological parameters: bifurcation ratio, leaf angle. Input: leaf:stem ratio from CR.

ACKNOWLEDGEMENTS

The ideas expressed here and much of the research upon which they were based were developed with the aid of National Science Foundation Grant GB35266.

REFERENCES

Albrecht, E. (1972). CO$_2$-gaswechsel und Hill-Aktivität von *Hedera helix* L. im Jahresverlauf unter Berücksichtigung klimatischer Faktoren [in German, English abstract]. *Photosynthetica* **6**, 240–246.

Allen, L. H., Jr., Steward, D. W. and Lemon, E. R. (1974). Photosynthesis in plant canopies: effect of light response curves and radiation source geometry. *Photosynthetica* **8**, 184–207.

Austin, R. B. (1972). The relationship between dry matter increment and sugar concentrations in beetroot leaves. *Photosynthetica* **6**, 123–132.

Baker, D. N., Hesketh, J. D. and Duncan, W. G. (1972). Simulation of growth and yield in cotton: I. Gross photosynthesis, respiration, and growth. *Crop Sci.* **12**, 431–435.

Bate, G. C. and Canvin, D. T. (1971). A gas-exchange system for measuring the productivity of plant populations in controlled environments. *Can. J. Bot.* **49**, 601–608.

Bate, G. C. and Canvin, D. T. (1972). Simulation of the daily growth of an aspen population from the measured CO$_2$-exchange rates of the components. *Can. J. Bot.* **50**, 205–214.

Beevers, H. (1970). Respiration in plants and its regulation. *In* "Prediction and Measurement of Photosynthetic Productivity" (I. Šetlik, ed.), pp. 209–214. Centre for Agricultural Publishing and Documentation, Wageningen, Netherlands.

Bidwell, R. G. S. (1973). A possible mechanism for the control of photo-assimilate translocations, pp. 77–89. *In* Proc. Res. Inst. Pomol., Skierniewice, Poland, Ser. E, Conf. and Symp. Nr. 3. Trans. Third Symp. on Accumulation and Translocation of Nutrients and Regulators in Plant Organisms. Warszawa, Jablonna, Skierniewice, Brezezna, Krakow, Poland.

Bidwell, R. G. S., Levin, W. B. (Turner) and Tamas, I. A. (1968). The effects of auxin on photosynthesis and respiration. *In* "Biochemistry and Physiology of Plant Growth Substances" (F. Wightman and G. Setterfield, eds.), pp. 361–376. Runge Press, Ottawa.

Björkman, O. and Holmgren, P. (1963). Adaptability of the photosynthetic apparatus to light intensity in ecotypes from exposed and shaded habitats. *Physiol. Plant.* **16**, 889–914.

Borchert, R. 1973. Simulation of rhythmic tree growth under constant conditions. *Physiol. Plant.* **29**, 173–180.

Borsdorf, W. 1967. Uber die Beziehungen zwischen Assimilations-intensitat und Ertrag bei Jungpflanzen einiger Pappelklone [in German, English summary]. *Züchter* **37**, 300–306.

Botkin, D. B., Woodwell, G. M. and Tempel, N. (1970). Forest productivity estimated from carbon dioxide uptake. *Ecology* **51**, 1057–1060.

Bourdeau, P. F. (1958). Relation between growth and unit rate of photosynthesis, pp. 58—62. *In* Proc. Fifth Northeast. For. Tree Improv. Conf. Orono, Maine.

Bourdeau, P. F. (1959). Seasonal variations of the photosynthetic efficiency of evergreen conifers. *Ecology* 40, 63—67.

Bourdeau, P. F. (1963). Photosynthesis and respiration of *Pinus strobus* L. seedlings in relation to provenance and treatment. *Ecology* 44, 710—716.

Brix, H. (1962). The effect of water stress on the rates of photosynthesis and respiration in tomato plants and loblolly pine seedlings. *Physiol. Plant.* 15, 10—20.

Brix, H. (1971). Effects of nitrogen fertilization on photosynthesis and respiration in Douglas-fir. *Forest Sci.* 17, 407—414.

Brix, H. and Barker, H. (1973). Rooting studies of Douglas-fir cuttings. Can. For. Serv., Pacific For. Res. Centre Inform. Rep. No. BC-X-87. 45 pp.

Brown, K. W. (1969). A model of the photosynthesizing leaf. *Physiol. Plant.* 22, 620—637.

Burkhalter, A. P., Robertson, C., and Reines, M. (1967). Variation in photosynthesis and respiration in southern pines. Georgia For. Res. Pap. 46. Georgia For. Res. Council, Macon, Georgia. 5 pp.

Campbell, R. K. and Rediske, J. H. (1966). Genetic variability of photosynthetic efficiency and dry-matter accumulation in seedling Douglas-fir. *Silvae Genet.* 15, 65—72.

Cartledge, O. and Connor, D. J. (1972). Field chamber measurements of community photosynthesis. *Photosynthetica* 6, 310—316.

Čatský, J., Nováková, J. and Šesták, Z. (1967). Daily carbon dioxide balance and its changes with age in a fodder cabbage plant grown in controlled conditions. *Photosynthetica* 1, 215—218.

Chang, C. S. and Huang, B. K. (1973). Plant growth simulation based on net carbon dioxide consumption. *Trans. Am. Soc. agric. Eng.* 16, 724—727.

Charles-Edwards, D. A. (1971a). A simple kinetic model for leaf photosynthesis and respiration. *Planta* 101, 43—50.

Charles-Edwards, D. A. (1971b). Photosynthesis and photorespiration in *Lolium multiflorum* and *Lolium perenne*. *J. exp. Bot.* 22, 663—669.

Chartier, P. (1970). A model of CO_2 assimilation in the leaf. *In* "Prediction and Measurement of Photosynthetic Productivity" (I. Šetlik, ed.), pp. 307—315. Centre for Agricultural Publishing and Documentation, Wageningen, Netherlands.

Clark, J. and Bonga, J. M. (1970). Photosynthesis and respiration in black spruce (*Picea mariana*) parasitized by eastern dwarf mistletoe (*Arceuthobium pusillum*). *Can. J. Bot.* 48, 2029—2031.

Connor, D. J., Tunstall, B. R. and van den Driessche, R. (1971). An analysis of photosynthetic response in a brigalow forest. *Photosynthetica* 5, 218—225.

Corley, R. H. V., Hardon, J. J. and Ooi, S. C. (1973). Some evidence for genetically controlled variation in photosynthetic rate of oil palm seedlings. *Euphytica* 22, 48–55.

Delaney, R. H. and Dobrenz, A. K. (1974). Yield of alfalfa as related to carbon exchange. *Agron. J.* 66, 498–500.

Dorman, K. W. (1952). Hereditary variation as the basis for selecting superior forest trees. U.S. For. Serv., Southeast. For. Exp. Sta., Sta. Pap. No. 15. 88pp.

Duncan, W. G. (1971). Leaf angles, leaf area, and canopy photosynthesis. *Crop Sci.* 11, 482–485.

Duncan, W. G., Loomis, R. S., Williams, W. A. and Hanau, R. (1967). A model for simulating photosynthesis in plant communities. *Hilgardia* 38, 181–205.

Eagles, C. F. (1974). Diurnal fluctuations in growth and CO_2 exchange in *Dactylis glomerata*. *Ann. Bot.* 38, 53–62.

Evans, L. T. and Dunstone, R. L. (1970). Some physiological aspects of evolution in wheat. *Aust. J. biol. Sci.* 23, 725–741.

Farmer, R. E., Jr. (1974). Juvenile growth of northern red oak: effects of light, temperature, and genotype. Final Report, Tennessee Valley Authority, Norris, Tennessee. 31pp.

Fryer, J. H. and Ledig, F. T. (1972). Microevolution of the photosynthetic temperature optimum in relation to the elevational complex gradient. *Can. J. Bot.* 50, 1231–1235.

Gatherum, G. E. (1965). Photosynthesis, respiration and growth of forest tree seedlings in relation to seed source and environment, pp. 10–18. *In* Proc. Fourth Central States For. Tree Improv. Conf. Lincoln, Nebraska.

Gatherum, G. E., Gordon, J. C. and Broerman, B. F. S. (1967). Effects of clone and light intensity on photosynthesis, respiration and growth of aspen-poplar hybrids. *Silvae Genet.* 16, 128–132.

Gauhl, E. (1969). Differential photosynthetic performance of *Solanum dulcamara* ecotypes from shaded and exposed habitats. *Carnegie Inst. Wash. Yb.* 67, 482–487.

Gopal, M. (1972). A mathematical approach to photosynthesis and growth of tree species. *Indian Forester* 98, 387–401.

Gordon, J. C. and Gatherum, G. E. (1968). Photosynthesis and growth of selected Scotch pine seed sources, pp. 20–23. *In* Proc. Eighth Lake States For. Tree Improv. Conf. U.S. For. Serv. Res. Pap. NC-23.

Halás, L. (1971). Effect of habitat irradiance on the year's course of photosynthetic rate of *Prunus laurocerasus* L. leaf discs. *Photosynthetica* 5, 352–357.

Hall, A. E. (1971). A model of leaf photosynthesis and respiration. *Carnegie Inst. Wash. Yb.* 70, 530–540.

Heichel, G. H. (1971). Confirming measurements of respiration and photosynthesis with dry matter accumulation. *Photosynthetica*, 5, 93–98.

2. Physiological Genetics, Photosynthesis and Growth Models

Helms, J. A. (1965). Diurnal and seasonal patterns of net assimilation in Douglas-fir, *Pseudotsuga menziesii* (Mirb.) Franco, as influenced by environment. *Ecology* **46**, 698–708.

Hesketh, J. D., Baker, D. N. and Duncan, W. G. (1971). Simulation of growth and yield in cotton: respiration and the carbon balance. *Crop Sci.* **11**, 394–398.

Huber, B. and Polster, H. (1955). Zur Frage der physiologischen Ursachen der unterscheidlichen Stofferzeugung von Pappelklonen. *Biol. Zentralbl.* **74**, 370–420.

Huber, B. and Rüsch, J. (1961). Uber den Anteil von Assimilation und Atmung bei Pappelblattern. *Ber. Deutsch. Bot. Ges.* **74**, 55–63.

Jahnke, L. S. and Lawrence, D. B. (1965). Influence of photosynthetic crown structure on potential productivity of vegetation, based primarily on mathematical models. *Ecology* **46**, 319–326.

Kaul, R. and Crowle, W. L. (1974). An index derived from photosynthetic parameters for predicting grain yields of drought-stressed wheat cultivars. *Z. Pflanzenzüchtung* **71**, 42–51.

Keller, T. (1973). CO_2 exchange of bark of deciduous species in winter. *Photosynthetica* **7**, 320–324.

Kramer, P. J. and Clark, W. S. (1947). A comparison of photosynthesis in individual pine needles and entire seedlings at various light intensities. *Plant Physiol.* **22**, 51–57.

Kramer, P. J. and Fiscus, E. L. (1972). Some problems of modeling water flow through plants. *In* "Modeling the Growth of Trees" (C. E. Murphy, Jr., J. D. Hesketh and B. R. Strain, eds.), pp. 99–106. Proc. Workshop on Tree Growth Dynamics and Modeling, Oak Ridge National Laboratory, Oak Ridge, Tennessee.

Krueger, K. W. and Ferrell, W. K. (1965). Comparative photosynthetic and respiratory responses to temperature and light by *Pseudotsuga menziesii* var. *menziesii* and var. *glauca* seedlings. *Ecology* **46**, 794–801.

Laisk, A. (1970). A model of leaf photosynthesis and photorespiration. *In* "Prediction and Measurement of Photosynthetic Productivity" (I. Šetlik, ed.), pp. 295–306. Centre for Agricultural Publishing and Documentation, Wageningen, Netherlands.

Larcher, W. (1969a). Die Bedeutung des Faktors "Zeit" für die photosynthetische Stoffproduktion [in German, English summary]. *Ber. Deut. Bot. Ges.* **82**, 71–80.

Larcher, W. (1969b). The effect of environmental and physiological variables on the carbon dioxide gas exchange of trees. *Photosynthetica* **3**, 167–198.

Ledig, F. T. (1969). A growth model for tree seedlings based on the rate of photosynthesis and the distribution of photosynthate. *Photosynthetica* **3**, 263–275.

Ledig, F. T. (1974). Photosynthetic capacity: developing a criterion for the early selection of rapidly growing trees. *In* "Toward the Future Forest: Applying Physiology and Genetics to the Domestication of Trees" (F. T. Ledig, ed.), pp. 19–39. Yale Univ., School For. and Environ. Stud., Bull. No. 85. New Haven, Connecticut.

Ledig, F. T. and Botkin, D. B. (1974). Photosynthetic CO_2-uptake and the distribution of photosynthate as related to growth of larch and sycamore progenies. *Silvae Genet.* 23, 188–192.

Ledig, F. T. and Perry, T. O. (1976). Variation in photosynthesis and respiration among loblolly pine progenies, pp. 120–128. *In* Proc. Ninth Southern Conf. For. Tree Improv. Knoxville, Tennessee.

Ledig, F. T. and Perry, T. O. (1969). Net assimilation rate and growth in loblolly pine seedlings. *Forest Sci.* 15, 431–438.

Ledig, F. T., Drew, A. P. and Clark, J. G. (1976). Maintenance and constructive respiration, photosynthesis, and net assimilation rate in seedlings of pitch pine. *Ann. Bot.* 40, 289–300.

Lister. G. R., Slankis, V., Krotkov, G. and Nelson, C. D. (1967). Physiology of *Pinus strobus* L. seedlings grown under high or low soil moisture conditions. *Ann. Bot.* 31, 121–132.

Liu, P., Wallace, D. H. and Ozbun, J. L. (1973). Influence of translocation on photosynthetic efficiency of *Phaseolus vulgaris* L. *Plant Physiol.* 52, 412–415.

Logan, K. T. (1971). Monthly variations in photosynthetic rate of jack pine provenances in relation to their height. *Can. J. For. Res.* 1, 256–261.

Logan, K. T. (1974). Photosynthetic efficiency of jack pine provenances in relation to their growth. *Can. For. Serv., Petawawa For. Exp. Sta. Inform. Rep. PS-X-48.* Chalk River, Ontario, Canada. 11pp.

Logan, K. T. and Krotkov, G. (1968). Adaptations of the photosynthetic mechanism of sugar maple (*Acer saccharum*) seedlings grown in various light intensities. *Physiol. Plant.* 22, 104–116.

Lommen, P. W., Schwintzer, C. R., Yocum, C. S. and Gates, D. M. (1971). A model describing photosynthesis in terms of gas diffusion and enzyme kinetics. *Planta* 98, 195–220.

Longman, K. A. and Coutts, M. P. (1974). Physiology of the oak tree. *In* "The British Oak" (M. G. Morris and F. H. Perring, eds.), pp. 194–221. E. W. Classey, Faringdon, England.

Loustalot, A. J., Gilbert, S. G. and Drosdoff, M. (1950). The effect of nitrogen and potassium levels in tung seedlings on growth, apparent photosynthesis, and carbohydrate composition. *Plant Physiol.* 25, 394–412.

Luukkanen, O. and Kozlowski, T. T. (1972). Gas exchange in six *Populus* clones. *Silvae Genet.* 21, 220–229.

Manley, S. A. M. and Ledig, F. T. (1976). Photosynthesis in black and red spruce and their hybrid derivatives: ecological isolation and hybrid inviability. *In litt.* (*Am. J. Bot.*).

McCree, K. J. (1970). An equation for the rate of respiration of white clover plants grown under controlled conditions, *In* "Prediction and Measurement of Photosynthetic Productivity" (I. Šetlik, ed.), pp. 221–229. Centre for Agricultural Publishing and Documentation, Wageningen, Netherlands.

McCree, K. J. (1974). Equations for the rate of dark respiration of white clover and grain sorghum, as functions of dry weight, photosynthetic rate, and temperature. *Crop Sci.* 14, 509–514.

McCree, K. J. and Troughton, J. H. (1966). Prediction of growth rate at different light levels from measured photosynthesis and respiration rates. *Plant Physiol.* 41, 559–566.

McGregor, W. H. D. and Kramer, P. J. (1963). Seasonal trends in rates of photosynthesis and respiration of loblolly pine and white pine seedlings. *Am. J. Bot.* 50, 760–765.

McGregor, W. H. D., Allen, R. M. and Kramer, P. J. (1961). The effect of photoperiod on growth, photosynthesis, and respiration of loblolly pine seedlings from two geographic sources. *Forest Sci.* 7, 342–348.

Moldau, H. (1971). Model of plant productivity at limited water supply considering adaptation. *Photosynthetica* 5, 16–21.

Monsi, M. and Murata, Y. (1970). Development of photosynthetic systems as influenced by distribution of matter. *In* "Prediction and Measurement of Photosynthetic Productivity" (I. Šetlik, ed.), pp. 115–129. Centre for Agricultural Publishing and Documentation, Wageningen, Netherlands.

Mooney, H. A., West, M. and Brayton, R. (1966). Field measurements of the metabolic responses of bristlecone pine and big sagebrush in the White Mountains of California. *Bot. Gaz.* 127, 105–113.

Moss, D. N. and Musgrave, R. B. (1971). Photosynthesis and crop production. *Adv. Agron.* 23, 317–336.

Negisi, K. (1966). Photosynthesis, respiration and growth in one-year-old seedlings of *Pinus densiflora, Cryptomeria japonica* and *Chamaecyparis obtusa. Bull. Tokyo Univ. For.* No. 62. Tokyo, Japan. 115pp.

Neuwirth, G. (1967). Gasstoffwechselökologische Untersuchungen an Larchen-Provenienzen (*Larix decidua* Mill.). I. Vergleichende Stoffwechselanalysen an Pfropflingen [in German, English abstract]. *Photosynthetica* 1, 219–231.

Parker, J. (1953). Photosynthesis of *Picea excelsa* in winter. *Ecology* 34, 605–609.

Parker, J. (1961). Seasonal trends in carbon dioxide absorption, cold resistance, and transpiration of some evergreens. *Ecology* 42, 372–380.

Perry, T. O. (1971). Winter-season photosynthesis and respiration by twigs and seedlings of deciduous and evergreen trees. *Forest Sci.* 17, 41–43.

Polster, H. and Weisse, G. (1962). Vergleichende Assimilationsuntersuchungen an Klonen verschiedener Larchenherkünfte (*Larix decidua* und *Larix leptolepis*) unter Freiland- und Klimaraumbedingungen. *Züchter* 32, 103–110.

Promnitz, L. C. and Rose, D. W. (1974). A mathematical conceptualization of a forest stand simulation model. *Angew. Bot.* 48, 97–108.

Rastorgueva, E. Y. (1966). Photosynthesis and respiration in advance growth of Siberian stone pine and fir on the northern spurs of the western Sayan. *In* "Physiology of Woody Plants of Siberia" (H. A. Khlebnikov *et al.*, eds), pp. 65–74. Acad. Sci. U.S.S.R. Siberian Branch, Trans. For. and Timber Inst., vol. 60. Translated from Russian by Israel Program for Scientific Translations, Jerusalem.

Reines, M. (1963). Photosynthetic efficiency and vigor in pines: variation. *In* "Proc. Forest Genetics Workshop" (J. W. Johnson, ed.), pp. 14–15. Sponsored Pub. No. 22, Southern For. Tree Improv. Comm. Macon, Georgia.

Robertson, C. F. and Reines, M. (1965). The efficiency of photosynthesis and respiration in loblolly pines: variation, pp. 104–105. *In* Proc. Eighth Southern Conf. For. Tree Improv. Savannah, Georgia.

Ross, Y. K. (1966). Mathematical description of plant growth. *Doklady Bot. Sci.* 171, 168–170.

Rüsch, J. (1959). Das Verhältnis von Transpiration und Assimilation als physiologische Kenngrösse, untersucht an Pappelklonen. *Züchter* 29, 348–354.

Rutter, A. J. (1957). Studies in the growth of young plants of *Pinus sylvestris* L. I. The annual cycle of assimilation and growth. *Ann. Bot.* 21, 399–426.

Saeki, T. and Nomoto, N. (1958). On the seasonal change of photosynthetic activity of some deciduous and evergreen broadleaf trees. *Bot. Mag., Tokyo* 71, 235–241.

Schulze, E. D. (1970). Der CO_2-Gaswechsel der Buche (*Fagus silvatica* L.) in Abhängigkeit von den Klimafaktoren im Freiland [in German, English summary]. *Flora* 159, 177–232.

Schultz, R. C. and Gatherum, G. E. (1971). Photosynthesis and distribution of assimilate of Scotch pine seedlings in relation to soil moisture and provenance. *Bot. Gaz.* 132, 91–96.

Schulze, E. D., Mooney, H. A. and Dunn, E. L. (1967). Wintertime photosynthesis of bristlecone pine (*Pinus aristata*) in the White Mountains of California. *Ecology* 48, 1044–1047.

Servaites, J. C. and Geiger, D. R. (1974). Effects of light intensity and oxygen on photosynthesis and translocation in sugar beet. *Plant Physiol.* 54, 575–578.

Sharpe, P. J. H. and DeMichele, D. W. (1974). A morphological and physiological model of the leaf. *Trans. Am. Soc. agric. Eng.* 17, 355–359.

Shiroya, T., Lister, G. R., Slankis, V., Krotkov, G. and Nelson, C. D. (1966).

Seasonal changes in respiration, photosynthesis, and translocation of the ^{14}C labelled products of photosynthesis in young *Pinus strobus* L. plants. *Ann. Bot.* 30, 81–91.

Smith, J. H. C. (1943). Molecular equivalence of carbohydrates to carbon dioxide in photosynthesis. *Plant Physiol.* 18, 207–223.

Smith, J. H. C. (1944). Concurrency of carbohydrate formation and carbon dioxide absorption during photosynthesis in sunflower leaves. *Plant Physiol.* 19, 394–403.

Sorensen, F. C. and Ferrell, W. K. (1973). Photosynthesis and growth of Douglas-fir seedlings when grown in different environments. *Can. J. Bot.* 51, 1689–1698.

Steinhübel, G. and Halás, L. (1969). Seasonal trends in rates of dry-matter production in the evergreen and winter green broadleaf woody plants. *Photosynthetica* 3, 244–254.

Sweet, G. B. and Wareing, P. F. (1966). Role of plant growth in regulating photosynthesis. *Nature, Lond.* 210, 77–79.

Tepper, H. B. (1967). The role of storage products and current photosynthate in the growth of white ash seedlings. *Forest Sci.* 13, 319–320.

Thornley, J. H. M. (1972a). A model to describe the partitioning of photosynthate during vegetative plant growth. *Ann. Bot.* 36, 419–430.

Thornley, J. H. M. (1972b). A balanced quantitative model for root : shoot ratios in vegetative plants. *Ann. Bot.* 36, 431–441.

Thornley, J. H. M. and Hesketh, J. D. (1972). Growth and respiration in cotton bolls. *J. Appl. Ecol.* 9, 315–317.

Tooming, H. (1967). Mathematical model of plant photosynthesis considering adaptation. *Photosynthetica* 1, 233–240.

Townsend, A. M., Hanover, J. W. and Barnes, B. V. (1972). Altitudinal variation in photosynthesis, growth, and monoterpene composition of western white pine (*Pinus monticola* Dougl.) seedlings. *Silvae Genet.* 21, 133–139.

Tranquillini, W. (1963a). Der Jahresgang der CO_2-Assimilation junger Zirben, pp. 501–534. *In* Forschungsstelle für Lawinenvorbeugung, Ökologische Untersuchungen in der subalpinen Stufe zum Zwecke der Hochlagenaufforstung. Teil II. Forstichlen Bundes-Versuchsanstalt Mariabrunn Österreichischer Agraverlag, Wien, Austria.

Tranquillini, W. (1963b). Die CO_2-Jahresbilanz und die Stoffproduktion der Zirbe, pp. 535–546. *In* Forschungsstelle für Lawinenvorbeugung, Ökologische Untersuchungen in der subalpinen Stufe zum Zwecke der Hochlagenaufforstung. Teil II. Forstichlen Bundes-Versuchsanstalt Mariabrunn Österreichischer Agraverlag, Wien, Austria.

Uhl, A. (1937). Untersuchungen über die Assimilationverhältnisse und die Ursachen ihrer Unterschiede in der Gattung *Pinus*. *Jb. wiss. Bot.* 85, 368–421.

Van Bavel, C. H. M. (1974). Soil water potential and plant behaviour: a case modeling study with sunflowers. *Oecol. Plant.* 9, 89—109.

Waggoner, P. E. (1969). Predicting the effect upon net photosynthesis of changes in leaf metabolism and physics. *Crop Sci.* 9, 315—321.

Wareing, P. F., Khalifa, M. M. and Treharne, K. J. (1968). Rate-limiting processes in photosynthesis at saturating light intensities. *Nature, Lond.* 220, 453—457.

Wilson, D. W. and Cooper, J. P. (1967). Assimilation of *Lolium* in relation to leaf mesophyll. *Nature, Lond.* 214, 989—992.

Wit, C. T. de, Brouwer, R. and Penning de Vries, F. W. T. (1970). The simulation of photosynthetic systems. *In* "Prediction and Measurement of Photosynthetic Productivity" (I. Šetlik, ed.), pp. 47—70. Centre for Agricultural Publishing and Documentation, Wageningen, Netherlands.

Zelawski, W. (1967). Variation in the photosynthetic efficiencies of trees with special reference to Scots pine (*Pinus silvestris* L.), pp. 515—535. *In* XIV. IUFRO-Kongress, vol. 3, section 22. Munich, West Germany.

Zelawski, W., Kucharska, J. and Kinelska, J. (1971). Relationship between dry matter production and carbon dioxide absorption in seedlings of Scots pine (*Pinus silvestris* L.) in their second vegetation season. *Acta Soc. Bot. Pol.* 40, 243—256.

Zelawski, W., Szaniawski, R., Dybcznyski, W. and Piechurowski, A. (1973). Photosynthetic capacity of conifers in diffuse light of high illuminance. *Photosynthetica* 7, 351—357.

3
Factors Influencing Net Photosynthesis in Trees: An Ecological Viewpoint

JOHN A. HELMS
Department of Forestry and Conservation, University of California, Berkeley, U.S.A.

I.	Introduction	55
II.	Plant Factors	56
III.	Environmental Factors	59
IV.	Interaction of Processes	63
V.	Comparison of Physiological Capacities of Tree Species in Particular Environments	68
References		73

I. INTRODUCTION

There are three kinds of questions that require answers in order to understand variation in physiological processes in trees. These are:

1. What factors limit process rates and what is the functional relationship between them?

2. What is the likely productivity of plants in a particular environment?

3. What is the likely response of plants to treatments which modify environments?

These kinds of questions can be approached from two complementary directions: firstly in the laboratory where complexity can be minimized, control over variation and interaction can be minimized, and where the degree of resolution of the problem can be high. This approach has proved to be especially fruitful in the study of agricultural plants which are relatively small, have a short life span, and are usually grown as crops under relatively uniform site conditions. Most of our knowledge of tree physiology has been developed using seedling material in laboratory studies. Secondly, the question can be addressed using field studies where complexity is monitored rather than controlled, where problems are resolved to a lesser degree of precision, but where a high degree of direct ecological relevance is maintained. This approach has special importance in forestry where the study organism is large, long lived, and grows on a wide array of site conditions. As pointed out by Jarvis (1970), studies on the effects of rate-limiting factors on photosynthesis can best be made in the laboratory; however, field studies must necessarily be used when information is required on the relation between physiological processes and canopy characteristics, and to test models under naturally varying conditions of the real world — which, after all, is basically what we are interested in.

In recent years, technological advances have tended to produce a merging of these two approaches. Phytotrons and controlled environment glasshouses can create any desired combination of environments. From the other direction, sophisticated field equipment is now being used which can provide the same level of precision as measures made in the laboratory. In addition, it is now possible to control conditions within the sampling chamber (Koch *et al.*, 1971). Thus the desirable situation is developing in tree physiological research where, through integrated approaches, hypotheses can be tested and relationships developed in the laboratory, validated in the field, and vice versa.

In this contribution, I shall explore some of the relationships involved in photosynthesis studies in trees. One example of my field analyses will be used to illustrate these relationships and to highlight the following questions: what are limiting factors; what is the likely productivity of plants; and what are predictable responses to treatments?

II. PLANT FACTORS

The influence of plant factors on photosynthesis has been well reviewed in such standard texts as Kramer and Kozlowski (1960). The topic continues to be of considerable importance, however, since plant factors have a major influence on measures of photosynthesis. Consequently, the type of leaf tissue analyzed and the intensity of sampling should be compatible with the type of research question asked. Indeed, the type of equipment available has a limiting effect on the kinds of research questions that can be appropriately asked. For example, decisions must be made on whether it is more appropriate to design the study to obtain estimates of variation in photosynthesis within or between age-classes of foliage, or within or between tree crowns.

The stage of phenological development of the plant has a major impact on rates of photosynthesis (Bormann, 1958; Kramer and Kozlowski, 1960). In general, cotyledonous material and young, fully grown tissues exhibit much higher photosynthetic efficiencies than do older leaves (Freeland, 1952; Gordon and Larson, 1968; Brix and Ebell, 1969; Higginbotham and Strain, 1974). This fact is of considerable importance when studying *Abies* which may have 7–12 age classes of foliage compared to some species of *Pinus* which may have only two. Part of this variation is due to the effects of mutual shading and differences in environment within tree crowns. Consequently it may be necessary to evaluate differences in the way in which foliage is displayed, differences in crown density, and position within the crown (Helms, 1970; Higginbotham and Strain, 1974).

Similarly, the photosynthetic performance of sun leaves and shade leaves is considerably different. Exposed sun leaves are smaller, thicker and more leathery compared with shade leaves. Logan (1970) demonstrated that photosynthesis of shade leaves of yellow birch was at a higher rate under conditions of low light intensity, also that light saturation was reached at lower light levels, and at lower rates of net photosynthesis, than in sun leaves. These results were similar to those reported earlier by Larcher (1969a). In natural stands of *Pseudotsuga menziesii*, Woodman (1971) showed that the most productive leaves were at the boundary between continuous sun and shade conditions in the upper crown. The position of foliage in the tree's canopy is clearly of great importance in the overall economy of the plant (Helms, 1970; Woodman, 1971). In some species the upper foliage can be exposed to the extent that it becomes chlorotic through photo-oxidation (Ronco, 1975). At the other extreme, in trees with dense foliage, the most shaded, oldest foliage can be below the light compensation point for much of the time and, when rates of respiration are high, can constitute a drain on the tree's carbohydrate resources (Helms, 1964).

Most models of photosynthesis and growth incorporate the factor of leaf area, or leaf area index (Zavitkovski *et al.*, 1974; Ledig, 1974). Recognition of potentially large differences in photosynthetic performance of sun leaves and shade leaves, and the differences due to position within the crown, would appear to necessitate the incorporation of factors that accommodate these differences.

The great variation in net photosynthesis within crowns, due to age of leaves, sun and shade adaptations and variations in light and temperature environment (Mitscherlich *et al.*, 1967), make the sampling problem acute, especially when monitoring equipment is limited.

The question whether different provenances of one species have inherently different photosynthetic capacities has been studied for some time. Results of these studies have been rather conflicting with some evidence supporting a correlation between genotype and growth (Campbell and Rediske, 1966; Gatherum *et al.*, 1967; Ludlow and Jarvis, 1971). Other studies have shown no relationship or inconclusive results (Krueger and Ferrell, 1965; Ledig and Perry, 1967; Gordon and Gatherum, 1968). Logan (1971) found with 7-year-old *Pinus banksiana* saplings that a correlation between rate of photosynthesis and tree height did not exist in data obtained before September, but after that date a positive correlation was found. Logan comments that in some experiments, differences in rates of photosynthesis are obscured by differences in leaf area development. Also, there are substantial differences in seasonal rates, and the ranking of provenances may vary during the season.

Conflicting evidence on the relationship between photosynthesis and growth is also partly attributable to the measurement of photosynthesis in terms of rates per unit of foliage, and growth in terms of one parameter such as height. By using growth analysis methods Ledig and Perry (1967) were able to explain differences in growth of *Pinus taeda* progeny in terms of variations in seasonal patterns of photosynthesis. The importance of measuring seasonal photosynthesis was again emphasized by Logan (1971). Further, to increase productivity it is not just a matter of selecting provenances on the basis of differences in photosynthetic efficiency but perhaps more importantly to select on the basis of inherent capacity to develop larger leaf area, to display that foliar surface more efficiently, and to reduce the time required to reach optimal leaf area index (Zavitkovski et al., 1974). In a similar vein, Dickman (1973) reminds us that relative growth of progeny also depends on the manner in which photosynthate is distributed and utilized by the plant. Differences in rates of respiration and photorespiration can also be an important factor causing differences in growth between provenances.

Another factor influencing photosynthetic capacity is the recognition that plants can be grouped into either C3 or C4 types depending upon differences in the way in which carbon is metabolized. C4 plants have been characterized as being 2—3 times higher in primary productivity; as having photosynthetic capacities of 50—80 mg CO_2 dm^{-2} h^{-1}; CO_2 compensation points of 0—10 ppm; no photorespiration; requiring less water; and commonly growing in stress environments such as tropics, arid areas, mountains and estuaries (Hatch et al., 1971; Black, 1971). Typical C4 plants are maize, sorghum, sugar cane, crabgrass, pigweed and saltbush. Dickman (1973) examined 14 conifers, 16 poplar clones and 39 hardwood species and found them to be typical C3 plants which are characterized by having relatively low photosynthetic capacities (10—35 mg CO_2 dm^{-2} h^{-1}), CO_2 compensation points between 30 and 70 ppm, and respiration stimulated by light.

In a recent article, Moore (1974) argues that there has been some oversimplification regarding the potential advantages of the C4 system. He recognizes the major significance of the C4 process as its capacity for operating at very low ambient CO_2 concentrations, and in the maintenance of growth in water-deficient conditions. However, Moore states that in crops with similar leaf area indices under the same growth conditions, the C4 plants may have only a 1.5-fold advantage at high light intensities, and that this declines with lower light availability. He concludes that if crop growth rates are measured rather than gas exchange under idealized conditions then one finds considerable variation among both C3 and C4 species and there is no evidence for consistently higher growth rates among C4 plants.

III. ENVIRONMENTAL FACTORS

Photosynthetic performance is obviously a function of the plant's environment. In field studies especially, selection of test material and location is important since these factors can largely determine the kinds and ranges of data obtained.

It is particularly important to recognize and evaluate site and productive potential. Elevation influences length of growing season; aspect and slope influence temperature regimes and water relations; and soil characteristics influence water-holding capacity and nutrition. Similarly, stocking levels directly influence available water, light intensity and levels of carbon dioxide concentration, temperature and humidity. It is these kinds of complexities, compounded by diurnal and seasonal movements of the sun, cloud and shadow patterns which necessitate that field studies of photosynthesis be addressed to simple questions. It is unfortunate that research questions are sometimes asked which cannot appropriately be answered given the equipment and resources available for sampling. Scaling up from branch data to give estimates of tree and stand productivity must be avoided when estimates of within-tree variation are not available. This problem can be avoided through the use of meteorological techniques which provide estimates of CO_2 flux for entire stands (Denmead, 1969; Denmead and McIlroy, 1971). However, the complications inherent in this approach may be just as unpalatable as those using sampling chambers.

Of the major nutrients, nitrogen and phosphorus appear to be of most importance in influencing forest productivity. Brix and co-workers (Brix and Ebell, 1969; Brix, 1971) have shown that the main effect of applying nitrogen fertilizers is to increase leaf area; however, Keller (1971) has observed direct influences on rates of photosynthesis.

The importance of light intensity is well documented (Kramer and Kozlowski, 1960; Šesták *et al.*, 1971). Light saturation point is commonly 1400–3000 ft-c for well exposed foliage (Walker *et al.*, 1972); however, in tree crowns where mutual shading occurs, rates of photosynthesis increase with increasing light intensity up to full sunlight (Kramer and Decker, 1944; Kramer and Kozlowski, 1960). In this regard, Dickman (1973) comments that it is unlikely that the total photosynthetic mechanism of most trees is saturated, even in full sun, except by seedlings and sparsely foliated trees. Of special importance in field analyses is the need to quantify the specific seasonal patterns of light attenuation in the particular crowns or canopies being studied.

3. Factors Affecting Net Photosynthesis

Responses of photosynthesis to temperature are difficult to define since they are usually tied to other factors through interactions. The temperature extremes are commonly cited as being between $-5°C$ and $35-40°C$ (Kozkowski and Keller, 1966; Pisek et al., 1969) with wide ranges in optimal temperatures, usually $18-25°C$, depending upon light intensity (Krueger and Ferrell, 1965; Brix, 1967; Larcher, 1969b). Recent field and laboratory studies on *Pseudotsuga menziesii* and *Pinus ponderosa* (Salo, 1974; Helms, unpublished) indicate, however, that maximum rates of net photosynthesis were recorded at the surprisingly low temperature of approximately $10°C$. In the field, higher temperatures of approximately $20°C$ are probably associated with adverse water relations, vapor pressure deficits, and partial stomatal closure. Another consideration, clearly shown by Rook (1969), is that the optimal temperature for photosynthesis in *Pinus radiata* seedlings is dependent upon the temperature at which the plant material is grown. With day and night temperatures of both $15/10°C$ and $24/19°C$, the optimum temperature for photosynthesis was $16°C$. However, with day/night temperatures of $33/28°C$, the optimum was $24°C$.

One of the main difficulties in interpreting the response of photosynthesis to temperature is that increases in temperature are associated with increases in vapor pressure deficit. Observed decreases in rates of net photosynthesis could therefore either be due to increases in temperature or increases in vapor pressure deficits, or both (P. G. Jarvis, personal communication). Consequently, in studies on photosynthesis, both these variables need to be monitored.

In ecological studies, it becomes important to measure not only the temperature of the air, but also the temperature of the leaf and soil. Knowledge of the temperature differences in the entire plant/environment system is necessary to describe metabolic activity which in turn influences plant water stress. Soil and root temperature influences water uptake, and the difference between leaf and air temperature has a marked influence on the vapor pressure gradient and the tendency for water to leave the system (Reed, 1972). An excellent review of methods of leaf temperature measurement has been made by Perrier (1971).

Water relations of plants probably have one of the largest influences on plant productivity and a substantial literature on the subject is available (Kozlowski, 1964 and Chapter 18). In areas which have no summer rainfall, adverse water relations are the prime cause of seedling mortality. In general, net assimilation rate is closely correlated with leaf water potential owing to the tendency for stomatal closure (Hodges, 1967; Kaufmann, 1968; Lopushinsky, 1969; Hinkley and Ritchie, 1970; Ludlow and Jarvis, 1971). In mature trees, the question of water relations differs from that in seedlings in two major respects: firstly, the sapwood constitutes a major water reservoir and Running *et al.* (1975) point out the importance of a high sapwood to leaf area ratio in trees growing in drought-prone areas. Secondly, tall trees are subject to an increasing water potential due to the action of the hydrostatic gradient. Barker (1973) measured increases in potential equivalent to the hydrostatic gradient in *Pinus ponderosa* and in *Abies concolor*. However, Tobiessen *et al.* (1971) found that the gradient in a tall *Sequoiadendron* was somewhat less than expected, about —0.8 bar per 10 m of height. The reasons for this observation have not been satisfactorily resolved.

Field studies of photosynthesis are particularly susceptible to changes in ambient levels of carbon dioxide concentration. Diurnal and seasonal fluctuations occur at all levels from the ground surface to above the tree canopy (Fig. 1). These fluctuations are due to changes in decomposition rates of organic matter in the soil and litter which follow changes in temperature and water content. They are also due to air movements within the forest canopy, and rates of carbon dioxide uptake and evolution. Photosynthesis in field conditions is commonly limited by carbon dioxide availability within dense plant foliage (Moss *et al.*, 1961; Hellmers, 1964; Zelitch, 1971), and process rates can be raised two to three times by increasing carbon dioxide concentrations (Hellmers and Bonner, 1960; Koch, 1969; Ludlow and Jarvis, 1971). Since carbon dioxide is denser than air, and since large amounts are evolved from forest soil, particularly after summer rains (De Sélm, 1952), under stable conditions higher concentrations can occur near the forest floor. This could be ecologically important for photosynthesizing plants in the understory. Similarly in dense canopies resulting from overstocking, and within crowns of trees, carbon dioxide deficiencies can occur under conditions otherwise conducive to active photosynthesis. In most weather conditions, however, there is vigorous mixing of air

3. Factors Affecting Net Photosynthesis 63

due to wind movement, and the concentrations in the canopy and in the atmosphere above it are very similar (Zelitch, 1971).

A final environmental factor to be mentioned in relation to field studies is the effect of animals, insects and diseases. These must be included as exogenous environmental variables because they are potentially rate limiting and process regulating through their actions as consumers of leaf biomass and storage carbohydrate.

IV. INTERACTION OF PROCESSES

Physiological performance reflects the capacity of a particular genotype to develop within a particular operating environment. The interrelationships involved need to be studied as a system in which the variables are largely inter-dependent. Photosynthesis cannot be evaluated in the absence of information on transpiration, respiration and, in particular, leaf resistance.

In general, net assimilation rates of conifers are commonly about the same as, or somewhat lower than, those for broadleaved species. Differences in reported rates are commonly due to differences in light intensity. Larcher (1969a) reports rates of 10–12 mg CO_2 dm^{-2} h^{-1}. Jarvis and Jarvis (1964) quote rates of 5–10 mg CO_2 dm^{-2} h^{-1} for deciduous broad-leaved trees and shrubs. Extremely high rates of 18 mg CO_2 dm^{-2} h^{-1} for *Pseudotsuga menziesii* (Krueger and Ferrell, 1965) and 17 mg CO_2 dm^{-2} h^{-1} for excised *Picea sitchensis* (Ludlow and Jarvis, 1971) have been recorded. Similarly, in a spherical chamber designed to eliminate all mutual shading, Żelawski *et al.* (1973) have obtained rates of 33 mg CO_2 g^{-1} h^{-1} for 5-week-old *Pinus sylvestris* seedlings. The high rates reached by fully exposed seedlings may, however, never be attained by the mature foliage of a mutually shaded crown. This again indicates that a prime purpose of field studies is to provide rate data which correspond to the sub-optimal conditions commonly prevailing in nature.

The importance of photorespiration has generally been underestimated in the past and has recently been brought to attention by Poskuta (1968), Decker (1970), Ludlow and Jarvis (1971), and Zelitch (1971). In many studies, respiration in the light is taken to be

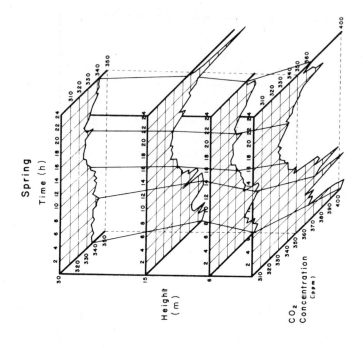

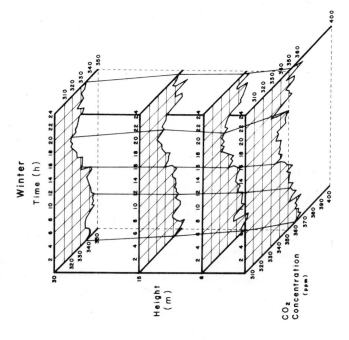

3. Factors Affecting Net Photosynthesis

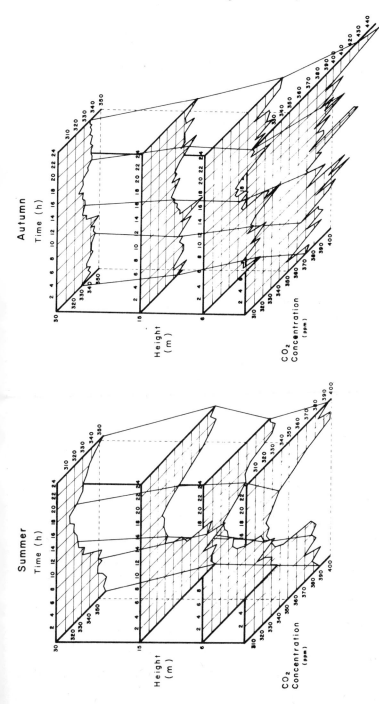

FIG. 1 Gradients in ambient carbon dioxide concentration within a mixed conifer forest in California in each season of the year.

equivalent to dark respiration. In *Fraxinus americana, Liriodendron tulipifera* and *Ginkgo biloba* this practice probably underestimates photorespiration by a factor of four (Zelitch, 1971). Zelitch quotes earlier work done by Żelawski (1967) indicating that photorespiration in *Pinus sylvestris* seedlings was 2.8 times greater than dark respiration. When considering productivity, total respiration becomes important because in forest trees, carbohydrates consumed in respiration may amount to 30–60% of that produced. Consequently, in operations such as seedling establishment, more important impact on growth and survival can probably be made by decreasing respiratory loss by providing shade than by any attempt to increase photosynthetic efficiency.

Together with radiation and temperature, one of the most important factors regulating photosynthesis is leaf resistance to carbon dioxide transfer. Total leaf resistance may be broken down into four components (Salo, 1974). These are: (1) cuticular; (2) stomatal; (3) boundary layer; and (4) mesophyll resistance. Cuticular resistance is very high and is considered to be infinite in normal leaves (Holmgren *et al.*, 1965). The boundary layer resistance is commonly regarded as negligible compared to other resistances in conifers because of the small size of the leaves (Gates, 1968; Jarvis, 1971; Zelitch, 1971; Running *et al.*, 1975). This is particularly true when using adequate flow rates in sampling chambers. The two component resistances of considerable rate-limiting importance in photosynthesis then are stomatal and mesophyll resistances (Chartier *et al.*, 1970; Ludlow and Jarvis, 1971; Neilson *et al.*, 1972; Dykstra, 1974). As reported by Ludlow and Jarvis (1971), the minimum stomatal resistance for carbon dioxide transfer in *Picea sitchensis* was 2.7 and 4.3 s cm^{-1} for glasshouse and forest material respectively. Values obtained for mesophyll resistance were 8.4 and 5.0 s cm^{-1} respectively. Extreme values of stomatal resistance are in the order of 20–30 s cm^{-1}. The measurement of leaf resistance is of prime importance in current analyses of photosynthesis and is obtained indirectly by calculation from measures of transpiration (Zelitch, 1971) or directly by use of a diffusion porometer. The porometer provides a measure of resistance to water flux and this is usually multiplied by a factor of 1.6–1.7 to convert to carbon dioxide transfer resistance (Jarvis, 1971; Zelitch, 1971).

Resistance to water movement is of critical importance since a doubling of resistance reduces transpiration by one half (Running *et al.*, 1975). Running and co-workers also showed that daily mean resistance is a function of pre-dawn moisture stress. As stress increases, leaf resistance increases until the stomata close and the final resistance is equivalent to that through the cuticle. Conifers tolerate greater water loss than broadleaved trees (Jarvis and Jarvis, 1963): for example, stomata of aspen and birch close at -5 and -10 bar respectively (Levitt, 1972). *Abies* stomata closed at -25 bar while those of *Pseudotsuga menziesii* were partially closed at -19 bar (Lopushinsky, 1969). Ecologically, differences in water potential at which stomata close can have an important bearing on the competitive relationship between different species growing on the same site. For example, Barker (1973) demonstrated that in a mixed conifer forest in California which experiences dry summers, lower leaf water potentials were developed in *Abies concolor* than in *Pinus ponderosa*. This indicates that stomatal closure in the more tolerant fir was not as effective as in the intolerant pine which is more capable of growing well in xeric sites.

Photosynthesis also decreases with decrease in leaf water potential and becomes zero in *Pinus contorta* seedlings when water potential reaches -15 bar (Dykstra, 1974). Cleary (1970) demonstrated that with *Pseudotsuga menziesii* photosynthesis declined almost linearly from a maximum when stress was less than -8 atm to 20% of maximum at -22 atm. In contrast, *Pinus ponderosa* showed little decline in photosynthesis until stress reached -15 atm and then declined abruptly with no measurable photosynthesis occurring below -20 atm water stress.

The driving force for transpiration is the gradient in vapor pressure from the leaf to the air when stomata are open. Ritchie's (1971) studies showed that vapor pressure deficits accounted for 56% of the variation in transpiration rate on a seasonal basis. Similarly, Reed (1972) showed that low vapor pressure deficits limited transpiration of *Pseudotsuga menziesii* in the spring, and increased stomatal resistance resulting from depleted soil water limited transpiration in the summer.

These factors integrating physiological response to environmental conditions are probably best focused through the development of conceptual and mathematical models. Considerable progress has been made in this area in the last 5 years, particularly through the activity and stimulation provided by the International Biological Programmes. The objective of these studies has been to evaluate and predict the productive capacity of individual branches, plants or communities. An evaluation of photosynthesis measurements used for ascertaining primary productivity is provided by Lange et al. (1970). Recent studies which have contributed to our understanding of productivity include those of Bray and Gorham (1964), Fujimori and Yamamoto (1967), Whittaker and Woodwell (1967), Botkin et al. (1970), Zavitkovski and Stevens (1972), Reed (1972), Webb (1972), Richardson (1974) and Reed et al. (1975). A good review of criteria used for selecting models was provided by Reed and Webb (1972). In many cases, the development of productivity models is hampered by the lack of adequate data. There is a clear need for more laboratory studies and particularly more field research aimed at providing appropriate process information suitable for analysis of problems of productivity.

V. COMPARISON OF PHYSIOLOGICAL CAPACITIES OF TREE SPECIES IN PARTICULAR ENVIRONMENTS

As an example of the kind of eco-physiological approach discussed above we may consider the results of field studies being done in California, aimed at determining the relative capacities of several species to produce carbohydrate and to maintain satisfactory water relations in natural environments. The research approach is to develop for each species under study a wide response surface quantifying the relation between net photosynthesis, respiration and transpiration, and controlling factors such as leaf resistance, light intensity, temperature and vapour pressure deficit.

Three species are being studied which grow in a natural mixture at 1300 m in the foothills of the Sierra Nevada mountains. The species are ponderosa pine (*Pinus ponderosa*), Douglas-fir (*Pseudotsuga menziesii*), and incense cedar (*Libocedrus decurrens*), which represent a wide range in relative tolerance.

The equipment used includes a Siemens assimilation chamber and a computer-controlled data acquisition system, and is mounted in a mobile trailer so that it can be taken to any desired field location (Fig. 2). The use of the Siemens system permits the development of a

FIG. 2 Field installation of equipment for monitoring gas exchange.

wide response surface by first monitoring physiological behaviour in relation to ambient conditions; then the foliar sample can be subjected to controlled levels of temperature and vapor pressure (Lange et al., 1969).

In this initial phase of the study, complexities of within-tree variation in processes were avoided by restricting sampling to the same relative location in the crowns of each tree. Preliminary data, obtained during the summer, are now available which permit a comparison of species performance (Fig. 3).

Temperature optima for net photosynthesis appear to be unexpectedly low for trees growing naturally in a hot, dry summer area. This may be partially due to the combined effects of increased foliar water stress, vapor pressure deficits, and leaf resistance as leaf temperatures increase above 15–20°C. Douglas-fir, which is the most shade tolerant of the three species, produced maximum rates of net photosynthesis of about 200 mg CO_2 m^{-2} h^{-1} at the low temperature of 8–12°C. The more intolerant pine exhibits similar maximal rates in the field at about 18°C and sustains higher rates of photosynthesis at higher temperatures than does Douglas-fir. Incense cedar, which has an unusually wide range of relative tolerance, exhibits photosynthetic rates intermediate between the other two species. It is interesting to note that the limited amount of data obtained on Douglas-fir photosynthesis in early winter shows that rates of photosynthesis at low temperature are much higher than during the summer, reaching average rates of about 310 mg CO_2 m^{-2} h^{-1}. This higher performance is probably associated with favourable water relations and low leaf resistance.

All species appear to become saturated at light intensities of about 1 cal cm^{-2} min^{-1}. The more intolerant pine produced more photosynthate at any given light level than did either incense cedar or Douglas-fir. The early winter Douglas-fir data indicate very rapid responses of photosynthesis to increases in light intensity. In fact, at any given level of light intensity, rates of net photosynthesis in winter were more than twice those obtained for this species in summer. In winter, a given level of light intensity is associated with lower temperature and lower vapor pressure deficits than the same light level in summer.

As leaf resistance increases, rates of net photosynthesis decrease

3. Factors Affecting Net Photosynthesis

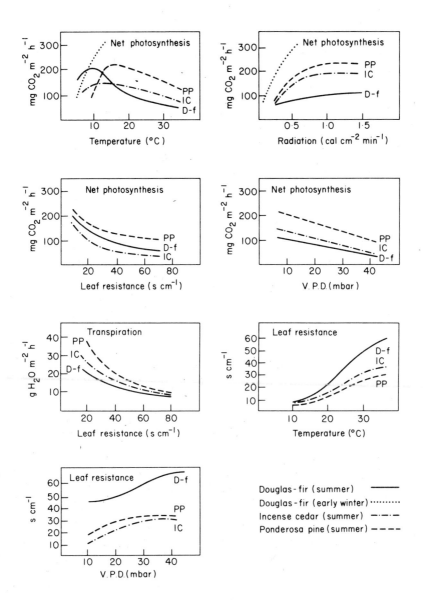

FIG. 3 Relationships between net photosynthesis, transpiration, and leaf resistance of naturally growing ponderosa pine (*Pinus ponderosa*), Douglas fir (*Pseudotsuga menziesii*), and incense cedar (*Libocedrus decurrens*) in terms of leaf temperature, light intensity, vapour pressure deficit and leaf resistance.

with the greatest reduction occurring between a leaf resistance of 0–20 s cm^{-1}. The more intolerant pine maintains higher rates of net photosynthesis at any given leaf resistance than does either Douglas-fir or incense cedar. A similar relationship is shown for each species with increases in vapour pressure deficit. Again, pine maintains higher rates of net photosynthesis at any given vapour pressure deficit than does incense cedar and Douglas-fir.

Rates of transpiration decline rapidly with increase in leaf resistance. At resistances of 80 s cm^{-1}, transpiration of all species dropped to about 10 g H$_2$O m^{-2} h^{-1}. Maximum rates of transpiration of about 40 g H$_2$O m^{-2} h^{-1} were measured at levels of leaf resistance less than 20 s cm^{-1}. At any given leaf resistance, ponderosa pine exhibited higher rates of transpiration than either Douglas-fir or incense cedar.

Leaf resistance of all species was markedly influenced by changes in leaf temperature from 10° to 35°C. Within this temperature range leaf resistance of all species increased from a common level of about 10 s cm^{-1} to an upper level markedly dependent upon species. Intolerant ponderosa pine was shown to have a lower leaf resistance at any given temperature than the other two species and to reach a maximum leaf resistance of about 25 s cm^{-1} at 35°C. The more tolerant Douglas-fir on the other hand reached a leaf resistance of about 60 s cm^{-1} at this high leaf temperature. The relationship between leaf resistance and vapor pressure deficit was similar in that Douglas-fir had markedly higher levels of leaf resistance with increasing vapour pressure deficits than the other two species.

These preliminary findings illustrate different physiological behaviour of trees of different relative tolerance growing side by side in the same field environment. The data indicate that the more shade-intolerant ponderosa pine has the greatest physiological capacity to exploit the hot, dry environmental conditions during summer. Douglas-fir and to a lesser extent incense cedar appear to be at a physiological disadvantage; this is borne out by general observations on survival and relative growth rate on exposed sites.

Currently, these data are being analyzed to describe quantitatively the relationships between processes, productivity, and environment. Similar response surfaces will be developed describing physiological performance in each season of the year and for other representative regions in the tree's crown. With this information simulations can be constructed which predict the likely response of vegetation to natural or induced changes in field environment.

REFERENCES

Barker, J. E. (1973). Diurnal patterns of water potential in *Abies concolor* and *Pinus ponderosa. Can. J. Forest Res.* 3, 556–564.

Black, C. C. (1971). Ecological implications of dividing plants into groups with distinct photosynthetic productive capacity. *In* "Advances in Ecological Research (J. B. Cragg, ed.), Vol. 7, pp. 87–114. Academic Press, London and New York.

Bormann, F. H. (1958). The relationships of ontogenetic development and environmental modification to photosynthesis in *Pinus taeda* seedlings. *In* "The Physiology of Forest Trees" (K. V. Thimann, ed.), pp. 197–215. The Ronald Press, New York.

Botkin, D., Woodwell, G. M., and Tempel, N. (1970). Forest productivity estimated from carbon dioxide uptake. *Ecology* 51, 1057–1060.

Bray, J. R. and Gorham, E. (1964). Litter production in forests of the World. *In* "Advances in Ecological Research" (J. B. Cragg, ed.), Vol. 2, pp. 100–158. Academic Press, London and New York.

Brix, H. (1967). An analysis of dry matter production of Douglas-fir seedlings in relation to temperature and light intensity. *Can. J. Bot.* 45, 2063–2072.

Brix, H. (1971). Effects of nitrogen fertilization on photosynthesis and respiration of Douglas-fir. *Forest Sci.* 17, 407–414.

Brix, H. and Ebell, L. F. (1969). Effects of nitrogen fertilization on growth, leaf area, and photosynthesis rate in Douglas-fir. *Forest Sci.* 15, 189–196.

Campbell, R. K. and Rediske, J. H. (1966). Genetic variability of photosynthetic efficiency and dry matter accumulation in seedling Douglas-fir. *Silvae Genet.* 15, 65–72.

Chartier, P., Chartier, M. and Čatský, J. (1970). Resistances for CO_2 diffusion and for carboxylation as factors in bean leaf photosynthesis. *Photosynthetica* 4, 48–57.

Cleary, B. D. (1970). "The effect of plant moisture stress on physiology and establishment of planted Douglas-fir and ponderosa pine seedlings". Ph.D. Thesis, Oregon State University, Corvallis.

Decker, J. P. (1970). Early history of photorespiration. *Bioeng. Bull.* No. 10. Eng. Res. Center, Ariz. State Univ., Tempe.

Denmead, O. T. (1969). Comparative micro-meteorology of a wheat field and a forest of *Pinus radiata. Agric. Meteorol.* 6, 357–371.

Denmead, O. T. and McIlroy, I. C. (1971). Measurement of carbon dioxide exchange in the field. In "Plant Photosynthetic Production, Manual of Methods" (Z. Šesták, J. Čatský, and P. G. Jarvis, eds.), pp. 467–516. Dr. W. Junk, The Hague.

De Sélm, H. R. (1952). Carbon dioxide gradients in a beech forest in central Ohio. Ohio J. Sci. 52, 187–198.

Dickman, D. I. (1973). Carbohydrate relations. In "Tree Physiology Colloquium", pp. 111–140. Coop. Extension, University of Wisconsin.

Dykstra, G. (1974). Photosynthesis and carbon dioxide transfer resistance of lodgepole pine seedlings in relation to irradiance, temperature, and water potential. Can. J. Forest Res. 4, 201–206.

Freeland, R. O. (1952). Effect of age of leaves upon the rate of photosynthesis in some conifers. Plant Physiol. 27, 685–690.

Fujimori, T. and Yamamoto, K. (1967). Productivity of Acacia dealbata stands. J. Jap. For. Soc. 49, 143–149.

Gates, D. M. (1968). Transpiration and leaf temperature. A. Rev. Plant Physiol. 19, 211–238.

Gatherum, G. E., Gordon, J. C. and Broerman, B. S. F. (1967). Effects of clone and light intensity on photosynthesis, respiration, and growth of aspen-poplar hybrids. Silvae Genet. 16, 128–132.

Gordon, J. C. and Gatherum, G. E. (1968). Photosynthesis and growth of selected Scotch pine seed sources. In "Proc. 8th Lake States For. Tree Impr. Conf." USFS Res Pap NC-23.

Gordon, J. C. and Larson, P. R. (1968). Seasonal course of photosynthesis, respiration, and distribution of ^{14}C in young Pinus resinosa trees as related to wood formation. Plant Physiol. 43, 1617–1624.

Hatch, M. D., Osmond, C. B. and Slatyer, R. O. (1971). "Photosynthesis and Photorespiration". Wiley-Interscience. N.Y.

Hellmers, H. (1964). An evaluation of the photosynthetic efficiency of forests. Q. Rev. Biol. 39, 249–257.

Hellmers, H. and Bonner, J. (1960). In "Photosynthesis limits of forest tree yields". Proc. Soc. Amer. For. Ann. Meeting 1959, pp. 32–35.

Helms, J. A. (1964). Apparent photosynthesis of Douglas-fir in relation to silvicultural treatment. Forest Sci. 10, 432–442.

Helms, J. A. (1970). Summer net photosynthesis of ponderosa pine in its natural habitat. Photosynthetica 4, 243–253.

Higginbotham, K. O. and Strain, B. R. (1974). The influence of canopy position and age of leaves on net CO_2 exchange in loblolly pine. In "Proc. 3rd. North American Forest Biology Workshop. Sept. 1974", pp. 357. Coll. For. and Nat. Res., Colo. State Univ., Fort Collins.

Hinkley, T. M. and Ritchie, G. A. (1970). Within-crown patterns of transpiration, water stress and stomatal activity in Abies amabilis. Forest Sci. 16, 490–492.

Hodges, J. D. (1967). Patterns of photosynthesis under natural conditions. *Ecology* 48, 234–242.

Holmgren, P., Jarvis, P. G. and Jarvis, M. S. (1965). Resistances to carbon dioxide and water vapour transfer in leaves of different plant species. *Physiol. Plant.* 18, 557–573.

Jarvis, P. G. (1970). Characteristics of the photosynthetic apparatus derived from its response to natural complexes of environmental factors. In "Prediction and Measurement of Photosynthetic Productivity". Proc. IBP/PP Tech. Meeting, Trebon, 1969.

Jarvis, P. G. (1971). The estimation of resistances to carbon dioxide transfer. In "Plant Photosynthesis Production: Manual of Methods", (Z. Šesták, J. Čatský, and P. G. Jarvis, eds), pp. 556–631. Dr. W. Junk, The Hague.

Jarvis, P. G. and Jarvis, M. S. (1963). The water relations of tree seedlings. 1. Growth and water use in relation to soil water potential. *Physiol. Plant.* 16, 215–235.

Jarvis, P. G. and Jarvis, M. S. (1964). Growth rates of woody plants. *Physiol. Plant.* 17, 654–666.

Kaufmann, J. R. (1968). Water relations of pine seedlings in relation to root and shoot growth. *Plant Physiol.* 44, 281–288.

Keller, Th. (1971). Der Einfluss der Stickstoffernährung auf den Gaswechsel der Fichte. *Alg. Forst- u. Jagdzt.* 142, 89–93.

Koch, W. (1969). Untersuchungen über die Wirkung von CO_2 auf die Photosynthese einiger Holzewächse unter Laboratoriumsbedingungen. *Flora* 158, 402–428.

Koch, W., Lange, O. L. and Schulze, E. D. (1971). Ecophysiological investigations on wild and cultivated plants on the Negev Desert. 1. Methods: A mobile laboratory for measuring carbon dioxide and water vapour exchange. *Oecologia (Berl.)* 8, 296–309.

Kozlowski, T. T. (1964). "Water Metabolism in Plants". Harper and Row, New York.

Kozlowski, T. T. and Keller, T. (1966). Food relations of woody plants. *Bot. Rev.* 12, 293–382.

Kramer, P. J. and Decker, J. P. (1944). Relation between light intensity and rate of photosynthesis of loblolly pine and certain hardwoods. *Plant Physiol.* 19, 350–358.

Kramer, P. J. and Kozlowski, T. T. (1960). "Physiology of Trees". McGraw-Hill, New York.

Krueger, K. W. and Ferrell, W. K. (1965). Comparative photosynthetic and respiratory responses to temperature and light by *Pseudotsuga menziesii* var. *menziesii* and var. *glauca* seedlings. *Ecology* 46, 794–801.

Lange, O. L., Koch, W. and Schulze, E. D. (1969). CO_2-gas exchange and water relationships of plants in the Negev Desert at the end of the dry period. *Ber. deut. bot. Ges.* 82, 39–61.

Lange, O. L., Schulze, E. D. and Koch, W. (1970). Evaluation of photosynthesis measurements taken in the field. *In* "Prediction and Measurement of Photosynthetic Productivity". Proc. IBP/PP Tech. Meeting, Trebon. 1969.

Larcher, W. (1969a). Physiological approaches to the measurement of photosynthesis in relation to dry matter production by trees. *Photosynthetica* 3, 150–166.

Larcher, W. (1969b). The effect of environmental and physiological variables in relation to dry matter production by trees. *Photosynthetica* 3, 150–166.

Ledig, F. T. (1974). Concepts of growth analysis. *In* "Proc. 3rd. North American Forest Biology Workshop". pp. 166–182. Coll. For. and Nat. Res., Colo. State Univ., Fort Collins.

Ledig, F. T. and Perry, T. O. (1967). Variation in photosynthesis and respiration among loblolly pine progenies. *In* "Proc. 9th Southern For. Tree Impr. Conf.", pp. 120–128.

Levitt, J. (1972). "Responses of Plants to Environmental Stress". Academic Press, New York and London.

Logan, K. T. (1970). Adaptations of the photosynthetic apparatus of sun- and shade-grown yellow birch (*Betula alleghaniensis* Britt.). *Can. J. Bot.* 48, 1681–1688.

Logan, K. T. (1971). Monthly variations in photosynthetic rate of Jack pine provenances in relation to their height. *Can. J. Forest Res.* 1, 256–261.

Lopushinsky, W. (1969). Stomatal closure in conifer seedlings in response to leaf moisture stress. *Bot. Gaz.* 130, 258–265.

Ludlow, M. M. and Jarvis, P. G. (1971). Photosynthesis in Sitka spruce. 1. General characteristics. *J. appl. Ecol.* 8, 925–953.

Mitscherlich, G., Künstle, E. and Lang, W. (1967). Ein Beitrag zur Frage der Beleuchtungstärke im Bestande. *Allg. Forst- u Jagdzt.* 138, 213–223.

Moore, P. D. (1974). Misunderstandings over C_4 carbon fixation. *Nature, Lond.* 252, 438–439.

Moss, D. N., Musgrave, R. B. and Lemon, E. R. (1961). Photosynthesis under field conditions III. Some effects of light, carbon dioxide, temperature, and soil moisture on photosynthesis, respiration, and transpiration of corn. *Crop Sci.* 1, 83–87.

Neilson, R. E., Ludlow, M. M. and Jarvis, P. G. (1972). Photosynthesis in Sitka spruce (*Picea sitchensis* (Bong.) Carr.) 2. Response to temperature. *J. appl. Ecol.* 9, 721–745.

Perrier, A. (1971). Leaf temperature measurement. *In* "Plant Photosynthetic Production, Manual of Methods" (Z. Šesták, J. Čatský, and P. G. Jarvis, eds), pp. 632–671. Dr. W. Junk, The Hague.

Pisek, A., Larcher, W., Moser, W. and Pack, I. (1969). Temperatur-abhängigkeit und optimaler Temperaturbereich der Netto-Photosynthese. *Flora* 158 B, 608–630.

Poskuta, J. (1968). Photosynthesis, photorespiration and respiration of detached spruce twigs as influenced by oxygen concentration and light intensity. *Physiol. Plant.* 21, 1129–1136.

Reed, K. L. (1972). A computer simulation model of seasonal transpiration in Douglas-fir based on a model of stomatal resistance. Ph.D. Thesis, Oregon State University, Corvallis.

Reed, K. L., Hamerly, E. R., Dinger, B. E. and Jarvis, P. G. (1975). An analytical model for field measurements of photosynthesis. Submitted to *J. appl. Ecol.*

Reed, K. L. and Webb, W. L. (1972). Criteria for selecting an optimal model: terrestrial photosynthesis. *In* "Proc. Research on Coniferous Forest Ecosystems — A Symposium", pp. 227–236. U.S. Forest Service, PNW For. and Range Expt. Sta.

Richardson, C. J. (1974). A net photosynthesis index — an indirect method of predicting photosynthetic capacity. *In* "Proc. 3rd. North American Forest Biology Workshop", p. 385. Coll. For. and Nat. Res., Colo. State Univ. Fort Collins.

Ritchie, G. A. (1971). Transpiration, water potential and stomatal activity in relation to microclimate in *Abies amabilis* and *Abies procera* in a natural environment. Ph.D. Thesis, University of Washington, Seattle.

Ronco, F. (1975). Diagnosis: "sunburned" trees. *J. For.* 73, 31–35.

Rook, D. A. (1969). The influence of growing temperature on photosynthesis and respiration of *Pinus radiata* seedlings. *N.Z. Jl. Bot.* 7, 43–55.

Running, S. W., Waring, R. H. and Rydell, R. A. (1975). Physiological control of water flux in conifers. *Oecologia (Berl.)* 18, 1–16.

Salo, D. J. (1974). Factors affecting photosynthesis in Douglas-fir. Ph.D. Thesis, University of Washington, Seattle.

Šesták, Z., Čatský, J. and Jarvis, P. G. (eds.). (1971). "Plant Photosynthetic Production, Manual of Methods". Dr. W. Junk, The Hague.

Tobiessen, P., Rundel, P. W. and Stecker, R. E. (1971). Water potential gradient in a tall *Sequoiadendron. Plant Physiol.* 48, 303–304.

Walker, R. B., Scott, S. R. M., Salo, D. J. and Reed, K. L. (1972). Terrestrial process studies in conifers: a review. *In* "Proc. Research on Coniferous Forest Ecosystems — A Symposium". pp. 211–225. U.S. Forest Service, PNW For. and Range Expt. Sta.

Webb, W. L. (1972). A model of light and temperature controlled net photosynthetic rates for terrestrial plants. *In* "Proc. Research on Coniferous Forest Ecosystems — A Symposium". pp. 237–242. U.S. Forest Service, PNW For. and Range Expt. Stn.

Whittaker, R. H. and Woodwell, G. M. (1967). Surface area relations of woody plants and forest communities. *Am. J. Bot.* 54, 931–939.

Woodman, J. N. (1971). Variation in net photosynthesis within the crown of a large forest-grown conifer. *Photosynthetica* 5, 50–54.

Zavitkovski, J. and Stevens, R. D. (1972). Primary productivity of red alder ecosystems. *Ecology* 53, 235–242.

Zavitkovski, J., Isebrands, J. G. and Crow, T. R. (1974). Application of growth analysis in forest biomass studies. *In* "Proc. 3rd. North American Forest Biology Workshop". Coll. For. and Nat. Res., Colo. State Univ., Fort Collins.

Żelawski, W., Szaniawski, R., Dybczynski, W. and Pierchurowski, A. (1973). Photosynthetic capacity of conifers in diffuse light of high illuminance. *Photosynthetica* 7, 351–357.

Zelitch, I. (1971). "Photosynthesis, Photorespiration, and Plant Productivity". Academic Press, New York and London.

4
Photosynthetic and Enzymatic Criteria for the Early Selection of Fast-Growing *Populus* Clones

J. C. GORDON AND L. C. PROMNITZ
Department of Forestry, University of Iowa, Ames, U.S.A.

I.	Introduction	79
II.	Choosing Physiological Variables	81
III.	Choosing Plant Material	82
IV.	Clonal Differences in Net Carbon Fixation	83
	A. Leaf Age	84
	B. Light Intensity	86
	C. Light Interception	87
V.	Clonal Differences in Peroxidase Activity	88
VI.	Clonal Differences in Nitrate Reductase Activity	91
VII.	Multivariate Methods	93
VIII.	Summary and Conclusions	95
Acknowledgement		96
References		96

I. INTRODUCTION

Populus clones are currently being examined in North America for use in intensive silvicultural systems because of their rapid growth, ease of propagation, and high utility for a variety of wood-fibre products (Schreiner, 1959; Cram, 1960; Larson and Gordon, 1969b; Dawson and Hutchinson, 1973). Because more *Populus* species and variants are available than can reasonably be field tested, a rapid technique for the selection of superior clones must be devised. A desirable technique must be simple and fast, in contrast to field-growth studies that might take from 3 to 20 years and would occupy large areas. If controlled-environment growth studies and physiological indicators can be used to select genotypes capable of rapid growth, field trials could be smaller, with attendant savings in time, effort and money.

The chances of successful early selection of poplar clones are enhanced by the genetic constancy of clonal material, by the great amount of knowledge about poplars and their culture already accumulated, and by the well defined cultural conditions and relatively short rotations used in the intensive silviculture systems now emerging (e.g. Larson and Gordon, 1969b). Intensive culture and short rotations particularly improve the chances of successful selection because environmental variation and time, the principal saboteurs of growth-predicting correlations, are both small.

Our ultimate objective is to construct indices based on controlled-environment growth and relatively simple physiological measurements that accurately predict field performance under specific cultural regimes. Our physiological approach has been to work the problem backwards; that is, to choose clones that differ markedly in juvenile growth rate, and to examine them systematically for differences in the physiological processes that we feel must be major determinants of productivity (Gordon, 1975). At the same time, we are studying growth of the same clones in a variety of controlled and field environments and determining what combinations of controlled-environment conditions and growth variables best predict field performance (Hennessey and Gordon, 1975). We also are developing mathematical models that will more fully describe and predict field growth for longer periods of time (Promnitz, 1975; Promnitz and Rose, 1974; Rose and Promnitz, 1975). These models are based on the physiological responses of known genotypes to major environmental determinants of productivity and can be continuously refined as new physiological information becomes available. We hope they will eventually replace conventional yield tables, at least for intensive silvicultural systems using short rotations.

This contribution discusses studies of physiological differences among genotypes showing different juvenile growth rates, with a brief treatment of how knowledge of these differences can be used to construct prediction indices. We believe, however, that all the components mentioned are necessary to the efficient practical utilization of physiological information on intensive silvicultural systems.

II. CHOOSING PHYSIOLOGICAL VARIABLES

Although it is true that, if prediction of growth is the only objective, there is no absolute need to choose variables causally related to growth rate, we think there are sound reasons for working with physiological processes that we have good reason to believe are causally related to growth and yield. Not only are such variables likely to be good predictors; knowing more about them may point the way to efficient yield improvement in the future.

On the basis of what is generally known about tree and stand growth, three broad aspects of tree physiology seem most likely to contain predictive variables:

1. photosynthesis and respiration, the processes directly determining net carbon fixation (approximating dry weight growth);

2. nitrogen uptake and metabolism, the processes responsible for the utilization of the nutrient most often limiting tree growth;

3. hormone metabolism, the basic system of meristem coordination and control.

Each of these physiological aspects can be described, at least theoretically, by many measured variables, and measurements can be made at different levels of organization (subcellular, cellular, tissue, organ, tree, or even stand) and at different places in the biochemical chain linking information stored in the genome to accomplished fact of form and function (nucleic acids, individual enzymes, systems of enzymes, fluxes of substrates, and products). It would be best, of course, to "read" base sequences in the DNA of individuals and thus to predict directly the genetic component of productivity. Some promising basic work on nucleic acids at the subcellular level in trees (e.g. Hall and Miksche, 1975) indicates that this may some day be possible, but currently it is not. Similarly, measurements of specific substrate and product fluxes for whole trees, as with whole-tree measurements of photosynthesis and respiration by gas exchange, are sometimes possible and desirable. They are also, very often, cumbersome and expensive. If thousands of genotypes must be screened rapidly, physiological measurements must be simple, cheap and relatively fast. We have therefore concentrated on measurements of gas exchange and specific enzymes at the tissue and organ levels, while being careful to relate those measurements to whole-tree and stand measures of growth.

Specifically, we have measured the following:

1. net photosynthesis, photorespiration and dark respiration (Domingo and Gordon, 1974; Dickmann et al., 1975), as well as the related measures of leaf physiological state and activity, CO_2 compensation point (Dickmann and Gjerstad, 1973), stomatal diffusion resistance, incorporation of ^{14}C-labelled photosynthate into leaf protein (Dickmann and Gordon, 1975), glycolate oxidase activity (Gjerstad, 1975), and plant shape (Max, 1975);

2. peroxidase activity, both as total activity in crude extracts and as isoenzymes resolved by acrylamide gel electrophoresis (Gordon, 1971; Wolter and Gordon, 1975; Wray, 1974);

3. nitrate reductase activity, assayed in excised tissues (Dykstra, 1974; Fasehun, 1975).

III. CHOOSING PLANT MATERIAL

Two sets of poplar clones were chosen which had different patterns and rates of growth when grown in a variety of environmental conditions (Wray, 1974). All were natural or artificial poplar hybrids; one group was relatively similar genetically, all being *Populus* x *euramericana,* and the other group was genetically diverse, including representatives of the poplar sections Leuce, Tacamahaca and Aegeiros (Table I). Henceforth, the clones will be referred to by their acquisition numbers as listed in the first column of Table I. For

TABLE I. Acquisition numbers, sex, botanical names and clone common names for the hybrid *Populus* clones for which physiological comparisons were made.

Acquisition number	Sex	Botanical name (parentage)	Clone name
5321	female	*Populus* x *euramericana*	Negrito de Granada
5323	male	*Populus* x *euramericana*	Canada Blanc
5326	male	*Populus* x *euramericana*	Eugenii
5328	male	*Populus* x *euramericana*	I 45/51
5377	male	*Populus* x *euramericana*	Wisconsin No. 5
5260	male	*Populus tristis* x *P. balsamifera*	Tristis No. 1
5339	male	*Populus alba* x *P. grandidentata*	Crandon

various experiments, group 1 included some or all of these clones: 5321, 5323, 5326, 5328 and 5377, whereas group 2 included 5377, 5360 and 5339. Note that, within group 1, clone 5321 was usually a relatively slow grower, while clones 5323 and 5326 were usually fast (Table II).

A final point is that all clones were selected by someone, at some location, for rapid growth and other desirable characteristics.

IV. CLONAL DIFFERENCES IN NET CARBON FIXATION

The clones compared were chosen, in part, because of differences in their rates of dry weight accumulation during early growth. Thus it was virtually certain that they differed in some aspects of the overall processes of carbon fixation, and the major questions remaining were: (1) what measures of photosynthetic and respiratory activities were best related to growth? and (2) did the differences in rates of dry weight accumulation persist, and were they, therefore, predictable from early differences in photosynthesis and respiration rates? The answer to (2) is not yet in, but Table II shows that early

TABLE II. Measures of growth of *Populus* clones for which physiological comparisons were made. Growth chamber growth is for 6, whereas glasshouse growth is for 8 weeks. Total height and top dry weight for these periods are presented, as well as total height and dry weight for 2 years' field growth for clones 5321, 5323, 5326 and 5377.

Clone	Growth Chamber		Glasshouse		Field	
	Height (cm)	Top Dry Weight (g)	Height (cm)	Top Dry Weight (g)	Height (cm)	Total Dry Weight (g)
5321	65	19	109	28	265	709
5323	76	22	161	63	294	987
5326	79	24	171	70	289	1129
5328	54	18	125	54	—	—
5377	76	25	170	67	259	949
5260	86	22	135	38	—	—
5339	52	16	—	—	—	—

differences resemble those observed after 2 years' growth in the field, which may be one-third to one-half of the anticipated rotation age.

A. Leaf Age

The answer to the first question demands detailed knowledge of leaf developmental patterns in relation to photosynthesis and respiration rates, because the aggregate photosynthetic capability of a tree depends upon the display and relative photosynthetic activity of all its leaves. These are strongly influenced, in turn, by leaf age and the environment under which leaves develop (Larson and Gordon, 1969a; Dickmann, 1971; Isebrands and Larson, 1973).

In general, young poplar leaves have high rates of dark respiration while they expand and reach maximum net photosynthetic rates after full expansion (Dickmann *et al.*, 1975). After a period of sustained high net photosynthetic rates, a gradual decline in photosynthetic activity begins, and just before leaf abscission, there is a rapid loss of photosynthetic capability and a sudden rise in CO_2 compensation point. Gjerstad (1975) has compared patterns of net photosynthesis at high and low oxygen levels in three of our selected clones to determine whether differences in leaf development rates could underlie the observed differences in clonal growth, and also to determine whether clones respond differently to the suppression of light respiration. He found differences among clones in leaf photo-

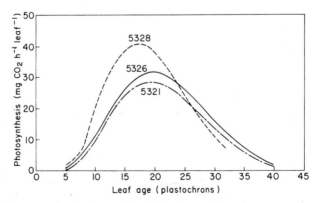

FIG. 1 Photosynthesis rates per leaf at normal (21%) O_2 concentrations for *Populus* × *euramericana* clones 5321, 5326, and 5328.

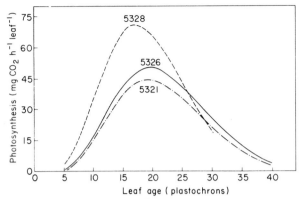

FIG. 2 Photosynthesis rates per leaf at low (2%) O_2 concentrations for *Populus* x *euramericana* clones 5321, 5326 and 5328.

synthetic rates with increasing leaf age measured at saturating light intensities at both normal (21%) and low (2%) oxygen concentrations (Figs 1 and 2). Table III indicates that clonal productivity could be expected to differ markedly in response to suppression of photorespiration. Clone 5328 had a faster rate of leaf development and senescence, and a greater response to low oxygen, than either 5321 or 5326.

Statistical analyses showed that clone and leaf age interacted strongly in their effects on photosynthesis, photorespiration, dark

TABLE III. Calculated net photosynthetic rates per plant when all leaves are exposed to saturating light intensities, and calculated photosynthetic gain due to the suppression of light respiration for *Populus* x *euramericana* clones 5321, 5326 and 5328.

Clone	Net photosynthesis (mg CO_2 h^{-1} $plant^{-1}$)		Increased net photosynthesis (%) due to suppression of photorespiration $\left(\dfrac{y - x}{x} \times 100\right)$
	21% O_2 (x)	2% O_2 (y)	
5321	600.5	948.0	57.9
5326	536.5	830.9	54.9
5328	666.1	1152.0	73.0

respiration, leaf diffusive resistance and CO_2 compensation concentration. Diffusive resistance and CO_2 compensation concentration, however, were poorly correlated with other measures of photosynthetic activity.

B. Light Intensity

The environment in which a leaf develops profoundly affects its photosynthetic properties, and the number of possible combinations of specific levels of environmental factors affecting leaf development is very large. To begin learning about clonal differences in growth and dry weight accumulation as affected by environmental factors, we grew four group 1 clones (5321, 5326, 5328, and 5323) at different shade levels in a glasshouse. The clones responded differently to shade (Table IV), so we measured photosynthetic responses to light intensity in leaves grown at different light intensities (Fasehun, 1975). In conventional forestry terms, clone 5321 was most "shade tolerant", whereas clones 5326 and 5328 were least "shade tolerant". Clone 5323 seemed to do relatively well under all light conditions. These differences were paralleled somewhat by differences in photosynthetic response to light intensity, although with important exceptions. For example, clone 5323, which grew relatively well under all light intensities, had a relatively

TABLE IV. Dry weight at the end of 6 weeks glasshouse growth for 4 *Populus* x *euramericana* clones grown at 3 light intensity levels (37, 75 and 100% of full sunlight).

Clone	Percentage of full sunlight		
	37%	75%	100%
	(total dry weight, g)		
5321	6.31	13.43	15.56
5323	5.54	15.65	26.12
5326	2.73	10.75	22.22
5328	4.46	10.36	31.68

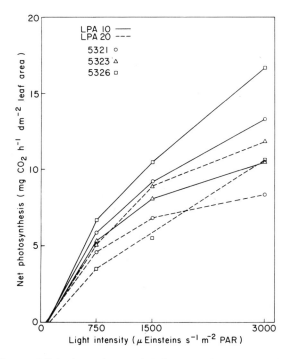

FIG. 3 Effects of light intensity on the photosynthetic rates of leaves of three *Populus* × *euramericana* clones previously grown in full sunlight. LPA = Leaf Plastochron Age. Clone 5321 was "shade tolerant", 5326 was "shade intolerant", and 5323 grew relatively well at all light intensities.

low photosynthetic rate at saturating light intensities in young leaves (LPA 10; i.e. leaf plastochron age = 10) but had a high rate in older leaves (LPA 20), in contrast to both "shade-tolerant" clone 5321 and "shade-intolerant" clone 5326 (Fig. 3). Also, clone 5321 showed the greatest adaptation to shade in terms of change in its photosynthetic response to light intensity. When grown under the densest shade, 5321 had a lower apparent light-saturation intensity than did 5323 or 5326 (Fig. 4).

C. Light Interception

The photosynthetic variables so far discussed were measured independently of leaf display. It may be, however, that plant shape, through its effects on mutual shading of leaves, is an important

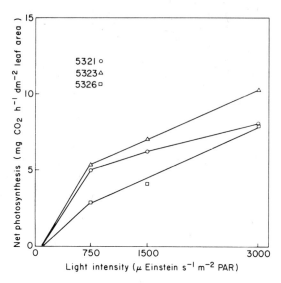

FIG. 4 Effects of light intensity on the photosynthetic rates of three *Populus* × *euramericana* clones previously grown in dense shade. (Compare Fig. 3).

determinant of photosynthetic capability differences among clones. The young trees we have used are relatively simple to describe in terms of shape, and yet the difficulties are formidable (Max, 1975). Leaf angle, however, has been an important determinant of the photosynthetic capability of annual crops and can be measured with relative ease. Figure 5 shows differences in leaf angle for clones 5321, 5323 and 5326. Clone 5323, which was a relatively good grower at all light intensities, and which performed well in dense stands in the field (Rose and Promnitz, 1975), had the most uniform leaf angle, and its change in leaf angle with plastochron age was characteristically different from that of clones 5321 ("shade tolerant") and 5326 ("shade intolerant"). Clones 5321 and 5326 differed from each other in leaf-angle pattern, and this may help to explain their differential performance at high light intensities.

V. CLONAL DIFFERENCES IN PEROXIDASE ACTIVITY

We chose peroxidases as the first enzyme systems to be examined because previous work implicated them in the control of growth rate

4. Photosynthetic and Enzymatic Criteria for Selection

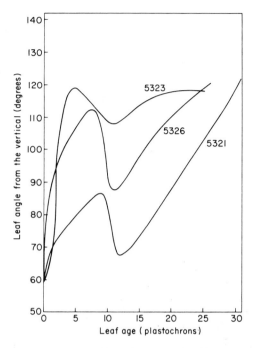

FIG. 5 Leaf angle from the vertical of leaves of different ages, for *Populus* x *euramericana* clones 5321, 5323, and 5326. An angle of 0° is vertical, leaf tip pointing straight up, 90° is horizontal and 180° is vertical, leaf tip pointing straight down.

in a number of higher plants (Galston and Davies, 1969), most probably through effects on hormone metabolism. This, together with simplicity of assay, led us to hypothesize that peroxidase activity and expression might be useful as growth predictors in rapid selection schemes. Before enzymatic criteria can be successfully used in this way, however, it is necessary to know whether peroxidase activity is, in fact, related to growth rate in poplars. Because peroxidase activity is invariably present in multiple forms in plant tissues, it is possible that the activity of only one or a few isoenzymes is quantitatively related to growth. Furthermore, because growth rate is strongly modified by environment, it was reasonable to suppose that those components of total peroxidase activity most directly related to control of growth would be those most easily observed by comparing genetically identical plant material growing under radically different environments.

We first measured peroxidase activity in *Populus tremuloides* callus cultures grown in different hormonal environments. Total peroxidase activity was linearly and positively related to growth (Fig. 6), but no qualitative isoenzyme differences consistently paralleled growth differences (Wolter and Gordon, 1975). In a second experiment, we examined whether photoperiodically induced differences in growth rate were paralleled by activity changes in any of the isoperoxidases of leaves and internodes of three genetically distinct poplar clones (5377, 5260, and 5339). In 5377 and 5260, one internode isoperoxidase showed large quantitative differences in response to photoperiod. One isoenzyme, located at R_f 0.4, stained more intensely in trees grown under the shorter photoperiod, and staining intensity differences at this location were observable just before height growth rates under the two photoperiods began to diverge. Thus encouraged, we analyzed the *Populus x euramericana* clones that differed in growth characteristics (5321, 5323 and 5326) although they were of similar parentage. Internodes of these clones showed differences in total peroxidase activity and in activity of the isoenzyme location which we suspected might be related to growth

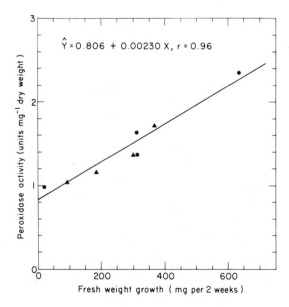

FIG. 6 Total peroxidase activity and fresh weight growth of *Populus tremuloides* callus cultures.

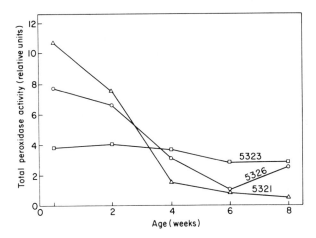

FIG. 7 Total peroxidase activity in the expanding internode above the first mature leaf as a function of age (0 weeks is a rooted tip cutting) for *Populus* x *euramericana* clones 5321, 5323 and 5326.

rate on the basis of the previous study. The most striking differences, however, were in changes in total peroxidase activity in expanding internodes with time (Fig. 7). Clone 5321 had initially high activity, but ultimately the lowest activity of the three, whereas activity in 5323 varied little over the course of the experiment. Obviously, although the initial and final ranking of the three clones by peroxidase activity parallels what we know about their growth-potential ranking, total peroxidase activity differences will have to be better understood before they can serve as a useful selection criterion for poplars.

VI. CLONAL DIFFERENCES IN NITRATE REDUCTASE ACTIVITY

As with peroxidase, our first step was to determine if clones differed in nitrate reductase activity. The question was: if the ability of clones to respond to a nitrate-rich environment is related to growth rate under intensive culture, as it should be, are levels of the inducible enzyme, nitrate reductase, related to this ability? If so, it could be a quick tool to select clones with maximal response to applied nitrate. Our first study compared nitrate reductase levels in two clones (5377 and 5260) subjected to a range of substrate nitrogen levels (Dykstra, 1974). Clone 5377 had approximately twice the nitrate reductase activity of 5260 at optimal substrate nitrogen levels (Fig. 8). In a field growth experiment conducted under high levels of nitrate fertilization, 5377 grew much more than 5260 (G. Dykstra, personal communication). Although far from conclusive, this result encouraged us to compare nitrate reductase activities in leaves of the genetically similar clones 5321, 5323 and 5326. In these clones, nitrate reductase activities were higher in the leaves than in the roots, and clone 5323, a superior performer in the field, had a 3-fold higher activity in young (LPA 10) leaves than did either 5321 or 5326 (Table V). In older leaves (LPA 20), both 5323 and 5326 had more than twice the nitrate reductase activity of clone 5321, a relatively poor performer in both controlled-environment and field tests. Nitrate reductase, however, is an inducible enzyme and thus highly dependent on substrate nitrate activity. Thus, assay conditions and methods will have to be standardized and improved before the growth predictive value of this enzyme can be firmly established.

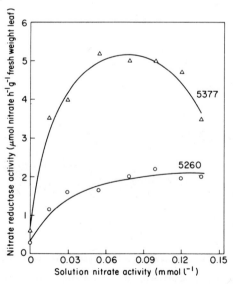

FIG. 8 Nitrate reductase activity in young leaves as a function of substrate nitrate activity for clones 5377 and 5260 of *Populus* x *euramericana*.

TABLE V. Nitrate reductase activities of clones 5321, 5323 and 5326 as affected by Leaf Plastochron-Age (LPA). Each value is the mean of four leaves.

Clone	Nitrate Reductase Activity	
	LPA 10	LPA 20
	(μ mol NO_2 h^{-1} dm^{-2} leaf)	
5321	9.0	7.7
5323	30.2	16.1
5326	10.4	18.8

VII. MULTIVARIATE METHODS

The class of statistical techniques commonly referred to as multivariate methods offer promise for the simultaneous accounting of many physiological measures (Pearce, 1969). Our main purpose is to discriminate among clones and thereby identify growth potential. The use of multivariate techniques for quantitative prediction of field performance through canonical correlations requires more extensive field information than is currently available, but rough rankings in terms of growth potential can be made.

To illustrate multivariate techniques, a canonical analysis was performed on eight photosynthetic and respiratory measures for clones 5321 and 5326. The first two canonical variates are presented in Table VI and represent independent, linear combinations of the original, non-independent variables. The result is a decrease in the dimensionality of the problem (from eight to two dimensions) while still accounting for 71% of the variability in the original data. Both canonical variables can be interpreted as contrasts between photosynthetic and respiratory activity, and when plotted for each clone and leaf age, differences in photosynthetic capacity become readily apparent (Fig. 9). Clones 5321 and 5326 have similar photosynthetic rates for all leaf ages (see Figs 1 and 2), but can be separated when all measures are taken simultaneously. Thus, the plot for clone 5326 encompasses clone 5321, and this, coupled with glasshouse and

TABLE VI. Results of canonical analysis of measures of gas exchange characteristics for clones 5321 and 5326 of *Populus* × *euramericana*.

	Canonical variable 1		Canonical variable 2	
	Coefficient	Correlation[a]	Coefficient	Correlation[a]
Photosynthesis (21% O_2)	−0.025	−0.418	−0.011	0.167
Photosynthesis (2% O_2)	−0.013	−0.360	0.015	0.318
Dark respiration (21% O_2)	0.105	0.551	0.031	−0.150
Dark respiration (2% O_2)	0.109	0.674	−0.158	−0.289
Apparent photo-respiration (21% O_2)	0.047	0.395	0.055	0.591
Apparent photo-respiration (2% O_2)	0.005	0.358	0.074	0.630
CO_2 compensation concentration	−0.0002	−0.028	−0.0008	−0.366
Diffusive resistance	−0.007	−0.118	−0.002	−0.321
Percent variation explained	44.24		27.03	

[a] Correlation coefficients between each canonical variable and the dependent variables.

growth chamber dry weight growth, would indicate the photosynthetic superiority of clone 5326. This superiority in dry weight productivity remains consistent in field performance after 2 years' growth (Table II).

Other variables, such as nitrate reductase and peroxidase activity, could be included in multivariate analyses to sharpen clonal comparisons. Of more practical importance, however, is the fact that controlled-environment growth measures can be combined with physiological variables to achieve maximum clonal discrimination. It is likely that this combination would result in accurate predictions of field performance under intensive culture.

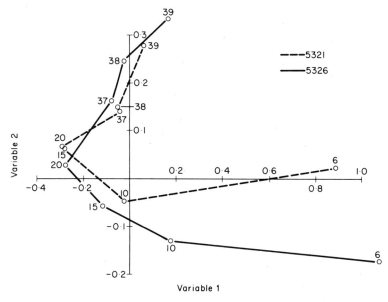

FIG. 9 Multivariate plot of canonical variables 1 and 2 derived from the gas exchange characteristics of leaves of different ages (numbered points indicate leaf plastochron ages) for clones 5321 and 5326 of *Populus* × *euramericana*. The difference between the plots indicates a general clonal difference, and the high negative values for leaves of clone 5326 indicate their photosynthetic superiority, in that photosynthesis exceeds respiration.

VIII. SUMMARY AND CONCLUSIONS

Rapid, early selection methods for *Populus* clones are needed to decrease the size of field trials. Physiological techniques that help discriminate clones could be important parts of these methods, if clones consistently differ in physiological characteristics related to yield. We have examined a variety of physiological variables in several hybrid *Populus* clones and found that they differ in (a) net photosynthesis and photorespiration rates as affected by leaf age and light intensity, (b) patterns of leaf orientation and (c) activities of peroxidase and nitrate reductase. The clonal differences in these variables seem fairly consistent, and any one of them could probably be used with some success to predict field performance under intensive culture. However, canonical variables that combine several physiological measurements with controlled-environment growth information will probably give the most reliable predictions of field growth.

ACKNOWLEDGEMENT

This contribution forms journal paper No. 3-8161 of the Iowa Agriculture and Home Economics Experiment Station, Projects 2033 and 1872, supported in part by the North Central Forest Experiment Station, USDA Forest Service.

REFERENCES

Cram, W. H. (1960). Performance of seventeen poplar clones in South Central Saskatchewan. *Forest Chron.* 36, 204—208.

Dawson, D. H. and Hutchinson, J. G. (1973). Farming for fiber. *Wisc. Conserv. Bull.* (May—June), pp. 24—26.

Dickmann, D. I. (1971). Photosynthesis and respiration by developing leaves of cottonwood (*Populus deltoides* Bartr.). *Bot. Gaz.* 132, 253—259.

Dickmann, D. I. and Gjerstad, D. H. (1973). Application to woody plants of a rapid method for determining leaf CO_2 compensation concentration. *Can. J. Forest Res.* 3, 237—242.

Dickmann, D. I. and Gordon, J. C. (1975). Incorporation of ^{14}C-photosynthate into protein during leaf development in young *Populus* plants. *Plant Physiol.*, 56, 23—27.

Dickmann, D. I., Gjerstad, D. H. and Gordon, J. C. (1975). Developmental patterns of CO_2 exchange, diffusion resistance, and protein synthesis in leaves of *Populus* x *euramericana*. *In* "Proc. of the Symposium on Env. and Biol. Control of Photosynthesis", in press.

Domingo, I. L. and Gordon, J. C. (1974). Physiological responses of an aspen-poplar hybrid to air temperature and soil moisture. *Bot. Gaz.* 135, 184—192.

Dykstra, G. F. (1974). Nitrate reductase activity and protein concentration of two *Populus* clones. *Plant Physiol.* 35, 632—634.

Fasehun, F. E. (1975). "Effects of light intensity on growth, photosynthesis and nitrate reductase activity in hybrid poplar". Ph.D. Dissertation, Iowa State University, Ames, Iowa.

Galston, A. W. and Davies, P. J. (1969). Hormonal regulation in higher plants. *Science* 163, 1288—1297.

Gjerstad, D. H. (1975). "Photosynthesis, photorespiration and dark respiration in *Populus* x *euramericana*: effects of genotype and leaf age". Ph.D. Dissertation, Iowa State University, Ames, Iowa.

Gordon, J. C. (1971). Changes in total nitrogen, soluble protein and peroxidases in the expanding leaf zone of eastern cottonwood. *Plant Physiol.* **47**, 595–599.

Gordon, J. C. (1975). The productive potential of woody plants. *Iowa State J. Res.* **49**, 267–274.

Hall, R. B. and Miksche, J. P. (1975). Determining conifer genome characteristics through computer analysis of DNA reannealing data. *In* "Proc. of the Third North American For. Biol. Workshop" (C. Reid and G. Fechner, eds.) pp. 375–376. Colorado State University.

Hennessey, T. C. and Gordon, J. C. (1975). A comparison of field and growth chamber productivity of three poplar clones. *In* "Proc. of the 9th Central States Forest Tree Imp. Conf." (C. Bey, ed.) USDA Forest Service. In press.

Isebrands, J. G. and Larson, P. R. (1973). Anatomical changes during leaf ontogeny in *Populus deltoides*. *Am. J. Bot.* **60**, 199–208.

Larson, P. R. and Gordon, J. C. (1969a). Leaf development, photosynthesis and ^{14}C distribution in *Populus deltoides* seedlings. *Am. J. Bot.* **56**, 1058–1066.

Larson, P. R. and Gordon, J. C. (1969b). Photosynthesis and wood yield. *Agric. Sci. Rev.* **7**, 7–14.

Max, T. A. (1975). Crown geometry, light interception and photosynthesis of selected *Populus* x *euramericana* clones: a modeling approach. Ph.D. Dissertation, Iowa State University, Ames, Iowa.

Pearce, S. L. (1969). Multivariate techniques of use in biological research. *Expl Agric.* **5**, 67–77.

Promnitz, L. C. (1975). A photosynthate allocation model for tree growth. *Photosynthetica* **9**, 1–15.

Promnitz, L. C. and Rose, D. W. (1974). A mathematical conceptualization of a forest stand simulation model. *Angew. Bot.* **48**, 97–108.

Rose, D. W. and Promnitz, L. C. (1975). Verification of a forest stand simulation model. *Angew. Bot.*, in press.

Schreiner, E. J. (1959). Production of poplar timber in Europe and its significance and application in the United States. Agriculture Handbook No. 150. USDA Forest Service, Washington, D.C.

Wolter, K. E. and Gordon, J. C. (1975). Peroxidases as indicators of growth and differentiation in aspen callus cultures. *Physiol. Plant.* **33**, 219–223.

Wray, P. H. (1974). Peroxidases and growth in hybrid poplar. Ph.D. Dissertation, Iowa State University, Ames, Iowa.

5
Variation in the Photosynthetic Capacity of *Pinus sylvestris*

WŁODZIMIERZ ŻELAWSKI
Institute of Silviculture, Warsaw Agricultural University, Poland.

I.	Introduction	99
II.	Photosynthesis	100
	A. Non-genetic Sources of Variation	100
	B. Inherent Differences	101
III.	Respiration, Photorespiration and Re-assimilation	103
IV.	Gas Exchange and Dry Matter Accumulation	104
Acknowledgements		105
References		105

I. INTRODUCTION

This contribution reviews work on photosynthetic productivity of Scots pine (*Pinus sylvestris*). It considers the large non-genetic sources of variation that occur in net photosynthesis, making it difficult to characterize genotypes by single measurements of gas exchange. Attempts to discern inherent differences in carbon fixation efficiency have shown a need for comprehensive models of dry matter production which incorporate, in particular, estimates of needle surface development and duration, and integrated changes in photosynthesis and respiration for the whole plant.

II. PHOTOSYNTHESIS

A. Non-genetic Sources of Variation

1. Seasonal Changes

Both juvenile and fascicular needles photosynthesize fastest per unit weight when they begin to elongate. Their photosynthetic rates decrease exponentially to about 50% of their former peak values when the needles are fully grown. However, this decrease is more than offset by their several-fold increase in weight, so that total photosynthesis per needle is greatest in late summer (Żelawski and Góral, 1966; Żelawski et al., 1969 and 1971; Żelawski, 1972; Żelawski and Łotocki, 1974). After the first frosts net photosynthesis is greatly depressed. Both juvenile and fascicular needles recover from this "winter depression" the next spring (Żelawski and Kucharska, 1967), but never return to their previous peak photosynthetic rates (Bourdeau, 1959; Parker, 1961, 1963; Polster and Fuchs, 1963; Żelawski et al., 1971; Żelawski and Łotocki, 1974). Nevertheless, fascicular needles continue to provide substantial amounts of assimilate in their second year (Stålfelt, 1924; Iwanoff and Kossowitsch, 1929; Freeland, 1952), and may photosynthesize at up to 65% of their first-year rates even into their third year, just before abscission (Żelawski and Łotocki, 1976). Needle retention is therefore very important for productivity (Żelawski and Łotocki, 1974).

2. Spatial Variation

The photosynthetic rates of needles within a crown depend, particularly, on their ages and local shade environments (Iwanoff and Kossowitsch, 1929; Helms, 1970). Scots pine needles grown in shade have relatively rapid photosynthetic rates per unit needle weight, but small rates per needle or per unit of chlorophyll, associated with anatomical modifications (Żelawski and Gowin, 1966; Żelawski and Żelawska, 1967; Żelawski and Kinelska, 1967; Żelawski et al., 1968). This species is considered shade-intolerant because it is unable to compensate for shade conditions by increasing its photosynthetic surface to any great extent (Żelawski et al., 1968). The photosynthetic characteristics of needles also differ along single shoots; top needles show lower rates than those at the middle and base (Żelawski et al., 1971). The age of the tree has little effect (Iwanoff and Kossowitsch, 1929; Helms, 1970).

3. Environmental Influences

The structure and photosynthetic capacity of pine needles can be greatly modified by changes, for example, in plant water or nutrient status. Scots pine seedlings grown under water stress have smaller needles, lower shoot to root ratios, lower photosynthetic activity (per needle) and higher respiration rates (per unit weight) than seedlings which are adequately watered (Żelawski et al., 1969). Increased nitrogen nutrition enhances photosynthesis, both directly and by modifying the structure of new needles (Zajączkowska, 1974a). Also, in young seedlings, ammonia-nitrogen can be more stimulatory than nitrate-nitrogen (Łotocki and Żelawski, 1973; Zajączkowska, 1973, 1974b) provided the soil is not waterlogged (Łotocki, 1975).

Because the environment has so much influence on needle structure and size, photosynthetic rates have only relative meaning and limited value for predicting productivity. This is well illustrated by work on pines (McGregor et al., 1961; Gatherum et al., 1967a, b; Sweet and Wareing, 1968; Schultz and Gatherum, 1971). It is also seen from experiments where X-ray treatment of seed had an effect on growth in which changes in photosynthesis or respiration were secondary, resulting from, rather than causing, differences in growth (Żelawski and Nalborczyk, 1971; Żelawski et al., 1974).

B. Inherent Differences

1. Maximum Photosynthetic Capacity

The potential photosynthetic capacities of tree genotypes may be measured under optimal conditions, overcoming effects of mutual shading among needles (Kramer and Decker, 1944; Kramer and

Clark, 1947). This is done using diffuse light of high illuminance and/or by using multi-directional light chambers. Under these conditions the photosynthetic rates of both juvenile (Żelawski *et al.*, 1973) and fascicular needles of Scots pine (Żelawski *et al.*, 1971) appeared to be much higher than had generally been assumed (Larcher, 1963, 1969b; Keller and Wehrmann, 1963). Individuals vary greatly in maximum photosynthetic capacity, but as yet little work has been done to evaluate the value of such measurements for selecting superior genotypes.

2. Ecotype Variation

Great difficulties are encountered in attempts to discern inherent differences in net photosynthesis or its component processes. First, there are the non-genetic sources of variation instanced above, secondly, there is the problem of finding a meaningful reference unit, and thirdly, the genotypes themselves may differ in phenology and so cannot easily be compared in the same physiological condition. Thus, lowland and highland ecotypes, or Polish and Turkish provenances, of Scots pine may differ in net photosynthesis when compared at the same chronological age, but when compared at the same stage of needle growth these differences disappear (Żelawski, 1967; Żelawski *et al.*, 1971; Al-Shahine, 1969). The "winter depression" of photosynthesis occurs earlier in highland than in lowland ecotypes, irrespective of difference in needle formation (Żelawski and Góral, 1966; Żelawski *et al.*, 1971).

Clearly, genotypic differences are relative and specific to given environments; ecotypes do not find optimal conditions in the same environment, and the most valuable approach has been to determine the nature of differing genotype—environment interactions. In this way inherent differences have been revealed in the shade-tolerance of seedlings of highland and lowland ecotypes of *Pinus sylvestris* (Żelawski *et al.*, 1968), and the photosynthesis light-response curves of provenances of *Pinus strobus* (Bourdeau, 1963) and *Pinus contorta* (Sweet and Wareing, 1968). Other instances of differing responses to frost, water stress, temperature and light are reviewed by Kozlowski and Keller (1966), Larcher (1969b) and elsewhere in this volume.

III. RESPIRATION, PHOTORESPIRATION AND RE-ASSIMILATION

Investigations on whole Scots pine seedlings show that dry matter accumulation is sometimes more closely correlated with rates of dark respiration than with photosynthetic rates. It seems particularly important to evaluate the balance of photosynthesis to respiration in differing genotypes.

Respiratory losses in trees are large. In whole seedlings of Scots pine the average ratio between daily total net photosynthesis (16 h) and respiration (24 h) varies from about 4 to 2 (Żelawski, 1974). In the second growing season the same ratios estimated for similar daily periods were from 4 to 6 (Żelawski et al., 1971). In mature forest stands respiratory losses amount to 25—60% of photosynthesis, depending on their age and site conditions (Polster, 1967).

Inherent differences seem to be more clearly discernible when the proportion of photosynthesis to respiration is considered. A lowland ecotype of Scots pine was found to have a significantly greater photosynthesis : respiration ratio compared with a highland one, while neither photosynthesis nor respiration alone revealed clear differences (Żelawski et al., 1971). The balance of photosynthesis to respiration, both in the above-ground parts of a seedling and in the whole plant, could be related to variation in productivity.

Determinations of CO_2-compensation points of Scots pine seedlings, and extrapolations to zero CO_2 concentrations, suggested that CO_2-evolution in light exceeded dark respiration by up to 200% (Żelawski, 1967). However, during the period of "winter depression" of photosynthesis the CO_2-compensation point was low, showing that CO_2 evolution in light was probably less than CO_2 evolution in darkness (Żelawski and Kucharska, 1967). The question arises whether the different rates of CO_2 evolution in light and dark result from photorespiration, as related to the glycolic acid pathway, or are the outcome of interactions between respiratory processes, in chlorophyllous and non-chlorophyllous tissues, and some re-assimilation of internal CO_2 within the tree.

Experiments with seedlings of *Pinus elliottii*, using $^{14}CO_2$, have shown that some respiratory CO_2 is re-assimilated in the young stem, where the bark contains chlorophyll (Żelawski et al., 1970). In larger trees, considerable quantities of respiratory CO_2 are transported to the crown in the transpiration stream, and it is possible that total dry matter production could be influenced by the extent to which this CO_2 was re-assimilated. However, some portion of this "internal" CO_2 is released into the atmosphere and this loss could erroneously be attributed to photorespiration.

IV. GAS EXCHANGE AND DRY MATTER ACCUMULATION

The amount of dry matter accumulated by a plant may be viewed as the difference between the integrated quantities of CO_2 photosynthesized and respired during its life (e.g. Larcher, 1969a; Ledig, Chapter 2). Integrals are not easily calculated using field data because photosynthesis fluctuates greatly under field conditions (e.g. Hodges, 1967). As an alternative, CO_2-exchange data were obtained for Scots pine seedlings in water culture in semi-controlled and continuously monitored diurnal environments imposed in a glasshouse. Estimates of dry matter production calculated from CO_2-exchange measurements and response curves agreed closely with direct determinations of seedlings dry weights (Żelawski, 1972).

This was the first step in building a model of productivity for Scots pine seedlings. To evaluate genotypic differences, more information is needed on relationships between the component processes of dry weight gain. Some progress in this field has already been made for pines (Van den Driessche and Wareing, 1966; Sweet and Wareing, 1966, 1968; Ledig, 1969; Ledig and Perry, 1969; Ledig, Chapter 2). Additionally, efforts are needed to define yield more precisely in terms of the quantity of cellulose compounds produced, and the proportion harvested (the "harvest index" or "yield coefficient"). Investigations using ^{14}C showed that less than 15% of the carbon fixed by Scots pine seedlings was incorporated into lignin and cellulose, and even less as the plants aged (Niziołek et al., 1969; Góral, 1973a, b).

A decrease in the net assimilation rate (NAR, on needle weight) of Scots pine seedlings during a season is not paralleled by a decrease in net photosynthetic rate because of an increase during the year in the plants' respiratory load. An index of "efficiency" which takes both photosynthesis and respiration into account is the ratio of the actual net assimilation rate (NAR_{true}) to the estimated capacity for assimilation (NAR_{cap}) calculated from gas exchange data. This ratio shows the fraction of the maximum rate of dry matter production, as estimated under the most favourable conditions, which is actually attainable under real conditions of growth. The differences between highland and lowland ecotypes of Scots pine were recognized better this way than by considering photosynthesis and dry matter production separately. Highland ecotypes had significantly larger ratios than lowland ones during the main part of the growing season. This did not mean that the highland ecotypes had greater photosynthetic efficiency, but that they responded better to the given growing conditions (Żelawski et al., 1971; Gowin, 1973).

It appears that the carbon fixation efficiency of genotypes can be characterized only by investigating the seasonal course of processes involved in dry matter production, applying various indices, and integrating with time. Single measurements, even of maximum photosynthetic capacities, do not allow one to predict the potential productivity of a tree.

ACKNOWLEDGEMENTS

Experimental work summarized in this review was carried out by the author and his co-workers within the following projects: U.S. Department of Agriculture (grant numbers FG-Po-167 and 240); Polish Academy of Sciences, Institute of Ecology (grant number 09.1.7.6.2.6.).

REFERENCES

Al-Shahine, F. O. (1969). Photosynthesis, respiration and dry matter production of Scots pine (*Pinus silvestris* L.) seedlings originating from Poland (Nowy Targ) and Turkey (Eskishaher). *Acta Soc. Bot. Pol.* 38, 355—369.

Bourdeau, P. F. (1959). Seasonal variations of the photosynthetic efficiency of evergreen conifers. *Ecology* 40, 63—67.

Bourdeau, P. F. (1963). Photosynthesis and respiration of *Pinus strobus* L. seedlings in relation to provenance and treatment. *Ecology* 44, 710—716.

Driessche, van den, R. and Wareing, P. F. (1966). Dry-matter production and photosynthesis in pine seedlings. *Ann. Bot.* 30, 673—682.

Freeland, R. (1952). Effect of age of leaves upon the rate of photosynthesis in some conifers. *Plant Physiol.* 27, 685—690.

Gatherum, G. E., Gordon, J. C. and Broerman, B. F. S. (1967a). Physiological variation in Scotch pine seedlings in relation to light intensity and provenance. *Iowa St. J. Sci.* **42**, 19—26.

Gatherum, G. E., Gordon, J. C. and Broerman, B. F. S. (1967b). Physiological variation in European black pine seedlings in relation to light intensity and provenance. *Iowa St. J. Sci.* **42**, 27—35.

Góral, I. (1973a). Distribution of radioactive products of photosynthesis in Scots pine (*Pinus silvestris* L.) seedlings during the first vegetation season. *Acta Soc. Bot. Pol.* **42**, 541—553.

Góral, I. (1973b). Incorporation of radioactive products of photosynthesis into lignin and cellulose of Scots pine (*Pinus silvestris* L.) seedlings. *Acta Soc. Bot. Pol.* **42**, 555—565.

Gowin, T. (1973). Growth of Scots pine (*Pinus silvestris* L.) seedlings of different provenience on comparative plantations in three regions of Poland. *Ekol. pol.* **21**, 309—321.

Helms, J. A. (1970). Summer net photosynthesis of Ponderosa pine in its natural environment. *Photosynthetica* **4**, 243—253.

Hodges, J. D. (1967). Patterns of photosynthesis under natural environmental conditions. *Ecology* **48**, 234—242.

Iwanoff, L. A. and Kossowitsch, N. L. (1929). Über die Arbeit des Assimilationsapparates verschiedener Baumarten. I Die Kiefer (*Pinus silvestris*). *Planta* **8**, 427—464.

Keller, Th. and Wehrmann, J. (1963). CO_2-Assimilation, Wurzelatmung und Ertrag von Fichten- und Kiefernsämlingen bei unterschiedlicher Mineralstoffernährung. *Mitt. schweiz. Anst. forstl. VersWesen* **39**, 217—242.

Kozlowski, T. and Keller, Th. (1966). Food relations of woody plants. *Bot. Rev.* **32**, 293—382.

Kramer, P. J. and Clark, W. S. (1947). A comparison of photosynthesis in individual pine needles and entire seedlings at various light intensities. *Plant Physiol.* **22**, 51—57.

Kramer, P. J. and Decker, J. P. (1944). Relation between light intensity and rate of photosynthesis of loblolly pine and certain hardwoods. *Plant Physiol.* **19**, 350—358.

Larcher, W. (1963). Die Leistungsfähigkeit der CO_2-Assimilation höherer Pflanzen unter Laborbedingungen und am natürlichen Standort. *Mitt. flor.-suz. ArbGemein.* **10**, 20—33.

Larcher, W. (1969a). Physiological approaches to the measurement of photosynthesis in relation to dry matter production by trees. *Photosynthetica* **3**, 150—166.

Larcher, W. (1969b). The effect of environmental and physiological variables on the carbon dioxide gas exchange of trees. *Photosynthetica* **3**, 167—198.

Ledig, F. T. (1969). A growth model for tree seedlings based on the rate of photosynthesis and the distribution of photosynthate. *Photosynthetica* 3, 263—275.

Ledig, F. T. and Perry, T. O. (1969). Net assimilation rate and growth in Loblolly pine seedlings. *Forest Sci.* 15, 431—438.

Łotocki, A. (1975). Wpływ aeracji podłoża i formy żywienia azotowego na produktywność fotosyntezy siewek sosny zwyczajnej (*Pinus silvestris* L.). [Effect of aeration and nitrogen form upon the photosynthetic productivity of Scots Pine (*Pinus silvestris* L.) seedlings.]. Ph.D. dissertation, Warsaw Agricultural University Faculty of Forestry.

Łotocki, A. and Żelawski, W. (1973). Effect of ammonium and nitrate source of nitrogen on productivity of photosynthesis in Scots pine (*Pinus silvestris* L.) seedlings. *Acta Soc. Bot. Pol.* 42, 599—605.

McGregor, W. M. H. D., Allen, R. M. and Kramer, P. J. (1961). The effect of photoperiod on growth, photosynthesis, and respiration of Loblolly pine seedlings from two geographic sources. *Forest Sci.* 7, 342—348.

Niziołek, S., Kączkowski, J. and Żelawski, W. (1969). Incorporation of assimilated $^{14}CO_2$ into cellulose and lignin of Scots pine (*Pinus silvestris* L.) seedlings. *Bull. Acad. pol. Sci. Cl. II Sér. Sci. biol.* 17, 363—367.

Parker, J. (1961). Seasonal trends in carbon dioxide absorption, cold resistance and transpiration of some evergreens. *Ecology* 42, 372—380.

Parker, J. (1963). Causes of the winter decline in transpiration and photosynthesis in some evergreens. *Forest Sci.* 9, 158—166.

Polster, H. (1967). Photosynthese, Atmung und Stoffproduktion. *In* "Gehölz Physiologie" (H. Lyr, H. Polster and H. J. Fiedler, eds.), pp. 197—237. Gustav Fischer Verlag, Jena.

Polster, H. and Fuchs, G. (1963). Winterassimilation und -Atmung der Kiefer (*Pinus silvestris*) im mitteldeutschen Binnenlandklima. *Arch. Forstw.* 12, 1011—1023.

Schultz, R. C. and Gatherum, G. E. (1971). Photosynthesis and distribution of assimilate of Scotch pine seedlings in relation to soil moisture and provenance. *Bot. Gaz.* 132, 91—96.

Stålfelt, M. G. (1924). Tallens och granens kolsyreassimilation och dess ekologiska betingelser. *Meddn. St. SkogsforskInst.* 21/5, 181—258.

Sweet, G. B. and Wareing, P. F. (1966). Role of plant growth in regulating photosynthesis. *Nature, Lond.* 210, 77—79.

Sweet, G. B. and Wareing, P. F. (1968). A comparison of the rates of growth and photosynthesis in first year seedlings of four provenances of *Pinus contorta* Dougl. *Ann. Bot.* 32, 735—751.

Zajączkowska, J. (1973). Gas exchange and organic matter production of Scots pine (*Pinus silvestris* L.) seedlings grown in water culture with ammonium or nitrate form of nitrogen. *Acta Soc. Bot. Pol.* 42, 607—615.

Zajączkowska, J. (1974a). Effect of the change of nitrogen source on photosynthesis and respiration of Scots pine (*Pinus silvestris* L.) seedlings. *Acta Soc. Bot. Pol.* 43, 93—101.

Zajączkowska, J. (1974b). Gas exchange and organic substance production of Scots pine (*Pinus silvestris* L.) seedlings grown in soil cultures with ammonium or nitrate form of nitrogen. *Acta Soc. Bot. Pol.* 43, 103—116.

Żelawski, W. (1967). Variation in the photosynthetic efficiencies of trees with special reference to Scots pine (*Pinus silvestris* L.). *Proc. XIV IUFRO-Congress*, Munich, pp. 515—535.

Żelawski, W. (1972). Uptake and evolution of carbon dioxide and accumulation of organic substance in Scots pine (*Pinus silvestris* L.) seedlings. *Bull. Acad. pol. Sci. Cl. II Sér. Sci. biol.* 20, 747—753.

Żelawski, W. (1974). Gas exchange processes in shoot and root of Scots pine (*Pinus silvestris* L.) seedlings as related to dry matter production. *Proc. II International Symposium Ecology and Physiology of Root Growth*. Academie-Verlag. Berlin, pp. 433—436.

Żelawski, W. and Gowin, T. (1966). Variability of some needle characteristics in Scots pine (*Pinus silvestris* L.) ecotypes grown on the comparative plantation. *Ekol. pol.* 14, 275—283.

Żelawski, W. and Góral, I. (1966). Seasonal changes in the photosynthesis rate of Scots pine (*Pinus silvestris* L.) seedlings grown from seed of various provenances. *Acta Soc. Bot. Pol.* 35, 587—598.

Żelawski, W. and Kinelska, J. (1967). Photosynthesis and respiration of Scots pine (*Pinus silvestris* L.) seedlings of various provenance grown under different light conditions. *Acta Soc. Bot. Pol.* 36, 713—723.

Żelawski, W. and Kucharska, J. (1967). Winter depression of photosynthetic activity in seedlings of Scots pine (*Pinus silvestris* L.). *Photosynthetica* 1, 207—213.

Żelawski, W. and Łotocki, A. (1974). Photosynthetic capacity of Scots pine (*Pinus silvestris* L.) needles from a forest subjected to chronic industrial pollution. *Bull. Acad. pol. Sci. Cl. II Sér. Sci. biol.* 22, 431—434.

Żelawski, W. and Łotocki, A. (1976). Ageing of assimilatory organs of conifers. Manometric measurements of photosynthesis and respiration. In preparation.

Żelawski, W. and Nalborczyk, E. (1971). Productivity of photosynthesis in Scots pine (*Pinus silvestris* L.) seedlings grown from seed irradiated by X-rays. *Acta Soc. Bot. Pol.* 40, 413—421.

Żelawski, W. and Żelawska, B. (1967). Some aspects of the effect of shade on growth of Scots pine (*Pinus silvestris* L.) seedlings of various provenances. *Ekol. pol.* 15, 107—114.

Żelawski, W., Kinelska, J. and Łotocki, A. (1968). Influence of shade on productivity of photosynthesis in seedlings of Scots pine (*Pinus silvestris* L.) during the second vegetation period. *Acta Soc. Bot. Pol.* 37, 505—518.

Żelawski, W., Kucharska, J. and Łotocki, A. (1969). Productivity of photosynthesis in Scots pine (*Pinus silvestris* L.) seedlings grown under various soil moisture conditions. *Acta Soc. Bot. Pol.* 38, 143–155.

Żelawski, W., Riech, F. P. and Stanley, R. G. (1970). Assimilation and release of internal carbon dioxide by woody plant shoots. *Can. J. Bot.* 48, 1351–1354.

Żelawski, W., Kucharska, J. and Kinelska, J. (1971). Relationship between dry matter production and carbon dioxide absorption in seedlings of Scots pine (*Pinus silvestris* L.) in their second vegetation season. *Acta Soc. Bot. Pol.* 40, 243–256.

Żelawski, W., Szaniawski, R., Dybczyński, W. and Piechurowski, A. (1973). Photosynthetic capacity of conifers in diffuse light of high illuminance. *Photosynthetica* 7, 351–357.

Żelawski, W., Łotocki, A. and Nalborczyk, E. (1974). Productivity of photosynthesis in Scots pine (*Pinus silvestris* L.) seedlings during their first and second growing season after X-irradiation of seed. USDA-Final report, pp. 15–33.

6
Relationship Between the CO_2 Compensation Point and Carbon Fixation Efficiency in Trees

OLAVI LUUKKANEN
Department of Silviculture, University of Helsinki, Finland

I.	Introduction	111
II.	Studies with Clonal Material	114
III.	Population Analyses	116
IV.	Summary and Conclusions	116
References		117

I. INTRODUCTION

Plants which photosynthesize can be classified into those with high and those with low CO_2 compensation points (Γ). The former group is the larger, and includes nearly all forest trees. They are characterized by (a) an inability to fix CO_2 via PEP carboxylase (the Hatch-Slack or C4-dicarboxylic acid cycle); (b) a requirement for a finite minimum concentration of CO_2 in the external atmosphere, associated with a large output of CO_2 in light; (c) an increase in photosynthesis with decreasing concentrations of O_2; and (d) low photosynthetic rates, especially at high temperatures (Jackson and Volk, 1970). The production of CO_2 in light (photorespiration in a broad sense) is associated with the presence of peroxisomes (Luukkanen, 1972) which utilize glycolate produced during photosynthesis, implying a close relationship between photosynthesis and photorespiration (Tolbert, 1971). Rates of photorespiration may be estimated as explained in Fig. 1 (see Forrester *et al.*, 1966).

Inherent variation in photorespiration is also associated with differences in leaf anatomy. High-Γ plants, which photorespire, do not have distinct vascular bundle sheaths, nor any large variation in

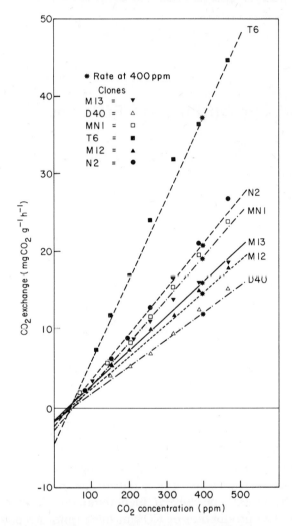

FIG. 1 The extrapolation method of estimating rates of photorespiration, applied to six *Populus* clones at 25°C. Net photosynthetic rates were measured at a known CO_2 concentration (400 ppm) and lines showing "carboxylation efficiencies" were extrapolated through Γ (CO_2 compensation point) to zero CO_2 concentration, where the rate was assumed to equal photorespiration. Points other than at 400 and zero ppm are checks for testing the linearity of the carboxylation efficiency curves. From Luukkanen and Kozlowski (1972), reproduced by permission of J. D. Sauerländer's Verlag, Frankfurt.

chloroplast structure. By contrast, low-Γ species generally have well developed vascular bundle sheaths which contain cells with large but weakly differentiated chloroplasts (Frederick and Newcomb, 1971). Most of their CO_2 uptake is fixed there, but not by the C4 pathway. C4 fixation occurs only in cells lying outside the bundle sheaths, which recycle respiratory CO_2, so giving apparently low rates of photorespiration (Hatch and Slack, 1970).

Zelitch and Day (1968) who found differences in photorespiration rates among varieties of tobacco (a high-Γ plant), demonstrated that they were associated with differences in amounts of glycolate oxidase. Also, in tobacco photorespiration rates were inversely correlated with rates of net photosynthesis.

Tolbert (1971) concluded that glycolate synthesis from carbohydrate reserves, and subsequent photorespiration, are unavoidable at high oxygen or low CO_2 concentrations, and at high light intensities. In the absence of CO_2, plants oxidize reserve foods to glycolate and subsequently to CO_2. When plants having high CO_2 compensation points are enclosed in an environment with a low concentration of CO_2 (or together with low-Γ species) they lose CO_2 continuously and eventually die. Thus, it may be possible to identify genotypes with lower than average Γ by observing their ability to survive at low CO_2 concentrations (Menz et al., 1969). Such screening for "photorespiration-defective" mutants in field crops has not, according to Ogren and Widholm (1970), given any confirmed positive results. However, intraspecific variation in Γ and photorespiration may differ between tree and other crop species having different genetic constitutions. In particular, mutation rates, which decrease with increasing genetic stability and inbreeding (Stebbins, 1967) may be greater in forest trees than in many field crops (see e.g. Sarvas, 1967).

Decker (1970) suggested using Γ as a means of selecting for rapid growth and indicated that photosynthetic and photorespiratory pathways might be under different genetic control and thus separable by breeding. He emphasized the role of Γ as a reliable indicator of the photosynthetic performance of entire plants or stands. This is in contrast to measurements of photosynthetic rate per unit of foliage, which are not always correlated with plant growth, particularly among unrelated species (see Ferrell, 1970).

Among forest trees, photorespiration has been investigated in

Pinus sylvestris (Żelawski, 1967), *Picea glauca* (Poskuta, 1968), *Pseudotsuga menziesii* (Brix, 1968) and *Populus* clones (Luukkanen and Kozlowski, 1972).

II. STUDIES WITH CLONAL MATERIAL

Six contrasting *Populus* clones were used to investigate variation in Γ, carbon fixation efficiency and rates of photorespiration and dark respiration (Luukkanen and Kozlowski, 1972). Particular attention was focused on relationships between Γ and net photosynthetic efficiencies, and the possibility that photosynthesis and photorespiration were under separate genetic control (Decker, 1970). However, the plant material did not include families for testing genetic parameters such as heritability.

The black poplar clones (within the species *Populus nigra* and *P. deltoides*) had higher CO_2 compensation points than three balsam poplar clones (representing two species, *P. trichocarpa* and *P. maximowiczii*). High photosynthetic rates were associated with low Γ and rapid photorespiration. However, when photorespiration rates were adjusted to average photosynthetic rates, using covariance analysis, these *relative* photorespiration rates seemed to be correlated positively with Γ and inversely with photosynthesis. That is, clones with the lowest measured rates of photorespiration had the fastest photorespiration rates in proportion to their rates of photosynthesis. High photosynthetic rates were observed in an interspecific cross, *P. maximowiczii* × *P. nigra*, which had the fastest rates of relative photorespiration over most of the temperature range (15—30°C) and also large Γ values at higher temperatures. This hybrid seemed to combine the properties of both parental species. A *P. nigra* clone included in the material had high Γ and high rates of relative photorespiration, but low photosynthetic rates. Two *P. maximowiczii* clones, on the other hand, had low Γ and low rates of relative photorespiration, but fast photosynthetic rates.

Overall a significant inverse relationship was found between estimates of Γ and net photosynthetic rates per unit of foliage, although it was strongly affected by temperature (Fig. 2). Thus Γ alone may provide a criterion for photosynthetic efficiency, although

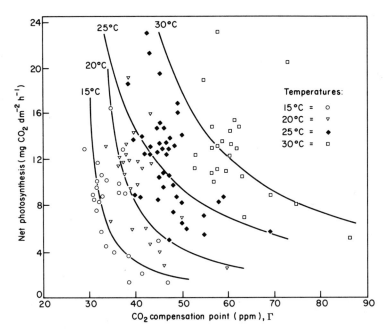

FIG. 2 Relationship between net photosynthesis and Γ at 15, 20, 25 and 30°C. Points are individual, varying observations for any of six *Populus* clones. From Luukkanen and Kozlowski (1972) reproduced by permission of J. D. Sauerländer's Verlag, Frankfurt.

further work is needed concerning the nature of the observed differences in relative photorespiration, albeit that they are in accordance with Decker's proposed model.

The relationship between Γ and net photosynthetic rates has also been investigated in four clones of *Picea abies* (Luukkanen, unpublished). A similar trend was found to that observed in the earlier work with *Populus* (Luukkanen and Kozlowski, 1972). In *P. abies* there appeared to be an inverse curvilinear relationship between Γ and rates of net photosynthesis per unit of foliage. The CO_2 compensation points of *P. abies* were also measured after subjecting the plants to water stress. Preliminary results indicated that Γ increased with increasing water stress and thus suggested that, at given temperatures and saturating light intensities, Γ, besides explaining part of the variation in photosynthesis, also provided a measure of the effects of water stress (see also Treganna et al., 1975).

III. POPULATION ANALYSES

In a recent study, two northern Finnish populations of *Picea abies* were found to have larger Γ values and higher rates of net photosynthesis per unit of foliage than a southern Finnish population (Pelkonen and Luukkanen, 1974). Values of Γ and net photosynthetic rates were not significantly related, taking the material as a whole, but seemed to be inversely related within each population (cf. Pelkonen, 1973). In view of the study of *P. abies* clones mentioned above, it is suggested that if Γ is used to predict net photosynthetic efficiency, better correlations may be found within genetically homogeneous groups of trees than, for instance, among a range of provenances or ecotypes.

IV. SUMMARY AND CONCLUSIONS

In some instances, the CO_2 compensation point (Γ) seems to be an inherent characteristic of forest trees, which is inversely related to their carbon fixation efficiency, although the relationship may be obscured by environmental factors.

It is well recognized that photosynthetic rates, when expressed per unit of foliage, are at best unreliable indicators of photosynthetic performance over long periods of time (Decker, 1955; Ferrell, 1970). Estimates of Γ may give more useful information about the plant's complex gas exchange processes and inherent differences in their responses to the environment. For instance, Γ values may reflect genetic variation in stomatal behaviour, which, in turn, can be correlated with differences in water balance (Luukkanen and Kozlowski, 1972).

In the author's view there is evidence to support large-scale screening for low CO_2 compensation points among trees using, for instance, the CO_2 starvation method already applied to agricultural plants. The speed and simplicity of such methods will appeal to any researcher who has attempted to combine conventional photosynthesis measurements with genetic studies.

REFERENCES

Brix, H. (1968). Influence of light intensity at different temperatures on rate of respiration of Douglas-fir seedlings. *Plant Physiol.* 43, 389–393.

Decker, J. P. (1955). The uncommon denominator in photosynthesis as related to tolerance. *Forest Sci.* 1, 88–89.

Decker, J. P. (1970). Photosynthetic efficiency, photorespiration and heterosis. *Arizona State Univ. Eng. Res. Center Bioeng. Bull.* 12.

Ferrell, W. K. (1970). Variation in photosynthetic efficiency within forest tree species. *Proc. First North Amer. Forest. Biol. Workshop.*

Forrester, M. L., Krotkov, G. and Nelson, C. D. (1966). Effect of oxygen on photosynthesis, photorespiration, and respiration in detached leaves. I. Soybean. *Plant Physiol.* 41, 422–427.

Frederick, S. E. and Newcomb, E. H. (1971). Ultrastructure and distribution of microbodies in leaves of grasses with and without CO_2-photorespiration. *Planta* 96, 152–174.

Hatch, M. D. and Slack, C. R. (1970). Photosynthetic CO_2 fixation pathways. *A. Rev. Pl. Physiol.* 21, 141–162.

Jackson, W. A. and Volk, R. L. (1970). Photorespiration. *A. Rev. Pl. Physiol.* 21, 385–432.

Luukkanen, O. (1972). Metsäpuiden fotosynteesin geneettinen vaihtelu. *Summary* Genetic variation of photosynthesis in forest trees. *Silva fenn.* 6, 63–89.

Luukkanen, O. and Kozlowski, T. T. (1972). Gas exchange in six *Populus* clones. *Silvae Genet.* 21, 220–229.

Menz, K. M., Moss, D. N., Cannell, R. Q. and Brun, W. A. (1969). Screening for photosynthetic efficiency. *Crop Sci.* 9, 692–694.

Ogren, W. L. and Widholm, J. M. (1970). Screening for photorespiratory-deficient soybean mutants. *Plant Physiol.* 46 (Suppl.), 7.

Pelkonen, P. (1973). Kolmen kuusimetsikön vapaapölytysjälkeläistöjen CO_2-aineenvaihdunnasta. [On the CO_2 Exchange in Open-Pollinated Progenies from Three Stands of Norway Spruce]. M.S. Thesis, University of Helsinki, Finland.

Pelkonen, P. and Luukkanen, O. (1974). Gas exchange in three populations of Norway spruce. *Silvae Genet.* 23, 160–164.

Poskuta, J. (1968). Photosynthesis, photorespiration and respiration of detached spruce twigs as influenced by oxygen concentration and light intensity. *Physiol. Plant.* 21, 1129–1136.

Sarvas, R. (1967). Pollen dispersal within and between subpopulations; role of isolation and migration in microevolution of forest tree species. *Proc. XIV IUFRO Congress*, pp. 332–345.

Stebbins, G. L. (1967). "Variation and Evolution in Plants". Columbia University Press, New York.

Tolbert, N. E. (1971). Microbodies — peroxisomes and glyoxysomes. *A. Rev. Pl. Physiol.* **22**, 45—74.

Treganna, E. B., Shikhobalov, V. V., Zabotin, A. I., Bassi, P. K., and Tarchevskii, I. A. (1975). Vliyanie vodnogo stressa na komponenty uglekislotnogo gasoobmena list'ev C_3 i C_4 rastenii. [Effect of water stress on components of CO_2 exchange in leaves of C_3 and C_4 plants]. *Abstr. XII Int. Bot. Congress*, p. 437.

Zelawski, W. (1967). A contribution to the question of the CO_2-evolution during photosynthesis in dependence of light intensity (*Pinus sylvestris* seedlings). *Bull. Acad. pol. Sci. Cl. II Sér. Sci. biol.* **15**, 565—570.

Zelitch, I. and Day, P. R. (1968). Variation in photorespiration. The effect of genetic differences in photorespiration on net photosynthesis in tobacco. *Plant Physiol.* **43**, 1838—1844.

SHOOT AND CAMBIAL GROWTH

7
Relationships Between Genetic Differences in Yield of Deciduous Tree Species and Variation in Canopy Size, Structure and Duration

R. E. FARMER, Jr.
Division of Forestry, Fisheries and Wildlife Development, Tennessee Valley Authority, Norris, U.S.A.

I.	Introduction	119
II.	Canopy Characteristics as Yield Components in Agronomic and Horticultural Crops	120
III.	Ecological Studies of Forest Canopy Structure	121
IV.	Shoot and Foliage Development on Individual Trees	122
V.	Genetic Variation in Shoot and Canopy Characteristics of Hardwoods	124
	A. *Juglans nigra*	124
	B. *Populus*	125
	C. *Quercus* and *Fagus*	127
	D. *Liquidambar styraciflua*	128
	E. *Acer*	129
	F. Other species	129
VI.	Conclusions	130
References		131

I. INTRODUCTION

Genetically controlled variation in the wood-producing capacity of deciduous trees can be separated into at least three components: (1) differences in net assimilation rate per unit of leaf or of chlorophyll; (2) differences in size, structure, and duration of the photosynthetic apparatus, i.e. the canopy; and (3) differences in distribution and use of photosynthate within the tree, as reflected in shoot : root and crown : bole ratios. The third component is closely tied to the first two through correlations and source–sink relationships which regulate plant structure. This contribution discusses variation in shoot and leaf characteristics which control canopy size, structure and duration.

II. CANOPY CHARACTERISTICS AS YIELD COMPONENTS IN AGRONOMIC AND HORTICULTURAL CROPS

Concern with the physiological basis of yield was probably first expressed by agronomists (e.g. Engledow and Wadham, 1923) who investigated the relation of yield per plant to other plant characters (i.e. components of yield), and the mode of inheritance of these characters. Agronomic workers initially studied such components as number and size of kernels in fruiting parts. Recently, attention has focused on the architecture and activity of the photosynthetic apparatus. Classical growth analysis (Evans, 1972) applied both to individual plants and to stands, has been widely exploited. Reviews by Watson (1952, 1956, 1971) suggest that variation in leaf area characteristics are mainly responsible for yield differences in field crops. Special attention has been given to the time when maximum leaf area indices (LAI) develop, canopy structure and source–sink relationships.

Loomis and Williams (1969) and Loomis et al., (1967) incorporated canopy characteristics (e.g. light extinction coefficients and leaf angles) in models of light distribution and stand production, and concluded that, other factors being constant, greatest efficiency will be obtained at low leaf angles and LAI values when light intensities are low, and at high leaf angles and LAI values when light intensities are high. A thorough analysis of canopy structure and duration in relation to Relative Growth Rate (RGR) and Net Assimilation Rate (NAR) was presented by Wallace et al., (1972). They noted that genetic variation in growth is usually caused by variation in many components of yield, and the large number of factors and complex interactions involved make it difficult to identify and genetically manipulate individual components. Thus it is unlikely that single factors will be highly correlated with yield across numerous genotypes, each of which may have a slightly different production strategy. Some modelling efforts (e.g. McKinion et al., 1974), are aimed at resolving this difficulty.

Recent research reveals that genetically controlled canopy characteristics have been associated with yield differences in some cases and unrelated in others. Perhaps the best known positive examples are the semi-dwarf rice and wheat varieties in which high leaf angles are believed to be partly responsible for high yield. Some reports for maize (e.g. Russell, 1972; Ariyanayagam et al., 1974) indicate no effect of leaf angle on yield. However, more generally positive effects of erect leaves and low-light extinction coefficients have been noted among varieties of forage species (e.g. Rhodes, 1971; Sheehy and Cooper, 1973). In an interesting study of tea foliage characteristics, Hadfield (1974) observed that varieties with semi-erect leaves maintained greater LAI values and responded better to nitrogen than varities with horizontal leaves.

III. ECOLOGICAL STUDIES OF FOREST CANOPY STRUCTURE

While few data on genetic variation in crown structure are contained in forest ecology literature, ecologists concerned with productivity have developed concepts, techniques, and information which will be useful in studying the contribution of this yield component. Relationships between leaf area and tree diameter were subjects of early forest hydrology investigations (e.g. Kittredge, 1944; Rothacher et al., 1954), and considerable recent attention has been devoted to estimating the biomass and productivity of forest canopies (for good entrances to the literature see especially Black, 1966; Madgwick, 1970; Satoo, 1970; and Zavitkovski et al., 1974). It has long been known that crown structure has a major effect on wood production (Matthews, 1963), and some work with conifers (e.g. Brown and Goddard, 1961) emphasizes individual tree variation in crown efficiency. Jahnke and Lawrence (1965) and others have shown that tree crowns which are extended vertically can be more productive than horizontally arranged photosynthetic surfaces, i.e. heightened cones intercept more light. Clearly, the complexity of the crowns, in terms of light utilization, increases with increasing vertical extension. The light extinction coefficient (K) has been used to quantify light absorption properties of such canopies and is thoroughly discussed by Saeki (1963). Papers by Anderson (1966) and Warren Wilson (1967) present methods of quantitatively characterizing light penetration.

Good illustrations of these investigative techniques, which are directly applicable to the study of natural variation in canopy characteristics, are found in papers by Ford and Newbould (1971), Kinerson *et al.* (1974), and Zavitkovski *et al.* (1974) which discuss vertical distributions of foliage in relation to season and ontogeny. Kinerson *et al.*, have developed a computer program which simulates spatial and temporal dynamics in a conifer canopy.

Some attention has been given to phenology and its relationship to duration of the photosynthetic surface (e.g. Lieth, 1971), but there are few recent references to natural variation within populations.

IV. SHOOT AND FOLIAGE DEVELOPMENT ON INDIVIDUAL TREES

The size, complexity and form of the photosynthetic apparatus of an individual tree depends on (1) the number of shoot growing points, (2) the rate of leaf production by these apices, and (3) the size, shape and arrangement of leaves (Wareing, 1966). These factors are fundamentally related to the degree to which photosynthate is reinvested in new shoot and leaf growth (i.e. leaf relative growth rate). In juvenile trees, the degree of this reinvestment may account for major differences in growth (Ledig, 1969). For example, growth analysis of some tropical deciduous trees (Coombe, 1960; Coombe and Hadfield, 1962) has revealed that rapid growth of these species does not lie in particularly efficient energy conversion but rather in their capacities for unrestricted leaf production and economical branching patterns. Thus, episodic growth controls, and growth correlations, are pertinent to the problem of canopy development, though control mechanisms are still unclear (Alvim, 1964; Borchert, 1973; Huxley and Van Eck, 1974).

In temperate-region deciduous trees, differences in Relative Growth Rate among species and treatments have been related to both NAR and Leaf Area Ratio (LAR), though some differences in

NAR are believed to be due to differences in crown form (Pollard and Wareing, 1968). Isebrands (in Zavitkovski *et al.*, 1974) noted the importance of the large leaf area per tree in the high productivity of hybrid *Populus*. Madgwick's (1971) work with *P. alleghaniensis* indicated that positive effects of nitrogen on RGR were due mainly to high NAR. Feedback mechanisms associated with source–sink relationships may affect NAR, making it more difficult to delineate yield components. Thus, a decrease in leaf surface may increase NAR values (Maggs, 1964; Sweet and Wareing, 1966). There is an extensive literature supporting the notion that photosynthetic rates may be controlled by the level of assimilate in leaves (Neales and Incoll, 1968). For example, long shoots in apple trees, which are presumably near active sinks, may have higher net photosynthetic rates than short shoots (Ghosh, 1973). Thus, the location of leaves within canopies may have effects which are not solely associated with light intensity. Such effects could account for the different NAR values reported for juvenile and mature trees, since short shoots may make up the bulk of the canopy in older trees (Pollard, 1970). These source–sink relationships may account for the generally negligible effects of minor defoliation by insects. However, major defoliations (and genetic variation in susceptibility to them) must be considered when evaluating canopy size and duration as yield components, because they can account for major increment losses (Franklin, 1970; Rafes, 1970). Priestley (1964) noted the importance of autumn foliage in providing storage carbohydrates and promoting root growth.

While much is known in a general way about shoot growth and its control (Kozlowski, 1964, 1971; Romberger, 1963), few deciduous species have received the detailed attention required to determine why genetic differences exist in the structure of the photosynthetic apparatus. The work of Ward (1964) on bud distribution and branching in red oak represents the type of approach which will reveal the nature and basis of leaf number and distribution. The application of the plastochron index (Erickson and Michelini, 1957) to shoot development in trees (Larson and Isebrands, 1971; Isebrands and Larson, 1973) also shows promise.

To summarize, one can conclude that the ideal individual deciduous tree, in terms of photosynthetic apparatus, is one with a large Leaf Area Ratio and a relatively long conical crown, with many small branches, and leaves arranged so that there are high leaf angles near the apex and low angles near the base. Such a tree will also be able to produce a large amount of frost-resistant foliage rapidly in early spring, expand it throughout a major portion of the growing season, and retain it until early winter in the face of insects and diseases.

V. GENETIC VARIATION IN SHOOT AND CANOPY CHARACTERISTICS OF HARDWOODS

Hardwood tree improvement programmes have not usually included analyses of the physiological basis of genetic variation in yield. During the past decade, however, many data on genetic differences in juvenile shoot growth have been published, especially those associated with North American provenances.

A. *Juglans nigra*

Some of the more complete observations of phenology and its relationship to growth differences have been made on *Juglans nigra* in which range-wide provenance tests were the first steps in breeding. In southern Illinois plantings (lat. 37° N), trees from southern locations (lat. 32° N) flush earlier, defoliate later, and are taller, larger, and branchier than local trees (Bey et al., 1971a). Correlations between growth and characters responsible for leaf area expansion and duration are strong, suggesting that phenological characteristics, with high heritabilities, may be major components of growth. In a similar study (Bey et al., 1971b) of trees from 32° N to 42° N, about 40% of variance in height was associated with latitude, with southern trees growing 134 days and northern trees 93 days. Seventy-seven percent of the variation in date of leaf fall was associated with latitude, and southern trees held their leaves after frost in late October. During the period of active growth in late May and June, trees from some southern sources were also growing more rapidly than those from the most northerly sources. Thus, it was difficult to distinguish an assimilation rate component from that of duration in this test. In other work with material from 37° N to 42° N, broad-sense heritabilities for foliation date and defoliation date were high, and late-foliating clones were also early defoliators (Beineke and Master, 1973). However, there was no correlation between foliation dates and height growths. Numbers of branches and resistance to anthracnose defoliation were under fairly strong genetic control, but no estimate was made of their relationship to growth.

The possible effect of flushing date may depend on where the trees are planted. For instance, the range of foliation time is less at northern sites (Bey, 1972), probably because all trees are completely chilled and in a state of imposed dormancy long before spring bud-burst. Moreover, autumn frost-damage to southern material at central locations (37° N) can be a problem (Williams et al., 1974) as can early spring frosts. Nevertheless, experience to date suggests that sources as far as 200 miles south of the planting site will grow larger than trees of local or more northerly origin (Bey, 1973); a substantial, but as yet inexactly determined, portion of this growth difference is due to highly heritable variation in duration of the growing period.

B. *Populus*

Since the work of Pauley and Perry (1954), Sylven (1940) and Vaartaja (1960), it has been apparent that racial differences in the period over which *Populus* shoots elongate and produce new leaf can be responsible for major differences in growth (see Ekberg et al., Chapter 11). The poor growth of some hybrid poplars in the southern United States (Maisenhelder, 1970) is partly due to their brief period of shoot elongation under the short photoperiods experienced there.

Both broad-sense and narrow-sense heritabilities for characters affecting leaf area and duration on juvenile eastern cotton-wood (*P. deltoides*) in Mississippi (i.e. foliation date, defoliation date, *Melampsora* rust resistance, number of branches) have all been moderate to high and generally higher than the heritability for growth (Wilcox and Farmer, 1967; Farmer, 1970a; Jokela, 1966). In some tests (Farmer, 1970a; Wilcox and Farmer, 1967), there were moderate genetic correlations between diameter growth and earliness of foliation (bud-break), number of branches, and resistance to rust which causes premature defoliation.

In the same population, major clonal differences in leaf weight and in the ratios of stem and root weights to leaf weight were noted by Farmer (1970b) in a pot study of first-year growth. Under good growing conditions there was a poor phenotypic correlation and no genetic correlation between stem and leaf weights; some clones with heavy stems had large stem : leaf ratios, suggesting genetic differences in stem yield efficiency per unit of foliage.

Published data from the several existing *Populus deltoides* provenance studies are few, but suggest that genetic differences in phenology and crown form, and their interactions with planting sites, may be related to growth. First-year growth in a single planting at Stoneville, Mississippi (34° N) of material from six populations along the Mississippi River (31°–45° N) has been evaluated by Rockwood (1968). In this test, Louisiana (31°) and Minnesota (45°) trees were the first to come into leaf in spring, and the Missouri (38° N) and southern Illinois (37° N) material flushed last. During the spring, rates of height increment were greatest for trees from 45° N, but by late summer trees from 31° N were making the most rapid shoot elongation. At the end of the season, the largest trees were those from 37° N, which had fewer and longer branches than those from other sources. While leaf area and distribution were not considered, the data suggest that trees from southern Illinois produced more leaves, and displayed them more effectively, than trees with other origins. Variation in chilling requirements and response to spring and summer temperatures, as well as to photoperiod toward the end of the growing season, appeared to have influenced growth patterns (see Ekberg *et al.*, Chapter 11; Thielges and Beck, Chapter 14).

In another test, Posey (1969) noted that *P. deltoides* saplings grown from seed collected at high elevations in western Oklahoma (36° N, 100° W) were much branchier than those from the Mississippi River (36° N, 91° W), though eastern material exhibited the greatest growth.

An example of breeding directly for leaf area duration is found in the work of Pryor and Willing (1965) who were concerned with development of *Populus* for low latitudes. Crosses of the Chilean semi-evergreen clone with a clone of *P. monilifera* indicated that the semi-evergreen character is under simple genetic control. The authors suggest that southern provenances of *P. deltoides* may be useful as breeding stock in hybridization work with the Chilean clone.

C. *Quercus* and *Fagus*

Russian workers (Luk'yanets, 1969; Shutgaev, 1969) concluded that the early flushing "form" of English oak (*Quercus robur*) exhibits better height growth (5–16%) than the late flushing form when they are grown in areas without late spring frost. The reverse holds true in frost-prone areas where the seed crop of the early flushing forms is smaller, presumably because of frost damage (Ievlev, 1969). However, Efimov (1969) observed a higher rate of photosynthesis (on a unit-leaf basis) for the late form. *Fagus sylvatica* has also been classified into early and late flushing forms (Mikulka, 1955; Schober and Seibt, 1971), and Krahl-Urban (1953) and Cieslar (1923) have published evidence suggesting that foliation time in oak and beech is under strong genetic control. In an early provenance test, Cieslar also noted that narrow-crowned trees of *Q. robur* with ascending branches were faster growing.

In North America, the few published data on leaf-area duration have come mostly from provenance tests with northern red oak (*Q. rubra*). Early results of a range-wide test (Kriebel, 1965) showed little relationship between time of height-growth cessation and total first-year height, though both varied significantly among provenances. However, Irgens-Möller (1955) reported major provenance differences in height due to cessation time when material was grown at 42° N; Georgia (34° N) plants grew twice the height of Maine (45° N) trees. In a 5-year-old study of material from 34° to 41° N planted in eastern Tennessee (36° N), the sequence of flushing date was predictably from southern to northern sources (Gall and Taft, 1973). Regressions of total height on date of bud-break and number of flushes per season indicated that these two independent variables accounted for only about 13% of 5-year height variance.

Genetic variation in time of bud-break of northern red oak has also been found among altitudinal provenances in the Southern Appalachian Mountains (37° N) (McGee, 1974); low-elevation provenances flushed earlier than high-elevation ones. However, this difference in leaf duration was positively correlated with a difference in growth only at a low-elevation site.

A provenance study of juvenile Shumard oak (*Q. shumardii*) by Gabriel (1958) illustrates a typical effect of genetic differences in growth periodicity at a northern site (40° N). In this test, Florida

(31° N) and Mississippi (33° N) trees continued shoot growth into late summer and were almost twice as tall as those from Tennessee (36° N) and Illinois (38° N). Winter dieback, however, reduced the height of Florida trees below those from Illinois.

D. *Liquidambar styraciflua*

Growth-chamber studies of sweetgum showed that 5-fold differences in height growth can be produced by manipulating the photoperiod and thermoperiod. Northern provenances consistently grew least (Williams and McMillan, 1971). Provenance differences occurred in both the rate and duration of growth, and photographs suggest that the larger plants had larger leaf area ratios. Within the southern United States, where sweetgum breeding is concentrated, results of regionwide provenance testing to-date (Sprague and Weir, 1973) indicate only that coastal-plain provenances may not be suitable for planting a few hundred miles north and east of origin, owing to their frost susceptibility. Some inheritance data for phenology and juvenile crown characteristics in a coastal-plain population (Wilcox, 1970) represent the kind of information which may ultimately be helpful in determining components of growth variation. Narrow-sense heritabilities for foliation dates ranged from 0.54 in a central Mississippi planting to 1.27* on the Gulf coast. Major variation was noted in branch length, diameter and angle (h^2 = 0.20, 0.34 and 0.90, respectively). The ratio of height to crown diameter was used as a measure of crown form and the range of family means in this character was 1.8—2.4 (h^2 = 0.56). However, except for branch length (r = 0.61) correlations between growth and these characteristics were not significant. After 6 years, little genetic control of branch characteristics was noted in this test (Mohn and Schmitt, 1973).

* Heritabilities for hardwood foliation date are commonly above unity when computed from open-pollinated progeny data using:

$$h^2 = \frac{4\sigma_F^2}{\sigma_F^2 + \sigma_{RF}^2 + \sigma_W^2}$$

where σ_F^2 = female variance, σ_W^2 = within-plot variance, and σ_{RF}^2 = replicate × female variance.

E. Acer

Variations in chilling requirement and flushing date among provenances of red (*Acer rubrum*) and sugar maple (*A. saccharum*) have been observed by several investigators (Perry and Wang, 1960; Kriebel, 1957; Olmsted, 1951). Bud-break in *Acer* follows a pattern similar to that noted for *Quercus* and *Juglans nigra* with southern trees usually beginning growth first. However, Kriebel and Wang (1962) noted that under some northern conditions, northern sources began growth first, presumably because they were capable of growth at lower spring temperatures than southern trees. Juvenile growth of sugar maple seedlings of southern origin was observed to be greater in midwestern plantings, probably due to their longer growth period (Kriebel, 1957); southern trees were also bushier, which suggests a greater leaf surface. In Wright's (1949) study of silver maple (*A. saccharinum*) from sources in Indiana, southern sources were branchier and more susceptible to winter dieback.

F. Other Species

Though it is known that there are provenance differences in photoperiodic response of yellow poplar (*Liriodendron tulipifera*) (Vaartaja, 1961) and that southern provenances are severely damaged by cold in the mid-western part of North America (Funk, 1958), a definitive field study of provenance differences in growth patterns has not been conducted. In two small studies (Sluder, 1960; Farmer et al., 1967) southern material started growing earlier and stopped later than other sources, but this was not associated with major differences in growth.

Provenance studies with *Platanus* (Schmitt and Webb, 1971), *Fraxinus* (Wright, 1944a, b), *Ulmus* and *Betula* (Vaartaja, 1959) also give evidence of major provenance differences in shoot growth periodicity and leaf area duration and their interaction with environmental factors, giving differences in yield. Recent data from *Platanus occidentalis* provenance tests in Mississippi indicate that trees from southern sources are more resistant to serious leaf and crown disease and that correlations between growth and crown characteristics are low.

VI. CONCLUSIONS

The above review indicates that most hardwood provenance, progeny and clonal tests to-date give evidence of major genetic differences in shoot elongation characteristics and leaf area duration; and some correlation analyses suggest they may be important components of yield. This is particularly so in tests of widely separated geographic sources growing in environments where their phenological differences are great. However, detailed information on variation in crown structure and its relationship to yield is not yet available. Presently unknown or poorly understood source—sink relationships probably have an effect on the balance of yield components which will be difficult to assess.

It is also apparent that while forest ecologists and silviculturists have given considerable attention to canopy structure and its relationship to light interception and yield, they have mostly ignored intraspecies variation in crown characteristics. Hardwood geneticists, on the other hand, have not yet incorporated studies of shoot growth and crown structure in their tree improvement programmes. Much of the information obtained to-date stems from "targets of opportunity" rather than planned data acquisition. For example, variation in phenology has been so striking and under such strong genetic control that it is hard to overlook. It has not, however, generally been considered in the context of yield component analysis. Other equally important components are probably being overlooked because we have not been analytically concerned with the physiological basis of yield. Annual measurements of progeny and provenance heights and diameters give only a fraction of the data necessary to assess the basis of yield differences.

If we wish to determine components of variation in yield of hardwoods, I believe we should first make better use of existing tests. This will entail more detailed observations of shoot and crown growth characteristics on at least a portion of test populations, perhaps material representing the range of yield variation. Procedures developed by ecologists to study canopy characteristics should be useful with both individual plants and stands. In particular, the methods of growth analysis, adapted for larger trees growing under field conditions, can be valuable in initial assessments of juvenile growth differences. Such studies should make it possible to delineate the relative effects of canopy characteristics and photosynthetic capacities. Existing field tests should be followed closely by intensive growth analysis of genetically variable material (Wareing, 1970). The investigations of Ledig and Perry (1969; see also Ledig, Chapter 2) with conifers are useful examples of this approach. Growth analysis procedures should also help determine the physiological basis of genotype—environment interactions which are proving to be important in breeding.

To be useful in breeding, this systematic deciphering must eventually enable one to identify components which account for appreciable yield differences. These components should be identified at an early age, and be correlated with yield throughout ontogeny, though this may be an unrealistic expectation given the complex relations among components. Whereas highly heritable phenological characteristics probably remain unchanged throughout the trees' lifetime, differences in crown form may change considerably as the trees develop. Inherent variation in crown development and its relationship to stand conditions will, therefore, require special attention in rotation-length experiments. Initially the most appropriate experimental species may be fast-growing trees such as *Populus deltoides* and *Platanus* which develop mature crowns in a few decades under normal plantation conditions. Growth studies, such as those described by Zavitkovski *et al.* (1974), expanded to include a range of genotypes and environmental conditions, should be most immediately fruitful.

REFERENCES

Alvim, P. de T. (1964). Tree growth periodicity in tropical climates. *In* "The Formation of Wood in Forest Trees" (M. H. Zimmerman, ed.) pp. 479—495, Academic Press, London and New York.

Anderson, M. C. (1966). Stand structure and light penetration II. A theoretical analysis. *J. appl. Ecol.* 3, 41—54.

Ariyanayagam, R. P., Moore, C. L. and Carangal, V. R. (1974). Selection for leaf angle in maize and its effect on grain yield and other characters. *Crop. Sci.* 14, 551—556.

Beineke, W. F. and Masters, C. J. (1973). Black walnut progeny and clonal tests at Purdue University. *Proc. 12th Southern Forest Tree Imp. Conf.* pp. 233—242.

Bey, C. F. (1972). Leaf flush in black walnut at several Midwest locations. *Proc. 19th Northeastern Forest Tree Imp. Conf.* pp. 47–51.

Bey, C. F. (1973). Growth of black walnut trees in eight Midwestern states – a provenance test. *USDA Forest Service, Res. Paper* NC-91, 7 pp.

Bey, C. F., Hawker, N. L. and Roth, P. L. (1971a). Variation in growth and form in young black walnut trees. *Proc. 11th Southern Forest Tree Imp. Conf.* pp. 120–127.

Bey, C. F., Toliver, J. R. and Roth, P. L. (1971b). Early growth of black walnut trees from twenty seed sources. *USDA Forest Service, Res. Note* NC-105, 4 pp.

Black, J. N. (1966). The utilization of solar energy by forests. *Forestry (Suppl.)* 39, 98–109.

Borchert, R. (1973). Simulation of rhythmic tree growth under constant conditions. *Physiol. Plant.* 29, 173–180.

Brown, C. L. and Goddard, R. E. (1961). Silvical considerations in the selection of plus phenotypes. *J. For.* 59, 420–426.

Cieslar, A. (1923). Untersuchungen über die wirtschaftliche Bedeutung der Herkunft des Saatgutes der Stieleiche. Centralbl. Ges. Forstw. 19, 97–149. Cited in Irgens-Möller, H. 1955. Forest genetics research: *Quercus* L. *Econ. Bot.* 9(1), 53–71.

Coombe, D. E. (1960). An analysis of the growth of *Trema guineensis*. *J. Ecol.* 48, 219–231.

Coombe, D. E. and Hadfield, W. (1962). An analysis of the growth of *Musanga cecropioides*. *J. Ecol.* 50, 221–234.

Efimov, Yu. P. (1969). Physiological characters of early and late forms of English oak (*Q. robur*). *In* "Proc. Conf. on Forest Tree Genetics, Selection and Seed Production". Trans. from Russian, Israel Program for Scientific Translation, Cat. No. 5500. pp. 39–40.

Engledow, F. L. and Wadham, S. M. (1923). Investigations on yield in the cereals. *J. agric. Sci., Camb.* 13, 390–439.

Erickson, R. O. and Michelini, F. J. (1957). The plastochron index. *Am. J. Bot.* 44, 297–305.

Evans, G. C. (1972). "The quantitative analysis of plant growth". University of California Press, Berkeley.

Farmer, R. E., Jr. (1970a). Genetic variation among open-pollinated progeny of eastern cottonwood. *Silvae Genet.* 19, 149–151.

Farmer, R. E., Jr. (1970b). Variation and inheritance of eastern cottonwood growth and wood properties under two soil moisture regimes. *Silvae Genet.* 19, 5–8.

Farmer, R. E., Jr., Russell, T. E. and Krinard, R. M. (1967). Sixth-year results from a yellow-poplar provenance test. *Proc. 9th Southern Conf. For. Tree Improvement* pp. 65–68.

Ford, E. D. and Newbould, P. J. (1971). The leaf canopy of a coppiced deciduous woodland. I. Development and structure. *J. Ecol.* 59, 843–862.

Franklin, R. T. (1970). Insect influences on the forest canopy. *In* "Analysis of Temperate Forest Ecosystems" (D. E. Reichle, ed.), pp. 86–99. Springer-Verlag, New York.

Funk, D. T. (1958). Frost damage to yellow-poplar varies by seed source and site. *U.S. For. Service Central States For. Exp. Sta., For. Res. Note* 115, 2 pp.

Gabriel, W. J. (1958). Genetic differences in juvenile Shumard oak. *USDA For. Service Northeastern For. Exp. Stn, For. Res. Note* 81, 3 pp.

Gall, W. R. and Taft, K. A., Jr. (1973). Variation in height growth and flushing of norhtern red oak (*Quercus rubra* L.). *Proc. 12th Southern For. Tree Improvement Conf.* pp. 190–199.

Ghosh, S. P. (1973). Internal structure and photosynthetic activity of different leaves of apple. *J. hort. Sci.* 48, 1–9.

Hadfield, W. (1974). Shade in Northeast Indian tea plantation. II. Foliar illumination and canopy characteristics. *J. appl. Ecol.* 11, 179–199.

Huxley, P. A. and Van Eck, W. A. (1974). Seasonal changes in growth and development of some woody perennials near Kampala, Uganda. *J. Ecol.* 62, 579–592.

Ievlev, V. V. (1969). Fruiting of early and late flushing forms of English oak from many years' observations in the Usman Forest, Voronezh region. *In* "Proc. Conf. on Forest Tree Genetics, Selection and Seed Production". Trans. from Russian, Israel Program for Scientific Translation, Cat. No. 5500. pp. 133–134.

Irgens-Möller, H. (1955). Forest tree genetics research: *Quercus* L. *Econ. Bot.* 9, 53–71.

Isebrands, J. G. and Larson, P. R. (1973). Anatomical changes during leaf ontogeny in *Populus deltoides*. *Am. J. Bot.* 60, 199–208.

Jahnke, L. S. and Lawrence, D. B. (1965). Influence of photosynthetic crown structure on potential productivity of vegetation, based primarily on mathematical models. *Ecology* 46, 319–326.

Jokela, J. J. 1966. Incidence and heritability of *Melampsora* rust in *Populus deltoides* Bartr. *In* "Breeding Pest Resistant Trees" (H. D. Gerhold et al., eds.), pp. 111–117. Pergamon Press, New York.

Kinerson, R. S., Higginbotham, K. O. and Chapman, R. C. (1974). The dynamics of foliage distribution within a forest canopy. *J. appl. Ecol.* 11, 347–353.

Kittredge, J. (1944). Estimation of the amount of foliage of trees and stands. *J. For.* 42, 905–912.

Kozlowski, T. T. (1964). Shoot growth in woody plants. *Bot. Rev.* 30, 335–392.

Kozlowski, T. T. (1971). "Growth and development of trees", Vol. I. Academic Press, New York and London.

Krahl-Urban, J. (1953). Hinweise auf individuelle Erbanlagen bei Eichen und Buchen. *Z. Forstgenet. ForstpflZücht.* **2**, 51–59.

Kriebel, H. B. (1957). Patterns of genetic variation in sugar maple. *Ohio Agr. Exp. Sta. Res. Bull.* 791. 56 pp.

Kriebel, H. B. (1965). Parental and provenance effects on growth of red oak seedlings. *Proc. 4th Central States For. Tree Improvement Conf.* pp. 19–25.

Kriebel, H. B. and Wang, C. W. (1962). The interaction between provenance and degree of chilling in budbreak of sugar maple. *Silvae Genet.* **11**, 125–130.

Larson, P. R. and Isebrands, J. G. (1971). The plastochron index as applied to developmental studies of cottonwood. *Can. J. Forest Res.* **1**, 1–11.

Ledig, F. T. (1969). A growth model for tree seedlings based on the rate of photosynthesis and the distribution of photosynthate. *Photosynthetica* **3**, 263–275.

Ledig, F. T. and Perry, T. O. (1969). Net assimilation rate and growth in loblolly pine seedlings. *Forest Sci.* **15**, 431–438.

Lieth, H. (1971). The phenological viewpoint in productivity studies. *In* "Productivity of Forest Ecosystems" (P. Duvigneaud, ed.), pp. 71–84. UNESCO, Paris.

Loomis, R. S. and Williams, W. A. (1969). Productivity and the morphology of crop stands: Patterns with leaves. *In* "Physiological Aspects of Crop Yield" (R. C. Dinauer, ed.), pp. 27–45. Amer. Soc. Agron. and Crop Sci.

Loomis, R. S., Williams, W. A. and Duncan, W. G. (1967). Community architecture and the productivity of terrestrial plant communities. *In* "Harvesting the Sun" (A. San Pietro, F. A. Greer and T. J. Army, eds.), pp. 291–308. Academic Press, New York and London.

Luk'yanets, V. B. (1969). The importance of selection of ecotypes and forms of English oak – results of experimental cultures. *In* "Proc. Conf. on Forest Tree Genetics, Selection and Seed Production". Trans. from Russian, Israel Program for Scientific Translation, Cat. No. 5500, pp. 81–82.

McGee, C. E. (1974). Elevation of seed sources and planting sites affects phenology and development of red oak seedlings. *Forest Sci.* **20**, 160–164.

McKinion, J. M., Hesketh, J. D. and Baker, D. N. (1974). Analysis of the exponential growth equation. *Crop Sci.* **14**, 549–551.

Madgwick, H. A. I. (1970). Biomass and productivity models of forest canopies. *In* "Analysis of Temperate Forest Ecosystems" (D. E. Reichle, ed.), pp. 47–54. Springer-Verlag, New York.

Madgwick, H. A. I. (1971). Growth of *Liriodendron tulipifera* seedlings with different levels of nitrogen supply. *Forest Sci.* **17**, 287–292.

Maggs, D. H. (1964). Growth-rates in relation to assimilate supply and demand. I. Leaves and roots as limiting regions. *J. exp. Bot.* **15**, 574–583.

Maisenhelder, L. C. (1970). Eastern cottonwood selections outgrow hybrids on southern sites. *J. For.* **68**, 300–301.

Matthews, J. D. (1963). Some applications of genetics and physiology in thinning. *Forestry* **36**, 172–180.

Mikulka, B. (1955). Spät- and frühtriebende Buchen in Sihlwald. *Schweiz. Z. Forstwes.* **106**, 666–670.

Mohn, C. and Schmitt, D. (1973). Early development of open-pollinated sweetgum progenies. *Proc. 12th Southern For. Tree Improvement Conf.* pp. 228–232.

Neales, T. F. and Incoll, L. D. (1968). The control of leaf photosynthesis rate by the level of assimilate concentration in the leaf: A review of the hypothesis. *Bot. Rev.* **34**, 107–125.

Olmsted, C. E. (1951). Experiments on photoperiodism, dormancy, and leaf age and abscission in sugar maple. *Bot. Gaz.* **112**, 365–393.

Pauley, S. S. and Perry, T. O. (1954). Ecotypic variation of the photoperiodic response in *Populus*. *J. Arnold Arbor.* **35**, 167–188.

Perry, T. O. and Wang, C. W. (1960). Genetic variation in the winter chilling requirement for date of dormancy break for *Acer rubrum*. *Ecology* **41**, 790–794.

Pollard, D. F. W. (1970). Leaf area development on different shoot types in a young aspen stand and its effect upon production. *Can. J. Bot.* **48**, 1801–1804.

Pollard, D. F. W. and Wareing, P. F. (1968). Rates of dry matter production in forest tree seedlings. *Ann. Bot.* **32**, 573–591.

Posey, C. E. (1969). Phenotypic and genotypic variation in eastern cottonwood in the southern great plains. *Proc. 10th Southern For. Tree Improvement Conf.* pp. 130–135.

Priestley, C. A. (1964). The importance of autumn foliage to carbohydrate status and root growth of apple trees. *Ann. Rep. East Malling Res. Stn*, pp. 104–106.

Pryor, L. D. and Willing, R. R. (1965). The development of poplar clones suited to low latitudes. *Silvae Genet.* **14**, 123–127.

Rafes, P. M. (1970). Estimation of the effects of phytophagous insects on forest production. *In* "Analysis of Temperate Forest Ecosystems" (D. E. Reichle, ed.), pp. 100–106. Springer-Verlag, New York.

Rhodes, I. (1971). The relationship between productivity and some components of canopy structure in ryegrass (*Lolium* spp.). II. Yield, canopy structure and light interception. *J. agric. Sci., Camb.* **77**, 283–292.

Rockwood, D. L. (1968). "Variation within eastern cottonwood along the course of the Mississippi River". Master's thesis. University of Illinois.

Romberger, J. A. (1963). Meristems, growth and development in woody plants. *USDA Tech. Bull.* No. 1292. 214 pp.

Rothacher, J. S., Blow, F. E. and Potts, S. M. (1954). Estimating the quantity of tree foliage in oak stands in the Tennessee Valley. *J. For.* **52**, 169–173.

Russell, W. A. (1972). Effect of leaf angle on hybrid performance in maize (*Zea mays* L.). *Crop Sci.* 12, 90–92.

Saeki, T. (1963). Light relations in plant communities. *In* "Environmental Control of Plant Growth" (L. T. Evans, ed.), pp. 79–94. Academic Press, New York and London.

Satoo, T. (1970). A synthesis of studies by the harvest method: Primary production relations in the temperate deciduous forests of Japan. *In* "Analysis of Temperate Forest Ecosystems" (D. E. Reichle, ed.), pp. 55–72. Springer-Verlag, New York.

Schmitt, D. M. and Webb, C. D. (1971). Georgia sycamore seed sources in Mississippi plantings: Site adaptability a key factor. *Proc. 11th Southern For. Tree Improvement Conf.* pp. 113–119.

Schober, R. and Seibt, G. (1971). Phenological observations on beech and spruce as a function of climate. *In* "Integrated Experimental Ecology" (H. Ellenburg, ed.), pp. 32–36. Springer-Verlag, New York.

Sheehy, J. E. and Cooper, J. P. (1973). Light interception, photosynthetic activity and crop growth rate in canopies of six temperate forage grasses. *J. appl. Ecol.* 10, 239–250.

Shutgaev, A. M. (1969). The characters of early-flushing and late-flushing forms of English oak. *In* "Proc. Conf. on Forest Tree Genetics, Selection and Seed Production". Trans. from Russian, Israel Program for Scientific Translation Cat. No. 5500. pp. 79–80.

Sluder, E. R. (1960). Early results from a geographic seed source study of yellow-poplar. *U.S. For. Service, Southeastern Forest Exp. Stn. Res. Note* 150. 2 pp.

Sprague, J. and Weir, R. J. (1973). Geographic variation of sweetgum. *Proc. 12th Southern For. Tree Improvement Conf.* pp. 169–180.

Sweet, G. B. and Wareing, P. F. (1966). Role of plant growth in regulating photosynthesis. *Nature, Lond.* 210, 77–79.

Sylven, N. (1940). Lang- och kortdagstyper av de suenska skogsträden. *Svensk PappTidn.* 43(17), 317–324; (18), 322–342; (19), 350–351.

Vaartaja, O. (1959). Evidence of photoperiodic ecotypes in trees. *Ecol. Monogr.* 29, 91–111.

Vaartaja, O. (1960). Ecotypic variation of photoperiodic response in trees especially in two *Populus* species. *Forest Sci.* 6, 200–206.

Vaartaja, O. (1961). Demonstration of photoperiodic ecotypes in *Liriodendron* and *Quercus*. *Can. J. Bot.* 39, 649–654.

Wallace, D. H., Ozbun, J. L. and Munger, H. M. (1972). Physiological genetics of crop yield. *Adv. Agron.* 24, 97–146.

Ward, W. W. (1964). Bud distribution and branching in red oak. *Bot. Gaz.* 125, 217–220.

Wareing, P. F. (1966). The physiologist's approach to tree growth. *Forestry (Suppl.)* 39, 7–18.
Wareing, P. F. (1970). Growth and its coordination in trees. *In* "Physiology of Tree Crops" (L. C. Luckwill and C. V. Cutting, eds), pp. 1–21. Academic Press, London and New York.
Warren Wilson, J. (1967). Stand structure and light penetration. III. Sunlit foliage area. *J. appl. Ecol.* 4, 159–165.
Watson, D. J. (1952). The physiological basis of variation in yield. *Adv. Agron.* 4, 101–145.
Watson, D. J. (1956). Leaf growth in relation to crop yield. *In* "The Growth of Leaves" (F. L. Milthorpe, ed.), pp. 178–191. Butterworths, London.
Watson, D. J. (1971). Size, structure and activity of the productive system of crops. *In* "Potential Crop Production: A Case Study" (P. F. Wareing and J. P. Cooper, eds), pp. 76–87. Heinemann, London.
Wilcox, J. R. and Farmer, R. E., Jr. (1967). Variation and inheritance of juvenile characters of eastern cottonwood. *Silvae Genet.* 16, 162–165.
Wilcox, Jr. R. (1970). Inherent variation in south Mississippi sweetgum. *Silvae Genet.* 19, 91–94.
Williams, G. J. III and McMillan, C. (1971). Phenology of six United States provenances of *Liquidamber styraciflua* under controlled conditions. *Am. J. Bot.* 58, 24–31.
Williams, R. D., Funk, D. T., Phares, R. E., Lemmien, W. and Russell, T. E. (1974). Apparent freeze damage to black walnut seedlings related to seed source and fertilizer treatment. *Tree Plrs' Notes, Wash.* 25, 6–8.
Wright, J. W. (1944a). Genotypic variation in white ash. *J. For.* 42, 489–495.
Wright, J. W. (1944b). Ecotypic differentiation in red ash. *J. For.* 42, 591–597.
Wright, J. W. (1949). Local genetic variation in silver maple. *J. For.* 47, 300–302.
Zavitkovski, J., Isebrands, J. G. and Crow, T. R. (1974). Application of growth analysis in forest biomass studies. *In* "Proc. 3rd North American Forest Biology Workshop" (C. P. P. Reid and G. H. Fechner, eds), pp. 196–226. Colorado State University, Fort Collins.

8
Tree Forms in Relation to Environmental Conditions: an Ecological Viewpoint

E. F. BRUNIG
Institute of World Forestry, University of Hamburg, Reinbek, F.R. Germany

I.	Introduction	139
II.	Tree Forms in Natural Environments	140
	A. Tropical Forests	140
	B. Temperate Forests	143
III.	Causal Relationships: Facts and Hypotheses	145
	A. Leaf Size and Orientation	146
	B. Crown Geometry	147
	C. Canopy Geometry	150
IV.	Crown Ideotypes	151
V.	Summary	155
References		155

I. INTRODUCTION

Natural forest sites which differ in soil type, climate or history, usually bear tree species which are locally adapted and have particular bole shapes, branching habits, and leaves of given sizes, colours and spatial arrangements, giving foliage canopies with certain architectural characteristics. The aim of this contribution is to examine such differences among primary forests, indicate their possible ecological significance, and arrive at some conclusions on the types of forest canopy architecture that might be best adapted to the varying site conditions encountered throughout the world.

II. TREE FORMS IN NATURAL ENVIRONMENTS

A. Tropical Forests

Nolde (1941) observed that trees with flat disc-type crowns occurred most frequently in hot, dry regions in tropical savanna areas, their frequency increasing proportionately with increasing solar radiation receipt. Spherical crown types were more common on more mesic* sites.

In eastern Australia evergreen rain forests and sclerophyllous, xeromorphic forests (adapted to tolerate drought, like eucalypts) occur on neighbouring sites within the same region, the former on the more nutrient-rich sites, which are less liable to water stress, the latter on poorer, drier sites (Webb, 1959, 1965). Poor mineral nutrition generally accentuates the sclerophyllous and xeromorphic properties associated with drought resistance, causing in particular, a decrease in leaf size and the formation of thicker leaf cuticles. Thus, Beadle (1966) showed that the degree of xeromorphy could be lessened in many taxa by adding phosphate and nitrate fertilizers. The formation of thickened main roots (lignotubers) on some eucalypt species, for example, is particularly common when the soils have low phosphorus contents (Beadle, 1968).

In equatorial, humid, evergreen forests of Sarawak (Borneo), tree stature, crown geometry and canopy architecture differ remarkably between stands growing on differing sites and soil types. For example, forests growing along a catena from mesic latosol to xeric podsol range from tall, large-leaved broad crowns in mesic conditions to short, small-leaved xeromorphic forms in xeric conditions (Fig. 1). Similar changes in vegetation occur along gradients of peat bog development on inland plateaus and terraces with extremely mineral-deficient and dry soils and along gradients of successional communities in deltaic peat swamp forests (Anderson, 1961; Brunig, 1970).

Primary humid, evergreen forests are patchworks of tree stands in different phases of successional development, ranging from pioneer

---

* "Mesic" refers to favourable, moist conditions; "xeric" to conditions where plants frequently suffer water stress, and may possess "xeromorphic" characters which prevent excessive water loss.

8. Tree Forms and Environmental Conditions

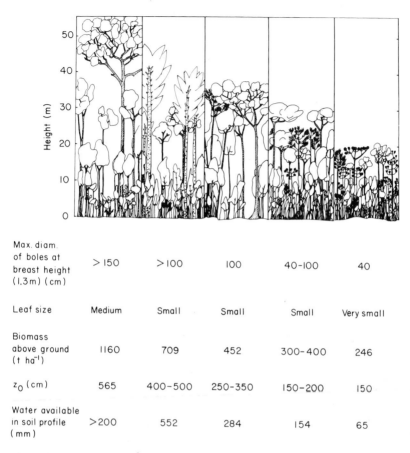

FIG. 1 Changes in stand architecture along a soil catena in Sarawak (Borneo) from mesic (wet) to xeric (dry) conditions (left to right). Tree heights, maximum diameters, and stand biomass decrease, accompanied by a change in crown architecture (leaf size, crown shape, and the estimator z_0 of aerodynamic roughness (Brunig, 1970)). The forest types are further described by Brunig (1974).

to overmature (Fig. 2). Short-lived pioneers on mesic sites are light-wooded and fast-growing, have large, thin leaves, often arranged in a single layer, are shade-intolerant and have poor apical dominance (Fig. 3A). Species typical of the next successional stage have pronounced apical dominance, more or less horizontal branching, often with long intervals between nodes and candelabra- or pagoda-shaped crowns (Fig. 3B). The leaves are often thick and intolerant of

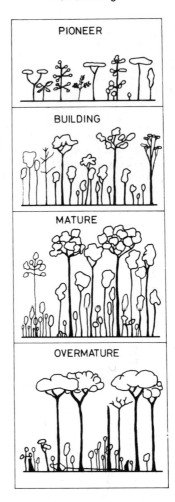

FIG. 2 Successional changes in stand architecture within a large gap in an equatorial, humid, evergreen forest. The gap is colonized by short-lived pioneer species which are gradually replaced by invading late-seral successional species in the "building" phase; these are replaced by mature-phase light-demanding species, and these finally by slow-growing, long-lived shade tolerant species with an accompanying change in tree form.

shade, although they differ greatly in size and shape. Typical genera of this "building" phase are *Anthocephalus* and *Ceiba*. Species in the next "mature" phase become relatively tall. The crowns of young trees are ellipsoid and multi-layered, but become hemispherical and often mono-layered at maturity. The leaves are xeromorphic, frequently upturned, leathery and highly reflectant, and are typically arranged in tufts (Fig. 3B) or spherical clumps (Fig. 3C). The final successional stage is marked by the emergence of slow-growing, heavy-wooded species with more diffusely arranged, smaller leaves, but otherwise resembling trees in the "mature" phase.

The understorey at all successional stages is multi-layered. Leaves may be larger or smaller than in the top canopies, depending on site conditions (Brunig, 1970, 1974), but are normally in more diffuse and random arrangements. In a non-successional Andean cloud forest the major differences between leaves in four canopy layers were in anatomy, although there were also some differences in leaf shape and size (Roth *et al.*, 1971).

Where soils occur within tropical rain forests which hold less than 100 mm water in the rooting zone, drought is a common, although intermittent event. On such sites, the successional sequence of species is very different (Brunig, 1971). Large-leaved pioneers are absent, and each successional stage has tree species with multi-layered crowns with small leaves, needles or phyllodes (Figs 1, and 3B). The understoreys are similarly xeromorphic and are often greatly separated from the top canopy layers.

B. Temperate Forests

In *Picea abies*, narrow, columnar crowns are generally associated with provenances from northerly latitudes, high altitudes and more xeric sites (Schmidt-Vogt, 1972), whereas broad crowns are associated with humid sites, deep shade and harsh tree-line conditions. A multitude of detailed crown shapes have been described, with branching habits varying from ascending to pendulous, but opinions differ on their relationships with site conditions. Alexandrov (1971) reported that upturned secondary branches on *P. abies* in the Rhodope mountains (Bulgaria) were associated with low air temperatures, high insolation, and xeric sites. Certain crown types seem to be associated with rapid increases in wood volume (Popov, 1970; Rubner, 1943; Schmidt-Vogt, 1972) and rapid rates of net photosynthesis (Neuwirth, 1968) but the causal relationships are obscure.

FIG. 3 Crown types from different sites and successional stages in equatorial, humid, evergreen forests.

A. *Cecropia sp.*, Venezuela; a mono-layered, large-leaved pioneer species on a mesic site. This crown type maximizes radiation interception at high sun elevations.

B. *Ploiarium alternifolium*, Sarawak; a late-seral to mature phase successional species on a humid, oligotrophic site, with upturned, reflectant leaves in tufts. To the left is a short-needled *Dacrydium pectinatum* with upright twigs. This crown type spreads the radiation load during the day and maximizes ventilation.

C. *Emmotum fulvum*, Venezuela; a light-demanding mature phase species showing composite-sphere crown geometry, with mesophyll, xeromorphic leaves, typical of emergent humid tropical forest species. This crown type receives an even radiative load during the day, intercepts light efficiently and has good ventilation. Oblique light penetrates effectively, but midday shading is heavy.

D. *Dacrydium pectinatum*, Bako National Park, Sarawak; a xeromorphic sclerophyll, which is present at all phases of succession on xeric and oligotrophic sites. This crown type maximizes convective heat transfer, and keeps plant temperatures close to ambient air temperatures.

8. Tree Forms and Environmental Conditions

Equally little is known about the reasons for differences in crown shape that occur within temperate pine species growing on different sites. Contrary to the situation in *Picea* and *Larix*, pines have relatively broad crowns when growing on xeric sites; narrow crowns are associated with mesic sites and high altitudes.

Horn (1971) described differences in crown geometry and leaf sizes in forest stands representing different successional stages in New Jersey, U.S.A. Short-lived pioneer species were soft-wooded, had small leaves, and multi-layered conical to weakly convex crowns. The ratios of crown heights to widths increased with age. In contrast, late successional species were heavy-wooded, and had two types of crowns; in the understoreys where most light comes from the zenith they were mono-layered, flat and spreading, whereas higher up in the canopies they filled out to tall ellipsoids. According to Horn, there can be intermediate types which are long-lived, with multi-layered canopies, and which persist by exploiting gaps in the forests. Assuming that "competition" for light is the primary determinant of crown architecture, these intermediate types should form young crowns which mimic the early successional habit, and later expand their crowns laterally in the top-canopy to dominate the gaps.

III. CAUSAL RELATIONSHIPS: FACTS AND HYPOTHESES

Energy is exchanged between plants and their environments mainly by radiation, evapo-transpiration, conduction and convection. The intensity of energy exchange depends on meteorological parameters, but can be influenced greatly by features of the leaf and canopy surfaces. Plant properties such as leaf sizes, shapes and optical properties are important, as well as leaf spacing, orientation and bulk physiological resistance to diffusion. Also, differences in exposure, which are influenced by canopy structure, have effects on the intensity of energy exchange both for individuals and whole stands (Brunig, 1970).

With high levels of net solar radiation, and correspondingly high bulk resistances, the "Bowen ratio" may rise above unity (i.e. sensible heat flux is greater than latent heat flux). In such conditions wind effectively decreases evapo-transpiration. Any geometric pattern of leaves and tree crowns which increases "diffusivity", ventilation and convective heat exchange will then decrease water stress and prevent the leaves from overheating. During humid periods, such "diffuse" leaf arrangements increase the rates of evapo-transpiration and allow rapid photosynthesis, whereas during periods of drought with high insolation they improve convective heat dissipation, and minimize the risks of leaf damage.

A. Leaf Size and Orientation

Almost all of the heat and energy transfer of a tree occurs through its leaves, and the leaf tissues and surface boundary layers exercise most resistance to transfer. Consequently, the size and reflectivity of the leaves have important effects on the extent to which plants can control their temperature. Parkhurst and Loucks (1972) predicted leaf sizes from models based on "water-use efficiency". Their scheme of expected trends in natural selection for leaf size agrees well with leaf size distributions that occur in natural, tropical, moist forests. High air temperature with low radiation flux densities favour large leaves, as happens in the understoreys of mesophyllous stands, whereas high radiation flux densities favour smaller leaves, as occur in the well illuminated understoreys of sclerophyllous stands on xeric sites, and generally in top canopies.

A simple way of capturing solar energy, with adequate supporting structure, is to arrange very large, entire leaves in a rosette or terminal whorl on a single light-wooded stem. But there are several disadvantages; the intercepted sunlight is distributed inefficiently, a great deal of water is required for transpirational cooling of large, entire leaves, and any deviation of the main stem from the vertical by branching makes it necessary to reinforce the structure, with an attendant energy cost. Trees with such photosynthetically inefficient, mono-layered crown forms are less able to afford this energy cost compared with multi-layered trees, especially if the trees are tall. Accordingly, large-leaved, mono-layered crown forms are typically found among short-lived early successional species on humid sites with high levels of incoming solar radiation and low storm risk.

Clearly, large, entire leaves are a disadvantage if the trees grow to great heights, or on sites both with occasional or seasonal periods of water stress and with high levels of solar radiation. The alternative is to have divided or small leaves, which have thinner atmospheric boundary layers, and are therefore more closely coupled to the temperatures of the ambient air.

When the leaves within tropical tree stands are steeply inclined, large areas of leaf surface are sunlit at high solar elevations, the flux density per unit leaf area is decreased, and the downward flux of radiation increased. Similarly, light penetrates the leaf canopies better when the leaves are clumped, especially at low sun angles. Also, erect leaves are more effectively cooled by wind, and when they are clumped there is increased air turbulence in the crowns. When individual needles, fascicles or groups of leaves are widely spaced there is increased turbulent exchange and lower diffusive resistance at higher temperatures. As already mentioned, these are crucial requirements in dry, sunny, still conditions, when the "Bowen ratio" rises above unity.

In temperate regions where the sun is at low elevations the maximum leaf surface is sunlit when the leaves are orientated more horizontally. However, the maximum area of sunlit leaf, which occurs at noon in midsummer, rarely exceeds twice the ground surface area, compared with about three times in tropical areas (Wilson, 1967).

Leaf bunching may have disadvantages for the plant, such as increased mutual shading. In pines, the net rate of photosynthesis per unit needle surface area is greater for two-needled than five-needled species (Kramer, 1957). Similarly, *Pinus taeda* seedlings with regularly spaced primary needles appear to be photosynthetically more efficient than older plants with secondary needles, which are more clumped (Borman, 1957). Leaf bunching also makes use of the structural elements of the tree less efficiently than random or regular distributions, because a given leaf surface area of bunched leaves intercepts less radiation, or alternatively, equal interception requires more leaf area and therefore more input into structure.

B. Crown Geometry

The variation in crown geometry in a successional series of broad-leaved stands was explained by Horn (1971) in terms of competition or "interference" for light alone, between successively invading species. This explanation is probably too simple. For immobile organisms to succeed, it is just as important to decrease the risks of damage by environmental factors such as drought, high temperatures, winds, snow, and so on, as to shade out competitors. Competitive advantages over other species can also, and possibly more effectively, be gained by adaptation to narrow ecological niches or by efficient breeding systems. This occurs, for instance, in tropical rain forests with respect to nutrients and drainage, and crown form seems more closely related to the efficient exploitation of light and water, and to the avoidance of damage, than to competition between succeeding species. In temperate forests Luganov and Solod'ko (1968) studied the length, width and surface angles of the crowns of six tree species at three latitudes in Siberia, and concluded that the angular distribution of sky light was among the most important factors which determine crown form.

Large, entire leaves can be effectively packed into monolayers to intercept maximum solar radiation without self-shading, but, as already mentioned, the total photosynthetic output of such canopies is less than that of multi-layered canopies of small leaves. However, canopies with small leaves need more complex supporting structures, and spreading the leaves over larger vertical distances introduces problems of self-shading. Photosynthesis is restricted severely when leaves are in the umbra of other leaves, and to a lesser degree when they fall in the penumbra (Fig. 4). Small, narrow leaves, which are inclined to the sun, produce shorter, narrower shadows, and enable greater leaf area indices to be adequately illuminated with greater total photosynthetic activity. However, to dispose the leaves in this way requires an extensive and complex branching system, which increases the amount of energy that has to be invested in canopy development. Therefore, radiation interception and utilization has to be much more efficient per unit of crown projected area in such small-leaved, multi-layered crowns, compared with large-leaved, mono-layered crowns.

The area of crown surface which a solitary tree exposes to direct sunlight varies with crown shape and sun altitude. A narrowly columnar crown of a solitary tree intercepts maximum radiation, and reflects least, at low sun elevations. However, in stands, the reverse may be true, owing to mutual shading. Thus, Thomas *et al.* (1972)

8. Tree Forms and Environmental Conditions

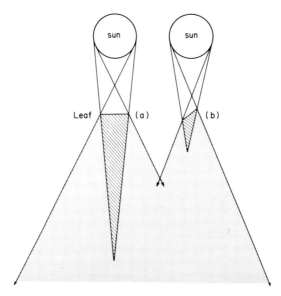

FIG. 4 Differences in the shadows cast by broad horizontal leaves (a) and narrow, inclined leaves (b). Adapted from Horn (1971). Leaf width ratio = 1 : 0.6; umbra ratio = 1 : 0.13.

calculated the proportion of solar radiation that was intercepted during the day by a dense pole stand of *Eucalyptus globulus* with narrow conical crowns on a humid site in southern India and found that minimum interception occurred at sunrise and maximum at noon.

A flat, disc-like crown intercepts maximum direct solar radiation at noon and least at sunrise, irrespective of whether the trees are solitary or in stands. The problem of dissipating excess heat at noon is solved by allowing rapid upward diffusion and leaving the greatest possible distance between the leaf layer and reflecting and re-radiating surfaces beneath. The major disadvantages are that solar radiation is poorly intercepted, and is also reflected greatly in the mornings and afternoons when moisture conditions for photosynthesis are most favourable, particularly for tall trees. Hemispherical and broadly conical crowns represent efficient compromises in these respects on humid sites. On xeric sites the flat disc-type crown is more efficient if excess heat loads at noon can be avoided both by holding the leaves vertically, and by having widely spaced individual or bunches of narrow leaves, needles or phyllodes, which allow rapid convective heat diffusion. Species with densely-leaved crowns cannot adopt this course, but must avoid overheating on xeric sites at noon by intercepting less solar radiation by arranging their crowns in narrow conical or columnar shapes. Such crowns are also efficient users of the environment within dense stands on humid sites where there is a high proportion of diffusive sky light and intermediate angles of direct solar radiation.

Evergreen conifers must endure wind, snow, freezing and drought, and accordingly have leaves that are thin and stiff, and branches which are flexible and able to do much of the wind and snow "load shedding". Deciduous, broad-leaved trees, on the other hand, are able to carry large surfaces of thin, flexible leaves; the leaf stalks and twigs absorb wind shocks while the main branches are less flexible than those of conifers (Banks, 1973). In both instances, the shape of the crowns, trunks and whole trees can be related to the mechanical stresses imposed by the environment and the trees' own structure. Wind-imposed stresses, and the general effects of aerodynamic features of tree crowns and canopies were used by Brunig (1970, 1973) to explain some differences in the structures of natural forests, and the risks of wind throw and drought damage in both natural and man-made forests.

Taking an isolated tree, Paltridge (1973) showed that considerations of leaf water potential alone might lead to the evolution of trees with hemispherical shapes, and isolated emergents on mesic tropical forest sites do have this shape. Paltridge also related the ontogenetic change in tree crown shapes, from pointed to broad, to shading effects in a stand, and the principle of maximizing instantaneous rates of net photosynthesis at all stages of growth.

C. Canopy Geometry

What is true for the crown is also true for the canopy. An open, irregularly broken canopy produces turbulent air, and rapid heat and vapour transfer both within itself and between it and the atmosphere. Rough canopies also intercept solar radiation more effectively than smooth surfaces; as a result, exchange intensities for radiation, heat, gases and water vapour are greater. A flat, uniform canopy is aerodynamically smooth; if its trees have large leaves, either water supplies must be ample or the leaves must have other efficient cooling mechanisms. The close interrelationship between tree form, canopy architecture and the resistance of trees and stands to drought, wind and storm has been described by Brunig (1970, 1971, 1973) and Farge (1974).

IV. CROWN IDEOTYPES

In summary: the known physio-ecological properties of leaves (see Gates, 1965, 1968) and the differing aerodynamic features of tree crowns and forest canopies with different geometries, are reflected in the differing leaf sizes and canopy architectures found among species along natural gradients of water supply, soil nutrient conditions, sun elevations, solar radiation and wind speeds. Where deviations occur these can invariably be explained by compensating plant and stand properties, such as leaf fluttering or leaf and canopy albedo (Brunig, 1970, 1973, 1974).

Table I and Fig. 5 present a scheme to distinguish and classify the major tree crown "ideotypes", that is, types considered ideally adapted to particular conditions of solar radiation, water supply and wind, and hence temperature. This scheme is tentative, purely descriptive and ecological, and is by no means complete; Halle and Oldeman (1970) distinguished 21 architectural crown types by branching habit for tropical tree species alone, and Letouzey (1969) described a multitude of architectural types of tree form for tropical Africa. The scheme in Fig. 5 aims at reducing this wealth of forms to a small number of ecologically meaningful prototypes. It does not take account of the evolutionary status of different crown forms, or the different seasonal growth patterns. The primary principles for classification are the aerodynamic properties of the leaves and the exposure of the mature leaves to radiation and wind, excluding such compensatory mechanisms as pigmentation, surface reflectivity and fluttering of leaves which shift leaf size tolerance limits further into hot and xeric ranges. The emphasis on leaf properties and exposure is justified because the leaf, acting like a "sun-paddle", is the primary motor of the tree's functions.

TABLE I. Classification scheme of crown form ideotypes

Predominating leaf size	Leaf form	Leaf distribution	Leaf orientation	Branching habit	Ideotype (see Fig. 5)	Example	Preferred site and phase
Large (macrophyll and larger)	entire, lobed, incised or densely spaced compound	whorl	flat to variable	single stem, vertical	T-d sc or twist (1)	palms	mesic, all phases
		whorl	flat to variable	forked with equal status, steep	umbel-d sc (2)	pandan, Cecropia, Musanga, Gmelina	mesic, pioneer
		whorl	flat to variable	complex, often flat	multi-layered ellipsoid, pagoda, umbrella (3)	Bombax, Ochroma, Ceiba	mesic, pioneer and building
		random	flat to variable	complex, flat to steep	multi-layered ellipsoid	Populus, Tectona	mesic, pioneer to building
Medium (mesophyll)	entire or widely spaced compound	whorl	flat to variable	complex, flat to steep	open-brunch (4)	Dyera spp., Terminalia spp.	mesic, building to mature
		whorl	erect	complex, flat to steep	candelabra (5)	Ploiarium alternifolium	xeric, all phases
		clumped	random to weakly erect	complex, flat to oblique	composite hemisphere (6)	Shorea spp., Emmotum fulvum, Quercus	mesic, building to mature
		random or spaced through crown	random	complex, oblique, multi-layered developing into monolayer	diffuse ellipsoid, umbrella (in dense mature to over-mature stands)	Fagus spp. Balanocarpus heimii	mesic, mature to overmature (tall trees) understorey (small trees)

Size	Leaf type	Arrangement	Leaf attitude	Branching	Crown shape	Example species	Ecology
Small (notophyll and smaller)	entire (or narrowly spaced compound)	random	random	complex	diffuse ellipsoid (7)	*Salix spp., Alnus spp.*	mesic, pioneer or understorey
		random or clumped	erect or drooping	complex, oblique to steep	multi-layered diffuse ellipsoid	*Eucalyptus, Betula spp.*	mesic, pioneer; xeric
				oblique to flat	umbrella (8)	*Eucalyptus, Acacia*	xeric, all phases
	needle — microphyll	random or clumped	erect in periphery	complex, oblique to steep	composite cone (9)	*Agathis spp., Phyllocladus spp.*	mesic to xeric, building to mature
	— leptophyll	spaced	erect near top	multi-layered drooping, flat or oblique	cone to column (10) (base:height ratio variable, mesic = broad xeric = narrow)	*Picea spp., Abies spp., Taxus spp.*	mesic-microtherm, building to mature
		clumped	erect near top	multi-layered drooping, flat or oblique; on xeric sites erect	narrow cone to umbrella, mesic = narrow, xeric = broad	*Pinus spp., Dacrydium spp.*	mesic-microtherm to xeric-megatherm
	scale	random	irregular, erect near top	multi-layered, flat to oblique	cone to column (11)	*Sequoiadendron, Chamaecyparis spp.*	mesic-microtherm to mesotherm, building to mature
	phyllode, phylloclade	random or clumped	irregular to erect near top and on xeric sites	multi-layered, flat to erect	cone (youth) to umbrella or disc broom stick (12)	*Acacia spp., Casuarina spp., Gymnostoma nobile*	xeric-megatherm, all phases

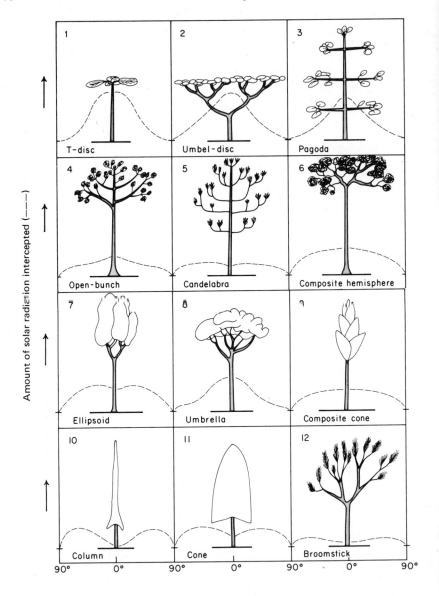

FIG. 5 Major crown-shape ideotypes according to the classification in Table I. The broken lines indicate the daily march of the total amount of solar radiation intercepted by the canopies at the equator, where the sun angle varies between 90° at sunrise and 0° at noon when the sun is overhead. The noon maxima give an indication of the relative rates of back-radiation that occurs on clear nights. "Normal" unimpaired foliage development is assumed.

V. SUMMARY

Examples are given of the different tree shapes and crown forms that occur naturally in different climatic and successional situations in tropical and temperate forests. Facts and hypotheses are reviewed on possible causal relationships between environmental conditions and tree form, and an attempt is made to distinguish the crown forms best adapted to particular situations ("ideotypes").

REFERENCES

Alexandrov, A. (1971). The occurrence of forms of Norway spruce based on branching habit. *Silvae Genet.* 20, 204–208.

Anderson, J. A. R. (1961). "The ecology and forest types of the peat swamp forests of Sarawak and Brunei in relation to their silviculture". Ph.D. Thesis, University of Edinburgh, Scotland.

Banks, C. C. (1973). The strength of trees. *J. Inst. Wood Science* 6, 44–50.

Beadle, N. C. W. (1966). Soil phosphate and its role in molding segments of the Australian flora and vegetation, with special reference to xeromorphy and sclerophylly. *Ecology* 47, 992–1007.

Beadle, N. C. W. (1968). Some aspects of the ecology and physiology of Australian xeromorphic plants. *Aust. J. Sci.* 30, 348–354.

Borman, F. H. (1957). The relationships of ontogenetic development and environmental modification to photosynthesis in *Pinus taeda* seedlings. In "The Physiology of Forest Trees" (K. V. Thimann, ed.), pp. 197–215. Ronald Press, New York.

Brunig, E. F. (1970). Stand structure, physiognomy and environmental factors in some lowland forests in Sarawak. *Trop. Ecol.* 11, 26–43.

Brunig, E. F. (1971). On the ecological significance of drought in the equatorial wet evergreen (rain) forest of Sarawak (Borneo). University of Hull, Department Geography, Miscellaneous Series 11, 66–67.

Brunig, E. F. (1973). Storm damage as a risk factor in wood production in the most important wood-producing regions of the earth. *Forstarchiv* 44, 137–140; *Mitt. BundForschAnst. Forst. u. Holzw.* 93, 17–34 and *Forestry Commission Translations*, H.M.S.O. p. 468.

Brunig, E. F. (1974). "Ecological studies in the Kerangas forests of Sarawak and Brunei". *Borneo Literary Bureau*, Kuching.

Farge, T. La (1974). Genetic differences in stem form of Ponderosa pine grown in Michigan. *Silvae Genet.* 23, 211–213.

Gates, D. M. (1965). Energy, plants and ecology. *Ecology* **46**, 1—13.
Gates, D. M. (1968). Energy exchange and ecology. *BioScience* **18**, 90—95.
Halle, F. and Oldeman, R. A. A. (1970). "Essai sur l'Architecture et la Dynamique de Croissance des Arbres Tropicaux". Masson et Cie., Paris.
Horn, H. S. (1971). "Adaptive Geometry of Trees". Princeton University Press, Princeton.
Kramer, P. J. (1957). Photosynthesis of trees as affected by their environment. *In* "The Physiology of Forest Trees" (K. V. Thimann, ed.), pp. 157—186. Ronald Press, New York.
Letouzey, R. (1969). "Description de différentes parties d'une plante". Centre Technique Forestier Tropical, Nogent-sur-Marne, pp. 52—119.
Luzganov, A. G. and Solod'ko, A. S. (1968). [Crown form and diffuse light.] *Lesn. Z., Archang.* **11**, 29—33 (In Russian).
Neuwirth, G. (1968). Photosynthese und Transpiration von Kamm- und Plattenfichten. *Arch. Forstw.* **17**, 613—620.
Nolde, I. v. (1941). Zur Entstehung von Flachkronen bei tropisch-afrikanischen Bäumen. *Kolonialforstl. Mitt.*, Reinbek **3**, 486—498.
Paltridge, G. W. (1973). On the shape of trees. *J. theor. Biol.* **38**, 111—137.
Parkhurst, D. F. and Loucks, O. L. (1972). Optimal leaf size. *J. Ecol.* **60**, 505—537.
Popov, P. P. (1970). [Ausnutzung der wertvollen Fichtenformen]. *Lesn. Khoz.* **23**, 84—85 (In Russian).
Roth, I. and Merida de Bifano, T. (1971). Morphological and anatomical studies of leaves of the plants in a Venezuelan cloud forest. *Acta biol. Ven.* **7**, 127—155.
Rubner, K. (1943). Die praktische Bedeutung unserer Fichtentypen. *Forstwiss. ZentBl.* **65**, 233—247.
Schmidt-Vogt, H. (1972). Studien zur morphologischen Variabilität der Fichte (*Picea abies* Karst.). 3. Gesetzmässigkeiten und Theorien. *Allgem. Forst-Jagdztg.* **143**, 133—144 and 221—240.
Thomas, P. K., Chandrasekhar, K. and Haldorai, B. (1972). An estimate of transpiration of *Eucalyptus globulus* from Nilgiris watersheds. *Indian Forester* **98**, 168—172.
Webb, L. J. (1959). A physiognomic classification of Australian rain forests. *J. Ecol.* **47**, 551—570.
Webb, L. J. (1965). The influence of soil parent materials on the nature and distribution of rain forests in South Queensland. Proc. Symp. Ecol. Res. in Humid Tropics Veget., pp. 3—14. UNESCO Sci. Coop. Off., S.E.A., Tokyo.
Wilson, J. W. (1967). Stand structure and light penetration. 3. Sunlit foliage area. *J. appl. Ecol.* **4**, 159—165.

9
Some Mechanisms Responsible for Differences in Tree Form

L. S. JANKIEWICZ
Vegetable Crops Research Institute, Skierniewice, Poland.
and Z. J. STECKI
Research Institute of Dendrology, Kórnik, Poland.

I.	Introduction	157
II.	Development and Relative Lengths of Branches	158
	A. Correlative Inhibition in Pines	158
	B. Growth Correlations in Deciduous Trees	159
	C. Effects of Gravity	160
	D. Acrotony and Basitony	161
III.	Direction of Branch Growth	162
	A. Branch Angles on Pines	162
	B. Branch Angles on Deciduous Trees	164
IV.	Ageing Phenomena	167
V.	Inherent Differences in Tree Form Among Pines and Poplars	168
References		169

I. INTRODUCTION

Well known differences in the dimensions and forms of tree crowns can be attributed to differences in rates of height growth, and to factors influencing (a) the numbers and relative lengths of branches, (b) the directions in which the branches grow, and (c) ageing phenomena. Aspects of these three component processes will be considered in this contribution, drawing mainly on the authors' work on *Pinus sylvestris*, *Populus* and *Malus*.

II. DEVELOPMENT AND RELATIVE LENGTHS OF BRANCHES

A. Correlative Inhibition in Pines

Branching and branch growth are closely controlled in a young pine. The leader bud is normally the largest, develops the most preformed short shoots, becomes the longest shoot, and will tend to bear most lateral branches. With some exceptions, there is a progressive regular decrease in these parameters for branches in lower positions in the crowns (for data see: Friesner, 1943; Friesner and Jones, 1952; Szymański and Szczerbinski, 1955; Kozlowski and Ward, 1961; Jankiewicz, 1967).

Apparent correlative inhibition exists within developing pine buds between the branch-forming long-shoot primordia at the tops of the buds and the needle-forming short-shoot primordia below, which lack active apical and subapical meristems. The latter appear to be inhibited, because they are able to develop when the apices of the long shoots are removed (see also Cannell *et al.*, Chapter 10).

Further growth correlations occur both among the long shoots, and between them and the terminal apex. Thus, when one of these structures is removed there is compensatory growth by those remaining. In *Pinus sylvestris*, decapitation and long-shoot bud thinning causes (a) needles on the remaining predetermined long shoots to grow longer, (b) more long-shoot buds to develop, and (c) more short shoots to develop on each long-shoot bud. Consequently, the numbers of branches and amounts of foliage produced the following year are much greater than on untreated shoots (Giertych, 1968; Jankiewicz and Westrom, 1977). *Pinus sylvestris* does not usually compensate bud damage immediately by forming proleptic shoots as does *Pinus strobus* (Little, 1970).

We, like others (Zimmermann and Brown, 1971) emphasize the importance of nutrition in correlative relationships in pine buds. Little (1970) stated that the long shoots in a whorl compete for dominance by competing for a common supply of nutrients transported from the preceding year's growth. Removing shoots decreases this competition. In an intact whorl of buds, the apices which (a) attract most nutrients, (b) develop most needles, and (c) produce the longest shoots the next year, are probably those that produce most hormones, particularly auxins. Terminal apices normally produce more auxin than long-shoot lateral buds, and exercise greatest "dominance" over them when the supply of nutrients is poor (Little, 1970). Also, terminal apices may have better vascular connections with internal nutrient sources than their lateral buds, and, in this regard, Little observed that shoots that become dominant within a tree invariably have large stem diameters relative to their neighbours.

B. Growth Correlations in Deciduous Trees

On deciduous trees, all leaves can subtend axillary buds which are capable of growing into branches. However, the buds may not develop until the year after they are formed, and may become either long shoots, short shoots, remain dormant or abort.

It is well known that growing terminal apices and young leaves inhibit the development of axillary buds on shoots of many deciduous trees, just as on herbaceous plants. As in pines, these "apical dominance" phenomena may be explained in terms of auxin production by the apices and diversion of nutrients away from the lateral buds (Wareing, 1970; Zimmermann and Brown, 1971). However, this hypothesis meets difficulties in instances where axillary buds develop precociously, forming sylleptic shoots at times when the main apices are growing and producing auxin most actively (Champagnat, 1965). It is more likely that several mechanisms act together to inhibit axillary buds; mechanisms based on nutrient diversion do not exclude simultaneous direct inhibition of buds by specific inhibitors or by auxins in supra-optimal concentrations. Infallibility is as important in living organisms as in space vehicles, and can be achieved in a similar way: by doubling or trebling mechanisms responsible for given functions.

Lateral buds that have been released from apparent apical dominance can develop in many ways, involving further control mechanisms. This occurs, for instance, in spring on 1-year-old shoots of *Malus* and *Populus* in Poland. Instead of distributing finite internal nutrient resources evenly to all lateral buds, a large part of the nutrients is directed to a few buds which develop into long shoots.

The remainder form short or dwarf shoots, or remain dormant. This growth pattern has obvious adaptive significance in that it enables the trees to build an adequate scaffold on which to display their leaves.

A cybernetic, positive feedback model was presented by Jankiewicz (1972) to explain this "vigour differentiating" mechanism, which applies equally to the process whereby some pine long shoots become dominant over their neighbours (Jankiewicz, 1967). It was proposed that a bud showing minute prevalence over other buds starts to develop a little faster during the spring and so produces slightly more auxin. Since auxin stimulates cambial activity, this bud develops better vascular connections with the main axis, which, in turn, gives it a favoured supply of nutrients, and hormones, including gibberellins and cytokinins from the roots. This may (a) further stimulate vascular development, since gibberellins and cytokinins act synergistically with auxin in promoting vascular development (Digby and Wareing, 1966; Pieniążek and Jankiewicz, 1966; Moraszczyk *et al.*, 1974); and (b) protect the favoured bud from inhibition by other apices (Plich *et al.*, 1975). Clearly, such a process is self-sustaining and would cause some buds to gain progressive advantage over others, enhancing small differences that exist when growth begins, as appears from observation. Additionally, it would mean that buds might be very susceptible to changes in their environment when they first start to grow. Again, this is supported by observation; for instance, some of the lateral buds on an apple tree can be induced to form short rather than long shoots by keeping them in the dark for only 3–5 days (Plich *et al.*, 1975).

C. Effects of Gravity

The development of branches on young deciduous trees depends markedly on their positions relative to gravity ("gravimorphism", Wareing and Nasr, 1958). The highest, upwardly directed shoots are often the dominant ones, possibly because they receive the greatest share of root-originated hormones and metabolites (Wareing, 1970). Similarly, buds on the lower sides of horizontally placed deciduous tree stems develop poorly compared with those on the upper sides (Smith and Wareing, 1964; Borkowska, 1966; Jankiewicz, 1971). Evidently, buds on the lower sides of such stems are inhibited in some way by the action of gravity. However, inhibition need not be sustained. Thus, when *Malus* or *Populus* trees were placed horizontally for only 1—5 days at the time of bud-break and then placed upright, buds on the previously lower sides of the stems grew weakly whereas those on the upper sides grew vigorously (Jankiewicz et al., 1967; Borkowska, 1968). Our interpretation was that even tiny differences in bud development set up during the 1—5 days were sufficient to cause major differences in subsequent growth owing to the action of positive feedback mechanisms described above (Borkowska and Jankiewicz, 1972). Such an interpretation helps to reconcile some apparently conflicting observations in the literature (e.g. Smith and Wareing, 1964; Mullins, 1965).

D. Acrotony and Basitony

The attention of French and Belgian workers (Champagnat, 1969; Champagnat et al., 1971; Crabbé and Arias, 1974; Barnola, 1970, 1972) was recently focused on the question why basal laterals often develop more than ones near shoot tips, or vice versa (basitony and acrotony, respectively). Their data showed that, in autumn, buds on the basal parts of woody shoots can exhibit weak dormancy and so tend to develop relatively fast (basitony) when the shoots are cut into single-node sections. On an intact shoot this tendency is often masked, because the basal buds may be inhibited by those growing distal to them. Nevertheless, in the case of raspberry (*Rubus idaeus*), buds on the lower parts of canes do show weak dormancy in autumn, and are able to swell considerably at low temperatures during the winter. In spring, when the more distal buds are released from dormancy and are able to develop, the basal buds are already well developed and can no longer be inhibited. They then give basal vegetative shoots which replace the mother shoots (Barnola, 1970).

This does not occur in *Malus* or *Populus*, on which all buds along a shoot have their chilling requirements satisfied well before the spring. In these instances, the main differences appear to be in the size of the buds and/or their abilities to attract assimilates. These species are usually acrotonous; the uppermost buds can be shown to attract ^{14}C-labelled carbohydrate reserves more readily than basal ones, even though the upper buds may initially be smaller (Borkowska,

1975). In this case, gravity may also play a part (Jankiewicz et al., 1967; Jankiewicz, 1971). Thus, placing a poplar tree in a horizontal position for only 2–7 days at bud-burst decreases its acrotonic tendency, while several weeks of such treatment can induce a wholly basitonic habit.

III. DIRECTION OF BRANCH GROWTH

Two features determine the direction in which a branch grows: first, the angle it makes with the main axis (the branch or crotch angle), and secondly, the extent to which the branch bends naturally. Both these features are under correlative control and have been investigated in *Pinus* by Münch (1938) and in *Malus* by Verner (1955). They speculated that the mechanism depended on different degrees of cell elongation on the upper and lower side of elongating branches, whereas Horsfall and Vinson (1938) proposed that the widening of the crotch angle in spring was caused by the growing trunk pushing aside the branch bases as it increased in diameter. Later work has revealed a complex situation (Jankiewicz, 1956, 1964, 1966, 1970, 1973; Little, 1967).

A. Branch Angles on Pines

Initially, pine branches on the main stem grow straight upwards, and can be shown to be negatively geotropic (Jankiewicz, 1966) (Fig. 1), but after 2–3 weeks, the bases of the branches, which have then completed their elongation, begin deviating gradually from the main axis (Little, 1970; Jankiewicz and Westrom, 1977). On the other hand, the distal parts of the branches, which are still elongating, continue to be negatively geotropic, and the whorl of branches looks like a candelabra (Fig. 1). When the branches stop elongating their tendency to be negatively geotropic is lost, but their vertical angles still continue widening for over a month (late June to early August — Jankiewicz, 1966).

There are probably two major forces that push young branch bases horizontally in spring and summer. First, there are mechanical pressures exerted by the newly formed tissues of the thickening branches on the main axis and vice versa, which can be shown to be

9. Differences in Tree Form

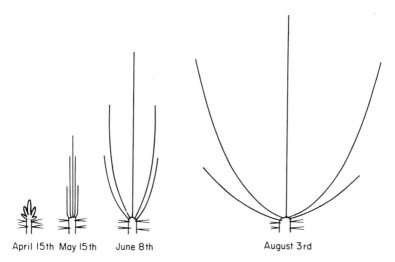

April 15th May 15th June 8th   August 3rd

FIG. 1 Changes in the positions of pine branches during their first season of growth.

at least 10 N cm^{-2}, sufficient to push a young non-lignified branch (Jankiewicz and Stępień, 1973; Jankiewicz, unpublished) (Fig. 3). Secondly, the formation of reaction wood in the crotch, possibly in response to auxin transported basipetally in the main stem, will tend to push the branches sideways (Fig. 4). There is indirect evidence for this phenomenon. Thus, when a tree is decapitated and partly disbudded, very little reaction wood is formed on the upper side of the remaining branch and it grows more upright (Jankiewicz, 1966) (Fig. 4). Additionally, there is the weight of the branches themselves, although this alone is not sufficient, because the basal parts of young branches can be shown to increase their crotch angles for at least 3 weeks in May—June after the branches have been cut off near the main stem.

Opposing all these forces is the pressure of reaction wood often formed on the lower sides of the branches. Also, as the branches get generally thicker and more lignified, they increasingly resist all the mechanical forces described. Thus, the rate of branch angle widening normally slows in summer and ceases altogether in early August.

The spatial positions of pine branches can change in their second or later growing seasons. Obviously, wet snow can bend branches down, and Jankiewicz (1966, 1971) (Fig. 2) showed that they do not return to their original positions after the thaw.

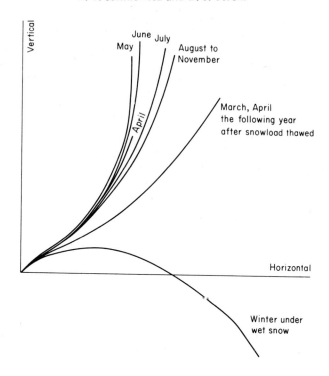

FIG. 2 Changes in the position of a pine branch during its third season of growth and after bearing a snowload.

Irrespective of snow bending, the middle and distal parts of mature branches move upwards slightly but significantly every spring, probably in response to reaction wood formation on their undersides, and move back downwards later in the year as they develop an increasing weight of newly formed needles. However, in the course of many years the bases of the branches move very little, become buried in the tissues of the trunk and only the more horizontal extremities of the branches are seen (Fig. 5).

B. Branch Angles on Deciduous Trees

The changes in branch angles on *Malus* and *Populus* and other trees with buds appressed to the stems are similar to those on pines. The crotch angles widen gradually in spring and early summer. When the

9. Differences in Tree Form

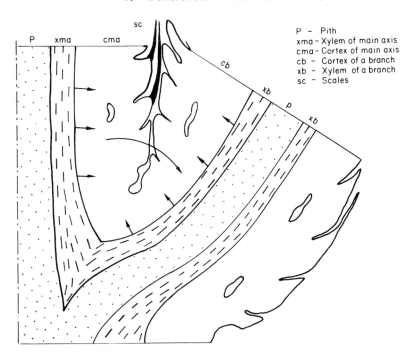

FIG. 3 Cross-section through the crotch of a young pine branch, indicating the forces causing angle widening. Small arrows mark the pressures exerted by new tissues produced by the cambium. The large arrow marks the resulting pressure acting on the branch base.

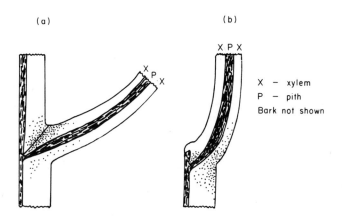

FIG. 4 Reaction wood (dotted) in the crotch of a 1-year-old pine branch: (a) control tree, (b) main stem decapitated and partly disbudded (Jankiewicz, 1966).

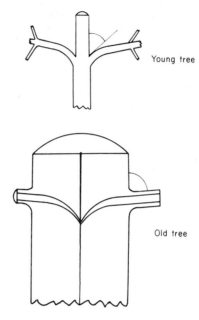

FIG. 5 The change in crotch angle at the bases of branches as they become progressively embedded in tissues of the main stem.

shoots are vigorous this widening occurs rapidly for a short period of time, whereas on weak shoots it occurs slowly over a longer period (Jankiewicz, 1973). The most probable explanation is that new tissues form faster in the crotches of vigorous branches, and produce larger forces which push the branches aside, but such branches also lignify and resist these forces sooner.

Paradoxically, dormant buds or weak shoots, possessing only 1—3 leaves, maintain very narrow angles with the main axis during the whole growing season, even though the axis thickens markedly. This can occur because dormant buds and weak shoots are connected to the stele of the main axis by very thin tracheids or vessels, reinforced with only helical thickenings, which can be easily stretched or broken. This occurs in response to pressures exerted by the thickening axis, so that the weak shoots tend to move aside rather than be pushed into more horizontal positions (Jankiewicz, 1957). Similar phenomena were described for tree leaves by Eames and McDaniels (1947).

On deciduous trees, negative geotropism, tending to cause upward growth, is often completely neutralized by geo-epinasty* causing downward growth in response to prolonged action of gravity (Kaldewey, 1962; Lyon, 1972). As a result, the branches may grow straight during the whole process of angle widening (Jankiewicz, 1964, 1971). The existence of negative geotropism can be revealed by decapitation, which removes the geo-epinastic tendency, and results in upward growth (Jankiewicz, 1964). In *Malus*, this effect of decapitation can be prevented by replacing the shoot tip with auxin (Verner, 1955; Jankiewicz *et al.*, 1961).

In *Fraxinus*, the buds are inserted nearly at right angles to the main stem and the young shoots grow out at wide angles. Consequently, the pressures exerted by tissues growing in the crotches probably have little or no influence on *Fraxinus* branch angles; these are determined more by negative geotropic and geo-epinastic responses.

In *Corylus* and *Tilia* the shoots show marked epinasty just after emerging from the buds and assume a hooked shape. Later on, they straighten at their bases so that the hook moves distally as the branches grow. This occurs because a "wave" of rapid extension growth progresses distally on the upper sides of the stems, tending to curve them downwards, to be followed by a similar "wave" of extension growth on the lower stem sides, which tends to bring them back to the vertical. Consequently, the shoots grow straight, but have hooks at the tips (Kaldewey, 1962).

IV. AGEING PHENOMENA

It is well known that, as trees age, their shoots eventually make progressively less growth, and terminal apices may exercise less dominance over lateral buds. Moorby and Wareing (1963) suggested that this was associated with a fast increase in numbers of developing shoot apices, compared with a slow increase in transporting capacity of the trunk and other skeletal parts of the tree. Consequently, there is increasing "competition" among apices for limited root-originated

---

* Epinasty is the greater growth of the upper side of a branch compared with the lower side, causing it to bend downwards.

metabolites. Many observations appear to support this view. For instance, when deciduous trees are heavily pruned, the remaining shoots are "rejuvenated", while, on pines, juvenile growth characters can persist when the trees are grown in dense stands because numerous apices on the older branches die off in the lower parts of the crowns.

Another cause of ageing in trees may be the increasing distance between root and shoot apices, with ever-increasing energy demands for translocation in both the xylem vessels, which are surrounded by living cells, and the phloem.

It is noteworthy that ageing phenomena always signal approaching death, and any decrease in shoot growth on trees must mean that fewer new leaves are produced which can support an increasing amount of scaffold tissue, leading eventually to internal starvation and death.

V. INHERENT DIFFERENCES IN TREE FORM AMONG PINES AND POPLARS

Differences in branching angles and crown forms are well known to silviculturists, and only a few examples are given here, drawn mainly from the authors' experience in Poland.

Mature *Pinus sylvestris* in Poland exhibits two inherently distinct crown types (Józefaciukowa and Ubysz-Borucka, 1972). Trees of Type A prevail in north-east areas. They are tall, thin-stemmed, with thick bark only on the lower parts, and their branches are thin and rise at acute angles in long, narrow crowns. Trees of Type B are commonest in the plains of south-west Poland. By contrast, they are shorter, have thick, panel-like bark, with thick branches normally set nearly at right angles to the stem in short, broad crowns. It has been stressed that elite trees of type A, which are considered the most desirable in Poland, differ markedly from the "plus" trees described in Scandinavian literature. The main difference is that Scandinavian plus trees have branches arising at wide angles whereas elite Polish trees of type A have acute-angled branches. The obtuse branch angles of Scandinavian trees may represent an adaptation to the heavier snow loads received there. Thus, trees have to be selected for differing "ideal" features in each territory, depending on the local environment.

Although the mechanisms influencing branch angles may not be strongly inherited among normally shaped Scots pines (Ehrenberg, 1963, 1966), some parents can produce a high proportion of progeny with abnormally formed whorls, forked or proleptic shoots, fasciations and other malformations (Ehrenberg, 1958). This is the case, for instance, among some local Norwegian sub-populations, which have had the better formed trees culled by man for centuries (Rudden, 1957).

In *Populus nigra italica* hybrids, crown form appears to be under polygenic control. Thus the F_1 progeny of fastigiate and broad-crowned parents exhibit the whole range of parent and intermediate crown forms (Pohl, 1964). However, in the section *Leuce* (aspen and silver poplars) the fastigiate character is inherited by nearly half the F_1 progeny, whereas the others are broad-crowned, suggesting single-gene inheritance (Stecki and Efros, 1968). An intriguing aspect of these inheritance relations is that all well known fastigiate poplars in section *Leuce* are males (*Populus tremula erecta* and *P. bolleana*). It is possible that females can be found in the variety *P. tremula pyramidalis*, but no data are available (Sokolov, 1951).

Evidently, the many mechanisms that influence crown form can be under the control of one or many genes, and, in either case, are often readily selected by tree breeders. Nevertheless, it is important to be aware that several independent processes are involved which could be inherited separately.

REFERENCES

Barnola, P. (1970). Recherches sur le déterminisme de la basitonie chez le framboisier (*Rubus idaeus* L.). *Ann. Sci. nat. bot., Paris* 12, 129–152.

Barnola, P. (1972). Etude expérimentale de la ramification basitone de sureau noir (*Sambucus nigra* L.). *Ann. Sci. nat. bot.* 13, 369–400.

Borkowska, B. (1966). Gravimorphism in young apple trees: sensitivity to gravity stimulus during dormancy and growth. *Bull. Acad. pol. Sci. Cl. V. Sér. Sci. biol.* 14, 563–567.

Borkowska, B. (1968). "Studia nad grawimorfizmem u jabloni. [Studies on gravimorphism in apple trees]". Doctoral Thesis. Agricultural University, Warsaw.

Borkowska, B. (1975). Distribution of stored C^{14} assimilates in apple trees during their spring development. Fruit Sci. Rep. 2, 1–11.

Borkowska, B. and Jankiewicz, L. S. (1972). The influence of gravity on bud development in apple trees and in poplars. *Acta agrobot.* 25, 185–193.

Champagnat, P. (1965). Rameaux courts et rameaux longs: problèmes physiologiques. In "Encyclopedia of Plant Physiology" (W. Ruhland, ed.), Vol. 15, pp. 1165–1171. Springer-Verlag, Berlin, Heidelberg and New York.

Champagnat, P. (1969). Le notion de facteurs de préséances entre bougeons. *Bull. Soc. bot.* 116, 323–348.

Champagnat, P., Barnola, P. and Lavarenne, S. (1971). Premières recherches sur le déterminisme de l'acrotonie des végétaux ligneux. *Ann. Sci. forest.* 28, 5–22.

Crabbé, J. and Arias, O. (1974). Rôle morphogène de la dormance chez les végétaux ligneux, spécialement espèces fruitières. *Proc. XIX Congrès Int. d'Horticult. Varsovie.* Vol. 1B, p. 494.

Digby, J. and Wareing, P. F. (1966). The effect of applied growth hormones on cambial divisions and the differentiation of cambial derivatives. *Ann. Bot.* 30, 539–548.

Eames, A. J. and McDaniels, L. H. (1947). "An Introduction to Plant Anatomy". McGraw-Hill, New York and London.

Ehrenberg, C. E. (1958). Über Entwicklungsanomalien in Kreuzungsnachkommenschaften bei *Pinus sylvestris* L. *Medd. Skogsforsk Inst., Stockholm* 48, 1–14.

Ehrenberg, C. E. (1963). Genetic variation in progeny tests of Scots pine (*Pinus sylvestris* L.). *Studia forest. suecica* 10, 1–135.

Ehrenberg, C. E. (1966). Parent-progeny relationship in Scots pine (*Pinus sylvestris* L.). *Studia forest. suecica* 40, 5–54.

Friesner, R. C. (1943). Correlation of elongation in primary, secondary and tertiary axes of *Pinus strobus* and *P. resinosa*. *Butler Univ. Bot. Stud.* 6, 1–9.

Friesner, R. C. and Jones, J. J. (1952). Correlation of elongation of primary and secondary branches of *Pinus resinosa*. *Butler Univ. Bot. Stud.* 15, 119–129.

Giertych, M. (1968). The process of crown restoration following debudding of *Pinus sylvestris* L. *Proc. third Symposium on Plant Growth Regulators*, Torun, Poland.

Horsfall, F. Jr and Vinson, C. G. (1938). Apical dominance in shoots and proximal in roots as related to structural framework in the apple. *Minnesota Agr. Exp. Sta. Res. Bull.* 293, 1–23.

Jankiewicz, L. S. (1956). The effect of auxins on crotch angles in apple trees. *Bull. Acad. Polon. Sci. Cl. II Sér Sci biol.* 4, 173–178.

Jankiewicz, L. S. (1957). "Wpływ auksyn na formowanie się kąta rozwidlenia u jabłoni. [The influence of auxins on crotch angle formation in apple trees]". Doctoral thesis, Szkoła Główna Gospodarstwa Wiejskiego, Warszawa.

Jankiewicz, L. S. (1964). Mechanism of the crotch angle formation in apple trees. 1. The crotches in the trees growing in a vertical and a horizontal position. *Acta Agrobot.* 15, 21—50.

Jankiewicz, L. S. (1966). Changes in angle width between the axis and a branch in young pines (*Pinus sylvestris* L.). *Acta Agrobot.* 19, 129—132.

Jankiewicz, L. S. (1967). Kształtowanie się pędu wegetatywnego i korelacje wzrostowe. *In* "Zarys fizjologii sosny zwyczajnej". [Formation of the vegetative shoot and growth correlations. *In* "Physiology of Pine Trees".] (S. Białobok and W. Żelawski, eds), pp. 223—246. P. W. N. Warszawa, Poznań.

Jankiewicz, L. S. (1970). Mechanism of crotch angle formation in apple trees. II. Studies on the role of auxin. *Acta Agrobot.* 23, 171—181.

Jankiewicz, L. S. (1971). Gravimorphism in higher plants. *In* "Gravity and the Organism" (S. A. Gordon and M. J. Cohen, eds), pp. 317—331. University of Chicago Press, Chicago.

Jankiewicz, L. S. (1972). A cybernetic model of growth correlations in young apple trees. *Biol. Plant.* (*Praha*) 14, 52—61.

Jankiewicz, L. S. (1973). Fizjologia wzrostu i rozwoju. *In* "Topole — *Populus* L." (Physiology of growth and development. *In* "Poplars — *Populus* L."] (S. Białobok, ed.), pp. 205—235. P. W. N. Warszawa, Poznań.

Jankiewicz, L. S. and Westrom, D. W. (1977). Factors involved in vertical branch angle formation in pine (*Pinus sylvestris.* L.). *Acta Agrobot.* 30, in press.
Proc. International Symposium on Biology of Wood Plants, pp. 377—381. Slovak Academy of Sciences, Bratislava.

Jankiewicz, L. S. and Westrom, D. W. (1977). Factors involved in vertical branch angle formation in pine (*Pinus sylvestris.* L.) *Acta Agrobot.* 30, in press.

Jankiewicz, L. S., Srzednicka, W. and Borkowska, B. (1967). Gravimorphism in the poplar (*Populus* sp.). *Bull. Acad. pol. Sci. Cl. V. Sér. Sci. biol.* 15, 11—15.

Jankiewicz, L. S., Szpunar, B., Barańska, H., Rumplowa, R. and Fiutkowska, K. (1961). The use of auxin to widen crotch angles in young apple trees. *Acta Agrobot.* 10, 151—171.

Józefaciukowa, W. and Ubysz-Borucka, L. (1972). Variation in habit forms of Scots pine (*Pinus sylvestris* L.) on the area of Poland. *Silvae Genet.* 21, 9—17.

Kaldewey, H. (1962). Plagio- und Diageotropismus der Sprosse und Blätter einschliesslich Epinastie, Hyponastie, Entfaltungsbewegungen. *In* "Encyclopedia of Plant Physiology" (W. Ruhland, ed.), Vol. 17, pp. 200—321. Springer-Verlag, Berlin, Heidelberg and New York.

Kozlowski, T. T. and Ward. R. C. (1961). Shoot elongation characteristics of forest trees. *Forest Sci.* 7, 357—368.

Little, C. H. A. (1967). "Some aspects of apical dominance in *Pinus strobus* L.". Ph.D. Dissertation, Yale University, Connecticut.

Little, C. H. A. (1970). Apical dominance in long shoots of white pine (*Pinus strobus*). *Can. J. Bot.* 48, 239—253.

Lyon, C. J. (1972). Growth responses of plants to gravity. *In* "Gravity and the Organism" (S. A. Gordon and M. J. Cohen, eds) pp. 427–437. University of Chicago Press, Chicago.

Moorby, J. and Wareing, P. F. (1963). Ageing in wood plants. *Ann. Bot.* **27**, 291–308.

Moraszczyk, A., Jankiewicz, L. S. and Plich, H. (1974). Role of growth regulators in apple internode elongation and in the formation of secondary structure. *Proc. 19th Inter. Hort. Congr.* **1A**, 416.

Mullins, M. G. (1965). Lateral shoot growth in horizontal apple stems. *Ann. Bot.* **29**, 73–78.

Münch, E. (1938). Untersuchungen über die Harmonie der Baumgestalt. *Jb wiss. Bot.* **89**, 581–673.

Pieniążek, J. and Jankiewicz, L. S. (1966). Combined effect of naphtaleneacetic acid and 6-benzylaminopurine on bud development and on initiation of cambial activity in dormant apple seedlings. *Bull. Acad. pol. Sci. Cl. V. Sér. Sci. biol.* **14**, 805–808.

Plich, H., Jankiewicz, L. S., Borkowska, B. and Moraszczyk, A. (1975). Correlations among lateral shoots in young apple trees. *Acta Agrobot.* **28**, 131–149.

Pohl, Z. (1964). Dziedziczenie piramidalnego pokroju korony u mieszańców *Populus piramidalis* Roz. [Inheritance of pyramidal tree form in hybrids of *Populus piramidalis* Roz.] *Arboretum Kórnickie* **9**, 199–222.

Rudden, T. (1957). Arvelige dverg former av furu (*Pinus sylvestris* L.) fra Skjak, Gudbrandsdal. *Medd. Det Norske Skogforsøksvesen* **48**, 419–443.

Smith, H. and Wareing, P. F. (1964). Gravimorphism in trees. 2. The effect of gravity on bud break in osier willow. *Ann. Bot.* **28**, 283–295.

Sokolov, S.Ja. (1951). Derevia i Kustarniki SSSR. [The trees and shrubs of the U.S.S.R.] *Izd. Akad. Nauk SSSR.* Moscow and Leningrad.

Stecki, Z. and Efros, S. V. (1968). K voprosu o nasledovanii formy krony u gibridov topolej podroda *Leuce. Lesovodstvo i Agrolesomelioracie, Kiev* **15**, 136–142.

Szymański, S. and Szczerbiński, W. (1955). Pączki jako wskaźnik potencjału życiowego młodej sosny. [The buds as an indicator of growth potential of young pines]. *Roczn. Sekc. Dendrol. P.T. Bot.* **10**, 275–304.

Verner, L. (1955). Hormone relations in the growth and training of apple trees. *Univ. Idaho agr. exp. Stn. Res. Bull.* **28**, 1–31.

Wareing, P. F. (1970). Growth and its co-ordination in trees. *In* "Physiology of Tree Crops" (L. C. Luckwill and C. V. Cutting, eds), pp. 1–21. Academic Press, New York and London.

Wareing, P. F. and Nasr, T. A. A. (1958). Gravimorphism in trees. Effects of gravity on growth, apical dominance and flowering in fruit trees. *Nature, Lond.* **182**, 379–381.

Zimmermann, M. H. and Brown, C. L. (1971). "Trees, Structure and Function". Springer-Verlag, Berlin, Heidelberg and New York.

10
An Analysis of Inherent Differences in Shoot Growth Within Some North Temperate Conifers

M. G. R. CANNELL
Institute of Terrestrial Ecology, Bush Estate, Penicuik, Scotland.
S. THOMPSON
Department of Forestry, University of Aberdeen, Scotland.
and R. LINES
Forestry Commission Northern Research Station, Roslin, Scotland.

I.	Introduction	173
II.	Apical Growth and Initiation of Stem Units	175
	A. Periods of Apical Meristematic Activity	176
	B. Environmental Responses of the Shoot Apices	177
	C. Inherent Variation	178
	D. Genetic Improvement	182
III.	Stem Unit and Shoot Elongation	183
	A. Factors Controlling Elongation	184
	B. Inherent Variation	185
IV.	Needle Elongation	189
	A. Control of Pine Needle Elongation	191
	B. Inherent Variation in Pine Needle Elongation	192
V.	Lateral Bud Production	193
	A. Lateral Bud Formation	194
	B. Some Factors Affecting Lateral Bud Numbers	196
	C. Inherent Variation in Branch Frequency	198
VI.	Implications for Tree Breeders	198
VII.	Summary	199
	References	200

I. INTRODUCTION

Most conifer breeders who select for wood yield begin by selecting progeny which grow rapidly in height. Before canopy closure height growth can be closely correlated with needle biomass, total dry matter production and stem volume per tree, and it is assumed that rapid growth then establishes a lasting benefit in yield per hectare.

This review analyses the activities of four centres of cell division and expansion in conifer shoots that directly influence the lengths of leaders and amounts of needle tissue that young conifers produce. The four growth centres are: (a) the shoot apices where new stem units (defined on p. 224; Doak, 1935) and potential needles are produced; (b) the sub-apical meristems which regulate the subsequent elongation of the internode at each stem unit; (c) the needles themselves; and (d) the sites where new lateral vegetative apices can form (see Fig. 1).

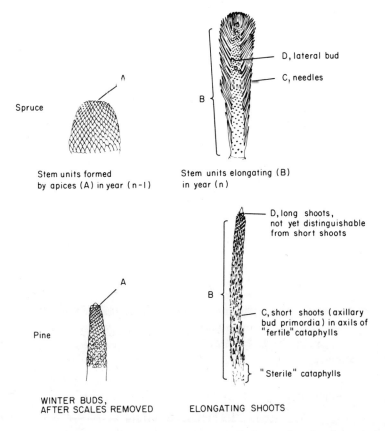

FIG. 1 Structure of pine and spruce buds and shoots showing four centres of cell division and growth. A. Apical meristems. B. Sub-apical, internode tissues. C. Needles. D. Lateral bud meristems.

Lanner (Chapter 12) gives an account of the various shoot growth patterns among temperate and tropical pine species. This contribution considers many genera, but focuses on north temperate conifers with single annual growth periods and "fixed" or "predetermined" habits of growth, including many important forest tree species in the genera *Pinus, Picea, Abies, Tsuga* and *Pseudotsuga*. In these conifers the activities of the four growth centres listed above are clearly phased, controlled and, we argue, consequently inherited independently. As a result, the usual measurements of shoot elongation, height growth and branching may not adequately describe the physiological possibilities for genetic gain; the component processes need to be understood.

In order to understand inherent differences in the activities of each growth centre, it is necessary to describe when each of them functions, and consider factors which may regulate their activity in north temperate climates. To provide a coherent account we have had to ignore features peculiar to particular species or situations.

II. APICAL GROWTH AND INITIATION OF STEM UNITS

Apical growth and needle initiation have rarely been measured on conifers and should not be confused with measurements of shoot elongation. These two growth processes involve different meristems, invariably occur over different periods of the year, and are controlled independently (Wareing, 1950; Romberger, 1963).

In first-year seedlings of many northern conifers, stem units (that is, needle internodes) elongate 2—4 weeks after they are formed at the apices, but in second-year and older plants a diminishing proportion of the stem units produced during a growing season is elongated in that season, and an increasing proportion is retained near the apex and overwintered in a bud (Fig. 2). This growth habit is presumed by us to be a strategy that enables (a) seedlings to compete with herbaceous plants, and (b) mature trees to protect their apices with scales and resinous materials and allow the production of initials to continue for the maximum period.

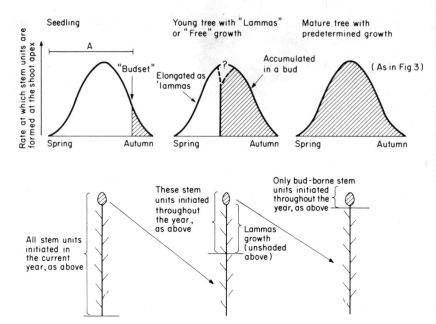

FIG. 2 Diagrammatic representation of the times when stem units are formed at the apices of many northern conifers with "predetermined" growth. Initiation rates are actually more intermittent (Sucoff, 1971). Also, some pines can move from stage 1 to stage 3 in one year, and fail to elongate a "rosette" of stem units in the first year.

A. Periods of Apical Meristematic Activity

Conifer apices are normally mitotically active throughout the growing season, on seedlings and mature trees alike. New initials can begin forming in April, before last year's overwintered bud elongates, and invariably continue to be formed after elongation ceases. The development of vegetative buds on mature trees has been described for *Abies concolor, Larix decidua, Pseudotsuga menziesii, Tsuga heterophylla, Picea glauca* and *Picea sitchensis* (Parke, 1959; Frampton, 1960; Owens and Molder, 1973a and b; Pollard, 1973; Cannell and Willett, 1975, respectively), and *Pinus ponderosa, P. lambertiana, P. densiflora, P. strobus, P. resinosa, P. banksiana* and *P. contorta* (Sacher, 1954; Hanawa, 1966; Owston, 1969; Sucoff, 1971; Curtis and Popham, 1972; Owens and Molder, 1976). The timing of events varies with species, ecotype and site, but the sequence is broadly similar. In March—April a wave of mitotic activity moves acropetally over the apical domes which, at first, produce scale initials or, in pines, "sterile" cataphylls that will not subtend axillary bud primordia (which are able to produce needles, cones or branches; see Fig. 1). During May, June and July mitotic rates increase to a maximum, the domes enlarge several-fold, become conical and zonate, and rapidly generate new needles or "fertile" cataphylls. In simple cases, this continues until August or October, when mitotic rates decrease basipetally, and initials that encroach on the diminishing domes of pines form a further set of "sterile" cataphylls. Few mitoses occur from late November to March, except in *Larix*, although DNA may be synthesized.

B. Environmental Responses of the Shoot Apices

In spring, the domes at northern conifer shoot apices become mitotically active, and produce new initials, as temperatures increase, provided they have received their chilling requirements. Close correlations have been found between measures of heat sum and the onset of initiation in *Pinus resinosa* (Sucoff, 1971). Also, heat sum is a useful measure of the onset of meristematic activity in buds on deciduous and fruit trees (see Landsberg, 1974). In all cases activity is apparently mediated by metabolites or growth regulators originating low in the trees (see Thielges and Beck, Chapter 14).

During the summer apical meristematic activity can be retarded or halted by water or nutrient stress and is rarely continuous, as Fig. 3 shows (Sucoff, 1971). Unstressed meristems are most active at high temperatures, and when metabolites are delivered to them rapidly. This usually occurs in mid-summer on conifers, when large amounts of soluble sugars are present in the shoots (Owens and Molder, 1973b).

In the autumn, apical growth slows and ceases, very probably in response to temperature-modified photoperiodic stimuli, at least in species where night-length has been shown to affect "bud-set" on first-year seedlings (i.e. length of period A in Fig. 2). The evidence for this is that "bud-set" on first-year seedlings, which is merely the time when stem unit *elongation* stops, also coincides with a slowing down of initiation at the apices (Burley, 1966; Pollard, 1974a), and photoperiodic ecotypes which set bud at different times probably stop initiation in the same relative sequence because they often initiate similar numbers of stem units after bud-set (Heide, 1974; Pollard and Logan, 1974). Nevertheless, it has yet to be demonstrated conclusively that the photoperiodic mechanism regulates apical growth in older trees, although it is certainly possible to double the number of needles produced per year by 5—6-year-old trees by giving them repeated short nights (*Pinus contorta*, Longman, 1960). Also, in *Picea sitchensis* and *Pinus contorta*, good correlations have been found between the time when initiation slows in autumn on young trees, and the latitude of seed origin (Cannell and Willett, 1975). Nevertheless, it is possible that water stress or poor apical nutrition may be involved, particularly where trees are adapted to periods of summer drought, or where well defined nodal diaphragms or cortical sclereids form at the bases of the buds in autumn (Pollard *et al.*, 1975, Venn, 1965).

C. Inherent Variation

Inherent differences in numbers of stem units, and hence needles, produced annually on conifer shoots reflect differences in the integral of initiation at the apical domes, that is, the areas under the curves in Figs 2 and 3. This is so irrespective of when the shoots elongate. The component processes are:

(1) number of initials $\propto$ duration $\times$ rate of initiation,

where

(2) rate of initiation $\propto \dfrac{\text{apical dome volume} \times \text{its relative growth rate}}{\text{dome volume taken per initial}}$.

1. Periods of Initiation

The period of the year when the shoot apices are active may be expected to vary among provenances in most northern conifers which have extensive geographic and ecological ranges, and this variation may often underlie provenance differences in height growth.

10. Inherent Differences in Conifer Shoot Growth

Ecotypes adapted to survive summer droughts may stop producing new stem units early in the summer, while northern or montane ecotypes may begin initiation after experiencing smaller heat sums in spring, but cease at longer daylengths, compared with southerly or lowland ecotypes (Stern and Roche, 1974). A latitudinal cline in apical activity of the latter sort has been demonstrated in 8–10-year-old trees as well as seedlings of *Picea sitchensis* (Burley, 1966; Pollard et al., 1975; Cannell and Willett, 1975; see Fig. 3). The timing of

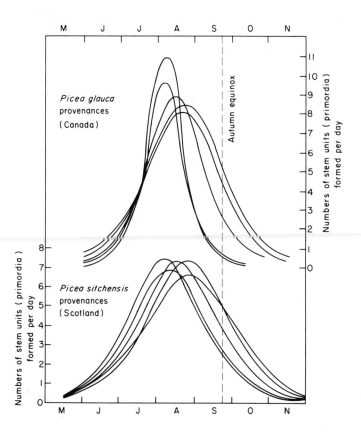

FIG. 3 Rates at which primordia are formed in buds on mature trees of differing provenances of *Picea glauca* in Canada and *Picea sitchensis* in Scotland. Data from Pollard (1973) and Cannell and Willett (1975). Progress curves have been smoothed using logistic functions. The provenances are (left to right): *Picea glauca* from Ontario region, numbers 2481, 2439, 2473, 2447 and 2571; *Picea sitchensis* from Kodiak and Juneau (Alaska), Queen Charlotte Is., Vancouver Is., and Oregon.

apical activity in species with more continental natural ranges may be less predictable; Pollard (1973) could find no obvious genecological pattern in the differing abilities of east Canadian *Picea glauca* provenances to continue initiation in late summer (Fig. 3).

In *Pinus contorta*, the rates as well as the periods of stem unit or cataphyll initiation vary among provenances (Fig. 4). In Britain, both Alaskan and Oregon coastal ecotypes produce cataphylls and axillary bud primordia faster than many inland ones, although some inland ecotypes can partially compensate for this by beginning sooner (Owens and Molder, 1976; Fig. 4).

2. Rates of Initiation

The volume of potentially meristematic tissue in the apical domes of conifer shoot apices can vary several-fold depending on their age, position and environment, and dome volumes often dominate potential initiation rates. Thus, *Picea abies* seedlings increase their dome volumes 20-fold, and initiation rates 5-fold, during the 10 weeks after germination (Gregory and Romberger, 1972); also large seedlings or branches have bigger domes which produce initials at faster rates than small domes (Tepper, 1963; Owston, 1969; Pollard 1974b). Seasonal changes in initiation rates, such as those shown in Fig. 3, are closely paralleled by changes in dome size. Over 80% of the variation in numbers of short-shoot primordia produced in branch buds on 14 contrasting provenances of *Pinus contorta* growing in Scotland could be explained by differences in the integral of dome diameter between June and October (Cannell, unpublished). The reasons why some genotypes can generate larger apical domes than others are unknown, but could involve differences in the rates at which carbohydrates, amino acids and other metabolites are supplied to the apical meristems.

It would be important if inherent differences existed in mitotic rates or the proportion of meristematic tissue in the apical domes. However, the limited evidence does not suggest this. At first, it seemed significant that the growth rates of seedlings of different provenances of *Picea sitchensis, Picea glauca* and *Pinus banksiana* were inversely related to differences in their mean nuclear volumes and DNA contents per cell (Burley, 1965; Miksche, 1968), but,

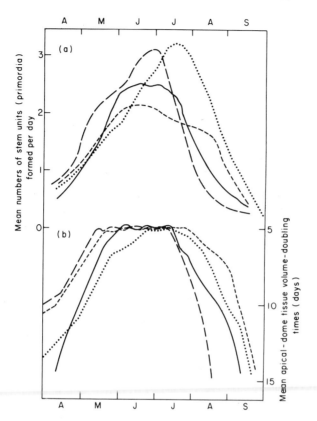

FIG. 4 (a) Rates at which primordia are formed in the terminal buds of 3-year-old *Pinus contorta* seedlings of contrasting provenance growing in Scotland; and (b) their apical-dome tissue volume-doubling times.
············· Hauser Dunes (Oregon coast) ———— Masset (Queen Charlotte Is.)
— — — Skagway (Alaskan coast) - - - - - - Cascade Mts (Oregon).
Prolonged activity by Cascade Mts provenance was associated with the development of a second "cycle" of primordia giving 2 branch whorls. Data are based on weekly samples of five buds per provenance, and are smoothed by taking 3-weekly moving means (Cannell, *in litt.*)

apparently, the period of DNA synthesis may be the same, and there is no direct link with growth rates (Miksche and Rollins, 1971). Recent data on the growth rates of apical dome tissues in terminal buds of seven contrasting provenances of *Pinus contorta* growing in Scotland showed that all reached minimum "dome-tissue volume-doubling times" of 120 ± 3 days in June–July, although their cataphyll initiation rates and dome diameters differed by up to 50% (Fig. 4 and unpublished results). In this study, 30% differences in short-shoot (fascicle) complements were, again, closely correlated with apical dome sizes attained during the year ($r = 0.92$), but were uncorrelated with relative rates of dome growth at any time or integrated over the year ($r = 0.17$).

The third component affecting initiation rates — the size of the initials — is probably of little importance. In *Picea abies* seedlings and mature *Pinus contorta* the amount of tissue assigned to each initial is largest when the initials are being produced fastest, and in any case, rarely exceeds 6% of the dome volume (Gregory and Romberger, 1972; Cannell, unpublished).

D. Genetic Improvement

Genotypes which initiate most stem units and produce most needles are potentially able to achieve the greatest annual height increments. These genotypes will be adapted in such a way that they (a) initiate primordia for the longest possible period, and (b) generate large domes capable of producing primordia rapidly. Both traits will not necessarily occur together, and so cannot be optimized by selecting for height growth alone. An approach may be to select for prolonged apical activity first, and then select for height growth.

If the starting and ending dates for apical activity are adaptive phenological traits, they can be expected to have sufficiently large additive genetic variances to respond well to selection in progeny tests (Hattemer, 1963; Stern and Roche, 1974; Ekberg *et al.*, Chapter 11). Selection among first-year seedlings for genotypes with suitably prolonged periods of height growth may be equivalent to selecting for prolonged apical activity, but this needs to be checked in subsequent years. The simplest meaningful observations on older progeny or "plus" trees are measurements of seasonal changes in apical dome diameters within, say, developing branch buds, which may reflect relative differences in the potential periods of initiation.

Selection for prolonged late-season initiation alone may not be sufficient, as delayed cessation in low latitude and low elevation ecotypes will often be associated with a late spring start (e.g. *Picea sitchensis*, Fig. 3; Stern and Roche, 1974). Consequently, hybridization with slow-growing northerly or montane races adapted to begin producing initials at small heat sums may prove profitable, provided, of course, there is no frost damage, and provided earliness in spring and autumn are not genetically linked, as may be the case, for example, in *Betula* spp. (Stern, 1964).

The potential "vigour" of some recurrently flushing southern pines, and hybrids between species with rapid early and rapid late season growth, appears to be due to sustained apical meristematic activity (*P. radiata*: Will, 1964; *P. taeda*: Woessner, 1972; *P. elliottii* x *P. caribea*: Slee, 1972). The basis of widely reported inter-racial hybrid vigour in conifers with wholly predetermined growth is unknown (Wright, 1963; Libby et al., 1969), but in most cases will be due to more needles being produced per year; it would seem important to know which components of duration and rate have been improved.

III. STEM UNIT AND SHOOT ELONGATION

Stem unit, internode, or sub-apical growth involves cell division as well as elongation, and differences in the lengths finally attained by stem units on woody plants are often due more to differences in cell numbers than to differences in cell lengths (Zimmermann and Brown, 1971, p. 16; Owens and Molder, 1973a; Lam and Brown, 1974). In northern conifers stem units rarely elongate until several weeks after they have been formed at the apical domes, even in seedlings. As trees age, this interval increases. "Lammas" or "free" growth is the elongation of stem units formed a few months beforehand, and the "spring flush" is the growth of stem units held over winter (Pollard and Logan, Chapter 13; see Fig. 5). As more and more stem units are overwintered, the annual period of *shoot* elongation shortens, because an increasing number of stem units elongate together (rather than successively) usually between mid-May and mid-July (Fig. 5). Speaking teleologically, shoot elongation need not occur throughout the available growing season, because it occurs merely to display the needles to the best advantage at the right time, whereas apical growth needs to occur throughout the year in order to generate as many new needles as possible.

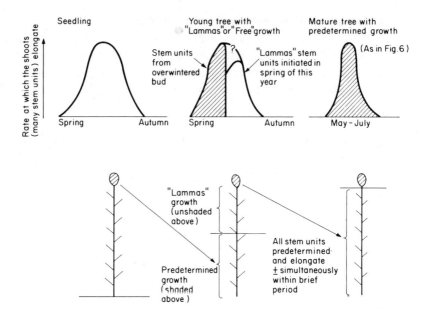

FIG. 5 Diagrammatic representation of the times when stem units elongate on many northern conifers with predetermined growth. Elongation is actually intermittent, and some pines have complex patterns of growth (see Lanner, Chapter 12).

A. Factors Controlling Elongation

Once chilling requirements have been met, sub-apical meristems become active as temperatures increase, in the same way as the apical meristems, but invariably at later dates. As with apical growth, the times when overwintered conifer buds begin to "flush" or "burst" have been closely linked with measures of heat sum (e.g. provenance studies of Langlet, 1936; Roche, 1969; growth models of Sarvas, 1966, 1973; Hari, 1972). However, it is doubtful whether trees integrate heat energy, and the response to temperature is not linear (Worrall and Mergen, 1967). What is certain is that photoperiodism is rarely involved.

Basal stem units elongate ahead of apical ones, and an elongating shoot consists of many stem units at different stages of elongation. Not surprisingly, the rate of shoot elongation fluctuates in response to current and recent weather, particularly as cells are both dividing and elongating, and young shoots import considerable quantities of carbohydrates from the rest of the tree (Kozlowski, 1964, 1971). Variations in the lengths finally attained by stem units reflect inherent differences in "potential", and the extent to which cell division has been checked or cell expansion irreversibly curtailed by premature lignification (e.g. by drought: Garrett and Zahner, 1973; Fig. 8).

The shoot as a whole stops extending when all its stem units have stopped elongating and no new ones start to elongate. In seedlings successive stem units may elongate until the sub-apical meristems are stopped by a temperature-modified photoperiodic stimulus, different in value to that which stops the apical meristems (recorded as "bud-set"; Wareing, 1950). In subsequent years an increasing proportion of stem units do not elongate, either because they are less sensitive to the prevailing daylength, or because they are inhibited by the presence of bud scales. Under field conditions, these sub-apical meristems require chilling before they can be activated by increased temperature. Thus, the time of bud-set occurs increasingly early in the year and is less affected by photoperiod. When all stem units are overwintered the concept of bud-set has no meaning, and, furthermore, the period and/or rate of shoot elongation becomes finite (Lanner, 1971b).

B. Inherent Variation

In *seedlings*, there are well known inherent differences in dates of flushing, critical photoperiods that stop elongation, and "lammas" shoot production, all of which have obvious, though temporary, effects on tree growth. In spite of intensive research, the question often remains whether these seedling differences, all of which involve *sub-apical* activity, are associated with lasting adaptations for brief or prolonged periods of *apical* activity which will influence potential growth when all stem units are overwintered. The case for an association between seedling height growth and apical activity has already been made. However, no such link can be assumed between "lammas" growth and apical activity certainly among progenies. An instance is shown in Table III where *Picea sitchensis* progenies which produced most needles (with parent J) did not produce most "lammas" (parent H).

On *mature* conifers, the variables of greatest interest are (1) the "flushing" date, (2) the period over which preformed stem units elongate, and (3) the extent to which individual stem units elongate.

1. Flushing

Species, provenances and individual trees often have well defined temperature requirements for flushing which are strongly inherited and may even be used to predict their likely dates of flushing from local meteorological data (e.g. in *Pseudotsuga menziesii*: Morris et al., 1957; Silen, 1962; Sziklai, 1966; Campbell, 1974). Various genecological patterns occur: ecotypes adapted to short continental or montane growing seasons often flush earlier than lowland ecotypes (e.g. *Pseudotsuga menziesii* and *Pinus contorta*), but maritime ecotypes from a range of latitudes may flush at the same time (e.g. *Picea sitchensis*: Lines and Mitchell, 1966). However, flushing dates may be of marginal interest to breeders, provided spring frosts are avoided, because they are often unrelated to total shoot growth among progenies, and may be inversely related among provenances (cf. Kleinschmit and Sauer, Chapter 30).

2. Periods of Shoot Elongation

The times when shoots elongate differ greatly between years and sites irrespective of genotype, but the relative differences in timing between genotypes are often the same (Fig. 6). These relative differences can be due to two factors: (a) inherent differences in response to the environment, such as spring temperatures, during the year when elongation occurs; and (b) differences in numbers of stem units originally present in the winter buds — it is important to realize that pre-formed shoots which have many stem units must elongate faster or over longer periods than ones with few stem units, provided stem unit lengths become similar (Lanner, 1968). Clearly, it is important to separate these two sources of variation because the first (a) involves stem unit growth in year n, whereas the second (b)

10. Inherent Differences in Conifer Shoot Growth

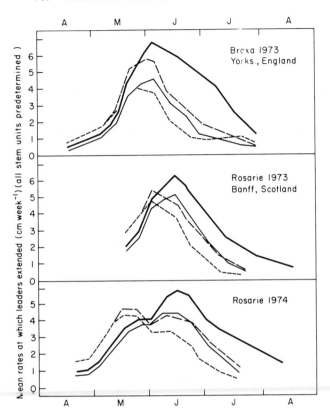

FIG. 6 Differing rates and periods of elongation of leading shoots on the same 10–11-year-old provenances of *Pinus contorta* growing at two sites in Britain. All stem units were preformed in year $(n-1)$.
──────── Mean of 3 provenances from Oregon, Washington and Vancouver Island coasts.
──────── Mean of 3 provenances from N. British Columbian and Alaskan coasts.
─ ─ ─ ─ Mean of 4 provenances from interior British Columbia.
- - - - - - Mean of 2 provenances from east of Rocky and Cascade Mountains.
Lengths were recorded weekly on 4 trees in each of 4 blocks per provenance and curves have been smoothed by taking 3-weekly moving means (full report *in litt.*)

involves apical growth in year $(n-1)$. Among mature north temperate conifers, particularly pines, the second component can be of over-riding importance, as shown by the fact that inherent differences in numbers of pre-formed stem units (often judged by bud lengths) are closely correlated with differences in rates or

TABLE I. Correlations (r) showing relationships between the numbers of pre-formed stem units on "predetermined" leaders of *Pinus contorta* provenances, and the rates and durations of leader elongation.

Trial location[a] (see also Fig. 6)	Range of stem unit numbers per leader	Correlations of numbers of stem units on the leaders with:	
		Maximum rates at which leaders elongated	Periods over which leaders elongated (May–July)
Rosarie, Scotland (1973)	306–540	0.96	0.84
Broxa, England (1974)	262–530	0.92	0.55

[a] Each trial had the same 14 provenances (12 degrees of freedom).

periods of shoot elongation (e.g. *Pinus sylvestris*: Wright and Bull, 1963; *P. resinosa*: Rehfeldt and Lester, 1966; *P. banksiana*: Teich and Holst, 1969; *P. taeda*: Boyer, 1970; *P. contorta*: Hagner and Fahlroth, 1974). Correlation and multivariate analyses suggested that 85–95% of the variation in shoot elongation rates and durations among 14 provenances of *Pinus contorta* growing in Britain (summarized in Fig. 6) were attributable to differences in numbers of stem units built into the buds at the start of the year (Table I). Thus, Oregon coastal provenances which extended their leaders fastest and longest (Fig. 6) had c. 473 stem units, whereas the slowest growing provenances had only c. 280 stem units. Apart from differences in starting dates, there was little evidence for inherent differences in the rates at which individual stem units elongated, although there were potentially important differences in their final lengths (range 1.02–1.42 mm).

3. Final Mean Stem Unit Lengths

Although height growth can probably be increased more readily by increasing the number of stem units than by increasing the extent to which they elongate (Table II), the latter can be a heritable trait of

10. Inherent Differences in Conifer Shoot Growth

TABLE II. Correlations (r) showing the relative importance of variation in numbers of stem units ($\triangleq$ needles) and lengths per stem unit, which affect the leader lengths of progenies of *Picea sitchensis* and provenances of *Pinus contorta* growing in Britain.

		Ranges of leader lengths (cm)	Correlations of leader lengths with:	
			Numbers of stem units on the leaders	Mean length per stem unit
As in Table III (47 d.f.)	*Picea sitchensis* 7 x 7 diallel progenies, Bush, Scotland	15—47	0.70	0.63
As in Fig. 6 (12 d.f.)	*Pinus contorta* 14 provenances, Rosarie, Scotland	34—57	0.84	—0.11
	Pinus contorta 14 provenances, Broxa, England	29—63	0.91	0.44

potential value. Thus, there was a significant positive correlation between mean stem unit lengths (distances between needles) and total leader lengths among 48 progenies in diallel cross trials of *Picea sitchensis* in Britain (Table II). The outstanding progenies were those which combined large needle numbers with long stem units (e.g. H x J, and H x G: Table III), and this combination may have been one cause of apparent "overdominance" for height growth in these trials (Samuel *et al.*, 1972). In *Pinus contorta*, similar opportunities for genetic gain may exist by crossing Oregon coastal provenances which produce many short stem units with interior British Columbian provenances which have fewer, but inherently longer stem units.

IV. NEEDLE ELONGATION

The growth of spruce and fir needles is determinate, depending particularly on the numbers of cell divisions at an early stage of differentiation, much like the growth of the internodes with which they are associated. Indeed, in these instances, provenances with long needles often have long internodes, associated with prolonged

TABLE III. Leader lengths, their components (needle numbers and stem unit lengths) and branch frequencies: means for all progenies of seven parents in a diallel cross of *Picea sitchensis* planted at two sites in Britain.

		A	C	D	G	H	J	K	GCA	SCA	
1.	Leader lengths	32	29	28	32	33	38	37	***	***	Bush, Scotland
2.	Fascicle numbers	563$_7$	558$_2$	520$_6$	555$_9$	526$_{23}$	650$_{10}$	577$_6$	**	N.S.	
3.	Stem unit lengths	56	52	53	57	62	58	64	**	*	
4.	Branch frequencies	32	33	30	35	30	37	36	*	N.S.	
1.	Leader lengths	38	—	36	41	43	43	38	***	*	Tywi, Wales
2.	Fascicle numbers	781$_{37}$	—	727$_{32}$	778$_{59}$	789$_{78}$	801$_{48}$	745$_{40}$	N.S.	N.S.	
3.	Stem unit lengths	49	—	49	53	55	54	51	**	N.S.	
4.	Branch frequencies	33	—	30	41	31	43	36	N.S.	N.S.	

1. Leader length (cm) is the product of fascicle numbers × mean length per stem unit.
2. Total number of fascicles (≙ stem units) occurring on the leader, subscript numbers refer to fascicles on lammas growth.
3. Mean distance between needle fascicles at mid-leader level (mm × 10^3).
4. Mean number of needle fascicles separating each branch bud on the leader.

GCA = general combining ability; SCA = specific combining ability.
Means are based on measurements of single 7-year-old (since germination) trees in 3 (Bush) or 4 (Tywi) replicate blocks per site.

periods when both extend (e.g. *Abies concolor*: Wright *et al.*, 1971; *Picea abies*: Baldwin *et al.*, 1973). The growth of pine needles is less determinate, as they grow from basal, intercalary meristems which arise on the flanks of the apical domes of short-shoot axillary bud primordia (for details see Sacher, 1955; Hanawa, 1967; Gabilo and Mogensen, 1973). Consequently, pine needles can grow throughout the summer and justify separate consideration here.

A. Control of Pine Needle Elongation

Even on northern pines with wholly preformed shoots, the needles may start growing in the autumn, particularly on young trees of "hard" pines. This growth occurs inside the scales on the basal parts of the winter buds, which also begin to elongate, and is a feature associated in British-grown *Pinus contorta* with interior or montane ecotypes. These same ecotypes remain ahead of southerly, coastal ones in both needle and stem unit elongation in the spring (Figs 6 and 7). The suggestion is that factors linked to heat sum regulate the onset of intercalary as well as sub-apical and apical meristematic activity in the buds, although not all at the same threshold levels.

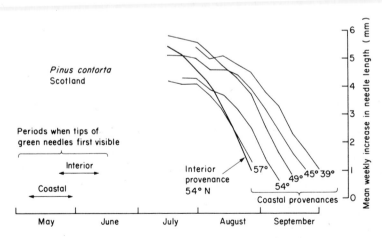

FIG. 7 Rates of needle elongation on five 7-year-old coastal provenances of *Pinus contorta* growing at Rosarie, Scotland. Needles were measured weekly in the middle of the leaders of 4 trees in each of 4 blocks. Data given are 3-weekly moving means.

Continued needle elongation during the spring and summer is sensitive to changes in the physical environment, for the same reasons as internode growth, but for almost twice as long a period. The important factors that can check or curtail needle elongation are poor sources of carbohydrate in the first few weeks (Gordon and Larson, 1968), water stress (Garrett and Zahner, 1973), mineral nutrient deficiencies and low temperatures. On the other hand, continued activity of the intercalary meristems enables pines to take advantage of favourable conditions and compensate for poor conditions the previous year. Thus, needles can become relatively long when few of them compete for abundant internal metabolites (Garrett and Zahner, 1973; Fig. 8) or when temperatures and other physical variables are favourable (Kienholz, 1934; Jensen and Gatherum, 1965; Larson, 1967). This facility enables pines to produce similar amounts of foliage in successive years, and means that needle length is a particularly plastic character.

If intercalary meristematic activity is not halted by stress conditions, it may be checked in late summer by temperature-modified photoperiodic stimuli as postulated for the apical meristems. Thus, young trees of *Pinus sylvestris* given long nights stop extending their needles sooner, and produce shorter needles, than trees given normal or short nights (Wareing, 1950). Also, latitudinal clines exist for the time of needle growth cessation in the same way as for height and apical growth cessation in first-year seedlings (*P. sylvestris*: Langlet, 1959; *P. contorta*: Fig. 7). However, pine needles do not grow indefinitely even in continuous long days, and the intercalary meristems are not reactivated in their second year.

B. Inherent Variation in Pine Needle Elongation

Needle lengths are often more unstable determinants of foliage production than stem unit numbers and lengths, since they depend on climatic conditions throughout the current and previous years (Fig. 8). Nevertheless, some factors regulating needle growth onset and cessation are clearly adaptive in an ecological sense, and needle lengths are often a heritable trait. In particular, provenances taken north from areas with long growing seasons often produce longer needles than provenances adapted to short growing seasons (e.g.

10. Inherent Differences in Conifer Shoot Growth

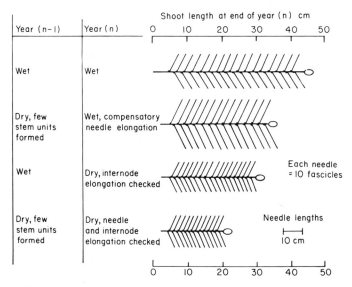

FIG. 8 Effects of water stress in June—July of the current and previous years on the growth of *Pinus resinosa* shoots and needles (After Garrett and Zahner (1973), by permission of the Ecological Society of America).

Pinus sylvestris: Langlet, 1959; *P. contorta*: Fig. 7; *P. ponderosa*: Squillace and Silen, 1962; Kung and Wright, 1972). However, short-season ecotypes sometimes compensate by extending their needles more rapidly, as do some interior provenances of *P. contorta* in Britain. At the progeny level there seems less likelihood that needle growth characteristics can be relied upon as phenological traits of value to increase foliage biomass.

V. LATERAL BUD PRODUCTION

It is principally by producing branches that young conifers multiply their foliage biomass and begin to make rapid radial growth (e.g. Rangnekar *et al.*, 1969). The numbers of branches on young trees commonly multiply annually 2-fold to 4-fold in the first 2—5 years (*Pinus resinosa*: Miller, 1965; *P. contorta*, *Picea sitchensis*: Cannell, 1974). The *actual* multiplication rate depends on the rate at which the leaders and branches increase in length, because long shoots tend to produce more branches than short ones. Consequently, total branch numbers may reflect growth rates as much as differences in branch formation (Cannell, 1974). The *potential* multiplication rate, of concern here, depends on factors regulating the numbers of lateral buds produced per unit length of shoot, that is, the branch frequency.

A. Lateral Bud Formation

Unlike most angiosperms, conifers do not, as a rule, form branch buds in the axils of all leaves, although pines can potentially form branches from each axillary bud primordium.

1. Lateral Buds on Pines

The sequence of appendages produced during a "cycle" of pine bud development (that is sterile zone to sterile zone) is almost always the same as shown in Fig. 1. To explain this development, it is necessary to hypothesize that the presence and fate of axillary bud primordia are (a) predestined when their subtending cataphylls are formed, and/or (b) depend on their positions relative to other structures after the cataphylls are formed. Evidence, in favour of hypothesis (a) is that the fate of primordia on many plants is influenced by the size and activity of the apical peripheral zones when they were formed (e.g. Allsopp, 1954). Thus, it is conceivable that cataphylls remain sterile (i.e. without axillary bud primordia) when they are produced on small apical domes, as at certain stages of seedling development (Riding, 1972) and at the start and end of each "cycle". The fates of axillary bud primordia depend more clearly on their positions, because all are morphologically similar, potentially able to form long shoots, but lacking cortical and rib meristem activity as long as the main apical meristem remains active (e.g. "foxtails": Lanner, 1971a), or branch buds are already present above them. However, this situation is not a simple case of apical dominance (see Zimmermann and Brown, 1971).

2. Lateral Buds on Spruces and Firs

Spruces and firs do not produce true axillary buds. Instead they form lateral bud meristems by de-differentiating areas of cortical tissue between a few of their needle primordia, long after the needles were initiated. These lateral bud meristems form when the sub-apical tissues of the parent buds begin to grow and the needle primordia separate, usually in March–April when there is an overwintering bud, 1–2 weeks before "bud-burst" (e.g. *Pseudotsuga menziesii*: Owens 1969; *Abies balsamea*: Powell, 1974). Local areas of metabolic activity can be detected 1–2 weeks before lateral buds become visible (Owens, 1969), but there is no evidence that their presence or positions are determined when the needle primordia are initiated. It appears that lateral bud meristems form only when the needle primordia have separated to a given degree. Thus, on large, overwintered buds of *Picea sitchensis* the lateral meristems can be seen forming acropetally, following a "wave" of bud expansion from base to tip, and as the bud swells further, additional lateral meristems can be added (Fig. 9).

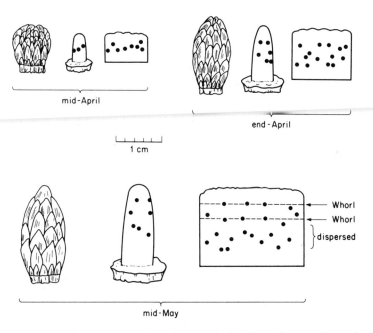

FIG. 9 Positions of lateral bud meristems at the time they are formed within swelling leader buds of *Picea sitchensis*. The buds are shown before and after bud scales and elongating needles have been removed, and the surfaces of the buds are displayed in two dimensions. The arrangements of points showing the positions of lateral bud meristems are non-random.

B. Some Factors Affecting Lateral Bud Numbers

On many northern conifers the mean number of lateral branches developing on each parent shoot is loosely related to the number of needles on that shoot, and hence its ability to supply metabolites to those branches when they begin to elongate. Linear regressions have been calculated between branch numbers and their parent shoot lengths over the range 5—40 cm for young *Pinus contorta* and *Picea sitchensis* in Scotland (Cannell, 1974), and others have noted close correlations between the height and numbers of branches on young conifers (e.g. Miller, 1965; Samuel *et al.*, 1972). It appears, consequently, that factors controlling branch bud formation can be linked with the size of the parent structures on which they form.

On pines, the numbers of axillary buds that escape short-shoot status seem to depend on the numbers in a "critical" zone below the diminishing dome at the end of each cycle. Small, slim buds with active sub-apical and rib meristems hold fewer axillary buds in this zone compared with large, wide buds which elongate very little. The former have on their flanks less closely packed primordia which are arranged in lower-order phyllotaxy spirals. Differences of this kind occur not only between buds of different sizes, but also between buds with the same needle complement on differing provenances, and progenies which produce different numbers of branches per unit number of needles or unit shoot length (e.g. *Pinus contorta*, Fig. 10).

The total number of branches produced per year by pines will depend not only on the numbers per growth cycle, but also the number of growth cycles. In the latter case, particularly, it will be influenced by factors affecting lateral bud abortion, and, in all cases, by any tendency to form female strobili instead of lateral branches.

The arrangement and numbers of lateral buds on single elongated shoots of many spruces and firs appears haphazard, but studies on the leaders and top terminal branches of *Picea sitchensis*, *Picea omorika* and *Pseudotsuga menziesii*, showed that in April, when the branch buds form, they are significantly evenly distributed over most of the surface of the expanding parent buds (Fig. 9). This tends to maximize both (a) the vertical distance between overlying branches, and (b) the difference in compass direction between branches in a whorl. No lateral buds are formed over the basal 10—20% of most shoots, and above that the numbers seem to be regulated by the size

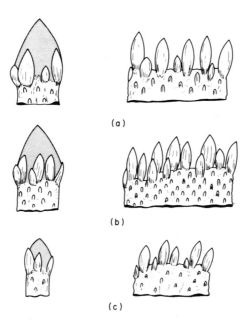

FIG. 10 Long-shoot lateral buds surrounding the leading buds on 3 contrasting genotypes of *Pinus contorta*. Scales (cataphylls) enclosing the lateral buds have been removed. (a) An interior Washington and coastal British Columbian hybrid, which produces relatively few lateral buds on long leading shoots; (b) an Oregon coastal provenance that produces relatively numerous lateral buds on long leading shoots; (c) an Alaskan coastal provenance that produces short leaders with proportionately fewer lateral buds when compared with (b) (Cannell, 1974).

and numbers of needle primordia on the parent buds and the degree of apparent mutual inhibition between lateral buds. All "available spaces" are occupied by lateral buds at the tops of the parent buds, mostly giving terminal branch "whorls".

Thus, the precise positions of lateral buds on spruces and firs are subject to chance, but their overall arrangement and numbers on the leading shoots are rather closely controlled. On the leaders of young trees some small buds on the basal parts of the parent buds can remain latent or abort, but most of the others survive to become branches of one kind or another. However, on branches of mature trees over half can become reproductive or abort (Owens, 1969).

Environmental effects on conifer lateral bud formation have not been closely studied, although controlled environment studies and field assessments on young *Picea sitchensis* and *Pinus contorta* in Scotland confirm that it is much less affected by temperature, daylength, and nutritional treatments than by apical and extension growth (see Cannell, 1974).

C. Inherent Variation in Branch Frequency

Inherent variation in the number of branch whorls produced on a given length of leader is well known in pine species which have polycyclic buds (e.g. *Pinus contorta*: Franklin and Callaham, 1970), and recurrent flushes of growth (e.g. *Pinus radiata*: Fielding, 1960). Inherent differences also occur in numbers of branches per whorl on *Pinus banksiana* (Schoenike et al., 1962) and *Pinus sylvestris* (Ehrenberg, 1963; Nilsson, 1968), and it would be interesting to see whether they could be predicted from differences in bud morphology as is sometimes the case on *Pinus contorta* (Fig. 10).

There are several reports of inherent differences in branch numbers on spruces and firs, but it is not always clear whether they reflect differences in branch frequency. Nevertheless, frequency differences almost certainly occur in such instances as *Picea abies*, where there is marked variation in branching habit (Hofmann, 1968; Schmidt-Vogt, 1972). The Scottish diallel progeny trials of *Picea sitchensis,* referred to earlier, show differences in mean distances, and numbers of needles, separating branch buds on the leaders. Branches tend to be widely separated on progenies of parents G, J and K and closer together on progenies involving parents D, H and A (Table III). Interestingly, the mean numbers of needles separating the branches is similar at the two sites, irrespective of differences in total needle number, reinforcing the view that inherent control processes affecting branch frequency can be separated from those affecting tree height growth and total branch number.

VI. IMPLICATIONS FOR TREE BREEDERS

Although the apical, sub-apical and intercalary meristems are all re-activated in spring by increasing temperature, or its integral heat-sum, and can all be stopped in late summer by lengthening nights, there is ample evidence that they are controlled separately. Consequently, needle numbers, stem unit lengths and needle lengths may vary independently, and wherever possible should be estimated as separate components of annual leader or shoot lengths and foliage biomass. Needle numbers will reflect the previous total activity of the apical meristems, which are otherwise difficult to study in breeding trials, unless close correlations can be established between apical dome diameters and plastochron durations. Emphasis often given to periods of shoot elongation may be better given to estimates of final mean stem unit numbers and lengths. Similarly, studies on branching should include estimates of branch frequencies, as the numbers of branches produced by many conifers are fundamentally related to the sizes and dimensions of the parent shoots on which they develop.

VII. SUMMARY

Inherent variation in shoot growth within species of north temperate conifers is considered in terms of the activities of four regions of cell division and elongation, which are apparently controlled, phased and probably inherited independently. They are: (a) the shoot apices where new stem units are formed; (b) the sub-apical meristems which regulate subsequent elongation of each stem unit; (c) the needles, particularly of pines; and (d) the sites where new lateral vegetative apices can form.

The periods and rates of initiation of stem units at the shoot apices are often the most important determinants of shoot lengths. The sizes of the apical domes, or meristematic "factories", reveal potential initiation rates. More estimates should be made of inherent differences in stem unit or needle numbers produced per year. When all stem units are pre-formed, inherent differences in shoot elongation phenology can reflect differences in numbers of stem units contained in the winter buds. However, some genotypes may be predisposed to extend individual stem units more than others, and this may be inherited separately from factors affecting stem unit numbers. Pine needle growth is influenced by climatic conditions through the current and previous years and can compensate for poor activity in the other meristems. It is consequently an unreliable phenological trait. Differences in branch frequency are considered to be as important as absolute numbers. Pines form many branches when many axillary buds occur at critical distances from the shoot apices. Spruces and firs form branch meristems in non-random arrays on the expanding buds, and potential numbers are determined by the sizes of the buds and degrees of apparent mutual inhibition between branch bud meristems.

REFERENCES

Allsopp, A. (1954). Juvenile stage of plants and the nutritional status of the shoot apex. *Nature, Lond.* 173, 1032–1035.

Baldwin, H. I., Eliason, E. J. and Carlson, D. E. (1973). IUFRO Norway spruce provenance tests in New Hampshire and New York. *Silvae Genet.* 22, 93–114.

Boyer, W. D. (1970). Shoot growth patterns of young loblolly pine. *Forest Sci.* 16, 472–402.

Burley, J. (1965). Karyotype analysis of Sitka spruce, *Picea sitchensis* (Bong) Carr. *Silvae Genet.* 14, 127–132.

Burley, J. (1966). Provenance variation in growth of seedling apices of Sitka spruce. *Forest Sci.* 12, 170–175.

Campbell, R. K. (1974). Use of phenology for examining provenance transfer in reforestation of Douglas fir. *J. appl. Ecol.* 11, 1069–1080.

Cannell, M. G. R. (1974). Production of branches and foliage by young trees of *Pinus contorta* and *Picea sitchensis*: provenance differences and their simulation. *J. appl. Ecol.* 11, 1091–1115.

Cannell, M. G. R. and Willett, S. C. (1975). Rates and times at which needles are initiated in buds on differing provenances of *Pinus contorta* and *Picea sitchensis* in Scotland. *Can. J. Forest Res.* (in press).

Curtis, J. D. and Popham, R. A. (1972). The developmental anatomy of long-branch terminal buds of *Pinus banksiana*. *Am. J. Bot.* 59, 194–202.

Doak, C. C. (1935). Evolution of foliar types, dwarf shoots, and cone scales of *Pinus. Illinois biol. Monogr.* 13(3).

Ehrenberg, C. E. (1963). Genetic variation in progeny tests of Scots pine (*Pinus sylvestris* L.). *Studia For. Suecica* No. 10.

Fielding, J. (1960). Branching and flowering characteristics of Monterey pine. *Bull. Aust. For. Timb. Bur.* No. 37.

Frampton, C. V. (1960). Some aspects of the developmental anatomy of the "long" shoot of *Larix decidua* Mill. with particular reference to seasonal periodicity. *New Phytol.* 59, 175–191.

Franklin, E. C. and Callaham, R. Z. (1970). Multinodality, branching and forking in Lodgepole pine (*Pinus contorta* var. *murrayana*, Engelm.). *Silvae Genet.* 19, 180–186.

Gabilo, E. M. and Mogensen, H. L. (1973). Foliar initiation and the fate of the dwarf-shoot apex in *Pinus monophylla. Am. J. Bot.* 60, 671–677.

Garrett, P. W. and Zahner, R. (1973). Fascicle density and needle growth responses of red pine to water supply over two seasons. *Ecology* 54, 1328–1334.

Gordon, J. C. and Larson, P. R. (1968). Seasonal course of photosynthesis, respiration, and distribution of ^{14}C in young *Pinus resinosa* trees as related to wood formation. *Plant Physiol.* 43, 1617–1624.

Gregory, R. A. and Romberger, J. A. (1972). The shoot apical ontogeny of the *Picea abies* seedling. I. Anatomy, apical dome diameter and plastochron duration. II. Growth rates. *Am. J. Bot.* 59, 587–597, 598–606.

Hagner, S. and Fahlroth, S. (1974). Om contortatallen och dess odlingsförutsättninger i Noorland. (On the prospects of cultivating *Pinus contorta* in northern Sweden). O.D.C. 651–656. *Sartryck ur Sveriges Skogsvardsforbunds Tidskriff* 4, 477–528.

Hanawa, J. (1966). Growth and development of the shoot apex of *Pinus densiflora*. I. Growth periodicity and structure of the vegetative shoot apex. *Bot. Mag., Tokyo* 79, 736–746.

Hanawa, J. (1967). Growth and development of the shoot apex of *Pinus densiflora*. II. Ontogeny of the dwarf shoot and the lateral branch. *Bot. Mag., Tokyo* 80, 248–256.

Hari, P. (1972). Physiological stage of development in biological models of growth and maturation. *Ann. Bot. Fenn.* 9, 107–115.

Hattemer, H. H. (1963). Estimates of heritability published in forest tree breeding research. *World Consult. on Forest Genetics and Tree Improvement.* FAO/FORGEN 63, 2a/3, 13 pp.

Heide, O. M. (1974). Growth and dormancy in Norway spruce ecotypes (*Picea abies*). I. Interaction of photoperiod and temperature. II. After-effects of photoperiod and temperature on growth and development in subsequent years. *Physiol. Plant.* 30, 1–12; 31, 131–139.

Hofmann, J. (1968). Ueber die bisherigen Ergebnisse der Fichtentypenforschung. *Arch. Forstw.* 17, 207–216.

Jensen, K. and Gatherum, G. E. (1965). Effects of temperature, photoperiod and provenance on growth and development of Scots pine seedlings. *Forest Sci.* 11, 189–199.

Kienholz, R. (1934). Leader, needle, cambial and root growth of certain conifers and their relationships. *Bot. Gaz.* 96, 73–92.

Kozlowski, T. T. (1964). Shoot growth in woody plants. *Bot. Rev.* 30, 335–392.

Kozlowski, T. T. (1971). "Growth and Development of Trees. I. Seed

Germination, Ontogeny and Shoot Growth". Academic Press, New York and London.

Kung, F. W. and Wright, J. W. (1972). Parallel divergent evolution in Rocky Mountain trees. *Silvae Genet.* 21, 77–85.

Lam, III O. C. and Brown, C. L. (1974). Shoot growth and histogenesis of *Liquidambar styraciflua* L. under different photoperiods. *Bot. Gaz.* 135, 149–154.

Landsberg, J. J. (1974). Apple fruit bud development and growth; analysis and an empirical model. *Ann. Bot.* 38, 1013–1023.

Langlet, O. (1936). Studien über die physiologische Variabilität der Kiefer und deren Zusammenhang mit dem Klima. Beiträge zur Kenntnis der Oekotypen von *Pinus sylvestris* L. *Meddl. Skogsfors Anst. Exkurshed.* 29, 219–254.

Langlet, O. (1959). A cline or not a cline – a question of Scots pine. *Silvae Genet.* 8, 13–22.

Lanner, R. M. (1968). "The pine shoot primary growth system". Ph.D. thesis., University of Minnesota.

Lanner, R. M. (1971a). Growth and morphogenesis of pines in the tropics. Selection and breeding to improve some tropical conifers. Section 22, 15th IUFRO Congress, Gainsville, Florida.

Lanner, R. M. (1971b). Comment on "Shoot growth patterns of young loblolly pine" by W. D. Boyer. *Forest Sci.* 16, 486–487.

Larson, P. R. (1967). Effects of temperature on the growth and wood formation of ten *Pinus resinosa* sources. *Silvae Genet.* 16, 58–65.

Libby, W. J., Stettler, R. F. and Seitz, F. W. (1969). Forest genetics and forest-tree breeding. *Ann. Rev. Genetics* 3, 469–494.

Lines, R. and Mitchell, A. F. (1966). Differences in phenology of Sitka spruce provenances. *Rep. Forest. Res., Lond.* H.M.S.O. 1173–1184.

Longman, K. A. (1960). "Problems of the phenology of flowering in forest trees". Ph.D. Thesis, University of Manchester.

Miksche, J. P. (1968). Quantitative study of intraspecific variation of DNA per cell in *Picea glauca* and *Pinus banksiana. Can. J. Genet. Cytol.* 10, 590–600.

Miksche, J. P. and Rollins, J. A. (1971). Constancy of the duration of DNA synthesis and per cent G and C in white spruce from several provenances. *Can. J. Genet. Cytol.* 13, 415–421.

Miller, W. E. (1965). Number of branchlets on red pine in young plantations. *Forest Sci.* 11, 42–49.

Morris, W. G., Silen, R. R. and Irgens-Möller, H. (1957). Consistency of bud-bursting in Douglas fir. *J. For.* 55, 208–210.

Nilsson, B. (1968). Studies of the genetical variation of some quality characters in Scots Pine (*Pinus sylvestris* L.). *Rapp. Uppsala Instn Skogsgenet, Skogshogsk* (Dept. For. Genetics, Royal Coll. of For., Stockholm) No. 3.

Owens, J. N. (1969). The relative importance of initiation and early development on cone production in Douglas fir. *Can. J. Bot.* 47, 1039–1049.

Owens, J. N. and Molder, M. (1973a). A study of DNA and mitotic activity in the vegetative apex of Douglas fir during the annual growth cycle. *Can. J. Bot.* 51, 1395–1409.

Owens, J. N. and Molder, M. (1973b). Bud development in western hemlock. I. Annual growth cycle of vegetative buds. *Can. J. Bot.* 51, 2223–2231.

Owens, J. N. and Molder, M. (1976). Development of long-shoot terminal buds of Pinus contorta. *Can. J. Forest Res.* (in press).

Owston, P. W. (1969). The shoot apex in eastern white pine: its structure, seasonal development, and variation within the crown. *Can. J. Bot.* 47, 1181–1188.

Parke, R. V. (1959). Growth periodicity and the shoot tip of *Abies concolor*. *Am. J. Bot.* 46, 110–118.

Pollard, D. F. W. (1973). Provenance variation in phenology of needle initiation in white spruce. *Can. J. Forest Res.* 3, 589–593.

Pollard, D. F. W. (1974a). Bud morphogenesis of white spruce, *Picea glauca*, seedlings in a uniform environment. *Can. J. Bot.* 52, 1569–1571.

Pollard, D. F. W. (1974b). Seedling size and age as factors of morphogenesis in white spruce, *Picea glauca* (Moench) Voss. buds. *Can. J. Forest Res.* 4, 97–100.

Pollard, D. F. W. and Logan, K. T. (1974). The role of free growth in the differentiation of provenances of black spruce, *Picea mariana* (Mill.) B.S.P. *Can. J. Forest Res.* 1, 308–311.

Pollard, D. F. W., Teich, A. H. and Logan, K. T. (1975). Seedling shoot and bud development in provenances of Sitka spruce, *Picea sitchensis* (Bong.) Carr. *Can. J. Forest Res.* 5, 18–25.

Powell, G. R. (1974). Initiation and development of lateral buds in *Abies balsamea*. *Can. J. Forest Res.* 4, 458–469.

Rangnekar, P. V., Forward, D. F. and Nolan, N. J. (1969). Foliar nutrition and wood growth in red pine: the distribution of radiocarbon photoassimilated by individual branches of young trees. *Can. J. Bot.* 47, 1710–1711.

Rehfeldt, G. E. and Lester, D. T. (1966). Variation in shoot elongation of *Pinus resinosa* Ait. *Can. J. Bot.* 44, 1457–1469.

Riding, R. T. (1972). Early ontogeny of seedlings of *Pinus radiata*. *Can. J. Bot.* 50, 2381–2387.

Roche, L. (1969). A genecological study of the genus *Picea* in British Columbia. *New Phytol.* 68, 505–554.

Romberger, J. A. (1963). Meristems, growth and development in woody plants. An analytical review of anatomical, physiological and morphogenic aspects. *U.S.D.A. Forest Service Tech. Bull.* 1293.

Sacher, J. A. (1954). Structure and seasonal activity of the shoot apices of *Pinus lambertiana* and *Pinus ponderosa. Am. J. Bot.* **41**, 749–759.

Sacher, J. A. (1955). Dwarf shoot ontogeny in *Pinus lambertiana. Am. J. Bot.* **42**, 784–792.

Samuel, C. J. A., Johnstone, R. C. B. and Fletcher, A. M. (1972). A diallel cross in Sitka spruce. Assessment of first year characters in an early glasshouse test. *Theor. appl. Genet.* **42**, 53–61.

Sarvas, R. (1966). Temperature sum as a restricting factor in the development of forest trees in the subarctic. *UNESCO Nat. Res. Org. Symp. on Ecology.* Paper no. 27, 79–82.

Sarvas, R. (1973). Investigations on the annual cycle of development of forest trees. Active period. *Metsäntutkimuslaitoksen Julkaisuja Comm. Instit. For. Fenniae* 76.

Schmidt-Vogt, H. (1972). Studien zur morphologischen Variabilität der Fichte (*Picea abies* L. Karst). *Allg. Forst- u. Jagdztg* **143**, 133–240.

Schoenike, R. E., Rudolph, T. E. and Jensen, R. A. (1962). Branch characteristics in a jack pine seed source plantation. *Minn. For. Note* No. 113.

Silen, R. R. (1962). A study of genetic control of bud bursting in Douglas fir, *J. For.* **60**, 472–475.

Slee, M. V. (1972). Growth patterns of Slash and Caribbean pines and their hybrids in Queensland. *Euphytica* **21**, 129–142.

Squillace, A. E. and Silen, R. R. (1962). Racial variation in ponderosa pine. *Forest Sci. Monogr.* No. 2.

Stern, K. (1964). Population genetics as a basis for selection. *Unasylva* **18**, 21–29.

Stern, K. and Roche, L. (1974). "Genetics of Forest Ecosystems". Springer-Verlag, Berlin, Heidelberg and New York.

Sucoff, E. (1971). Timing and rate of bud formation in *Pinus resinosa. Can. J. Bot.* **49**, 1821–1832.

Sziklai, O. (1966). Variation and inheritance in some physiological and morphological traits in Douglas fir. *U.S. For. Service Res. Paper* No. 6 pp. 62–67.

Teich, A. H. and Holst, M. J. (1969). Breeding for height growth of *Pinus banksiana* Lamb. *2nd World Consult. on For. Tree Breeding.* FAO/IUFRO pp. 131–138.

Tepper, H. B. (1963). Dimensional and zonational variation in dormant shoot apices of *Pinus ponderosa. Am. J. Bot.* **50**, 589–596.

Thompson, S. (1974). "Shoot elongation and dry matter production in two contrasted provenances of *Pinus contorta*". Ph.D. thesis University of Aberdeen.

Venn, K. (1965). Nodal diaphragms in *Picea abies* (L.) Karst and other conifers. *Meddr. norske Skogsfors Ves.* **20**, No. 73.

Wareing, P. F. (1950). Growth studies in woody species. I. Photoperiodism in first-year seedlings of *Pinus sylvestris*. II Effect of daylength on shoot growth in *Pinus sylvestris* after the first year. *Physiol. Plant.* 3, 258–279, 300–314.

Will, G. M. (1964). Dry matter production and nutrient uptake by *Pinus radiata* in New Zealand. *Commonw. For. Rev.* 43, 57–70.

Woessner, R. A. (1972). Growth patterns of one-year-old loblolly pine seed sources and interprovenance crosses under contrasting edaphic conditions. *Forest Sci.* 18, 205–210.

Worrall, J. and Mergen, F. (1967). Environmental and genetic control of dormancy of *Picea abies*. *Physiol. Plant.* 20, 733–745.

Wright, J. W. (1963). Hybridization between species and races. *Unasylva* 18, 30–39.

Wright, J. W. and Bull, W. I. (1963). Geographic variation in Scotch pine. *Silvae Genet.* 12, 1–25.

Wright, J. W., Lemmien, W. A. and Bright, J. N. (1971). Genetic variation in southern Rocky Mountain white fir. *Silvae Genet.* 20, 148–150.

Zimmermann, M. H. and Brown, C. L. (1971). "Trees, Structure and Function". Springer-Verlag, Berlin, Heidelberg and New York.

11
Inheritance of the Photoperiodic Response in Forest Trees

INGER EKBERG, INGEGERD DORMLING, GÖSTA ERIKSSON
Department of Forest Genetics and the Phytotron, Royal College of Forestry, Stockholm, Sweden
and DITER VON WETTSTEIN
Department of physiology, Carlsberg Laboratory, Copenhagen, Denmark

I.	Introduction	207
II.	Time of Growth Cessation Inferred from Height Measurements	208
III.	Time of Growth Cessation in *Picea abies* Determined by the Critical Night Length for Bud-Set	214
IV.	Discussion and Conclusions	216
V.	Summary	217
	Acknowledgements	218
	References	218

I. INTRODUCTION

In forest trees the photoperiod exerts an important control over the times of height growth cessation, winter bud formation, leaf abscission and in some instances bud-break, flower initiation and seed germination (Wareing, 1956; Nitsch, 1957; Downs, 1962; Flint, 1974; Nienstaedt, 1974). Best studied are the effects of differing photoperiods on the times of height growth cessation, bud formation and dormancy onset. As is well known for photoperiodic phenomena in general, photoperiodic responses in trees are determined by the length of the night, i.e. the diurnal number of dark hours (Wareing, 1950; Olson and Nienstaedt, 1957; Nitsch, 1962; Dormling et al., 1968). The phytochrome pigment, with its red and far-red absorbing forms, induces the response (Downs, 1962).

Temperate tree species have become genetically differentiated into populations that are adapted to the local climatic conditions at different latitudes and altitudes (Cieslar, 1895; Engler, 1908). Latitude and altitude provenances of *Populus, Picea, Pinus, Larix* and *Betula* display distinct clinal patterns of variation in photoperiodic responses such as terminal bud formation, young trees from northerly or high altitude regions being adapted to stop height growth at longer daylengths than plants from southerly or lowland regions (Sylvén, 1940; Langlet, 1943; Johnsson, 1951; Pauley and Perry, 1954; Vaartaja, 1959; Holzer, 1966; Dormling *et al.*, 1968; Morgenstern, 1969; Håbjørg, 1972; Magnesen, 1972; Dormling, 1973; Heide, 1974; Holzer and Nather, 1974). Height growth cessation in response to declining photoperiods is evidently an indirect adaptation to the environment, which allows trees to enter the resting stage before the occurrence of the first autumn frost.

Forest tree breeding programmes might be greatly aided by information on the mode of the inheritance of characters which determine the annual growth rhythm. The inheritance of the photoperiodic response of growth cessation can be studied most directly by determining the critical night length for bud-set (CN) in progenies of diallel crosses between trees displaying differences in this trait. Information can also be gained from the results of inter- and intraprovenance crosses involving provenances from different latitudes and/or altitudes tested at one locality or several localities with different natural photoperiodic regimes. Differences in the heights of the parents and progenies will provide acceptable evidence as long as the differences in plant height are attributable primarily to differences in the duration of growth (cf. Eriksson and Gagov, 1976).

This contribution reviews results of the few known investigations on the inheritance of photoperiodic controls over height growth cessation, indicated by (a) height measurements in progeny trials and (b) determinations of the critical night length for bud-set. All latitudes referred to below are in the northern hemisphere.

II. TIME OF GROWTH CESSATION INFERRED FROM HEIGHT MEASUREMENTS

Sylvén (1940), working with *Populus tremula*, was the first to report provenance variation in the effect of photoperiod on the time of height growth cessation. He made crosses between and within Swedish provenances growing at eight different latitudes varying between 56° N and 66° N. The heights of 18 different progenies were recorded at three localities. Plant heights attained after one growing season are shown in Figs 1 and 2 for some selected combinations, and can be largely attributed to differences in shoot elongation during summer and autumn. The most striking apparent photoperiodic effect was observed for the progenies of northern provenance crosses cultivated at the southerly location, Ekebo; they stopped growing in height very early in the year, becoming no more than 5 cm in height, and assuming rosette habits (e.g. Fig. 2, top left). With a few exceptions, plant heights increased with increasing latitude of test locality. At Ekebo (56°00') and Söråker (62°30') the growth of hybrids was generally intermediate between that of their parent provenance half-sib progenies, whereas at Vittjärv (65°50') the hybrids were mostly superior to both their intraprovenance half-sibs. The hybrids of two crosses between extremes (65°50' x 55°54'; 62°30' x 58°42') were twice as tall as the means of their respective intraprovenance half-sibs (Figs 1 and 2).

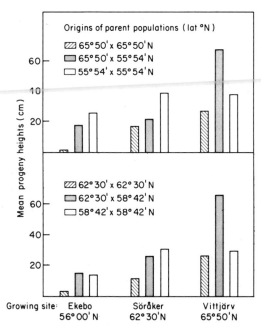

FIG. 1 Mean heights per plant, after one growing season, of *Populus tremula* progenies obtained from crosses within and between latitude populations, when tested at three sites. (Derived from data of Sylvén, 1940)

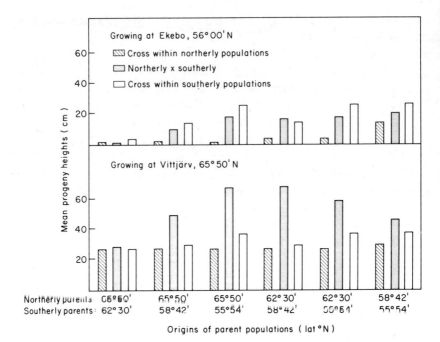

FIG. 2 Mean heights per plant, after one growing season, of *Populus tremula* progenies obtained from crosses within and between latitude populations, when tested at two sites. (Derived from data of Sylvén, 1940)

Northern intraprovenance progenies at the northern site, Vittjärv, stopped growing in height at the end of August and beginning of September, whereas southern and central Swedish progenies continued to grow until later into the autumn. Sylvén (1940) observed that these progenies were damaged by autumn frosts. His data illustrate the existence of hybrid vigour for height growth resulting from crosses between parents with northerly and southerly origins growing at Vittjärv (Fig. 2). It would be of interest to determine whether the superior growth attained by these hybrids under northern, long-day conditions was due to (a) faster growth rates than their southern half-sibs, or (b) greater frost resistance, which enabled them to continue to grow after the first autumn frost.

Johnsson (1956), also working with *Populus tremula* in Sweden, observed that F_1 hybrids between different latitude populations were intermediate between their respective intraprovenance half-sibs (Fig. 3). Grehn (1952) obtained similar results from crosses between

11. Inheritance of the Photoperiodic Response

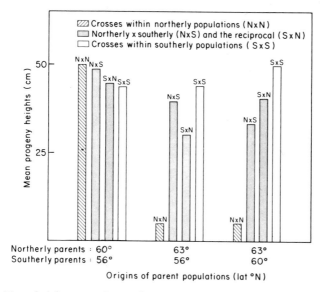

FIG. 3 Mean heights per plant, after one growing season, of *Populus tremula* progenies obtained from crosses within and between different Swedish latitude populations when grown at Ekebo (56°). (Derived from data of Johnsson, 1956)

Populus tremula populations from latitudes 54° and 64°, but, in contrast to Johnsson's results the hybrids grown at latitude 54° were all rosette plants.

Data presented by Hoffman (1953) on the heights of progenies resulting from crosses between one mother tree of *Populus tremuloides* at Boston, USA (latitude 42°) and different father trees of *Populus tremula* from latitudes 63°, 58° and 49°, showed a strong influence of the pollen parent origin on progeny height. At latitude 42° the progeny with the northern-most pollen parent stopped growing first.

By making interprovenance crosses of *Populus trichocarpa* involving trees from latitudes 58° and 47°, Pauley and Perry (1954) found that the date of height growth cessation of F_1 plants was, again, intermediate between the dates of their parents, when grown at latitude 42° (Fig. 4). Under artificial long-day conditions, corresponding to those at latitude 58°20′, growth stopped about one month later.

Following crosses between different latitude populations of *Betula pubescens* growing in Sweden, Johnsson (1951) reported a strong

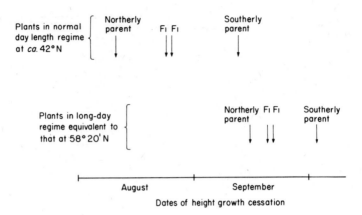

FIG. 4 The dates when height growth ceased in *Populus trichocarpa* F_1 progenies derived from northerly (58°30′) and southerly (46°50′) parent clones, recorded in normal and artificial long-day regimes at latitude 42°. (Derived from data of Pauley and Perry, 1954)

influence of latitude of origin on the growth of F_1 hybrids. At 7–8 years of age the volumes of the hybrids, grown at Ekebo (56°), were least for progenies resulting from crosses between northern populations and greatest for progenies from crosses between southerly populations (Table I). The northern-parent progenies stopped growing early in the year in response to the relatively long nights at Ekebo.

The results of a half-diallel cross involving two plus trees from

TABLE I. Mean stem volumes of 7–8-year-old hybrids obtained from crosses between different Swedish latitude populations of *Betula pubescens*. The hybrids were grown at Ekebo (56°). (After Johnsson, 1951)

Latitude of parent population origins (° N)	Volumes of hybrids after 7–8 years (dm^3 tree^{-1})
64° × 61°	0.44
64° × 60°	0.96
64° × 56°	2.05
60° × 56°	2.77
56° × 61°	2.80

11. Inheritance of the Photoperiodic Response

southern Sweden (E 4008 and E 4015 from c. 58°) and two plus trees from northern Sweden (BD 4016 and BD 4018 from c. 66°) serve as a possible example of differing photoperiodic responses in *Pinus sylvestris* (Ehrenberg, 1973). The progenies were tested at the Bogesund trial area, at c. 59°. Following two growing seasons the F_1 progenies were intermediate in height between their intraprovenance half-sibs (Table II). This agrees with Nilsson's (1969) finding that hybrids between Scots pine provenances from southern and northern Sweden are not usually superior in height to intraprovenance progenies. Also, Hagner (1966) suggested that such hybrids were intermediate in their degrees of lignification in autumn, which implies that they were intermediate in their times of height growth cessation.

Finally, Nilsson (1973) working with *Picea abies*, reported intermediate height growth for young hybrids resulting from crosses between provenances of central European and Swedish origins. Furthermore, Nilsson and Andersson (1969) observed that, as a rule, the hybrids were intermediate between their parent populations in time of bud-set in autumn, when grown at the latitude of the Swedish provenances. As expected, the earliest bud-set was observed on the Swedish progenies.

TABLE II. Heights, after two growing seasons, of hybrids between two northern (BD 4016 and BD 4018) and two southern (E 4008 and E 4015) plus trees of *Pinus sylvestris*, and their four F_1 hybrids. The plants were cultivated in a plastic greenhouse at Bogesund, 59°N. (After Ehrenberg, 1973)

Parentage of progenies		Mean progeny heights (cm)
Plus tree identity numbers	Latitudes of origin	
BD 4016 x BD 4018 (northern intraprovenance cross)	66°25' x 66°28'	15.4
BD 4018 x E 4015	66°28' x 58°07'	17.3
BD 4016 x E 4015	66°25' x 58°07'	18.5
E 4008 x BD 4018	58°07' x 66°28'	16.4
E 4008 x BD 4016	58°07' x 66°25'	18.8
E 4008 x E 4015 (southern intraprovenance cross)	58°07' x 58°07'	21.1

III. TIME OF GROWTH CESSATION IN *PICEA ABIES* DETERMINED BY THE CRITICAL NIGHT LENGTH FOR BUD-SET

The critical night length for bud-set (CN) has been defined as the number of dark hours inducing apical bud formation on 50% of the plants (Dormling *et al.*, 1974). The CN is determined graphically from diagrams of the type shown in Figs 5 and 6.

Clones of *Picea abies* of Swedish and French origin, known to exhibit short and long CN values, respectively, were crossed in all combinations (Swedish x Swedish, French x Swedish and French x French). Seed of six sets of progeny of each of the three types of cross were germinated and grown for 11 weeks at 20°C in continuous light. Thereafter the progenies were given various night lengths for 6

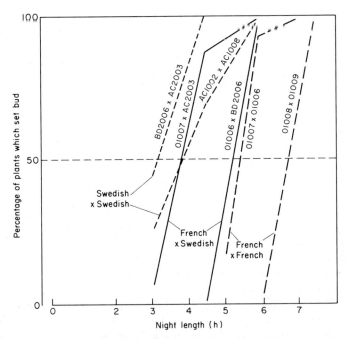

FIG. 5 Effects of 6 weeks in differing night-length regimes on the percentage bud-set among seedling progenies of *Picea abies*, resulting from crosses between and within French and Swedish provenances. Only the two most extreme of six progenies are shown for each type of cross. The critical night length for bud-set (CN) is determined from these curves, as the value giving 50% bud-set.

weeks. The following night lengths were used: 0.0, 2.0, 3.0, 4.0, 4.5, 5.0, 6.0, 7.0, 7.5 and 8.0 h. The percentage of plants which set terminal buds was recorded after 6 weeks' treatment (week 17) (see also Dormling *et al.*, 1974).

The effects of the differing night lengths on the percentage of seedlings which set terminal buds after 6 weeks, are shown in Fig. 5 for the two extreme progenies from each type of cross. Considerable variation occurred among progenies within each type of cross, even to the extent that some cross categories partially overlapped. However, the critical night length for the hybrid progenies of French x Swedish clones were clearly intermediate between progenies resulting from the intraprovenance crosses (Fig. 5).

Six of the 18 progenies tested represented a half-diallel crossing scheme (Fig. 6). Again, the variation in CN values between individual progenies was large, and only the mean value of the hybrids, French x Swedish, was intermediate between the parental combi-

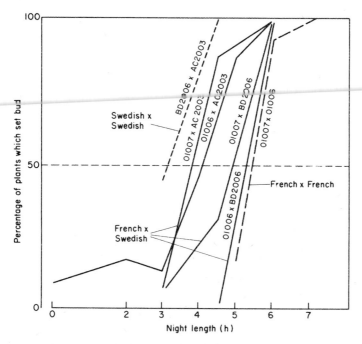

FIG. 6   Effects of 6 weeks in differing night-length regimes on the percentage bud-set among two intraprovenance seedling progenies of *Picea abies* and their hybrids (French x Swedish) in a small diallel cross.

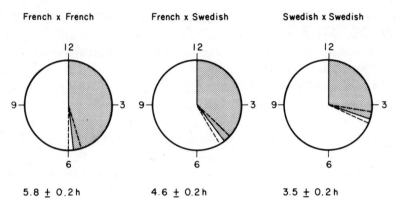

FIG. 7 The critical night length for bud-set (CN)* obtained by pooling data obtained for three types of progeny from intra- and interprovenance crosses of *Picea abies*. The standard errors observed with 6 tested groups of progeny are indicated by stippled lines.
* See Fig. 5.

nations. This study revealed the possibility that forest tree breeders may select hybrids with suitable photoperiodic characteristics.

The mean CN values for the three different types of progeny are given in Fig. 7. Analysis of variance revealed that the CN values of the hybrids were significantly different from those of the parental combinations ($P < 0.01$).

IV. DISCUSSION AND CONCLUSIONS

Data reported in this paper show that the critical night length that stops height growth and induces bud-set on hybrids between provenances adapted to widely different photoperiodic regimes have been either intermediate between that of their intraprovenance half-sibs or close to one of the half-sibs. However, it should be noted that the hybrids of *Populus tremula* growing at Vittjärv (65° 50′) involving a southern parent did not allow specific conclusions to be drawn about their photoperiodic responses (Figs 1 and 2). The data on hybrids, and the well documented clinal variation of the photoperiodic response, suggest a multiple factor inheritance (cf. Pauley and Perry, 1954). Furthermore, the phytotron study on the mode of inheritance of critical night-length values for bud-set revealed great variation in times of height growth cessation and bud-set between individual plants within full-sib families (Eriksson and Gagov, 1976) which, again, may imply that CN values are under multifactorial genetic control. More information would be gained if hybrids were compared with their individual parents instead of their intraprovenance half-sibs. This may be accomplished by comparing parental clones with clones taken from their hybrid progenies. Such comparisons are currently being made by the authors. Nonetheless, F_2 segregations will have to be analysed in order to obtain conclusive evidence on the numbers and types of genes that determine responses to differing photoperiodic regimes in trees.

Whaley (1965) distinguished two types of genes that control the photoperiodic responses for flower initiation and growth characters in higher plants: (1) genes which determine the presence or absence of a photoperiodic requirement; and (2) genes which determine the critical length of the dark period required. Three examples of the first type of genes are the recessive Maryland mammoth mutations, causing a short-day requirement for flowering of *Nicotiana tabacum* (Lang, 1948), the recessive *matura-a* mutations in barley, which remove the long-day requirement for heading in certain barley varieties (Dormling *et al.*, 1966; Gustafsson and Dormling, 1972), and the dominant allele that confers a long-day requirement for flowering in *Lathyrus odoratus* (Little and Kantor, 1941). The second type of genes was found in early, intermediate, late and ultra-late phenotypes obtained from crosses between certain strains of a *Sorghum* variety (Quinby and Karper, 1945). In this variety three independent maturity genes were shown to influence the critical night length for floral initiation.

Genetic adaptation to photoperiod is of great significance for growth in temperate forest trees. Further investigations on the mode of inheritance of the photoperiodic response are needed for the development of efficient breeding programmes.

V. SUMMARY

Tree provenances vary in time of height growth cessation in response to increasing night lengths. The inheritance of this photoperiodic response was reviewed for (a) ten progeny trials where heights have been recorded for progenies resulting from crosses between photoperiodic ecotypes (provenances from different latitudes) and (b) controlled environment studies on the critical night-length for bud-set of *Picea abies* inter- and intraprovenance progenies. In most cases the responses of hybrids have been intermediate between those of parents or intraprovenance half-sibs, but individual progenies with responses close to either intraprovenance half-sibs have been obtained. These observations and the well documented clinal variation of photoperiodic ecotypes suggest a multiple factor inheritance for photoperiodic responses.

ACKNOWLEDGEMENTS

The phytotron study on *Picea abies* was supported by a grant from the Knut and Alice Wallenberg Foundation, which is gratefully acknowledged. We are greatly indebted to Mr Kjell Lännerholm and the personnel of the phytotron for excellent technical assistance.

REFERENCES

Cieslar, A. (1895). Ueber die Erblichkeit des Zuwachsvermögens bei den Waldbäumen. *Zentbl. ges. Forstw.* **21**, 7–29.

Dormling, I. (1973). Photoperiodic control of growth and growth cessation in Norway spruce seedlings. *IUFRO Division 2, Working Party 2.01.4 Growth Processes Symposium on Dormancy in Trees*, Kórnik, September 5–9, 1973, 16 pp.

Dormling, I., Gustafsson, Å., Jung, H. R. and Wettstein, D. von. (1966). Phytotron cultivation of Svalöf's Bonus barley and its mutant Svalöf's Mari. *Hereditas* **56**, 221–237.

Dormling, I., Gustafsson, Å. and Wettstein, D. von. (1968). The experimental control of the life cycle in *Picea abies* (L.) Karst. I. Some basic experiments on the vegetative cycle. *Silvae Genet.* **17**, 44–64.

Dormling, I., Ekberg, I., Eriksson, G. and Wettstein, D. von. (1974). The inheritance of the critical night length for budset in *Picea abies* (L.) Karst. *Proceedings, Joint IUFRO Meeting, S.02.04.1–3* Stockholm 1974, Session VI, pp. 439–448.

11. Inheritance of the Photoperiodic Response

Downs, R. J. (1962). Photocontrol of growth and dormancy in woody plants. *In* "Tree Growth" (T. T. Kozlowski, ed.), pp. 133—148. Ronald Press, New York.

Ehrenberg, C. (1973). The effect of increased heterozygosity in pine plants (*Pinus sylvestris* L.). *IUFRO Division 2 Working Party S 2.03.05 Symp. on Genet. of Scots Pine*, Warsaw 8.X.1973 — Kórnik 18.X.1973. 15 pp.

Engler, A. (1908). Tatsachen, Hypothesen und Irrtümer auf dem Gebiete der Samenprovenienzfrage. *Forstwiss. Zent. bl.* 30.

Eriksson, G. and Gagov, V. (1976). Growth differences within full-sib families of *Picea abies* (L.) Karst. *Stud. for. suec.* 130, 1—25.

Flint, H. L. (1974). Phenology and genecology of woody plants. *In* "Phenology and Seasonality Modeling" (H. Lieth, ed.), pp. 83—97. Springer Verlag, Berlin, Heidelberg and New York.

Grehn, J. (1952). Über Spaltungserscheinungen und photoperiodische Einflüsse bei Kreuzungen innerhalb der Sektion *Populus* Leuce Duby. *Z. Forstgenet. Forstpfl. zücht.* 1, 61—69.

Gustafsson, Å. and Dormling, I. (1972). Dominance and overdominance in phytotron analysis of monohybrid barley. *Hereditas* 70, 185—216.

Hagner, M. (1966). [Methods for evaluation of hardiness of pine from Sweden and a study of the inheritance of hardiness]. M.Sc. thesis (in Swedish), Royal College of Forestry, Stockholm.

Heide, O. M. (1974). Growth and Dormancy in Norway Spruce Ecotypes (*Picea abies*). I. Interaction of Photoperiod and Temperature. *Physiol. Plant.* 30, 1—12.

Hoffmann, D. (1953). Die Rolle des Photoperiodismus in der Forstpflanzenzüchtung. *Z. Forstgenet. Forstpfl. zücht.* 2, 44—47.

Holzer, K. (1966). Die Vererbung von physiologischen und morphologischen Eigenschaften bei der Fichte. I. Sämlungsuntersuchungen. *Mitt. forstl. Bund. vers. anst. Mariabrunn* 71. Wien.

Holzer, K. and Nather, J. (1974). Die Identifizierung von forstlichem Vermehrungsgut. *In* "100 Jahre Forstliche Bundesversuchsanstalt". Pp. 13—42. Forstliche Bundesversuchsanstalt, Wien.

Håbjørg, A. (1972). Effects of photoperiod and temperature on growth and development of three latitudinal and three altitudinal populations of *Betula pubescens* Ehrh. *Meld. Norg. Landbr. høgsk.* 51, 1—27.

Johnsson, H. (1951). Avkomneprövning av björk — preliminära resultat från unga försöksplanteringar. *Svensk Papp. tidn.* 11—12, 379—393, 412—426.

Johnsson, H. (1956). Heterosiserscheinungen bei Hybriden zwischen Breitengradrassen von *Populus tremula*. *Z. Forstgenet. Forstpfl. zücht.* 5, 156—160.

Lang, A. (1948). Beiträge zur Genetik des Photoperiodismus. I. Faktorenanalyse des Kurztagcharakters von *Nicotiana tabacum* "Maryland—Mammut". *In*

"Vernalization and Photoperiodism — A symposium" (F. Verdoorn, ed.) pp. 175–189. Chronica Botanica, Waltham, Massachussetts.

Langlet, O. (1943). Photoperiodismus und Provenienz bei der gemeinen Kiefer (*Pinus silvestris* L.). *Medd. St. Skogsförs. anst.* 33, 295–330.

Little, T. M. and Kantor, J. H. (1941). Inheritance of earliness of flowering in the sweet pea. *J. Hered.* 32, 379–383.

Magnesen, S. (1972). Ecological Experiments Regarding Growth Termination in Seedlings of Norway Spruce. 3. Effect of daylength. Supplementary experiments with 53 seed lots. *Meddr. Vestland. forstL Forsstn.* 52, 273–317.

Morgenstern, E. K. (1969). Genetic Variation in Seedlings of *Picea mariana* (Mill.) BSP. I. Correlation with Ecological Factors. *Silvae Genet.* 18, 151–161.

Nienstaedt, H. (1974). Genetic variation in some phenological characteristics of forest trees. In "Phenology and Seasonality Modeling" (H. Lieth, ed.), pp. 389–400. Springer Verlag, Berlin, Heidelberg and New York.

Nilsson, B. (1969). Försök med provenienshybrider av gran (*Picea abies* (L.) Karst.). *Royal College of Forestry, Department of Forest Genetics, Research Notes* 5, 1–10.

Nilsson, B. (1973). Recent results of inter-provenance crosses in Sweden and the implications for breeding. *Royal College of Forestry, Department of Forest Genetics, Research Notes* 12, 1–7.

Nilsson, B. and Andersson, E. (1969). Spruce and pine racial hybrid variations in Northern Europe. *Royal College of Forestry, Department of Forest Genetics, Research Notes* 6, 1–10.

Nitsch, J. P. (1957). Photoperiodism in woody plants. *Proc. Am. Soc. hort. Sci.* 70, 526–544.

Nitsch, J. P. (1962). Photoperiodic regulation of growth in woody plants. In International Horticultural Congress 15 (Nice, 1958) Proc. "Advances in Horticultural Science and their Applications" (J. C. Garnaud, ed.), Vol. III, pp. 14–22. Pergamon Press, Oxford.

Olson, J. S. and Nienstaedt, H. (1957). Photoperiod and Chilling Control Growth of Hemlock. *Science* 125, 492–494.

Pauley, S. S. and Perry, T. O. (1954). Ecotypic variation of the photoperiodic response in *Populus*. *J. Arnold Arbor.* 35, 167–188.

Quinby, J. R. and Karper, R. E. (1945). The inheritance of three genes that influence time of floral initiation and maturity date in milo. *J. Am. Soc. Agron.* 37, 916–936.

Sylvén, N. (1940). Longday and shortday types of Swedish forest trees. *Svensk Papp. tidn.* Årg. 43. *Medd. Fören. Växtförädling Skogsträd.* (English summary), pp. 351–354.

Vaartaja, O. (1959). Evidence of photoperiodic ecotypes in trees. *Ecol. Monogr.* 29, 91–111.

Wareing, P. F. (1950). Growth studies in woody species. I. Photoperiodism in first-year seedlings of *Pinus silvestris*. *Physiol. Plant.* 3, 258–276.

Wareing, P. F. (1956). Photoperiodism in woody plants. *A. Rev. Pl. Physiol.* 7, 191–214.

Whaley, W. G. (1965). The interaction of genotype and environment in plant development. *In* "Handbuch der Pflanzenphysiologie" Band XV Differenzierung und Entwicklung Teil 1 (W. Ruhland, ed.), pp. 74–99. Springer-Verlag, Berlin, Heidelberg and New York.

12
Patterns of Shoot Development in *Pinus* and Their Relationship to Growth Potential

RONALD M. LANNER
Department of Forestry and Outdoor Recreation, Utah State University, Logan, U.S.A.

I.	Introduction	223
II.	Diversity of Developmental Patterns in Pine Shoots	224
	A. Development of the Shoot Axis	224
	B. Development of Lateral Structures	230
III.	The Developmental Pattern as an Adaptive Strategy	233
IV.	The Developmental Pattern as a Growth-Limiting Factor	236
V.	Summary	240
Acknowledgements		241
References		241

I. INTRODUCTION

The growth of a pine shoot is often represented as the springtime elongation of a bud laid down the preceding summer. This is a gross oversimplification, as we should intuitively expect. *Pinus* is, after all, a genus of about a hundred species distributed from the tropics to the Arctic, and from sea level to the peaks of high mountains. A developmental pattern that is successful in one species may not work well in another species growing elsewhere.

The details of pine shoot development have recently become of more importance to foresters, for three reasons. First, recent research has shown that in north temperate pine species, the potential length of a spring shoot is already determined the preceding summer by the number of stem units* laid down in the winter bud. Therefore, it becomes necessary to know when the critical stem units are being formed, in order to be able to manipulate the bud-formation process for greater yields.

Secondly, in order to measure the influence of environmental factors on height growth, those factors must be considered during the period that is critical for bud formation. Developmental studies of individual species are needed if this period is to be identified.

Finally, exotic pines are being planted on a large scale in many tropical and subtropical areas. An understanding of a species' developmental pattern can aid in interpreting its behaviour in a new area, and in predicting its behaviour where it has not yet been planted (Lanner, 1972).

This contribution will attempt to systematize the diverse patterns of shoot development found in pines — or at least those well enough described to demand recognition — and it will examine those patterns as adaptive strategies and as processes setting limits on potential growth.

II.   DIVERSITY OF DEVELOPMENTAL PATTERNS IN PINE SHOOTS

The diversity of developmental patterns is clearly seen in two areas of shoot morphogenesis. The first concerns the time of development of the shoot axis; the second is the timing of the development of structures arranged on the axis.

A.   Development of the Shoot Axis

All shoot axes are formed by the activity of apical meristems, but in *Pinus* there is unusual flexibility in the time relations of initiation and elongation. In this section a classification scheme is presented in

* After Doak (1935), a stem unit is "an internode, together with the node and nodal appendages at its distal extremity". It is, therefore, a portion of the shoot bearing a primary scale and its axillary structure, if any. Stem unit number can be arrived at by counting the fascicles, lateral buds, and sterile scales on a length of stem.

the hope that it will differentiate between currently known developmental patterns and others yet to be described (Table I). The classification scheme should be regarded as tentative and subject to change as new information comes to light.

The basis of this scheme is the degree to which the annual shoot is formed by the processes of "fixed" growth on the one hand, and "free" growth on the other. By *annual shoot* is meant that part of a shoot that is added during an annual cycle of growth in temperate or seasonal areas, or during a calendar year in non-seasonal tropical areas. By *fixed growth* is meant the elongation of pre-formed (predetermined) stem units after a rest period, typified by the elongation of a winter bud to form a spring shoot. By *free growth* is meant the elongation of a shoot due to the simultaneous initiation and elongation of new stem units (see Pollard and Logan, Chapter 13). It is typified by the growth of a seedling hypocotyl during its first growing season, but also includes the growth of temporary non-resting buds of summer shoots, because, at least in slash pine (*P. elliottii*), the apices of such buds continue to initiate new stem units concurrently with elongation at the bud base.

The fixed and free growth modes resemble the growth "*modalités*" of Debazac (1968), but they allow for more flexibility in classification because they can be combined in various ways. Furthermore, bud or shoot complexity is not a factor in this initial classification, thus allowing the categories to be widely embracing. The developmental patterns discussed here can be viewed as being along a continuum, the extremes of which represent all growth by free growth, and all growth by fixed growth.

It is not thought that a separate mode or pattern can be specified for *lammas growth*. In the absence of morphogenetic studies to the contrary, it must be assumed that lammas growth is merely the early elongation of stem units that would otherwise form part or all of a resting bud, with or without the concomitant formation of new stem units at the apex. Thus, it represents a "borrowing" of part of the spring shoot, to use Rudolph's (1964) term. This is accomplished by elongation in the basal part of the winter bud. While usually only a minor part of the winter bud elongates, up to 15 cm of such growth has been reported in *P. banksiana* (Rudolph, 1964), and over 32 cm in *P. attenuata* (Lanner, 1963). In the latter case lammas growth was

TABLE I. Developmental modes and patterns of the annual shoot in *Pinus*

Mode I. The annual shoot is elaborated entirely by free growth

  A. the annual shoot originates as an embryonic plumule

    1. there is a single cycle of free growth followed by formation of a resting bud . . . . *Seedling pattern 1*

    2. additional cycles of free growth occur by formation of temporary non-resting buds that elongate while stem unit initiation is progressing . . . . *Seedling pattern 2*

  B. the annual shoot does not originate as a plumule

    1. there is a single cycle of free growth which often continues for several years . . . . *Foxtail pattern*

    2. several cycles of free growth occur by formation of temporary non-resting buds that elongate while stem unit initiation is progressing . . . . *Caribaea pattern*

Mode II. The annual shoot is elaborated by both free and fixed growth

  A. free-growth components (summer shoots) constitute a significant part of the annual shoot, sometimes exceeding the fixed-growth component (spring shoot) in length

    1. the fixed-growth component (spring shoot) is monocyclic . . . . . *Elliottii pattern*

    2. the fixed-growth component (spring shoot) is polycyclic . . . . . *Echinata pattern*

  B. free-growth components (summer shoots) usually constitute only a minor part of the annual shoot . . . . *Piñon pattern*

Mode III. The annual shoot is elaborated entirely by fixed growth

  A. the annual shoot is monocyclic . . . . *Resinosa pattern*

  B. the annual shoot is polycyclic

    1. all cycles of the shoot are initiated in the bud consecutively . . . . *Contorta pattern*

    2. the final shoot cycle is initiated during the winter, long after earlier cycles . . . . *Pinaster pattern*

12. Patterns of Shoot Development in *Pinus*

so pronounced in some half-sib families as to make their height–growth curves essentially bimodal. As viewed here, lammas growth can occur in any of the patterns of Modes II and III of Table I.

1. *Seedling pattern 1* is the familiar pattern of seedling development. The growth cycle begins with seed germination and elongation of the hypocotyl, followed by generation of an axis bearing primary leaves. Some of the primaries may subtend axillary short shoots or even long shoots during the first year. Late in the growth cycle scales accumulate at the apex to form a resting terminal bud. An example is found in *P. banksiana* (Riding, 1967; cited in Kozlowski, 1971).

2. *Seedling pattern 2* differs from the preceding by the formation and rapidly ensuing elongation of non-resting terminal buds, as in *P. palustris* seedlings (Wakeley, 1954). It is, in effect, seedling pattern 1 with the addition of summer shoots.

3. *Foxtail pattern* is not a normal developmental pattern, but occurs in some genotypes of some species when grown in tropical areas where, for long periods of time, neither temperature nor water stress are growth-limiting. Foxtailing may follow upon seedling growth, without an intervening resting-bud stage, or it may occur after a resting-bud has elongated. It is typified by long branchless leaders covered year-round with needles grading into primordia at the apex. Such growth may continue for several years and produces shoots lacking defined annual rings (Lanner, 1966; Kozlowski and Greathouse, 1970). The pattern is especially common on *P. radiata* and *P. caribaea*, but occurs in several other species as well.

4. *Caribaea pattern* is characterized by an annual cycle that includes the overlapping initiation and elongation of several temporary non-resting buds, without a defined growing season or dormant period. It is exemplified by *P. caribaea* var. *hondurensis*, growing in Puerto Rico, as described by Chudnoff and Geary (1973). They recorded six growth flushes in a 12 month period, each of which was

sigmoid in pattern. The flushes were due to elongation of buds that were forming during the preceding flush and, therefore, result in summer shoots. During the 12 month period there were five periods of relatively little shoot elongation, but the longest of these was only about one month, and the aggregate period was only about 14 weeks. Thus, it is assumed that summer shoots were initiated in a manner similar to those of *P. elliottii*, cited below. Cambial growth was year-round, without a well defined annual ring.

5. *Elliottii pattern*. In this the annual shoot consists of a monocyclic* spring shoot emanating from a winter bud, and one or more summer shoots. The summer shoots form a significant portion of the annual shoot, but the spring shoot is usually the longest single shoot segment. It is exemplified by saplings of *P. elliottii* in Florida up to about 6 or 8 years in age (Lanner, unpublished data). Summer shoots in such trees are formed by the elongation of temporary non-resting buds that elongate while their apices are still initiating new stem units. *P. taiwanensis* and *P. palustris* appear to follow this pattern in youth.

6. *Echinata pattern*. Although no dissection studies have been made to determine the nature of summer bud development in these species, they are assumed to be similar to the elliottii pattern, but with polycyclic spring shoots, e.g. sapling shoots of *P. echinata* and *P. rigida*, as described by Tepper (1963). *P. taeda* appears to follow this pattern during its sapling stage (Griffing and Elam, 1971).

7. *Piñon pattern* is characterized by the addition of a minor summer shoot to the spring shoot. Its morphogenesis has been described by Lanner (1970) for piñon (*P. edulis*), singleleaf piñon (*P. monophylla*) and Parry piñon (*P. quadrifolia*). Presumptive evidence

---

* The terms "monocyclic" and "polycyclic" are used in preference to "uninodal" and "multinodal", for reasons explained elsewhere (Van Den Berg and Lanner, 1971; Lanner and Van Den Berg, 1973).

indicates its occurrence elsewhere in the subgenus *Strobus*: *P. gerardiana, P. armandii, P. cembroides, P. pinceana, P. nelsonii, P. ayachahuite, P. bungeana* and *P. strobus* (Lanner, 1970).

In the piñons it is limited to vigorous shoots, and commonly occurs on those bearing seed cones.

8. *Resinosa pattern* is the one most familiar to foresters of north temperate lands. A monocyclic spring shoot develops from a winter bud, and then a new winter bud forms. It is exemplified by *P. resinosa*, whose bud morphogenesis has recently been investigated by Sucoff (1971), but also characterizes *P. densiflora* (Hanawa, 1966), *P. lambertiana, P. ponderosa* (Sacher, 1954), *P. sylvestris* (Mi'halevskaja, 1963), *P. nigra* and others.

9. *Contorta pattern.* Like the resinosa pattern, all growth is from preformed winter buds, but the spring shoot is polycyclic, e.g. young *P. contorta* var. *latifolia* (Van Den Berg and Lanner, 1971), and probably *P. banksiana* (Curtis and Popham, 1972). The stem units constituting all of the shoot cycles are laid down in the terminal bud in a continuous process.

10. *Pinaster pattern.* In this pattern the stem units of the second shoot cycle are initiated in the winter, long after those of the first cycle. It has been described only in *P. pinaster*, where it is not of universal occurrence (David, 1968; Debazac, 1968).

The above-named patterns are not intended to categorize species throughout their life cycles, or under all conditions of growth. For example, leaders of slash pines fit seedling pattern 2 during the seedling stage; and the elliottii pattern until perhaps 10 years of age. The leaders of older trees have been observed to follow the contorta pattern or the resinosa pattern, contrary to the frequent observation in other species where polycyclism characterizes the most vigorous growth stage. First-order branches of saplings may fit the elliottii pattern or the resinosa pattern. Shoots bearing seed cones often fit the contorta pattern.

Shoots of most soft pines, including the piñons, usually follow the resinosa pattern. *Pinus contorta* in vigorous growth follows the contorta pattern, but as growth rate declines with age, the resinosa pattern becomes common. Morphological details of most species are insufficient to allow their ready classification, but it is apparent from Table I that pine shoot morphogenesis is far more varied than is usually stated.

B. Development of Lateral Structures

This discussion will relate only to the primary scales, short shoots, and needles of the vegetative annual shoots that form after completion of the seedling stage. For simplicity, only specific epithets will be used (see Table II).

Pines show variability in the dates of initiation of lateral structures, the amount of time elapsing between inception and maturity of these structures, and the coordination of initiation with other growth activities of the tree. These will be briefly examined to point out the diversity within the genus *Pinus*.

1. Scales

In all pines so far studied, basal sterile scales are the first lateral structures produced by renewed apical activity after a dormant period. These are followed by scales that will subtend axillary structures, and by terminal sterile scales that will form a membraneous cover around the dormant apical meristem. All but the latter will become scarious external scales of the winter bud or of a temporary non-resting bud shortly after they are formed. The terminal sterile scales will become scarious in the next growth cycle. Terminal sterile scales of year n are continuous with basal sterile scales of $n + 1$, and cannot be distinguished from them on the mature shoot. Additional short series of sterile scales have recently been reported at intermediate positions in monocyclic shoots of *P. contorta* var. *contorta* (Owens and Molder, 1973). This is a novel finding that should be further studied.

Scale initiation is reported to begin just prior to shoot elongation (*sylvestris, banksiana, pinaster*) or concurrently with it (*strobus, lambertiana, resinosa, densiflora, contorta* var. *contorta*).

TABLE II. Timing of morphogenetic events in the development of the normal annual shoot of several species of pines (see footnotes 1, 2, 3 and 4)

Species	Shoot type	Earliest Initiation of Scales	Short shoots	Needles	Period of Elongation	Reference
elliottii	spring shoot	—	July, n	Oct., n	March–May, $n + 1$	Lanner, unpublished
	summer shoot	—	April, $n + 1$	April, $n + 1$	April–July, $n + 1$	Lanner, unpublished
edulis	spring shoot	—	July, n	May, $n + 1$	May–June, $n + 1$	Lanner, 1970
	summer shoot	—	May, $n + 1$	June, $n + 1$	June, $n + 1$	Lanner, 1970
strobus	spring shoot	May, n	July, n	Sept., n	May–July, $n + 1$	Owston, 1969
	summer shoot	—	May (?), $n + 1$	May (?), $n + 1$	June, $n + 1$	Owston, 1968
lambertiana[5]	spring shoot	April, n	May, n	August (?), n	April–May, $n + 1$	Sacher, 1954, 1955
resinosa	spring shoot	May, n	June, n	spring, $n + 1$	April–July, $n + 1$	Sucoff, 1971
densiflora	spring shoot	April, n	July, n	fall, n	April–June, $n + 1$	Hanawa, 1967
sylvestris	spring shoot	April, n	July, n	August, n	May–June, $n + 1$	Mi'halevskaja, 1963
banksiana	spring shoot	April, n	June, n	May, $n + 1$	May–June, $n + 1$	Curtis and Popham, 1972
contorta var. *latifolia*	spring shoot	—	May, n[6] July, n[7] August, n[8]	August, n[6] Oct., n[7] Nov., n[8]	May–June, $n + 1$	Van Den Berg and Lanner, 1971
contorta var. *contorta*	spring shoot	March, n	May, n	August, n	March–Aug, $n + 1$	Owens and Molder, 1973
pinaster	spring shoot	April, n[6] Dec. (?), n[7]	July, n[6] Dec., n[7]	Jan., $n + 1$[6] Feb., $n + 1$[7]	May–July, $n + 1$ March–Aug., $n + 1$	David, 1968; Debazac. 1968
pumila	spring shoot	June, n	July, n	Sept., n	June–Aug., $n + 1$	Hanawa, 1972

1 the annual shoot consists of a spring shoot with or without summer shoots
2 "n" refers to the year of initiation of a winter bud that will form the spring shoot
3 most studies cited were on leaders or first-order branches of vigorous young trees, and may not apply in detail to all growth stages or experimental conditions
4 uncertain data indicated by (?)
5 growing outside its natural range
6 on first cycle of polycyclic shoot
7 on second cycle of polycyclic shoot
8 on third cycle of polycyclic shoot

2. Short Shoots

The newly initiated short-shoot primordium can be identified visually by the appearance of a meristem axillary to a scale. In all cases of spring-shoot development studied, such meristems are formed in year n, the year of bud initiation (Table II). In summer shoots they are of necessity formed in year $n + 1$. Therefore, the developmental cycle of a short shoot on a summer shoot is much shorter than that of a short shoot formed on a spring shoot.

When no summer shoots are formed, short-shoot primordia in newly forming winter buds may be initiated while elongation of the current spring shoot is still progressing (*resinosa* [according to Sucoff, 1971], *contorta* var. *latifolia* and var. *contorta*); or upon completion of elongation (*strobus, lambertiana, ponderosa* [Sacher, 1954], *resinosa* [according to Duff and Nolan, 1958]).

When a summer shoot does form, the initiation of short-shoot primordia which will be present on the spring shoot is necessarily delayed until after the period of current spring-shoot elongation (*edulis, elliottii*). When polycyclic winter buds are formed, short-shoot primordia appear first in the lower cycle, then at upper cycles (*contorta* var. *latifolia, pinaster*): thus the "development gradient" described in *P. contorta* (Van Den Berg and Lanner, 1971).

The presence of needle primordia is clear evidence that all the structures of the short shoot have been laid down; but incomplete short shoots, i.e. those without their full complement of scales, can be identified by scale counts in dissected material. Such data have shown that incomplete short shoots overwinter in terminal buds not only in piñon pines, which are monocyclic (Lanner, 1970), but in the upper parts of polycyclic *contorta* buds as well (Van Den Berg and Lanner, 1971). Unpublished data show the same to be true in *elliottii* saplings.

3. Needles

The appearance of needle primordia is extremely variable. The simplest case is their inception in all the short shoots of the terminal bud during the year of bud initiation, n. Some species in which needle primordia are present in the winter bud are *strobus* (Owston, 1969); *lambertiana, ponderosa* (Sacher, 1954); *sylvestris, sibirica*

(Minina and Piskunova, 1963); *nigra* and *taeda* (Doak, 1935); and *mugo, cembra,* and *flexilis* (Van Den Berg and Lanner, 1971).

In another group of species some needle primordia are present in winter buds, but others are initiated in winter or spring of year $n + 1$. These include the polycyclic *contorta* var. *latifolia* in which lower cycle needles are autumnal while upper cycle needles tend to be vernal, and *densiflora*, a monocyclic species. Hanawa and Sasaki (1970) originally reported that in the 5-needled species *pumila*, two needle primordia are formed in the fall of year n, and the other three in spring of year $n + 1$; but later Hanawa (1972) retracted this statement, and reported that all the needle primordia are present in September. An unusual situation obtains in *pinaster*, in which needle primordia form in the winter.

Needle initiation is vernal in spring shoots of several pines, requiring a very rapid growth cycle between inception and elongation, similar to that of needles on summer shoots. These include *resinosa* according to both Duff and Nolan (1958) and Sucoff (1971), though the present author has observed needle primordia in winter buds of Minnesota samples. Other species with vernal needles are *edulis* and other piñon species, and *banksiana*.

Needles of summer shoots grow rapidly from newly initiated primordia, at least in most species. A peculiar exception is *P. clausa*. In northern Florida this pine was observed overwintering with summer shoots that were bare of needles from September until March. The needles finally emerged in mid-March, concurrently with those of the elongating spring shoot. It is not known when the primordia of these late-maturing needles were initiated.

The needles of lammas shoots must also undergo a rapid cycle of inception and growth if they are to appear in the summer or fall of n, though they often fail to attain full size (Tepper, 1963; Rudolph, 1964). In *P. attenuata*, needles of lammas shoots may not elongate until spring of $n + 1$ (Lanner, 1963), but the date of needle inception in this species is not known.

III. THE DEVELOPMENTAL PATTERN AS AN ADAPTIVE STRATEGY

Pines are pioneers or seral species with a limited capacity to persist in understories. They bear abundant seed only when they have outpaced competing vegetation, and their seed production is positively correlated with vigorous crown development. Therefore, if a pine is to succeed it must form long annual shoots, at least in youth.

But environments place constraints on growth, especially by imposing climatic barriers. Frosts and droughts are the major barriers requiring plants to compromise the ideal for the feasible. In colonizing areas where cold or dry conditions are encountered, pines have evolved diverse strategies of shoot development.

Mild conditions favour free growth and the absence of a dormancy structure; harsh conditions do not. Thus, free growth is replaced by fixed growth in proportion to the severity of the environmental conditions to which a species is exposed.

Seedling growth is necessarily free growth, and the longer the favourable season, the longer it can safely continue. *P. palustris* seedlings, for example, may grow as long as 200 days from germination to winter bud set, a growth period that permits the flushing of one or more summer shoots (Wakeley, 1954).

Foxtailing is expressed only under conditions of continuously high moisture availability and mild temperatures, features of a very permissive environment. The free growth of foxtails is expressed only by some genotypes, and even in these, internal stresses seem to develop that eventually cause a lapse into the normal growth habit. The most interesting thing about foxtails is that they illustrate the capacity for free growth in species whose natural areas of distribution are too restrictive to permit it, i.e. summer drought in the range of *P. radiata*. The foxtail habit cannot be regarded as adaptive in that species because it is not part of the normal phenotypic expression. Actually it results in a relatively unstable tree with fewer branches on which to bear cones; and its growth rate is no greater than that of normally branched trees (Lanner, 1966).

The caribaea pattern features frequent flushes of free growth instead of one flush of indefinite duration. Normal branching occurs, so a stable structure is formed. This pattern permits advantage to be taken of favourable conditions at any time of the year, but can only succeed in the absence of frost and prolonged drought.

The elliottii pattern combines the opportunism of the free-growth summer shoot habit with the security of the fixed-growth spring shoot habit. This permits maximum advantage to be taken of an environment with long growing seasons that are separated by short, sometimes quite frosty, winters. The length of an annual shoot is determined almost equally by prior-year and current-year conditions, thus buffering the tree against the effects of an unfavourable year. Further, the number of summer shoots on an annual shoot can vary from one to six or more, providing a wide range of response to environmental conditions. It is notable that in *P. elliottii* the proportion of the annual shoot contributed by summer growth declines with increasing latitude of seed source (Bengtson et al., 1967). The elliottii habit permits pines to grow very rapidly to cone-bearing size. The echinata pattern is a variant of the elliottii.

It is difficult to view the piñon pattern as adaptive, especially in pines such as *P. strobus* where it occurs only occasionally. It is seldom important as a determinant of shoot length. Perhaps it merely represents an ability that is not selected against because it imposes no serious disadvantage on the bearer. In the case of the piñon pines, it may have once been advantageous in a subtropical climate.

The resinosa and contorta patterns are the most rigid, in keeping with the environments in which their exemplars are found. In cold climates where late or early frosts can pose a great threat, free growth cannot be risked (except in seedlings where there is no alternative and where the period of exposure is short). Students of northern species have frequently commented on the growth rhythm of trees of this pattern. Kramer (1943), for example, called this one of the most interesting and puzzling problems in tree physiology:

> "Why should most trees resume growth in the spring before the danger of frost is past, then cease growth in midsummer or at least weeks before the first killing frost of autumn? Cessation of growth in midsummer does not have any obvious survival value to the species, but would rather serve to decrease its ability to compete with species having longer growing seasons."

An answer to this question would stress the importance to the species of fully utilizing the time available for winter bud formation, the process that determines the potential length of next year's growth in shoots of the resinosa and contorta patterns. This strategy assures the unfolding of a shoot of maximum length during the rapid spring shoot flush.

IV. THE DEVELOPMENTAL PATTERN AS A GROWTH-LIMITING FACTOR

The focus of this book is on yield. As foresters, we know many ways to increase the length of the average annual shoot, thereby increasing long-term rates of height growth and wood yield. But after choosing the best genotypes, and bringing the various environmental factors to optimum levels, what will finally set limits on potential shoot increment? I will argue here that the ultimate factor imposing limits on annual shoot growth may be the relative inflexibility of the shoot development pattern.

The potential length of an annual shoot component, whether a spring or summer shoot, is largely a function of the number of stem units it contains. Within a class of similar shoots (same species or provenance, age class, crown position) differences in length are due mainly to differences in stem unit number, rather than stem unit length. An example of this effect is shown in Fig. 1. In three provenance-test plantations of *P. sylvestris* (planted in Michigan, Minnesota, and Missouri) there were much greater differences in rates of height-growth than in stem-unit lengths. Results were consistent and statistically significant in all plantations. Thus variation in shoot length was due overwhelmingly to differences in stem unit number, though stem unit length also varied systematically (Fig. 1).

Because *P. sylvestris* follows the resinosa pattern, all its stem units are formed the year before they elongate, so stem unit number reflects conditions of the year prior to elongation.

In species bearing summer shoots, the number of stem units in the summer shoot portion of the annual shoot reflects conditions of the current growth year. In *P. elliottii*, stem units on the first summer shoot are slightly longer than those on spring shoots or subsequent summer shoots (unpublished data), but the range of variation is not great.

Stem unit length can be seriously retarded, as by spring drought or root damage. It does not appear to be easily lengthened much beyond its norm, except perhaps by very intensive fertilizing and watering, measures not often employed under forest conditions. It therefore appears that significant increases in shoot length must be brought about mainly by increasing the number of stem units in the annual shoot.

12. Patterns of Shoot Development in *Pinus*

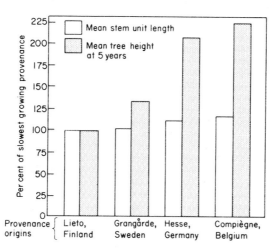

FIG. 1 Relative constancy of stem unit length on shoots of *Pinus sylvestris* provenances with varying growth rates. Data are means of trees growing in three plantations in the mid-western United States.

Stem unit number can be increased in either of two ways: *increasing the rate of stem unit initiation* (i.e. shortening the plastochron, or time interval between initiation of consecutive primordia), or *increasing the length of the initiation period*.

Few detailed plastochron analyses have been made in *Pinus*: the difficulties of measuring the plastochron in a bud that does not have a predictable number of components are indeed formidable. Hanawa (1966) estimated the average plastochron on *P. densiflora* as 6—8 h. Sucoff (1971) has made some estimates for *Pinus resinosa*, a species whose narrow range of genetic variation makes statistical estimates of the plastochron relatively feasible. According to Sucoff, the plastochron varies from more than a day (during the spring period of sterile scale initiation) to a midsummer value of 2½—3 h, but he pointed out that the plastochron varied greatly from week to week. Cannell and Willett (1975) have made estimates for *P. contorta* provenances which are discussed elsewhere in this volume. However, there have apparently been no attempts at influencing the plastochron experimentally, and we cannot yet predict whether shoot growth can be increased by shortening it.

The second alternative is to increase the period during which stem units are initiated. The assumption here is that the process in nature "runs down" because of depletion of some factor that can be supplied by the forester. This is probably what happens when we irrigate, weed or thin a stand — measures that increase the availability of soil water and its solutes. It is possible that the use of fertilizers also lengthens the "initiation season".

Indirect evidence that shoot growth potential is correlated with approximate length of the stem unit initiation period is presented in Fig. 2. Here we can see great differences in the period of initiation in annual shoots of four pine species.

In *P. caribaea* var. *hondurensis* the initiation of new stem units on a single study tree appears to have been year-long, with the annual shoot reaching a length of about 220 cm (Chudnoff and Geary, 1973). This tree was able to grow by forming and elongating summer shoots all year long in a climate not requiring dormancy. In *P. elliottii* the formation of the winter bud took 5 months; and summer shoot activity an additional 5.2 months; thus, the total initiation period was 10.2 months. The average annual shoot of 26 trees studied was 119 cm. In *P. edulis* winter bud formation and summer shoot activity combined totalled less than 3 months with annual shoots averaging about 20 cm in length. The annual shoot of *P. contorta* was about 45 cm long and resulted from an initiation period of about 3.5 months (Van Den Berg and Lanner, 1971).

In the last three species a degree of inaccuracy is introduced by excluding from the initiation period that time during which basal sterile scales of the spring shoot were initiated. In *P. elliottii* and *P. edulis* the initiation period is probably somewhat exaggerated by combining formation and elongation of the summer shoot.

The important question raised by these data is whether the periods of no initiation can be significantly shortened or eliminated without prejudice to survival or normal development. In *P. caribaea* var. *hondurensis* it is clearly impossible to do so, because initiation of stem units is already a year-long activity.

Can the inactive period of *P. elliottii* be significantly shortened? This species appears to have an inactive period of about 2.5 months (Fig. 2), but this is deceptive. Actually, the short shoots within the terminal buds are still developing, and they do not all have their full complements of scales and needles until late March, when rapid shoot elongation is already underway. In 1974 northern Florida experienced an unseasonably warm January (about 7°C above the normal)

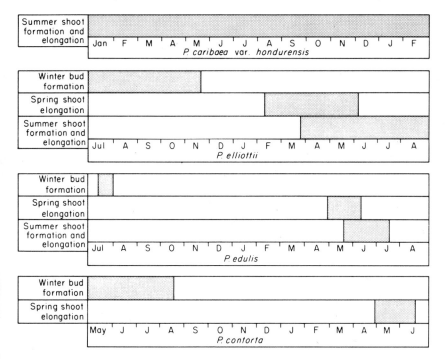

FIG. 2 Contrasting periods of activity in the formation of the annual shoot in several pines. The shaded bars denote periods of morphogenesis and elongation of stem units.

but these trees showed no signs of elongating their terminal buds even after rainstorms. Their failure to do so may have been due to an unfulfilled chilling requirement, or perhaps a necessity to attain a certain level of bud maturity. This commonly observed inability to exploit good growth conditions by prematurely breaking dormancy limits us in altering the normal developmental rhythm. When that rhythm is clearly adaptive, as in northern species that must go dormant early to escape frost, the likelihood of greatly lengthening the initiation season appears dim. Selection for early flushing lines may prove useful, but pine species do not usually show a great range of variation in this trait.

I do not wish to leave the impression that alteration of the normal developmental rhythm cannot be successful. Although it may not often work, sometimes it works very well. Some years ago I studied growth habits of pines growing in Hawaii (Lanner, 1966). Despite the short photoperiods and mild climate, many northern species developed normally and followed a seasonal pattern not very different from that of their homelands. The following species were not only normal in their morphology, but were successfully regenerating themselves without man's intervention: *Pinus coulteri, P. strobus, P. pinaster, P. contorta* and *P. sylvestris*. Normal growth rhythms, including a dormant period, were observed in *P. patula* and *P. taeda*. But there was one notable exception. In *P. radiata* plantations, there were trees that were dormant and trees that were active, at all times of the year. Each genotype was "on its own", liberated by the tropical climate, and able to grow at a much greater rate than in its California homeland. Detailed shoot development studies have not been made in California, and were not made in Hawaii, but I suspect that the increased growth was due to a switch from the contorta pattern to the elliottii or even the caribaea pattern. In addition, many trees entered the foxtail pattern. There may be other species with similar plasticity, but we know few details of such growth responses (Lanner, 1972).

V. SUMMARY

The development of pine shoots is not the stereotyped affair it is often represented to be, but is richly diverse. Two main categories of annual shoot growth are recognized: *free growth*, in which the initiation and elongation of stem units proceed contemporaneously; and *fixed growth*, in which a dormant period intervenes between the initiation and elongation of stem units. These are combined variously into three developmental modes and ten developmental patterns. The range of variation is from the entirely free growth of the seedling axis to the entirely fixed growth of so-called preformed shoots in northern pines.

The developmental modes and patterns are viewed as adaptive strategies imposed by natural selection acting through climatic constraints. Thus free growth in trees beyond the seedling stage is related to mildness of the climate.

Finally, the developmental pattern is examined as a factor that, when it is rigid, may resist attempts to enhance growth; and that, when it is flexible, may permit great gains in growth rate. Detailed developmental data of many more species are needed before we can claim an understanding of shoot growth throughout the genus *Pinus*.

ACKNOWLEDGEMENTS

I wish to thank Dr Ray E. Goddard of the School of Forestry and Conservation, University of Florida, at Gainesville for the opportunity to study slash pine; and Mr Harm R. Kok for assistance in the field.

REFERENCES

Bengtson, G. W., McGregor, W. H. D., and Squillace, A. E. (1967). Phenology of terminal growth in slash pine: some differences related to geographic seed source. *Forest Sci.* 13, 402–412.

Cannell, M. G. R. and Willett, S. C. (1975). Rates and times at which needles are initiated in buds on differing provenances of *Pinus contorta* and *Picea sitchensis* in Scotland. *Can. J. Forest Res.* 5, 367–380.

Chudnoff, M. and Geary, T. F. (1973). Terminal shoot elongation and cambial growth rhythms in *Pinus caribaea*. *Commonw. For. Rev.* 52, 317–324.

Curtis, J. D. and Popham, R. A. (1972). The developmental anatomy of long-branch terminal buds of *Pinus banksiana*. *Am. J. Bot.* 59, 194–202.

David, R. (1968). Trois aspects de la physiologie du pin maritime: la genése de la racine et de la tige, la formation de l'oléorésine. *In* "Colloque sur la Physiologie de l'Arbre, 1966" *Mém. Soc. bot. Fr.* 1968, 137–164.

Debazac, E. F. (1968). Les modalités de la croissance en longueur chez les pins. *In* "Colloque sur la Physiologie de l'Arbre, 1966". *Mém. Soc. bot. Fr.* 1968, 3–14.

Doak, C. C. (1935). Evolution of foliar types, dwarf shoots, and cone scales of *Pinus. Illinois biol. Monogr.* 13, 1–106.

Duff, G. H. and Nolan, N. J. (1958). Growth and morphogenesis in the Canadian forest species. III. The time scale of morphogenesis at the stem apex of *Pinus resinosa* Ait. *Can. J. Bot.* 36, 687–706.

Griffing, C. G. and Elam, W. W. (1971). Height growth patterns of loblolly pine saplings. *Forest Sci.* 17, 52–54.

Hanawa, J. (1966). Growth and development in the shoot apex of *Pinus densiflora*. I. Growth periodicity and structure of the terminal vegetative shoot apex. *Bot. Mag. (Tokyo)* 79, 736–746.

Hanawa, J. (1967). Growth and development in the shoot apex of *Pinus densiflora*. II. Ontogeny of the dwarf shoot and the lateral branch. *Bot. Mag. (Tokyo)* **80**, 248–256.

Hanawa, J. and Sasaki, Y. (1970). A preliminary study on the shoot apex of *Pinus pumila*. *Bull Fac. Ed.* Hirosaki Univ., No. 23B: 12–16.

Hanawa, J. (1972). Growth and development in the shoot apex of *Pinus pumila*. *Bull. Fac. Ed.* Hirosaki Univ., No. 27B: 79–90.

Kozlowski, T. T. (1971). "Growth and Development of Trees", Vol. 1. Academic Press, New York and London.

Kozlowski, T. T. and Greathouse, T. E. (1970). Shoot growth and form of pines in the tropics. *Unasylva* **24**, 2–10.

Kramer, P. J. (1943). Amount and duration of growth of various species of tree seedlings. *Plant Physiol.* **18**, 239–251.

Lanner, R. M. (1963). Growth and cone production of knobcone pine under interrupted nights. *U.S. Forest Service Res. Note* PSW-38.

Lanner, R. M. (1966). The phenology and growth habits of pines in Hawaii. *U.S. Forest Service Res. Paper*, PSW-29.

Lanner, R. M. (1970). Origin of the summer shoot of pinyon pines. *Can. J. Bot.* **48**, 1759–1765.

Lanner, R. M. (1972). Growth and morphogenesis of pines in the tropics. *In* "Selection and Breeding to Improve Some Tropical Conifers". (J. Burley and D. G. Nikles, eds), Vol. 1, pp. 126–132. Commonwealth Forestry Institute, Oxford, and Department of Forestry, Queensland.

Lanner, R. M. and Van Den Berg, D. A. (1973). The vegetative buds and shoots of lodgepole pine. *In* "Management of Lodgepole Pine Ecosystems" (D. M. Baumgartner, ed.), Vol. 1, pp. 68–85. Washington State University, Pullman, Washington.

Mi'halevskaja, O. B. (1963). [The development of pine bud habit under conditions of the Moscow region]. *Biull. Glavnogo Bot. (Akad. Nauk SSSR)* **48**, 61–68 (In Russian).

Minina, E. G. and Piskunova, T. M. (1963). The activity of buds of the Siberian stone pine as the most important stage in the processes of its fruit bearing. *In* "Fruiting of the Siberian Stone Pine in East Siberia" (A. P. Shimanyuk, ed.), transl. by Raya Karschon. U.S. Dept. Comm., TT-65-50123 (1966), 21–34.

Owens, J. N. and Molder, M. (1973). Development of long-shoot terminal buds of *Pinus contorta* ssp. *contorta*. *In* "Management of Lodgepole Pine Ecosystems" (D. M. Baumgartner, ed.), Vol. 1, pp. 86–104. Washington State University, Pullman, Washington.

Owston, P. W. (1968). Multiple flushing in eastern white pine. *Forest Sci.* **14**, 66–67.

Owston, P. W. (1969). The shoot apex in eastern white pine: its structure, seasonal development, and variation within the crown. *Can. J. Bot.* **47**, 1181–1188.

Pollard, D. F. W. and Logan, K. T. (1974). The role of free growth in the differentiation of provenances of black spruce *Picea mariana* (Mill.) B.S.P. *Can. J. Forest Res.* 4, 308–311.

Rudolph, T. D. (1964). Lammas growth and prolepsis in jack pine in the Lake States. *Forest Sci. Monogr.* 6, 1–70.

Sacher, J. A. (1954). Structure and seasonal activity of the shoot apices of *Pinus lambertiana* and *Pinus ponderosa*. *Am. J. Bot.* 41, 749–759.

Sacher, J. A. (1955). Dwarf shoot ontogeny in *Pinus lambertiana*. *Am. J. Bot.* 42, 784–792.

Sucoff, E. (1971). Timing and rate of bud formation in *Pinus resinosa*. *Can. J. Bot.* 49, 1821–1832.

Tepper, H. B. (1963). Leader growth of young pitch and shortleaf pines. *Forest Sci.* 9, 344–353.

Van Den Berg, D. A. and Lanner, R. M. (1971). Bud development in lodgepole pine. *Forest Sci.* 17, 479–486.

Wakeley, P. C. (1954). "Planting the Southern Pines". Forest Service USDA, *Agric. Monogr.* 18, 1–233.

13

Inherent Variation in "Free" Growth in Relation to Numbers of Needles Produced by Provenances of *Picea mariana*

D. F. W. POLLARD and K. T. LOGAN
Canadian Forest Service, Petawawa Forest Experiment Station, Chalk River, Ontario, Canada

I.	Introduction	245
II.	Free Growth on 4-year-old Seedlings of Different Provenance	246
III.	Photoperiod as a Factor Regulating Free Growth in Seedlings	247
IV.	Free Growth on 12-year-old Trees	248
V.	Discussion and Conclusions	250
References		250

I. INTRODUCTION

Northern conifers exploit two modes of shoot growth. In the more common mode new leaf or stem unit initials are assembled on a primordial shoot and extended together after a period of dormancy; this growth may be termed "predetermined" (or "fixed" — see Lanner, Chapter 12). In the other mode, new stem units are initiated and extended without interruption on a continuously expanding stem; this has been termed "free" growth (Jablanczy, 1971).

In the first year, free growth is the only mode of growth from germination to budset. Free growth may also appear in the second and subsequent years, augmenting the spring flush of "predetermined" growth. It has been reported in *Abies* and *Picea* species on trees up to 10 years old (Nienstaedt, 1966; Jablanczy, 1971), and is also an important component in the development of long shoots in *Larix* (Clausen and Kozlowski, 1970).

This contribution describes free growth in black spruce, *Picea mariana*. Its purpose is to demonstrate that free growth occurs only in the early years of seedling development, depending on the environment, and that it is an important factor accounting for differences in provenance growth in this species.

Three investigations are described. The first examined the contributions made by the free and predetermined modes of growth in the development of 4-year-old seedlings in a provenance nursery trial. This was followed by a controlled environment study to test the effect of photoperiod on free growth in second-season seedlings of several provenances. In the third investigation, 12-year-old saplings of several provenances were examined for evidence of free growth.

II. FREE GROWTH ON 4-YEAR-OLD SEEDLINGS OF DIFFERENT PROVENANCE

Four-year-old seedlings of 10 provenances spanning about 20° of latitude and 61° of longitude, between Ontario and Alaska, were selected from a nursery trial at Chalk River, Ontario (46°00' N, 77°26' W). The seedlings were divided into two groups. The first group was lifted from the nursery bed before bud break, potted, and flushed in 8 h photoperiods in a growth room. The purpose of this treatment was to prevent the onset of free growth after the spring flush so that only predetermined needles emerged. The second group of seedlings was left undisturbed in the nursery to flush and express their subsequent free growth potential (measured as newly-formed needles). The numbers of needles on leading shoots were compared at the end of the growing season (Table I).

Differences between needle numbers developing (i) on the nursery seedlings and (ii) in the growth rooms, were assumed to be needles initiated during the period of free growth in the nursery. Significant differences occurred for most provenances and were greatest for provenances from southern latitudes (Table I). Correlations between initial seedling heights and numbers of predetermined needles were weak ($r = 0.56$), whereas correlations with numbers of "free"-grown needles, and total needles, were strong ($r = 0.79$ and 0.91, respectively).

The small amount of variation among the predetermined foliage complements indicated that the wide range of mean heights of the

TABLE I. Numbers of predetermined (overwintered) and "free" grown needles (current-year's complement) extended by 4-year-old *Picea mariana* seedlings of 10 provenances growing in a nursery (after Pollard and Logan, 1974).

Provenance			Number of needles appearing in 1973			Height before flush May 1973 cm
Accession numbers	Lat. N	Long. W	Total	Predetermined[1]	Free grown[2]	
6901	45°10'	77°10'	486a	354a	132*	34.1
6949	45°44'	89°03'	460ab	337a	123*	36.6
6908	48°59'	80°31'	433abc	342a	91*	33.5
6909	49°45'	85°05'	434ab	348a	86*	29.8
6924	50°53'	93°44'	431ab	324a	107*	33.0
6937	51°06'	80°52'	418bcd	362a	56	33.0
6963	51°38'	100°47'	402cd	368a	34	27.8
6970	56°03'	108°42'	407cd	347a	60	24.6
6986	56°37'	121°28'	374d	337a	37	18.2
7006	64°44'	148°19'	260	233	27	11.0

[1] Predetermined needles were initiated in 1972, held overwinter, and "flushed" in spring 1973; their numbers were estimated by lifting seedlings and flushing them in short days which prevented "free" growth.
[2] Needles both initiated and extended in 1973.
* Difference between total and predetermined needle numbers significant at $P = 0.05$.
Numbers associated by common letters are not significantly different (Duncan's test, $P = 0.05$).

provenances could be attributed largely to the variable amounts of free growth accumulating each year. The free growth complements were not large; the highest proportional increase was 37% of the number of predetermined needles (provenance 6901 from Hastings County, Ontario). Although much of the height growth on 4-year-old seedlings was predetermined in overwintering buds, provenance differences appeared to result from different capabilities for free growth.

III. PHOTOPERIOD AS A FACTOR REGULATING FREE GROWTH IN SEEDLINGS

The previous experiment raised the question whether free growth on northern provenances had been inhibited by the comparatively short summer days at Chalk River. For example, mid-summer daylength at Chalk River is 15.7 h, compared to 22.3 h at Bonanza Creek, Alaska.

Six of the 10 provenances of black spruce used in the first experiment were raised from seed in 300 ml plastic pots filled with peat and vermiculite mix (3 : 1 by volume) and fed three times daily with Ingestad (1967) solution. Seedlings were reared in a growth room under 18 h photoperiods at 25°C constant temperature. After 12 weeks the photoperiod was decreased to 8 h and temperature to 20°C. Terminal buds were initiated within a few days. Eight weeks later (week 20) the seedlings were transferred to a dark, cold room (3°C) for 6 weeks.

After chilling (week 26) seedlings were divided into four matched groups (30 per provenance per treatment) and transferred to one of four growth cabinets held at 25°C constant temperature and 8, 12, 16 or 20 h photoperiods. The seedlings began to flush a few days later and, with the exception of those under 8 h photoperiods, were still growing 8 weeks later (week 34).

Seedlings subject to 8 h photoperiods formed bud scales immediately after flushing and did not enter free growth. Free growth was resumed in all the other photoperiod treatments. Needle counts indicated that all provenances supplemented their preformed complements (as estimated from the 8 h treatment) by about 25–40% (Table II). The results demonstrate that northern provenances are capable of entering free growth under suitable environmental conditions. Indeed in the controlled environments all provenances entered free growth in photoperiods as short as 12 h. The effects of longer photoperiods were inconsistent, probably because the seedlings grew close to the overhead lamps and received excessively high temperatures.

IV. FREE GROWTH ON 12-YEAR-OLD TREES

A subsequent investigation aimed to establish whether free growth was a source of provenance variation, affecting height growth of 12-year-old black spruce. Ten trees of each of six provenances were selected from a field trial at Chalk River, Ontario. Free growth was calculated as the difference between numbers of needle primordia in the winter buds in spring and the total needle complement at the end of summer. Because the trees were part of a long-term field trial,

TABLE II. Effects of different photoperiods on numbers of needles (predetermined plus free-grown) appearing on seedlings of 6 provenances of *Picea mariana* in their second growing season.

Photoperiod (h)	Provenance numbers and latitudes of origin						Mean
	6901 45°10'	6908 48°59'	6909 49°45'	6924 50°53'	6970 56°03'	6986 56°37'	
8	228a	222a	208a	230a	206a	198a	215
12	315b	276b	268b	281b	251b	236b	271
16	288b	283b	272b	300b	237b	218ab	266
20	275b	261b	279b	290b	250b	225b	263
Mean	277	261	251	275	236	219ab	

a, b, refer to Duncan's test, see Table I.

non-destructive sampling techniques were employed; predetermined foliage complements were estimated using the regular arrangement of needle primordia at bud break. Complements were calculated as the product of the numbers of columns of primordia (contact parastichies) and the average numbers of primordia in two columns. Total foliage complements for the season were calculated from shoot lengths and numbers of needles stripped from the middle 5 cm sections of the shoots. A separate study of leaders from 12 comparable trees revealed that total complements were underestimated by an average of 2.5% by this method. Free growth was assumed to have occurred if the total season's foliage exceeded the predetermined complement by 20%.

The total foliage complement was generally similar to the primordial count. Only one provenance, 5098 from Potter, Ontario (48°42' N, 80°50' W) showed significant free growth. Individual trees with free growth were also found in provenances from Chalk River (5025, 46°00' N, 77°26' W) and Moosonee (5118, 51°17' N, 80°40' W). This general lack of free growth supported the conclusion of Jablanczy (1971) that free growth disappears between the fifth and tenth years, and all the variation in needle complements thereafter is due to differences in numbers of predetermined needle initials. In this study predetermined complements ranging from 423 to 668 per leader, were significantly correlated with both current height increment and initial height ($r = 0.68$ and 0.42, respectively).

V. DISCUSSION AND CONCLUSIONS

In young seedlings, "free" growth after the spring flush contributes significantly to total height growth; seedlings not entering free growth are at a disadvantage. While the evidence presented is not conclusive, there are strong indications that the prevailing photoperiod and temperature can determine whether or not free growth begins. When grown at southern latitudes, southern provenances are more likely to enter free growth than are northern provenances.

As black spruce seedlings develop, their propensity for free growth diminishes, and time formerly spent in free growth becomes available for the development of a primordial shoot for the following year. However, good correlation may not exist between (a) the amount of predetermined plus free growth on young seedlings and (b) the amount of wholly predetermined growth on older trees. Changes in rank are observed in provenance trials over a period of years, e.g. Teich and Khalil (1973), and these may partly result from provenance variation in the two modes of growth. At the seedling stage, the rank of a provenance will reflect its capacity for free growth, and the effect of accumulated annual increments resulting from free growth may persist for many years. If the more mature mode of predetermined growth is of similar rank, then the provenance will retain its rank in the trial. If, however, there is not a close relationship between the two modes of growth, predetermined growth will gradually obscure the effects of earlier free growth and will lead to a new rank for the provenance. Nevertheless, because large differences in height growth can occur as a result of free growth over the first 5–10 years, the effect of differences in amounts of predetermined growth may take several decades to become fully manifest.

REFERENCES

Clausen, J. Johanna and Kozlowski, T. T. (1970). Observations on growth of long shoots in *Larix laricina. Can. J. Bot.* 48, 1045–1048.

Ingestad, T. (1967). Methods for uniform optimum fertilisation of forest tree plants. *Proc. XIV IUFRO Congress*, München. III, pp. 275–278.

Jablanczy, A. (1971). Changes due to age in apical development in spruce and fir. *Can. For. Serv. Bi-month. Res. Notes* 27, 10.

Nienstaedt, H. (1966). Dormancy and dormancy release in white spruce. *Forest Sci.* **12**, 374—384.

Pollard, D. F. W. and Logan, K. T. (1974). The role of free growth in the differentiation of provenances of black spruce *Picea mariana* (Mill.) B.S.P. *Can. J. Forest Res.* **4**, 308—311.

Teich, A. H. and Khalil, M. A. K. (1973). Predicting potential increase in volume growth by progeny testing white spruce plus trees. *Can. For. Serv. Bi-month. Res. Notes* **29**, 27—28.

14
Control of Bud Break and Its Inheritance in *Populus deltoides*

BART A. THIELGES and RICHARD C. BECK
Department of Forestry and Wildlife Management, Louisiana State University, Baton Rouge, U.S.A.

I.	Introduction	253
II.	Materials and Methods	254
III.	Results and Discussion	254
IV.	Summary	258
References		258

I. INTRODUCTION

Applications of kinetin and gibberellic acid have been effective in stimulating vegetative bud break in some woody species (Pieniazek, 1964; Wareing, 1965) and Eagles and Wareing (1963) have suggested that the increased activity of endogenous gibberellins in response to winter chilling initiates spring growth or bud break in many woody plants of the Temperate Zone. A recent study by Lavender *et al.*, (1973) demonstrated that gibberellic acid, applied exogenously, stimulated bud break in chilled seedlings of Douglas-fir (*Pseudotsuga menziesii*). Their study also provided evidence that, in nature, spring growth may be initiated by gibberellin-like substances exported from the roots to vegetative buds and that warming soil temperatures may therefore govern the initiation of bud activity in the spring.

Our studies on the breaking of dormancy in cuttings of eastern cottonwood (*Populus deltoides*) have yielded results that lend support to the hypothesis that root activity stimulates bud break in species that require a winter chilling period to break bud dormancy.

II. MATERIALS AND METHODS

One hundred branch-wood cuttings were obtained from each of 23 cottonwood clones of northern origin (Ohio, Minnesota, Missouri, Illinois and Kansas) growing in a field test at Wooster, Ohio. Cuttings were taken in late September, after the trees had become dormant but before they had been exposed to low temperatures for any appreciable time. The "unchilled" control treatment was established by immediately planting 10 cuttings per clone, five in a natural photoperiod (9.5—10.5 h over the course of the experiment) and five in an extended (16 h) photoperiod. The other cuttings were stored at 1.6°C; 10 cuttings were removed per clone at intervals of *c.* 10 days, and five were planted in each photoperiod. Glasshouse temperatures were controlled at 24°C for 16 h and 15°C for 8 h in both photoperiods.

Cuttings were planted in pure sand in *c.* 0.5 l plastic cartons. The use of flexible containers provided an easy means of observing the process of root development on individual cuttings; the carton was squeezed, causing a fissure in the sand from which the cuttings could be readily extracted for examination. This procedure enabled the basal ends of the cuttings to be observed. Cuttings were examined every third day and dates of callus formation, rooting and bud break were recorded for each cutting.

III. RESULTS AND DISCUSSION

A previous publication on some aspects of this study (Chandler and Thielges, 1973) reported a shorter interval to bud break with increased chilling time for all 23 clones, under strong genetic control. This finding corroborated earlier studies which indicated that, in northern latitudes, plants which have been adequately chilled initiate shoot growth in the spring in response to increasing temperatures and not in response to lengthening photoperiods (Wareing, 1956; Worrall and Mergen, 1967; Lavender *et al.*, 1970).

In the 2300 cottonwood cuttings, observable callus formation (either lenticular or cambial) and/or rooting usually preceded bud break (Fig. 1). Increased periods of chilling accelerated the growth responses and decreased the time differential between root formation

14. Control and Inheritance of Bud Break

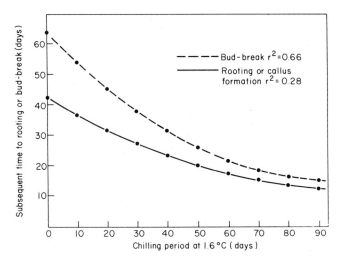

FIG. 1 Mean effects, on 24 *Populus deltoides* clones, of differing periods of cold storage, at 1.6°C, on the subsequent rates of rooting (———) and bud break (– – –) when grown at 24/15°C (day/night). Values of r^2 are significant at $P < 0.05$.

and bud break. Correlation coefficients for the positive relationship between callus and/or rooting and bud break were significant at all chilling times. These developmental observations suggest that root activity *per se* may stimulate bud break. If so, the broad-sense heritability (repeatability) is quite high when calculated for both variables. Heritabilities, shown in Table I, were calculated as:

$$h^2 = \frac{\sigma_c^2}{\sigma_c^2 + \sigma_{c \times p}^2 + \sigma_e^2}$$

where σ_c^2 = clonal variance

$\sigma_{c \times p}^2$ = clone × photoperiod interaction variance

σ_e^2 = environmental variance (residual or error variance)

As chilling time increased, h^2 values increased significantly due to a marked decrease in environmental variance. The stimulatory effects of extended photoperiod were negated as chilling time increased and,

TABLE I. Broad-sense heritabilities (h^2) for bud break and root initiation in 24 *Populus deltoides* clones previously chilled at 1.6°C for 0–90 day periods.

	Periods of chilling (days)									
	0	10	20	30	40	50	60	70	80	90
h^2 bud-break	0.51	0.84	0.86	0.99	0.98	0.98	0.97	0.96	0.96	0.98
h^2 callus formation/ root initiation	0.33	0.73	0.82	0.89	0.82	0.92	0.92	0.95	0.94	0.90

after about 30 days of chilling, the clone x photoperiod interaction became almost negligible. Thus, as environmental conditions (especially exposure to cold temperature) were satisfied, the high degree of genetic control of rooting and bud break became more evident.

Other studies have provided a physiological basis for the hypothesis that bud break depends on root activity. Michniewicz and Kriesel (1970) examined the levels of auxins, gibberellins and inhibitors in the basal ends of black poplar (*Populus nigra*) cuttings before and during adventitious root development. They detected gibberellins in the basal portions of cuttings only after root emergence and, as root development proceeded, gibberellin levels increased greatly to a maximum activity level several hundred times greater than that of auxins. A subsequent experiment (Michniewicz and Kriesel, 1972) showed that, in *Salix viminalis* cuttings, root formation preceded bud break and was accompanied by a 10-fold increase in gibberellin levels in the developing roots. The authors concluded that the adventitious roots were the major site of gibberellin synthesis because, although gibberellins were detected at very low levels in dormant buds, they were extracted in appreciable quantities from the buds only after root activity began. These findings were not discussed from the standpoint of gibberellins synthesized in developing roots stimulating bud break, but the data clearly indicate a temporal correlation between root and bud development and gibberellin activity, which can be interpreted as evidence for the export of root-synthesized gibberellins to vegetative buds prior to their development. These data are in close agreement with the findings of Lavender et al., (1973) who detected a marked increase in gibberellin levels in the xylem sap of Douglas-fir seedlings as soil temperatures (and, therefore, root activity) increased to a peak just prior to bud break. Bachelard and Wightman (1974) have also suggested a mechanism involving whole-plant growth coordination based on their studies of gibberellins, cytokinins and inhibitors in *Populus balsamifera*.

Observations of the temporal relationship of root development and bud break in other species provide additional indirect evidence to support the hypothesis that root activity stimulates bud break once the dormant condition of the vegetative bud is relieved. In other experiments with well chilled cuttings of species that are more difficult to root, viz. *Quercus rubra, Acer saccharum, Pinus strobus* and *Picea abies*, bud break was delayed, even within clones, over an unnaturally long period of time.

Experiments with *Picea abies* suggest that the delay in bud break may be due to a lack of adventitious root formation or active root growth. Cuttings taken from field-grown *Picea abies* in mid-winter and placed in a warm glasshouse broke bud very slowly over a period of several months (Mergen et al., 1964). On the other hand, when already rooted ramets of several of the same clones were exposed to very similar treatment, these broke bud in 16–30 days (Worrall and Mergen, 1967).

These observations are compatible with the theory that, once the dormant condition of the vegetative bud is broken, bud break is stimulated by growth substances synthesized in the roots and exported to the buds. In species that are easy to root, such as cottonwood and *Salix*, adventitious roots are initiated rapidly in plants that have been adequately chilled and bud break follows shortly thereafter. With cuttings of species that are more difficult to root (such as *Picea abies*), even though the material has been adequately chilled, bud break is significantly delayed or inhibited because the process of root formation (and synthesis and export) is much slower.

The observation that some chilled cuttings will eventually break bud in the absence of observable root formation does not invalidate the hypothesis. As discussed previously, external application of gibberellic acid to vegetative buds will stimulate bud break. Michniewicz and Kriesel (1972) found gibberellins present at low levels in dormant buds of unrooted *Salix* cuttings. Therefore, it seems reasonable to assume that these endogenous compounds would stimulate bud break if the buds did not become desiccated.

Thus, the hypothesis does not preclude bud break initiated by the activity of endogenous compounds in non-dormant (chilled) bud. Rather, it would attribute a significant stimulatory effect to additional gibberellins synthesized by an actively growing root system and exported to the bud. Lavender *et al.* (1973) demonstrated the potential magnitude of this stimulatory action: they shortened the time to break bud by 2 weeks by spraying GA on buds of chilled *Pseudotsuga menziesii* seedlings growing in cold (5°C) soils. This GA-induced response was the same as that obtained by growing the seedlings in warmer (20°C) soil which also resulted in a 2-week reduction in time of bud break. No response to GA treatment was observed in seedlings grown in the warmer soil.

The fact that bud break apparently occurs only when there is an actively growing root system or, in the case of unrooted cuttings, adventitious root initials, suggests that, once true dormancy or rest is alleviated by exposure to low temperatures, bud break is stimulated by growth substances originating in the roots. Our results with cottonwood have shown that these processes are under strong genetic control.

IV. SUMMARY

Studies on cottonwood clones indicate that apical bud dormancy is more effectively relieved by low temperatures than by extended photoperiods. Once dormancy is relieved bud break appears to be stimulated by growth substances synthesized in the roots in response to increased temperatures. At this stage, the heritabilities of the root initiation and leafing-out responses are high.

REFERENCES

Bachelard, E. P. and Wightman, F. (1974). Biochemical and physiological studies on dormancy release in tree buds. III Changes in endogenous growth substances and a possible mechanism of dormancy release in overwintering vegetative buds of *Populus balsamifera. Can. J. Bot.* 52, 1483–1489.

Chandler, J. W. and Thielges, B. A. (1973). Chilling and photoperiod affect dormancy of cottonwood cuttings. *Proc. South. Forest Tree Improvement Conf.* 12, 200–205.

14. Control and Inheritance of Bud Break

Eagles, C. F. and Wareing, P. F. (1963). Dormancy regulators in woody plants. Experimental induction of dormancy in *Betula pubescens*. *Nature, Lond.* **199**, 874–875.

Lavender, D. P., Hermann, R. K. and Zaerr, J. B. (1970). Growth potential of Douglas-fir seedlings during dormancy. *In* "Physiology of Tree Crops" (L. C. Luckwell and C. V. Cutting, eds), pp. 209–222. Academic Press, London and New York.

Lavender, D. P., Sweet, G. B., Zaerr, J. B. and Hermann, R. K. (1973). Spring shoot growth in Douglas-fir may be initiated by gibberellins exported from the roots. *Science* **182**, 838–839.

Mergen, F., Burley, J. and Yeatman, C. W. (1964). Variations in growth characteristics and wood properties of Norway spruce. *Tappi* **47**, 499–504.

Michniewicz, M. and Kriesel, K. (1970). Dynamics of auxins, gibberellin-like substances and growth inhibitors in the rooting process of black poplar cuttings (*Populus nigra* L.). *Acta Soc. Bot. Pol.* **39**, 283–390.

Michniewicz, M. and Kriesel, K. (1972). Dynamics of gibberellin-like substances in the development of buds, newly formed shoots and adventitious roots of willow cuttings (*Salix viminalis* L.). *Acta Soc. Bot. Pol.* **41**, 301–310.

Pieniazek, J. (1964). Kinetin-induced breaking of dormancy in 8-month old apple seedlings of "Antonovka" variety. *Acta agrobot.* **16**, 297–306.

Wareing, P. F. (1956). Photoperiodism in woody plants. *A. Rev. Pl. Physiol.* **7**, 191–214.

Wareing, P. F. (1965). Dormancy in plants. *Sci. Prog.* **53**, 529–537.

Worrall, J. G. and Mergen, F. (1967). Environmental and genetic control of dormancy in *Picea abies*. *Physiol. Plant.* **20**, 733–745.

15

The Leaf–Cambium Relation and Some Prospects for Genetic Improvement

PHILIP R. LARSON
USDA Forest Service, North Central Forest Experiment Station, Rhinelander, Wisconsin, U.S.A.

I.	Introduction	261
II.	Origin and Development of the Cambial System	262
	A. Origin and Development of the Cambium	262
	B. The Primary–Secondary Transition	268
III.	Resumption of Cambial Activity Following Dormancy	271
IV.	Species Distinctions	273
V.	Some Prospects for Genetic Improvement	275
VI.	Summary	279
References		280

I. INTRODUCTION

Wood is not only the most massive biological product of the cambium but also the most valuable economic product of the tree. It is therefore not surprising that improvement of cambial or volume growth should be the ultimate objective of most tree-breeding programmes. However, neither cambial growth nor any other growth attribute can be genetically improved without some basis for identifying selection criteria. For the overall improvement of volume growth, selection criteria based on relatively broad correlations between crown and stem may be adequate. But for the improvement of specific xylem characteristics and wood quality, more definitive criteria are required. Therefore an evaluation of cambial development as it relates to leaf development may reveal prospects for deriving selection criteria that are more meaningful than those presently available.

When we examine the origin and subsequent development of the cambium we find that throughout ontogeny it exists in close association with the development of foliar organs. This association begins with the cotyledons in the embryo and it continues with every leaf primordium produced thereafter by the buds. Within a growing bud the ontogenetic sequence usually proceeds continuously and uninterruptedly, but closer examination reveals that it can nonetheless be observed as a series of apparently discrete morphogenetic events. Each event pertaining to the cambium of a leaf trace can be correlated with other events occurring either within the primordium to which that trace leads or within the antecedent leaf from which that trace arose. This intimate relation between leaf and cambial development, established in the bud, continues throughout all subsequent stages of cambial development and cambial activity.

The leaf–cambium relation as it develops throughout ontogeny will be the subject of this chapter. Descriptions of the developmental events that follow are based on experimental data and observations. Many of the conclusions derived from the data, however, must be viewed as tentative and in some cases theoretical.

II. ORIGIN AND DEVELOPMENT OF THE CAMBIAL SYSTEM

A. Origin and Development of the Cambium

The literature on the procambium, the cambium, and their derivatives is both vast and exceedingly complex. Much of this research relates to forest trees, but it has been subjected to numerous interpretations and is therefore not easily summarized. In order to emphasize only those ontogenetic events essential to the leaf–cambium relation, I shall consider some recent research conducted in our laboratory using *Populus* as a model system for discussion (Larson, 1974; 1975, 1976). One stringent requirement of this system has been the adoption of the Leaf Plastochron Index (LPI) for maintaining uniformity in all plant material (Larson and Isebrands, 1971).

The procambial system, which precedes the cambium, conforms to a precise but complex phyllotaxy and its integrated development

15. Leaf–Cambium Relation and Genetic Improvement

assures that each leaf primordium will be served by an adequate amount of vascular tissue. Procambial development is relevant to this discussion because the blocking out of pattern by the procambium determines subsequent organization of the entire shoot (Wetmore et al., 1964).

The procambial system in *Populus* consists of an aggregation of leaf traces, three of which lead to each leaf primordium. The perpetuating members of this system are the central (C) traces. Each C-trace arises from an older C-trace below and develops acropetally toward the prospective site of a new primordium. Each C-trace gives rise in turn to lateral traces destined for adjacent leaf primordia. For example, in Fig. 1, the C-trace of LPI 6 (large arrow) gives rise to trace $-7C$, and the latter to lateral traces $-9R$ and $-10L$. In this system of trace designations, the term antecedent leaf will refer to the leaf 13 plastochrons older from which the C-trace of the respective primordium arose; i.e. LPI 6 is the antecedent leaf for LPI -7. Actually, each lateral trace leading to a primordium also has an antecedent leaf, but these will not be considered in the present discussion.

Once a procambial trace and the primordium to which it leads have attained certain states of development, then further trace development proceeds as outlined in Table I. Within each procambial trace protophloem differentiates continuously and acropetally from older protophloem below, whereas protoxylem differentiates discontinuously and bidirectionally from a point of origin at the primordium base. About this same time subsidiary procambial bundles also begin to develop basipetally from the primordium base using the original acropetal trace as a template in their descent. Thus, the pattern of procambial development in every trace consists of an acropetal component leading to the leaf primordium followed by a basipetal component emanating from the primordium base.

By examining this developmental sequence in terms of morphological events and physiological correlates some interesting relationships are revealed. For example, the acropetal initiation of a new C-trace from that of LPI -9 and the basipetal development of subsidiary procambial bundles in the C-trace of LPI -10 (Table I) approximately coincide with the rapid increase in photosynthesis and the beginning of photosynthate export, most of which is upward, by antecedent leaves LPI 3 and 4 (Table II). These data suggest that

TABLE I. Morphological events relating to procambial and cambial development in a *Populus deltoides* seedling bud with 5/13 phyllotaxy.

Primordia (Leaf Plastochron Index)	Morphological[a] event	Antecedent leaf[b] (Leaf Plastochron Index)
−11	*Acropetal* immature protophloem in C-trace. Lateral procambial traces enter lamina. Incipient basipetal procambial bundles.	2
−10	Protophloem of C-trace extends well into lamina. Protoxylem, discontinuous, appears in C-trace. *Basipetal* procambial bundles developing.	3
−9	Protophloem developed in lateral traces. *Acropetal* initiation of new C-trace (−22C).	4
−8	Protoxylem, discontinuous, appears in laterals.	5
−7	Protoxylem continuous in C-trace. *Acropetally* developing cambium in C-trace.	6
−6	Basipetal procambial bundles well developed. Cambium developed in all bundles of trace. *Basipetal* metaxylem discontinuous, in C-trace. Beginning of rapid internode elongation.	7
−5	Metaxylem continuous in C-trace. Phloem developing in daughter C-trace (−18C).	8

[a] Each morphological event occurred either within the C-trace or the lateral traces of the respective primordium. Italicized events indicate rhythmic pattern of acropetal–basipetal development. Data from Larson (1975).

[b] Antecedent leaf refers to the leaf 13 plastochrons older from which the C-trace of the respective primordium arose. See Table II for antecedent leaf correlates.

events occurring during physiological maturation of the antecedent leaves may influence events occurring during development of their daughter primordia and the traces that unite them.

While the foregoing morphogenetic events are proceeding, a cambium begins to develop within the initiating layer of each procambial bundle. The procambium is therefore the precursor of the cambium. Because the cambium develops acropetally and laterally from existing cambium below, it unites all the procambial bundles into a continuous vascular cylinder. Metaxylem is a product of the cambium but its differentiation within a trace does not begin until the cambium has advanced to the primordium base. Morphogenetic correlations can also be noted between the developing primordia and their traces and the antecedent leaves during cambial development. For example, the appearance of a cambium in the

TABLE II. Gas exchange in and ^{14}C translocation from *Populus deltoides* antecedent leaves at different developmental stages.

LPI[a]	Lamina length[a] (cm)	Lamina area[a] (cm^2)	Gas exchange[b] (mg CO$_2$ h^{-1} dm^{-2})		^{14}C translocated[c] from treated leaf		
			Photosynthesis	Respiration	Total %	Upward %	Downward %
0	2.0	0.8	−8.4	−13.0	—	—	—
1	2.7	2.8	0.1	−5.0	—	—	—
2	4.2	7.8	6.6	−1.7	—	—	—
3	5.9	17.1	11.2	−1.9	1.6	33.8	66.2
4	7.7	32.3	13.7	−1.6	5.9	81.7	18.3
5	9.2	47.6	12.6	−1.6	27.5	83.0	17.0
6[d]	9.4	52.9	12.6	−1.7	28.2	70.4	29.6
7	9.5	56.8	—	—	35.0	71.0	29.0
8	9.4	56.2	11.1	−1.4	39.1	50.4	49.6

[a] LPI (Leaf Plastochron Index) based on 16-leaf *P. deltoides* seedling. Data from Larson and Isebrands (1971).
[b] Based on 13-leaf *P. deltoides* seedlings, infrared gas analysis. Data from Dickmann (1971).
[c] Based on 16-leaf *P. deltoides* seedlings; total represents the amount of ^{14}C translocated in 24 h following a 1 h feeding of $^{14}CO_2$ to a single leaf. Data from Larson and Dickson (1973).
[d] LPI 6 considered first mature leaf although its marginal meristems are still expanding. It is subtended by the first mature internode and the primary–secondary transition zone.

C-trace of LPI −7 followed closely by the initiation of basipetal metaxylem in the C-trace of LPI −6 (Table I), coincide with maturation of antecedent leaf LPI 6 (Table II). Not only is a leaf at LPI 6 structurally and functionally mature, but the primary–secondary transition zone also occurs in the internode that it subtends (Larson and Isebrands, 1974).

The data of Table III and Fig. 1 reinforce these correlations and

TABLE III. Percentage of ^{14}C imported by developing primordia within *Populus deltoides* seedling buds when antecedent leaves were photosynthetically fed $^{14}CO_2$.

Primordia LPI	Lamina length[a] $\mu m \times 10^{-1}$	Antecedent Leaf LPI		
		6.0[a]	7.0[b]	8.0[b]
		% ^{14}C recovered (dpm mg^{-1})[c]		
0	200	0.5	2.9	8.0
−1	133	3.8	20.9	0.6
−2	102	14.6	0.2	12.2
−3	85	0.3	21.4	2.7
−4	66	17.5	3.8	6.1
−5	54	1.9	1.8	12.1
−6	39	4.1	21.9	0.9
−7	29	24.4	1.1	20.2
−8	18	2.5	9.8	8.8
−9	13	22.0	11.6	5.1
−10	8	8.4	4.6	23.3
Total		100.0	100.0	100.0
Total (dpm mg^{-1} × 10^{-5})		3.01	9.76	4.23

[a] Mean of six trees.
[b] Mean of two trees.
[c] Primordia were isolated under a dissecting microscope following 4 h of $^{14}CO_2$ treatment; ^{14}C recoveries determined by liquid scintillation spectrometry. Underlined values for LPI 6 refer to data plotted in Fig. 1; double underline identifies primordia that arose directly from the trace of the antecedent leaf. Because of phyllotaxis, each antecedent leaf exports to a different series of primordia. Comparable values in each series are identified by diagonal lines.

15. Leaf—Cambium Relation and Genetic Improvement

show that antecedent leaves do indeed provide developing primordia with essential requirements for growth. When antecedent leaf LPI 6 (Table III) was fed $^{14}CO_2$ photosynthetically, primordium LPI −7 received a high percentage of the labelled photosynthate. This pattern also held for LPI −6 and LPI −5 when their antecedent leaves LPI 7 and LPI 8, respectively, were fed $^{14}CO_2$. By comparing Tables I and III, one can see how each developing primordium is provided with photosynthates necessary for its development. Because of the phyllotactic arrangement of vascular traces, each primordium is served by a number of leaves although one leaf may predominate at any one point in time. As an example, LPI −3 received a high

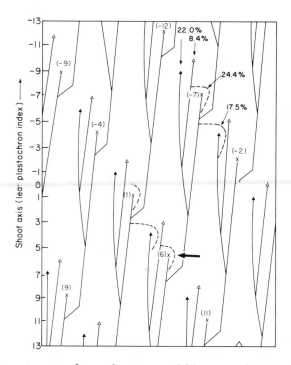

FIG. 1 Arrangement of vascular traces within part of a *Populus deltoides* seedling bud, showing the movement of ^{14}C-labelled photosynthate from a leaf at LPI 6 (large arrow). Percentages indicate the proportions of labelled photosynthate found in different primordia distal to the leaf at LPI 6 (e.g. 24.4% at LPI − 7) x = central traces (C) whose LPI values are indicated; △ and ▲ = lateral traces (R and L); dashed lines indicate phloem bridges created by basipetally developing subsidiary trace bundles that were split during their descent by existing traces below. Data from Table III, and see Larson, 1975.

amount of labelled photosynthate from LPI 7 but relatively little from LPIs 6 and 8.

The foregoing discussion by no means conveys a complete picture of the complexities of cambial origin and development within a bud. However, the morphological events and their physiological correlates strongly suggest that antecedent leaves exert some form of regulatory control not only on the vascular traces to which they give rise but also on the primordia to which the traces lead. The morphological events noted in Table I appear rhythmic, with acropetal progressions preceding basipetal ones. Such rhythmicity further suggests that the ontogenetic sequence may be rather precisely programmed with each event preparing the way for the next. Interestingly, in every case the acropetal event involved cambial or phloic development, whereas the basipetal event involved the development of xylem.

B. The Primary—Secondary Transition

In contrast to the origins of procambium and cambium which take place within a developing bud, the primary—secondary transition occurs in the internode that has just ceased elongating and that subtends the first mature leaf (Table II).

At the transition the cambium undergoes several important changes (Larson and Isebrands, 1974). The tempo of cambial activity increases markedly and the production of xylem derivatives is greatly accelerated. A matrix of fibres is produced instead of the poorly differentiated parenchymatous cells found interspersed with metaxylem vessels in the primary body. And, secondary xylem vessels differentiate rather than metaxylem vessels although the vessel system remains continuous and functional throughout.

An intriguing question concerns the nature and origin of the acropetally directed mechanism that induces all tissues to become secondary. Although it is most certainly correlated with leaf and internode maturation, it is not possible to say which of many events might be causal.

An important aspect of interest to cambial growth is the relative production of fibres and vessels following appearance of the transition. A careful analysis of these two tissue systems suggests that they may be segregated on the basis of origin and function. For example, the major product of the secondary cambium in *Populus* is a matrix of structural fibres. Developmental events suggest that metabolites and hormones received by the secondary cambium may exert a mass effect resulting in the differentiation of fibres. Even though fibre differentiation can be sectorially localized to reflect metabolic gradients to some extent, no one sector can be attributed to a single leaf. On the contrary, vessel differentiation occurs in a strictly polar gradient from a point source, the leaf basal meristem. During its descent, the vessel-forming stimulus induces cambial derivatives to differentiate as a series of metaxylem vessels in the primary zone that are interconnected via anastomoses with a similar series of secondary vessels in the secondary zone. The latter are embedded in the fibre matrix, but they are nonetheless developmentally and functionally continuous throughout both the primary and secondary zones.

The manner in which vascular continuity is maintained through the transition zone is schematically presented in simplified form in Fig. 2. At the developmental stage shown in Fig. 2A, procambial trace −13C has developed acropetally from parent trace 0C. Trace −13C contains protophloem throughout most of its length but no protoxylem. Parent trace 0C, however, contains continuous protoxylem that differentiated in the original procambial trace component and metaxylem that differentiated from later-formed cambial derivatives. When the metaxylem vessels of trace 0C encountered the transition zone at the level of LPI 6, they were directed radially outward to overlie the secondary xylem produced by the cambium below this point. But to the interior nearest the pith the original trace 0C still contains all proto- and metaxylem elements formed prior to advance of the transition zone. In other words, only those vessels will appear in the secondary xylem that were produced by a developing leaf after the transition zone has advanced to a particular level in the stem. Moreover, each vessel that appears in the secondary xylem will be continuous with one or more metaxylem vessels in the primary zone.

Three plastochrons later (Fig. 2B) the apex has advanced three nodes beyond LPI −13 and the transition zone has progressed upward to LPI 3. In trace −13C a continuous file of protoxylem elements has developed in the original procambial component and subsidiary trace bundles consisting entirely of procambium have

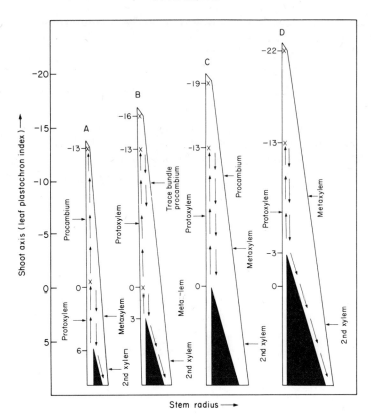

FIG. 2 Scheme showing the acropetal development of procambial traces, and the subsequent basipetal formation of trace bundles from a single leaf trace with "leaf plastochron index" value −13. See text for explanation of A, B, C and D. (From Larson, 1974).

begun to develop basipetally. Trace 0C now has but a short segment of metaxylem and the remainder consists entirely of secondary xylem.

After three additional plastochrons (Fig. 2C) LPI −13 lies six nodes below the apex and the transition zone has advanced to the level of LPI 0. The basipetally developing trace bundles of −13C have now descended to the transition zone. Although metaxylem is initiated at the primordium base, its maturation is discontinuous and it is frequently first observed in lower levels of the trace. At this stage it would not be possible to segregate the acropetal and basipetal components of trace −13C in the absence of a developmental study.

No new vessels will arise from LPI 0 following arrival of the transition zone, and those that have recently matured will appear far removed radially from their sister vessels in the primary zone.

After a total of nine plastochrons have elapsed (Fig. 2D) the transition zone has advanced to the level of LPI -3 and secondary xylem overlies the lower part of trace -13C. At this position LPI -13 is a rapidly enlarging primordium, its C-trace is well developed, and it now consists of basipetally differentiating metaxylem vessels that are shunted radially outward and dispersed laterally as they encounter the transition zone. Trace -13C would thereafter repeat the process described for trace 0C (Fig. 2A).

The sequence of developmental events presented in Fig. 2 supports the hypothesis that the main product of the secondary cambium is a matrix of fibres, formation of which is regulated principally by the maturing and recently matured leaves. The vessel-forming stimulus, produced by the developing and expanding leaves, induces metaxylem vessels in the primary zone and secondary vessels in the secondary zone. Thus, each developing leaf regulates formation of the water-conducting vessel system that will eventually serve it upon maturity. Although there are interesting ramifications of this problem, for present purposes it appears that the major function of the secondary cambium is to initiate production of structural fibres and that of the primary cambium is to initiate production of vessels and thereby provide for vascular continuity.

III. RESUMPTION OF CAMBIAL ACTIVITY FOLLOWING DORMANCY

It has long been known that the resumption of normal cambial activity, following dormancy, begins at the base of a reactivated bud (Jost, 1893). Despite the work of Priestley (Priestley, 1932; Priestley et al., 1933) and subsequent investigators (Reinders-Gouwentak, 1965), however, it is significant that few attempts have been made to isolate, either anatomically or physiologically, the precise location within the bud where reactivation of cambial activation actually begins.

From an examination of the literature we might postulate that resumption of cambial activity represents a recapitulation of events that occurred during development of the seedling bud. That is, the primordia would be reactivated first and these in turn would activate the basipetal components of the vascular traces associated with them. Tepper and Hollis (1967) reported that in *Fraxinus* mitotic figures first appeared in the leaf primordia and somewhat later in the procambial and cambial cells at the bud base. Owens (1968) observed enzymatic activity in the primordia and procambial strands of *Pseudotsuga* about 2 weeks prior to bud swelling. Bud swelling was not caused by internode elongation but by growth of the primordia, and this growth activity preceded bud burst by about 2 months.

Preparatory activities also occur within the cambium beneath the barely perceptible swelling bud. Fine structural changes have been found to occur in the cambial zone of *Pseudotsuga* long before noticeable growth initiation (Srivastava and O'Brien, 1966), and observations suggest that phloem activation progresses basipetally from the buds of *Vitis* (Esau, 1948). In both angiosperms (Davis and Evert, 1968) and gymnosperms (Alfieri and Evert, 1968) phloem activity preceded that of xylem by one to several months. As a generalization, it appears that phloem activation approximately coincides with bud swelling and new xylem production with bud burst and shoot elongation.

The successive stages of bud and shoot development, and consequently the reactivation of cambial activity, have also been correlated with the seasonal rise and fall of endogenous growth-regulating substances. For example, a growth inhibitor, presumably abscisic acid, was found to build up gradually in bud and cambial tissues with the onset of fall dormancy in both angiosperms and gymnosperms (Michniewicz and Galoch, 1972; During and Alleweldt, 1973; Davison and Young, 1974; Jenkins and Shepherd, 1972), while that of growth promoters, such as indole acetic acid, remained low (Shepherd and Rowan, 1967; Aldén, 1971; Kamieńska, 1971; Sladký, 1972). Prior to bud break the level of growth inhibitors fell while that of growth promoters rose. The actual evocator of resumed bud activity is not known, although gibberellin-like substances originating in the roots have been implicated (Hoad and Bowen, 1968; Michniewicz and Kriesel, 1972; Lavender *et al.*, 1973; Thielges and Beck, Chapter 14) as have cytokinins (Hewett and Wareing, 1973). In spite of specific uncertainties, there is little doubt that cambial reactivation is under hormonal control (Eschrich, 1970), and that auxin is necessary for the initiation of xylem differentiation (Roberts, 1969).

The objective of this discussion is not to elucidate the hormonal regulatory mechanisms involved in cambial reactivation, but rather to reveal correlative morphogenetic events and processes between leaf and cambial development. In angiosperms and gymnosperms there is a similar reactivation pattern that is strongly reminiscent of that in the actively growing *Populus* bud. An acropetal movement of hormone in some way releases the bud from dormancy and the primordia are activated. Subsequent hormonal interactions result in reactivation of the cambium to produce first phloem, and eventually xylem. Again, a cyclical acropetal–basipetal sequence of events can be observed, with each stage preparing the way for the next. Although the resumption of cambial activity following dormancy cannot be causally related to antecedent leaves, available evidence does suggest that the newly activated primordia within the bud will exert a decisive influence on the cambium during its subsequent development.

IV. SPECIES DISTINCTIONS

The similarities in cambial processes that prevail among species and species groups are far greater than the apparent differences. The ontogenetic sequence of primordial and procambial development is essentially the same in both angiosperms and gymnosperms. Procambial strands develop acropetally toward the prospective sites of new primordia, and once union is established, protophloem and protoxylem differentiate as described for *Populus* and other species (Sterling, 1945; Gunckel and Wetmore, 1946).

The most notable differences among species groups occur during the resumption of cambial activity following dormancy, and these differences may lie more in the way that the renewed cambial activity is manifested than in the underlying processes. From the work of Priestley and others (Reinders-Gouwentak, 1965) we know that, once initiated, cambial activity proceeds rapidly down the stems of ring-porous and much more slowly down those of diffuse-porous angiosperms. Moreover, single vessels may extend from the terminal shoot to ground level in ring-porous species, whereas they are much shorter and of determinate length in diffuse-porous species (Greenidge, 1952). To account for the differential rate of cambial reactivation, Imagawa and Ishida (1972) suggested that in ring-porous species the first few cells nearest the ring boundary overwintered in an immature state and therefore differentiated into the first earlywood vessels. Similarly, Digby and Wareing (1966) proposed that in *Ulmus* a precursor present in the cambium was rapidly converted to auxin at bud burst and initiated cambial activity almost simultaneously throughout the stem. Reactivation patterns in gymnosperms resemble, in general, those of the diffuse-porous angiosperms.

The similarities between angiosperms and gymnosperms persist into secondary development. Both the fibre matrix in angiosperms and the tracheid matrix in gymnosperms might conceivably be visualized as mass products of the cambium that develop under the influence of all leaves but with no one leaf predominating. Superimposed on this fibre matrix in angiosperms would be the strongly polar vessel-forming stimulus originating at the basal meristem of each leaf. Disposition of vessels within the fibre matrix of the secondary body would depend on the relative rates of cambial activity at successively lower positions in the stem (Larson, 1974). Evidence indicates that under normal circumstances mature angiosperm leaves produce neither vessels (Elliott, 1933) nor high levels of auxin (Eliasson, 1969; Kamieńska, 1971). When these latter observations are considered in relation to differences in seasonal leaf production, the types of vessel patterns observed in diffuse- and ring-porous woods may be partially accounted for.

A similar scheme might be proposed for gymnosperms. Earlywood tracheids are the principal conducting elements in the xylem and they may be considered functionally analogous to the vessels of angiosperms. An earlywood-forming stimulus produced by the developing needles would be propagated downward to induce earlywood tracheid differentiation in cambial derivatives below. However, because of differences either in the nature of the stimulus or in response of the cambium, the entire circumferential zone of derivatives would differentiate into earlywood tracheids. Like angiosperm leaves, mature gymnosperm needles neither add new tracheids to their traces (Priestley and Scott, 1936; Elliott, 1937) nor do they produce auxin (Egierszdorff and Tomaszewski, 1973). Consequently, in the absence of an earlywood-forming stimulus, latewood tracheids usually follow the cessation of shoot extension and needle elongation.

In both angiosperms and gymnosperms transitional forms of conducting elements occur. Gradients of stimuli undoubtedly arise because of interactions among leaves or needles in successive stages of maturation, and because of changes in foliar growth patterns induced by fluctuating environmental factors. These gradients could conceivably be responsible for transitional forms, such as vessel-like fibre tracheids in angiosperms (Jayme and Azzola, 1964) and earlywood-like, transition latewood tracheids in gymnosperms (Larson, 1972). Transitional elements are particularly common in young, vigorous trees.

The foregoing scheme has been synthesized from evidence obtained from numerous sources, but it is by no means conclusive. The scheme does indicate very clearly, however, that the leaves perform decisive roles in regulating and maintaining cambial activity in both angiosperms and gymnosperms throughout the growing season.

V. SOME PROSPECTS FOR GENETIC IMPROVEMENT

The purpose of the foregoing discussion has been to demonstrate what appears to be an ineluctable relation between cambial and leaf development throughout all stages of ontogeny. To emphasize this relation adequately and forcefully, it has been necessary to pursue ontogeny briefly from a relatively fundamental beginning. The evidence from this approach indicates that these tree parts are so intimately related developmentally that it is often difficult to distinguish between them. Although well separated both spatially and temporally in an older tree, it is clear that the cambium and its products are but functional extensions of the leaf. With this relation in mind, I shall explore a few ways in which it might be used to establish new selection criteria for use in either wood quality or wood volume improvement work.

Earlier I suggested that the differentiation products of cambial activity might be segregated into matrix tissue and induced elements, both of which are presumably regulated in their development by the leaves but by different metabolic pathways. Considering the matrix first, it is known that not only the number of cells produced but also the structure of their walls can be influenced by the foliar organs; that is, differentiation of earlywood and latewood, transition wood, and juvenile wood. This influence is exerted primarily by seasonal growth patterns inherent in leaf development. However, superimposed on these patterns are effects of prevailing environmental and growth conditions that are also largely perceived by the foliar organs.

Seasonal maintenance of the fibre matrix might be visualized by comparing it to an *in vitro* tissue culture. Although the cambium exhibits totipotency when isolated, it is totally dependent on other organs, primarily the leaves, for its *in vivo* growth requirements. Thus, the structural characteristics exhibited by a matrix cell might reflect the nature of the metabolites available to it at the time of differentiation. Availability would depend, in addition to the season and overall foliar vigour, on the position of each cell in the stem relative to the crown and consequently on cambial age.

The latter variables suggest further that the characteristics imparted to a xylem matrix cell may be strongly influenced by polarity gradients between crown and root. The relative metabolic gradients of sucrose and auxin occurring during an early stage of differentiation may determine whether a xylem or phloem cell is produced (Rier, 1970), while those occurring during a later stage may determine the radial diameter and wall thickness of a matrix cell (Gordon and Larson, 1968). The latter characteristics may be further influenced by radial gradients reflecting the time a matrix cell remained in the xylem maturation zone (Wodzicki and Peda, 1963). In addition to auxin gradients from the foliar organs, auxin may also be produced by cell autolysis of matrix cells within the zone of differentiation (Sheldrake, 1973). Although cell autolysis could provide an important supplementary auxin source for cambial maintenance, auxin from the leaves would still be required for cambial regulatory processes.

The induction of vascular tissue in the fibre matrix, as noted earlier, would arise directly from the developing leaves. Once induced, perpetuation of a vessel to lower stem levels could conceivably be by homogenetic induction (Lang, 1973) with the required auxin arising from cell autolysis (Sheldrake, 1973). It is therefore evident that the determination of cell type is not a property residing solely in the cambium.

The segregation of cambial products into matrix tissue and induced elements, which must at this time be considered conjectural, does have practical implications for genetic improvement. I noted earlier that the vessel-forming stimulus in angiosperms is presumably derived from developing leaves and that mature leaves do not ordinarily produce this stimulus. On the contrary, photosynthates required for the continued maintenance of the fibre matrix resulting from cambial activity are derived almost exclusively from the mature leaves. Therefore to produce wood with a high fibre : vessel ratio, which is desirable for most commercial purposes, one might select for a balance between leaf production and the retention of mature leaves on the stem. That is to say, the faster that new leaves are produced and the greater the number of leaves, the greater the number of vessels that will appear in wood of the stem. Conversely, the longer that mature leaves can be maintained on the stem in a functional condition, the greater the production of fibres.

In a similar way, the earlywood : latewood ratio of gymnosperms may be regulated by the relation between needle elongation and maturity. As long as new needles are being produced or needles are elongating, the earlywood-forming stimulus is produced. Mature needles do not produce this stimulus, but they do produce photosynthates required for maintenance of the tracheid matrix. Consequently, a decline in shoot and needle growth and a shift toward needle maturity increases the tendency toward the formation of thick-walled, latewood tracheids. Because of these relationships, it is possible for a given quantity of foliage to produce either a relatively large volume of low density wood composed mostly of earlywood, or a smaller volume of high density wood composed mostly of latewood. The many variables that contribute to both the fibre : vessel ratio in angiosperms and the earlywood : latewood ratio in gymnosperms are under genetic control and subject to improvement. Nonetheless, the regulating mechanism must be sought first in the relation between leaf and cambial development in the elongating shoot, and secondly between mature leaf performance and cambial growth on the stem.

The leaf–cambium relation also influences wood volume production in some indirect and subtle ways. The previous discussion stressed the close relation that exists between leaf primordia and the cambium during sequential stages of bud development. The number of buds on a tree, the number of leaf primordia within a bud, and the vigour of these primordia are all determined to a large extent well in advance of their appearance as photosynthesizing organs. Most buds form the season prior to opening, and because buds develop over a long period of time they are consequently exposed to a wide range of both predictable and unpredictable environmental influences. Furthermore, because trees have adapted to meet most of these exigencies, there undoubtedly exists a large genetic base that could be tapped for the improvement of cambial development by the selection of bud traits. To take advantage of these genetic potentials, however, one must have knowledge of how leaf and cambial development are related, and how this relationship is modified by both the natural environment and imposed growth conditions.

The relation between leaf development following bud break and cambial activity during the season of growth also influences wood volume production. Because trees are conservative organisms, they do not fully utilize the growing season and individual trees often differ in timing either of the initiation or of the cessation of cambial activity. By exploiting individual tree variability it should be possible not only to extend the period of cambial growth, but to do so during that part of the growing season when thick-walled cells are normally produced. These thick-walled fibres and tracheids contribute to the formation of high specific gravity wood. Such wood is desired for many commercial uses and it is therefore an important objective of many wood-quality improvement programmes.

There are undoubtedly many aspects of cambial development that could be improved by selection and breeding, and we might suggest many methods for improvement (Larson and Gordon, 1969; Wareing and Matthews, 1971). However, if the thesis pursued in this chapter is correct, that the determination of cambial growth characteristics is regulated by the foliar organs, then perhaps attempts to improve these characteristics genetically should be based on selection criteria related to the foliar organs. It is without question that a desired characteristic can generally be improved simply by selecting for that characteristic. If total wood volume is the improvement objective, for example, a satisfactory level of improvement could probably be attained by selecting for stem volume growth with only limited attention to other stem and crown attributes. Success would be a natural consequence of the inherently strong relation between overall stem and crown development. There is an equally strong relation between specific aspects of cambial development — wood quality — and foliar development, but these are manifested in much more subtle ways. Therefore, if a specific xylem characteristic is the improvement objective it might be most efficient to select for highly correlated foliar characteristics. Such an approach presupposes a knowledge of how each xylem characteristic is related to foliar development and how each modification of this relation by genetic selection might alter other growth processes. We are rapidly acquiring an understanding of the leaf—cambium relation through the research efforts of numerous investigators, but unfortunately our ability to express these relations in terms of meaningful selection criteria is still inadequate.

VI. SUMMARY

Studies of *Populus* seedling buds and shoots indicate that throughout ontogeny the cambium and the foliar organs develop in a close association. Although development of the cambium and its derivatives proceeds continuously and uninterruptedly in a growing bud, a series of rhythmic morphogenetic events can nonetheless be identified. These events oscillate between acropetal and basipetal phases of differentiation, with each event preparing the way for the next. Analysis of these patterns suggests that each acropetal event is related to a particular stage of physiological development attained by a specific antecedent leaf below, whereas each basipetal event represents the assertion of increasingly greater regulatory control by the primordium itself. Resumption of growth by dormant buds proceeds through a series of ontogenetic events that in many respects represents a recapitulation of events that occur in seedling buds. Similarly, buds and shoots of both angiosperm and gymnosperm trees proceed through an analogous series of ontogenetic events, suggesting that species differences may reside more in the way that cambial development is manifested on the stem than in the fundamental processes that precede them in the buds and shoots. Some prospects for identifying selection criteria based on the leaf—cambium relation are discussed.

REFERENCES

Aldén, T. (1971). Seasonal variations in the occurrence of indole-3-acetic acid in buds of *Pinus silvestris*. *Physiol. Plant.* 25, 54–57.

Alfieri, F. J. and Evert, R. F. (1968). Seasonal development of the secondary phloem in *Pinus*. *Am. J. Bot.* 55, 518–528.

Davis, J. D. and Evert, R. F. (1968). Seasonal development of the secondary phloem in *Populus tremuloides*. *Bot. Gaz.* 129, 1–8.

Davison, R. M. and Young, H. (1974). Seasonal changes in the level of abscisic acid in xylem sap of peach. *Plant Sci. Letters* 2, 79–82.

Dickmann, D. I. (1971). Photosynthesis and respiration by developing leaves of cottonwood (*Populus deltoides* Bartr.). *Bot. Gaz.* 132, 253–259.

Digby, J. and Wareing, P. F. (1966). The relationship between endogenous hormone levels in the plant and seasonal aspects of cambial activity. *Ann. Bot.* 30, 607–622.

During, H. and Alleweldt, G. (1973). Der Jahresgang der Abscisinsaüre in vegetativen Organen von Reben. *Vitis* 12, 26–32.

Lgierszdoiff, N. and Tomaszewski, M. (1973). Auxin dependent accumulation of photosynthates in cambium and wood formation in Scots pine. *Proc. Res. Inst. Pomol., Poland. Ser. E*, 3, 181–189

Eliasson, L. (1969). Growth regulators in *Populus tremula*. I. Distribution of auxin and growth inhibitors. *Physiol. Plant.* 22, 1288–1301.

Elliott, J. H. (1933). Growth and differentiation in the vascular system during leaf development in the dicotyledon. *Proc. Leeds phil. lit. Soc.* 2, 440–450.

Elliott, J. H. (1937). The development of the vascular system in evergreen leaves more than one year old. *Ann. Bot.* 1, 107–127.

Esau, K. (1948). Phloem structure in the grapevine, and its seasonal changes. *Hilgardia* 18, 217–296.

Eschrich, W. (1970). Biochemistry and fine structure of phloem in relation to transport. *A. Rev. Plant Physiol.* 21, 193–214.

Gordon, J. C. and Larson, P. R. (1968). The seasonal course of photosynthesis, respiration, and distribution of ^{14}C in young *Pinus resinosa* trees as related to wood formation. *Plant Physiol.* 43, 1617–1624.

Greenidge, K. N. H. (1952). An approach to the study of vessel length in hardwood species. *Am. J. Bot.* 39, 570–574.

Gunckel, J. E. and Wetmore, R. H. (1946). Studies of development in long shoots and short shoots of *Ginkgo biloba* L. II. Phyllotaxis and the organization of the primary vascular system; primary phloem and primary xylem. *Am. J. Bot.* 33, 532–543.

Hewett, E. W. and Wareing, P. F. (1973). Cytokinins in *Populus* × *robusta*: changes during chilling and bud burst. *Physiol. Plant.* 28, 393–399.

Hoad, G. V. and Bowen, M. R. (1968). Evidence for gibberellin-like substances in phloem exudate of higher plants. *Planta* 42, 22–32.

15. Leaf—Cambium Relation and Genetic Improvement

Imagawa, H. and Ishida, S. (1972). Study on the wood formation in trees. III. Occurrence of the overwintering cells in cambial zone in several ring-porous trees. *Res. Bull. Coll. exp. Forests, Hokkaido Univ.* 29, 207—221.

Jayme, G. and Azzola, F. K. (1964). Zur Morphologie der Tracheiden im Rotbuchenholz (*Fagus sylvatica* L.). *Holzforschung* 18, 9—14.

Jenkins, P. A. and Shepherd, K. R. (1972). Identification of abscisic acid in young stems of *Pinus radiata* D. Don. *New Phytol.* 71, 501—511.

Jost, L. (1893). Ueber Beziehungen zwischen der Blättentwickelung und der Gefässbildung in der Pflanze. *Bot. Zeit.* 51, 89—138.

Kamieńska, A. (1971). Auxin content in developing leaves of black poplar trees (*Populus nigra* L.). *Roczn. Naukro ln* Ser. A. 97, 13—21.

Lang, A. (1973). Inductive phenomena in plant development. *Brookhaven Symp. Biol.* 25, 129—144.

Larson, P. R. (1972). Evaluating the quality of fast-grown coniferous wood. *Proc. 63rd West. Forestry Conf.*, pp. 146—152. Seattle, Wash.

Larson, P. R. (1974). Development and organization of the vascular system in cottonwood. *In* "Proc. 3rd North American Forest Biology Workshop" (C. P. P. Reid and G. H. Fechner, eds), pp. 242—257. Colorado State University, Fort Collins.

Larson, P. R. (1975). Development and organization of the primary vascular system in *Populus deltoides* according to phyllotaxy. *Am. J. Bot.* 62, 1084—1099.

Larson, P. R. (1976). Development and organization of the secondary vessel system in *Populus grandidentata*. *Am. J. Bot.* 63, 369—381.

Larson, P. R. and Dickson, R. E. (1973). Distribution of imported ^{14}C in developing leaves of Eastern cottonwood according to phyllotaxy. *Planta* 111, 95—112.

Larson, P. R. and Gordon, J. C. (1969). Photosynthesis and wood yield. *Agr. Sci. Rev.* 7, 7—14.

Larson, P. R. and Isebrands, J. G. (1971). The plastochron index as applied to developmental studies of cottonwood. *Can. J. Forest Res.* 1, 1—11.

Larson, P. R. and Isebrands, J. G. (1974). Anatomy of the primary—secondary transition zone in stems of *Populus deltoides*. *Wood Sci. & Tech.* 8, 11—26.

Lavender, D. P., Sweet, G. B., Zaerr, J. B. and Hermann, R. K. (1973). Spring shoot growth in Douglas-fir may be initiated by gibberellins exported from the roots. *Science, N.Y.* 182, 838—839.

Michniewicz, M. and Galoch, E. (1972). Dynamics of endogenous inhibitor of abscisic acid properties in the development of buds, newly formed shoots and adventitious roots of willow cuttings (*Salix viminalis* L.). *Bull. Acad. pol. Sci. Cl. II Sér. Sci. biol.* 20, 333—337.

Michniewicz, M. and Kriesel, K. (1972). Dynamics of gibberellin-like substances in the development of buds, newly formed shoots and adventitious roots of willow cuttings (*Salix viminalis* L.). *Acta Soc. Bot. Pol.* 61, 301—310.

Owens, J. N. (1968). Initiation and development of leaves in Douglas fir. *Can. J. Bot.* **46**, 271–278.

Priestley, J. H. (1932). The growing tree. *Forestry* **6**, 105–112.

Priestley, J. H. and Scott, L. I. (1936). A note upon summer wood production in the tree. *Proc. Leeds phil. lit. Soc.* **3**, 235–248.

Priestley, J. H., Scott, L. I. and Malins, M. E. (1933). A new method of studying cambial activity. *Proc. Leeds phil. lit. Soc.* **2**, 365–374.

Reinders-Gouwentak, C. A. (1965). Physiology of the cambium and other secondary meristems of the shoot. *Handb. PflPhysiol.* **15**, 1077–1105.

Rier, J. P. (1970). Chemical basis for vascular tissue differentiation in plant tissues. *Trans. N.Y. Acad. Sci.* **32**, 594–599.

Roberts, L. W. (1969). The initiation of xylem differentiation. *Bot. Rev.* **35**, 201–250.

Sheldrake, A. R. (1973). The production of hormones in higher plants. *Biol. Rev.* **48**, 509–559.

Shepherd, K. R. and Rowan, K. S. (1967). Indoleacetic acid in cambial tissue of radiata pine. *Aust. J. biol. Sci.* **20**, 637–646.

Sladký, Z. (1972). The role of endogenous growth regulators in the differentiation processes of walnut (*Juglans regia* L.). *Biol. Plant.* **14**, 273–278.

Srivastava, L. M. and O'Brien, T. P. (1966). On the ultrastructure of cambium and its vascular derivatives. I. Cambium of *Pinus strobus* L. *Protoplasma* **61**, 257–276.

Sterling, C. (1945). Growth and vascular development in the shoot apex of *Sequoia sempervirens* (Lamb.) Endl. II. Vascular development in relation to phyllotaxis. *Am. J. Bot.* **32**, 380–386.

Tepper, H. B. and Hollis, C. A. (1967). Mitotic reactivation of the terminal bud and cambium of white ash. *Science, N.Y.* **156**, 1635–1636.

Wareing, P. F. and Matthews, J. D. (1971). Physiological and genetical factors determining productivity of species. *Proc. 15th IUFRO Congress*, pp. 136–143. Gainesville, Florida.

Wetmore, R. H., DeMaggio, A. E. and Rier, J. P. (1964). Contemporary outlook on the differentiation of vascular tissues. *Phytomorphology* **14**, 203–217.

Wodzicki, T. and Peda, T. (1963). Investigation on the annual ring of wood formation in European silver fir (*Abies pectinata* DC.). *Acta Soc. Bot. Pol.* **32**, 609–618.

16
Predicting Differences in Potential Wood Production From Tracheid Diameters and Leaf Cell Dimensions of Conifer Seedlings

M. P. DENNE
Department of Forestry and Wood Science, University College of North Wales, Bangor, Wales.

I.	Introduction	283
II.	Tracheid Diameter	284
III.	Leaf Cell Dimensions	286
IV.	Some Implications	286
V.	Conclusions	288
	Acknowledgements	289
	References	289

I. INTRODUCTION

Many years ago an association was established between tree vigour and stem anatomy for apple trees (Beakbane and Thompson, 1939). A linear relationship was found between the vigour of the scions and the proportion of wood to bark in the rootstocks, together with significant differences in the size and frequency of vessels (Beakbane, 1953). Recently, Doley (1974) compared the effects of vigorous and dwarfing rootstocks on scion wood anatomy in apples, and showed that the cross-sectional areas of vessels and fibres were closely related with ring widths. Leaf structure has also been shown to be associated with growth potential in both apples and plums (Beakbane, 1969; Watkins *et al.*, 1971). Curiously, these relationships do not seem to have been exploited in plants other than fruit trees.

II. TRACHEID DIAMETER

It may be possible to select for increased wood production indirectly by selecting for either rapid or prolonged shoot growth. The rates at which leading shoots of *Picea abies* seedlings elongated have been shown to be positively correlated with their rates of wood production, and their durations of shoot growth positively correlated with their durations of wood production (Denne, 1973). In this study, rates of shoot elongation varied independently from their duration, although this may not always be the case (Worrall, 1973). Importantly, the mean radial diameters of tracheids, measured in comparable parts of the growth rings of *P. abies* seedlings (i.e. in early or late wood), were found to be correlated with the rates of both shoot elongation and wood production, but not with the duration of either (Denne, 1973). This finding has been reinforced by recent work on *Pinus sylvestris*. Partial correlations showed that the mean radial diameters of comparable tracheids produced by transplants in their third year were greatest in plants which extended their leading shoots rapidly, irrespective of their periods of elongation (Table I). Furthermore, the diameters of tracheids which were produced when these transplants were first-year seedlings were also correlated with the rates at which their shoots elongated in their

TABLE I. Partial correlations describing relationships between the rates and durations of shoot elongation on 3-year-old transplants of *Pinus sylvestris* and the mean radial diameters of earlywood tracheids.

	Partial correlation coefficient
Third-year earlywood tracheid diameter on shoot elongation rate (eliminating duration)	0.45[a]
Third-year earlywood tracheid diameter on duration of shoot elongation (eliminating rate)	0.18 n.s.
First-year seedling tracheid diameter on shoot elongation rate (eliminating duration)	0.34[a]
First-year seedling tracheid diameter on duration of shoot elongation (eliminating rate)	0.13 n.s.

135 values   [a] significant at $P < 0.05$.

16. Predicting Potential Wood Production from Anatomy

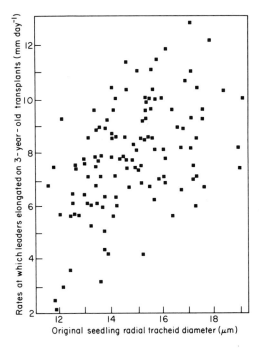

FIG. 1 Relationship between the rate at which the leading shoots elongated on 3-year-old *Pinus sylvestris* transplants, and the diameters of their early-wood tracheids when 1-year-old seedlings.

third year (Fig. 1), but, again, not with their durations of shoot growth (Table I).

Thus, it may be possible to predict rates of both shoot growth and wood production from the diameters of tracheids in seedlings. Obviously this is impracticable if the leader has to be destroyed, but fortunately there is a reasonable correlation between the mean diameters of tracheids taken from leading and lateral shoots. Given the association between rates of shoot elongation and tracheid diameters (Table I), and knowing the total shoot increment, it should be possible to estimate the duration of shoot growth. Similarly, it should also be possible to predict the rate of wood production from tracheid diameters, and hence to estimate the duration of cambial activity. A tree having a small mean tracheid diameter is likely to have a slow rate of cambial activity; if it also has wide rings, this would indicate a long growing season.

III. LEAF CELL DIMENSIONS

Casual field observations suggest that leaf sizes vary greatly between plants, irrespective of the lengths of the shoots on which they are borne. However, data from 3-year-old *P. sylvestris* transplants showed that leaf lengths were significantly correlated with rates of shoot elongation ($r = 0.34$), but not with durations of elongation ($r = 0.10$). Further preliminary data are shown in Table II, derived from 20 first-year seedlings of *Picea sitchensis* grown in 16h photoperiods, using means of five leaves from each plant. Cell dimensions were determined from the leaf surface, the outlines of the palisade cells being visible through the epidermis. These data show that rates of shoot elongation were positively correlated with leaf lengths, leaf widths and cell widths; the correlation between rates of growth and palisade cell widths looks particularly promising and may have some predictive value.

IV. SOME IMPLICATIONS

Some of the more interesting correlations between the growth characteristics of seedlings and those of older plants are summarized in Fig. 2. From this, it seems likely that potential rates of wood production could be predicted from the rates of shoot elongation of seedlings, or from seedling tracheid diameters, or leaf characteristics. The idea of using cell dimensions to predict rates of wood production may perhaps seem too esoteric to be practicable for selection on a large scale. As John Evelyn (1664) said, after

TABLE II. Coefficients correlating rates of shoot elongation with four leaf characteristics, using first-year seedlings of *Picea sitchensis*.

Correlations with rates of shoot elongation			
Leaf length	Leaf width	Mean epidermal cell width	Mean palisade cell width
0.60	0.47	0.59	0.72

20 values; all correlations significant at $P < 0.05$.

16. Predicting Potential Wood Production from Anatomy

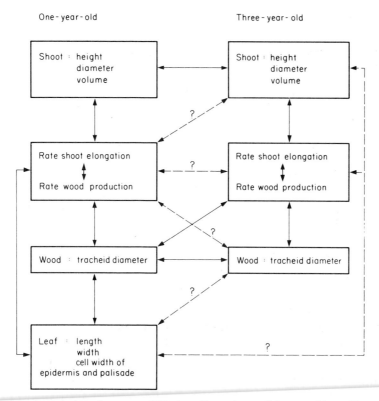

FIG. 2   Associations between differing dimensions of 1-year-old seedlings and 3-year-old transplants of *Picea sitchensis*.
Solid arrows: correlation coefficients significant at $P < 0.05$;
Broken arrows, with question marks: correlation coefficients not yet tested, but assumed to be significant.

discussing the merits of selecting seed from "... such as are found most solid and fair":

> And this shall suffice, though some might haply recommend to use a more microscopical examen, to interpret their most secret schematisms, which were an overnicety for these great plantations.

However, selection for wood production should go hand in hand with selection for wood quality. A measure of wood structure is already included as an estimate of potential wood quality in the selection of plus trees for breeding (Fletcher and Faulkner, 1972). If it is possible to predict rates of wood production from the dimensions of wood and leaf cells (Fig. 2), it should also be possible to predict potential tracheid diameters from the original seedling tracheid diameters, or from seedling leaf characteristics, or from rates of seedling growth.

An increase in wood production is generally considered to be accompanied by a decrease in specific gravity. This has given rise to some concern that a decrease in wood density may follow selection for increased wood production. Since mean tracheid diameter appears to be linked with the rate of cambial activity rather than with its duration, then selection for more rapid wood production would be likely to result in decreased wood density. On the other hand, increasing the period over which the wood is produced would not be likely to affect its density. Selection for prolonged cambial activity might be expected to increase wood density by increasing the proportion of late wood in the growth ring, but there is some evidence that this does not happen (Denne, 1973); presumably this is because the period of cambial growth is increased both before and after shoot elongation stops, so the proportion of early to late wood remains unchanged. Hence it should be possible to achieve both rapid height growth and high wood density by selecting trees with both prolonged cambial activity and small radial tracheid diameters. An alternative possibility would be to increase the proportion of late wood by selecting trees which continue cambial activity after their shoots have stopped elongating (see Larson, Chapter 15). Further work is needed to find a method of selecting for this feature in first-year seedlings.

In some circumstances it may be useful to select for decreased wood production. For example, it would be profitable to produce dwarfing rootstocks when tailoring trees for restricted areas, as for street planting, or for high density housing areas. Also, just as dwarfing rootstocks are used to promote early fruiting in apples, it may be feasible to screen rootstocks to be used for conifer seed orchards, selecting anatomical features associated with lack of vigour, and hence, hopefully, with early flowering.

V. CONCLUSIONS

It seems likely that potential rates of cambial activity can be predicted from mean tracheid diameters or leaf characteristics of first-year conifer seedlings. Using these features together with ring widths, it should be possible to estimate the duration of cambial activity. This would allow selection for rate of wood production separately from length of growing season. Also, it may be feasible to predict potential tracheid diameters from seedling tracheid or leaf dimensions, or from seedling growth rates.

ACKNOWLEDGEMENTS

This work was supported by a grant from the Natural Environment Research Council. Thanks are due to Mr M. J. Allen and Miss Lynda Hughes for technical assistance.

REFERENCES

Beakbane, A. B. (1953). Anatomical structure in relation to rootstock behaviour. *Proc. XIIIth int. hort. Cong.* 1952, 1, 152–158.

Beakbane, A. B. (1969). A new series of potential plum rootstocks. *A. Rep. East Malling Res. Stn* for 1968, 81–83.

Beakbane, A. B. and Thompson, E. C. (1939). Anatomical studies of stems and roots of hardy fruit trees. II. The internal structure of the roots of some vigorous and some dwarfing apple rootstocks, and the correlation of structure with vigour *J. Pomol.* 17, 141–149.

Denne, M. P. (1973). Tracheid dimensions in relation to shoot vigour in *Picea*. *Forestry* 46, 117–125.

Doley, D. (1974). Effects of rootstocks and interstock on cell dimensions in scion stems of apple (*Malus pumila* Mill.). *New Phytol.* 73, 173–194.

Evelyn, John (1664). "Silva, or a discourse of forest trees". (1786 Edition)

Fletcher, A. M. and Faulkner, R. (1972). "A plan for the improvement of Sitka spruce by selection and breeding". Research and Development Paper No. 85. Forestry Commission, H.M.S.O. London.

Watkins, R., Beakbane, A. B. and Werts, J. M. (1971). Selection techniques. *A. Rep. East Malling Res. Stn* for 1970, 98.

Worrall, J. (1973). Seasonal, daily, and hourly growth of height and radius in Norway spruce. *Can. J. Forest Res.* 3, 501–511.

17
Probable Roles of Plant Hormones in Regulating Shoot Elongation, Diameter Growth and Crown Form of Coniferous Trees

RICHARD P. PHARIS
Biology Department, University of Calgary, Alberta, Canada

I.	Introduction	291
II.	Shoot Elongation	292
III.	Xylem Growth	296
IV.	Tree Form	299
V.	Conclusions	301
VI.	Summary	301
References		302

I. INTRODUCTION

The interface between genetics and physiology of woody perennial plants covers a diverse spectrum of growth and differentiation processes. Two of the most important are (a) promotion of precocious flowering and enhancement of seed yield, and (b) early recognition of functional traits which characterize genetically superior individuals. The latter is the subject of this conference, and will be given emphasis in this contribution.

Three of the growth and differentiation traits of interest to the forester are stem elongation, diameter growth and tree form. Each is the end result of a number of physiological processes, most of which involve, at one stage or another, the action of plant growth hormones; it is the role of these hormones in coniferous trees that will be examined here. Special consideration will be given to the possible use of hormone levels, balances, or rates of metabolism for parental selection and progeny testing.

II. SHOOT ELONGATION

Whereas there is an extensive literature on the effects of the physical environment on shoot elongation in trees (see Zahner, 1968; Kozlowski, 1971), it is less apparent to the casual reader that growth hormones play a role. Even so, a recent review of gibberellin (GA) effects on conifers noted (a) over 20 references, concerned with some 18 conifer species, where exogenous GAs have increased elongation or endogenous GAs were implicated in elongation, and (b) a number of GAs (GA_1, GA_2, GA_3, GA_4, GA_7, GA_{34}) which have been characterized from conifer tissue by definitive means (Pharis and Kuo, 1976).

Unfortunately, the possible role(s) of auxin in shoot elongation is less clear, although indole-3-acetic acid (IAA) has been characterized from shoots of *Pseudotsuga menziesii* by "gas chromatography-mass spectrometry" (DeYoe and Zaerr, 1975) and *Pinus sylvestris* by a diverse combination of chromatographic methods (Alden and Eliasson, 1970; Alden, 1971). Numerous reports of IAA-like sub stances in conifers also exist (Giertych, 1964; Shepherd and Rowan, 1967; Wodzicki, 1968; Hashizume, 1969; Little and Bonga, 1974). It is possible that auxins increase cell elongation and division in trees, as they do in flower stalks of herbaceous plants, and are presumed to do in stems of herbaceous dicotyledons (Sachs, 1965). However, evidence that auxin acts directly on cell division and elongation in conifer shoots is currently lacking, as, indeed, it is on many other *in vivo* plant systems. Thus, excised or decapitated tissues are generally required to demonstrate auxin effects on elongation in angiosperms.

However, IAA and extractable conifer auxins will stimulate growth of excised *Pinus thunbergii* hypocotyl sections (Hashizume, 1969) and endogenous auxins are correlated, although not always positively, with shoot elongation in conifers (Giertych, 1964; Hashizume, 1969; Alden, 1971; Little and Bonga, 1974). Reports that applied IAA or other auxins stimulate shoot elongation in conifers are rare, although Wort (1975) has obtained significant increases in shoot elongation of *Pseudotsuga menziesii* with potassium napthenate, a synthetic auxin, and Table I illustrates a rather "back-handed" effect of IAA on lateral shoot elongation of *Pinus radiata* when the IAA is applied to the terminal shoot in the presence of $GA_{4/7}$. One attractive interpretation of the data in Table I is that

TABLE I. Effects of exogenous applications of $GA_{4/7}$ on the elongation of terminal and lateral branches of 3-year-old *Pinus radiata* seedlings. Plants were grown in the field at Rotorua, New Zealand.

Treatment[a]	Growth over 56 days, March–April (±S.E.)		
	Terminal Shoot (cm)	Lateral Shoots (cm/branch)	Lateral Shoots (cm/plant)
Retarded leader +$GA_{4/7}$	15.0 ± 4.20	13.3 ± 1.50	90 ± 9.3
Retarded leader +$GA_{4/7}$ + IAA	10.5 ± 2.10	15.6 ± 1.64	108 ± 11.3
Retarded leader (control)	12.6 ± 1.82	11.9 ± 1.85	67 ± 10.3
Normal leader (control)	18.2 ± 2.54	11.3 ± 0.64	154 ± 8.7

[a] 200 μg $GA_{4/7}$ (c. 30% A_4) and/or 10 μg IAA applied to each of 11–18 replicate plants in 80% ethanol every two weeks to the terminal shoot.
"Retarded leader" refers to a growth habit, common for young seedlings of *Pinus radiata* during their first 3 or 4 growing seasons, whereby terminal shoot elongation is initially retarded, then totally inhibited, while growth of the lateral shoots at the first node subtending the terminal continues unabated. One or more lateral shoots subsequently become dominant (Weston, 1957). (Data taken from Pharis and Sweet, unpublished research results).

IAA levels are already inhibitory in the "retarded" terminal, but still below optimal in the lateral shoots. Thus, although auxins are implicated in shoot elongation of conifers, ascertaining their exact role(s), or even correlating their concentrations or metabolism with the elongation process must await future research.

The possible roles of other growth regulators, including abscisic acid (ABA), in internode elongation of conifers are even more tenuous, although ABA, by implication from its effects in other plant species (Milborrow, 1974) and known presence in conifers (Milborrow, 1969; Lenton et al., 1971; Jenkins and Shepherd, 1972; Little et al., 1972; Zaerr et al., 1973), may well be a native hormone which regulates shoot elongation in conifers.

Thus, for the known growth hormones, it would appear that only GAs can currently be considered with certainty to play a controlling, or "limiting factor" role in shoot elongation of conifers. The roles of other growth regulators remain unclear, and their general use in correlative analyses of inherent differences in height growth may be questionable. What then is the specific evidence for the involvement of GAs in shoot elongation of conifers?

Exogenous application of certain of the GAs is known to increase shoot elongation in at least 17 conifer species, including 13 species of Pinaceae (Pharis and Kuo, 1976) and this is shown for *Pinus radiata* in Table I and *Cupressus arizonica* in Table II. Although such evidence does not prove that GAs regulate internode elongation in conifers, it is striking that: (a) increased amounts of endogenous GAs in *Cupressus arizonica* seedlings are closely correlated with the promotion of height growth, both being the result of increasing photoperiod length (Table III); (b) correlations occur between the concentration of a GA_3-like substance in the xylem sap of some tree species and vegetative bud break (Lavender *et al.*, 1973; Thielges and Beck, Chapter 14), and (c) increases in GA-like substances occur during bud break and shoot elongation of some species of the Pinaceae (see Pharis and Kuo, 1976). Additionally, there is evidence that shoot elongation in *C. arizonica* is decreased in proportion to the degree to which endogenous GA production is inhibited by AMO-1618 (Kuo, 1973; Kuo and Pharis, 1975), a growth retardant known to inhibit endogenous GA biosynthesis (Lang, 1970; Kuo and Pharis, 1975). Exogenous applications of GA_3 overcome the growth

TABLE II. Effects of weekly soil drenches of 100 ml of aqueous GA_3 solutions on the height growth of 6-month-old pot-grown *Cupressus arizonica* seedlings. Plants were grown in a glasshouse under 18 h photoperiods in full sunlight, supplemented with incandescent light. The period of 75 days was 7 May–21 July, Calgary, Canada.

GA_3 treatment (mg/litre)	Height growth (mm)		
	Days 0–25	Days 26–50	Days 51–75
0	44	46	69
10	38	53	78
100	50	123	134

Data adapted from Glenn (1973).

TABLE III. Effect of photoperiod on both height growth over a 25-day period, and on endogenous GA_3 levels, in 6-month-old *Cupressus arizonica* seedlings. Plants were grown 75 days prior to and during the treatments in growth chambers at 8000 $\mu W/cm^2$ light intensity. Gibberellins were extracted immediately after measuring elongation.

Photoperiod (h)	Height growth (mm)	Endogenous GA_3 levels[a] ($\mu g/25$ g dry wt tissue)
4	7	1.42
16	49	9.70

[a] GA_3 characterized by "gas chromatography–mass spectrometry" (Durley, Pharis, Morf and Glenn, unpublished research results) and measured by the dwarf rice, Tan-ginbozu assay (Murakami, 1966). Data adapted from Glenn (1973).

retardation induced by AMO-1618 in both *C. arizonica* and young seedlings of *Picea abies* (Dunberg and Eliasson, 1972; Dunberg, 1974).

Responses to growth retardants or exogenous GA applications have generally been restricted to species within the Cupressaceae and Taxodiaceae, to pines which flush several times during the growing season, or to very young seedlings (see Pharis and Kuo, 1976). In members of the Pinaceae, like *Pinus sylvestris*, which flush once each season, the number of cells present in the internode is apparently fixed the previous year (Kozlowski and Keller, 1966), and other endogenous factors may override the primary effect of GAs on shoot elongation. These factors could be the presence of inhibitory compounds such as ABA, and/or the rapid detoxification of "excesses" of promotive hormones by enhanced metabolism, thus preventing an effect even on the next year's growth.

Finally, rates of hormone metabolism (synthesis, interconversion, conjugation) may be good indicators of inherent differences in growth potential. Wample *et al.* (1975) have shown that *Pseudotsuga menziesii* shoots rapidly metabolize (^3H)-GA_4 to GA_{34}, GA_2, and six other unknown acidic radioactive substances, the presence and amounts of each depending on the stage of ontogeny. The possibility that rates of metabolism of potent growth hormones such as the auxins or GAs are correlated with inherent differences in shoot elongation bears close investigation, particularly as radioactive hormones could be used as an effective and rapid assay for rates of metabolism.

Given then, that (a) exogenous GAs will increase height growth of many conifers, (b) substances known to inhibit GA biosynthesis often retard growth, and (c) levels of endogenous GAs (or rates of GA metabolism?) are correlated with shoot elongation, one could reasonably conclude that GAs do indeed play a positive, perhaps often limiting, role in shoot elongation of most, if not all, conifers. Unfortunately the mechanism(s) by which GAs influence shoot growth remain even more of a mystery for trees than for herbaceous angiosperms, where little is still known (Jones, 1973) beyond their effect on inducing a "new", sub-apical meristem in which cell division and elongation are increased in pith, cortical and vascular tissues (see Sachs, 1965). Similar studies on trees remain undone, although the anatomical effects of GA_3 on the apical and branching meristems of two conifers have been described (Owens and Pharis, 1967, 1971). The biochemistry of GA_3-induced cell division and elongation, even in herbaceous angiosperms, still remains at a correlative stage. Increases in RNA synthesis, enzyme synthesis, and even the production of nucleolar bodies have been noted in meristematic cells shortly after GA_3 application. However, such correlations mean little without a thorough knowledge of the early kinetics of GA_3-induced growth (see Jones, 1973).

III. XYLEM GROWTH

The diameter of a tree, for all practical purposes, depends on the number and size of xylem cells, derived from a circumferential cambial meristem. The cambial zone can be thought of as including dividing cambial initials and dividing phloem and xylem mother cells. "Cambial activity", as used here, includes both cell division and enlargement; "differentiation" of xylem cells includes enlargement (primarily radial) and maturation (Wilson *et al.*, 1966), but only enlargement will be considered.

There is little doubt that cambial activity is under hormonal control, specifically of auxin (see reviews by Romberger, 1963; Westing, 1965, 1968; Roberts, 1969; Kozlowski, 1971; Torrey *et al.*,

1971; and papers by Zajeczkowski, 1973; Wodzicki and Wodzicki, 1973; Little, 1975). Other hormones that have been implicated as native growth regulators of cambial activity of conifers are gibberellins (Wareing et al., 1964; Farrar and Fan, 1970) and more recently, ABA (Little and Eidt, 1968, 1970; Little and Bonga, 1974; Little, 1975).

Xylem differentiation (as defined above) also appears to be governed primarily by auxin although GA_3 (Wareing et al., 1964; Wardrop and Davies, 1964; DeMaggio, 1966; Hejnowicz and Tomaszewski, 1969; Farrar and Fan, 1970), ABA (Little and Eidt, 1968; 1970; Jenkins, 1973; Little, 1975) and even cytokinins are implicated (see Westing, 1965, 1968; Hejnowicz and Tomaszewski, 1969). Gibberellins are especially promotive in the presence of IAA (Wareing et al., 1964; Hejnowicz and Tomaszewski, 1969), although high GA does not always promote early differentiation of xylem in preference to phloem (DeMaggio, 1966). Exogenous ABA decreases xylem cell radial growth in *Pinus radiata* and *Abies balsamea*, where it is apparently native (Little and Eidt, 1968; Jenkins, 1973; Little, 1975), and this decrease also occurs with drought stress, which is known to increase ABA levels in plants (Hiron and Wright, 1973). Exogenous GA_4 will increase xylem cell radial growth in *Pinus radiata* and ABA will enhance the conversion of (^3H)-GA_4 to more polar metabolites in this species, although neither has thus far been shown to significantly counteract the other's effects when applied simultaneously (Jenkins and Pharis, unpublished). However, Little (1975) has shown that ABA will antagonize an IAA-induced increase in cambial activity in *Abies balsamea*.

Thus, auxins, GAs, ABA and cytokinins are all implicated in the control of cambial activity and subsequent differentiation in conifers. From the work of Little (1975) it could be speculated that conifers may antagonize the positive growth effect of native auxin *in situ* with ABA, and given the growing list of plant systems where GA—ABA antagonisms have been noted (see Milborrow, 1974) conifers may use ABA to counteract both IAA and GA, perhaps regulating "dormant" and "active" periods of cambial activity in this manner.

Westing (1965, 1968) has concluded, on the basis of numerous published reports, that increased auxin levels and rapid growth are indeed positively correlated, although how directly remains open to question since most of the estimates of auxin level are based on indirect or rather crude techniques. More recently, however, work by Wodzicki and Wodzicki (1973) and Nix and Wodzicki (1974) can be interpreted to indicate that the level of an inhibitory substance (ABA?) is the key factor regulating cambial activity, not levels of auxin. Hence, additional work is necessary to establish just how closely any or all of the above hormones are correlated with rate of diameter growth. Such research could well be most usefully done on material where "rapid" or "slow" inherent rates of growth were already established.

It is now possible, with care, to quantify ABA by physical methods (see Milborrow, 1974) and GAs can be bioassayed with reasonable accuracy (Crozier et al., 1970; Bailess and Hill, 1971). Physical methods such as gas chromatography (Cavell et al., 1967) are not, in my opinion, suitable for the quantitative estimation of the nanogram amounts of GAs normally found in vegetative tissues of higher plants, even after purification (Crozier et al., 1969; Glenn et al., 1972). The method of Knegt and Bruinsma (1973) offers some hope that IAA, if characterized as being native by physical methods, can be quantitatively assayed as indolo-α-pyrone by fluorescence. Also, Bridges et al. (1973) have shown "single-ion-scanning", using "gas chromatography—mass spectrometry" of crude extracts of coleoptile tissue, to be a potentially useful method of quantifying IAA, although it remains to be seen whether it can be applied to extracts of all plants. Nevertheless, other auxins will still have to be bioassayed, a straight growth assay with excised coleoptile segments being perhaps the most satisfactory (Sirois, 1966). Although high pressure liquid chromatography may become a useful technique to detect cytokinins in plant extracts routinely, it would appear that purification and chromatography on polyvinylpyrolidone (Glenn et al., 1972; Thomas et al., 1975) and subsequent verification of cytokinin activity on soya bean callus (Miller, 1965), followed by more rapid assays (Fletcher et al., 1973), currently offers the best hope of accurately quantifying endogenous cytokinin levels. It should be emphasized that where a definitive form of identification is used only once to verify the presence of a plant hormone, and other more rapid but less definitive methods are subsequently used to quantify the hormone, purification and chromatography methods must be highly reproducible and rigorously adhered to.

If then, after appropriate assay, significant correlations are found to exist between diameter growth and the above hormones, then long-term testing for "juvenile : mature" correlations can begin.

IV. TREE FORM

The form of a tree depends upon branch angle, length, and diameter in relation to both vertical and circumferential growth of the main stem. It is the end result of interactions between environment and so-called "correlation effects". Environmental influences include gravity, light intensity, photoperiod, temperature and mineral nutrition. Correlation effects involve not only a response to gravity by the branch itself, via the branch apical and cambial meristems, but also a signal from the terminal shoot (see Champagnat, 1965; Phillips, 1969, 1975; Westing, 1965, 1968; Zimmerman and Brown, 1971).

The form of a conifer is of practical importance to foresters for many reasons:

(1) Trees with small, short branches of modest diameter are not only better from a self-pruning aspect, but may allow more stems to be grown per hectare in more photosynthetically efficient stands.

(2) Photosynthate used in the radial growth of lateral branches rather than the bole is, in general, lost to human use, although this may change as intensive utilization becomes more widely practised.

(3) The stimulus which causes the formation of undesirable compression wood in large vigorous branches may extend downward into the vertical bole (see Westing, 1965, 1968).

(4) A tendency toward large vigorous branches in the upper crown may result in excessive growth where they join the main stem, and in very rapidly growing trees compression wood may form around the main stem (Westing, 1965, 1968).

(5) The desirable conical shape of young trees depends on control of lateral shoot growth by the terminal shoots, a situation which does not exist, for instance, in a high percentage of young *Pinus radiata* seedlings in New Zealand, which are said to have "retarded leaders" (see footnote to Table I).

Most of the above aspects of tree form appear to be under genetic control (Bailey *et al.*, 1974), but may also be caused, or modified, by the environmental factors mentioned above. Additionally, the juvenile—mature transition may include not only changes in leaf morphology, flowering and ability to root, but also changes in many aspects of tree form. Finally, compression wood formation, or the absence thereof, appears to be almost inseparably linked with most factors that modify tree form (Westing, 1965, 1968). It would appear, then, that the form of a tree is a "quantitative" expression of many interacting factors and that tree form can be modified via many physiological mechanisms.

What are these mechanisms? First and foremost, circumferential growth in the main stem and branches responds to "signals" from the terminal shoot, apparently of an auxin nature (see Westing, 1965, 1968). However, the cambial meristem itself may be able to produce growth promotive substances, and cambial activity in lateral branches can be controlled not only by a signal from the terminal shoot (or superior branches), but also from their own apices. The possibility that the signal from terminal shoots may be the absence of a substance(s)" should not be overlooked; discussions of "hormone-directed transport" or "nutrient diversion theories" are given by Phillips (1969, 1975) and Westing (1965, 1968). Additionally, the cambial meristems of both lateral shoots and main stems seem able to perceive gravity independently of the apical meristems (Westing, 1965, 1968). Gibberellins have also been implicated, either alone or in conjunction with auxin (see Westing, 1968; Pharis and Kuo, 1976), and would appear to be limiting and controlling factors in many conifer species.

Although auxins are associated with many aspects of tree form, there are only a few instances where exogenous application of auxins, or compounds known to affect auxin action or movement, have modified tree form (Westing, 1965, 1968; McGraw, 1973). In contrast, the use of GAs, compounds known to inhibit GA biosynthesis, or environmental treatments such as short days, which decrease GA levels, have produced significant changes in the form of conifers (see Pharis and Kuo, 1976). These changes are expressed most obviously by species in the Cupressaceae and Taxodiaceae, but are also apparent in the Pinaceae. For most species tested, GAs can be thought of as factors which decrease apical control in the intact plant, but for some, such as *Sequoia sempervirens*, GAs enhance control by the apex over lateral buds.

However, I believe that the evidence still points towards correlative control via auxin, perhaps with GAs participating. But, the correlative signal perceived by the meristem is often so weak that exogenous GA_3 can override it, resulting in a significant change of tree form (Pharis *et al.*, 1972; Glenn, 1973; Kuo, 1973; McGraw, 1973). Westing (1965, 1968) also cites several instances where signals from the roots, or lower parts of the tree, seem necessary for a full geotropic response, or have modified tree form. These could either by cytokinins or GAs since both are known to move upward in woody shoots (Reid and Burrows, 1968; Sembdner *et al.*, 1968; Lavender *et al.*, 1973).

Thus, as might be expected from their apparent roles in diameter growth, hormones of the auxin, GA and cytokinin classes are implicated also in apical control. However, this control is made more complex by the dependence of tree form not only on correlative signals from the apex (and/or roots?), but also by the apparently independent response of the cambial meristem to gravity (see Jankiewicz and Stecki, Chapter 9).

V. CONCLUSIONS

How then may the physiologist and forest tree geneticist best interact to utilize the above information for practical purposes? Firstly, correlative studies should be done using material where differences in shoot elongation, diameter growth and tree form are known to be under genetic control. These studies should investigate the quantities, balance and metabolism of native growth regulators, as is appropriate for the trait in question. If it can be established that significant correlations exist between the desired trait and the level, balance or metabolism of growth regulators, then selection techniques using one or more of these hormonal parameters can begin. Experiments must also be initiated to determine the earliest age at which the "assay" may be used as an indicator of later performance. Success, then, could result in significant and rapid genetic gains through both parental selection and progeny testing by chemical and/or bioassay methods.

VI. SUMMARY

Analyses of endogenous hormones and their correlations with shoot elongation, xylem growth and tree form, together with information on responses to applied hormones, or compounds known to affect hormone levels or actions, can tell us which hormones play major roles in conifer growth and development. Gibberellins play a dominant role in longitudinal shoot growth, although auxins are also thought to be involved. Cambial activity and xylem cell radial growth appear to be regulated, under certain circumstances, by auxin, abscisic acid and gibberellins.

Tree form is not only a measure of the ability of the tree to respond to gravity, but also of the response of lateral branches to a "signal" from the terminal shoot. Both auxins and gibberellins are implicated in controlling tree form and, for some conifers at least, gibberellins appear to be a limiting factor in the differential growth that governs this response.

Future research that ascertains those desirable growth or developmental traits which are closely correlated with inherent differences in hormone levels, balance or metabolism could well provide criteria for selecting parents and progeny using appropriate physical or bioassay methods.

REFERENCES

Alden, T. (1971). Seasonal variation in the occurrence of indole-3-acetic acid in buds of *Pinus silvestris*. *Physiol. Plant.* 25, 54–57.

Alden, T. and Eliasson, L. (1970). Occurrence of indole-3-acetic acid in buds of *Pinus silvestris*. *Physiol. Plant.* 23, 145–153.

Bailess, K. W. and Hill, T. A. (1971). Biological assays for gibberellins. *Bot. Rev.* 37, 437–479.

Bailey, J. K., Feret, P. P. and Bramlett, D. L. (1974). Crown character differences between well-pruned and poorly-pruned Virginia pine trees and their progeny. *Silvae Genet.* 23, 181–185.

Bridges, I. G., Hillman, J. R. and Wilkins, M. B. (1973). Identification and localisation of auxin in primary roots of *Zea mays* by mass spectrometry. *Planta* 115, 189–192.

Cavell, B. D., MacMillan, J., Pryce, R. J. and Sheppard, A. C. (1967). Thin-layer and gas–liquid chromatography of the gibberellins; direct identification of the gibberellins in a crude plant extract by gas–liquid chromatography. *Phytochemistry* 6, 867–874.

Champagnat, P. (1965). Physiologie de la croissance et l'inhibition des bourgeons: Dominance apicale et phénomènes analogues. *In:* "Encyc. Pl. Physiol. XV/I" (W. Ruhland, ed.), pp. 1106–1164. Springer-Verlag, Berlin.

Crozier, A., Aoki, H. and Pharis, R. P. (1969). Efficiency of counter-current distribution, Sephadex G-10, and silicic acid partition chromatography in the purification and separation of gibberellin-like substances from plant tissue. *J. exp. Bot.* **20**, 786–795.

Crozier, A., Kuo, C. C., Durley, R. C. and Pharis, R. P. (1970). The biological activities of 26 gibberellins in nine plant bioassays. *Can. J. Bot.* **48**, 867–877.

DeMaggio, A. E. (1966). Phloem differentiation: induced stimulation by gibberellic acid. *Science, N.Y.* **152**, 370–372.

DeYoe, D. R. and Zaerr, J. B. (1975). Indole-3-acetic acid in Douglas-fir: a definitive analysis by gas–liquid chromatography and mass spectrometry. *In* "Proceeding Third N.A. Forest Biol. Workshop", (C. P. P. Reid and G. H. Fechner, eds). College of Forestry, Fort Collins, Colorado, USA. (abstr.).

Dunberg, A. (1974). Occurrence of gibberellin-like substances in Norway spruce (*Picea abies* (L.) Karst.) and their possible relation to growth and flowering. *Stud. For. Suec.* **111**, 1–62.

Dunberg, A. and Eliasson, L. (1972). Effects of growth retardants on Norway spruce (*Picea abies*). *Physiol. Plant.* **26**, 302–305.

Farrar, J. L. and Fan, W. Y. (1970). Induction of tracheid development in decapitated Jack Pine seedlings (*P. banksiana* Lamb.) with indole-3-acetic acid and gibberellic acid. *Annu. Rep. Fac. of For. and Dept. of Bot.*, Univ. of Toronto p. 8.

Fletcher, R. A., Teo, C. and Ali, A. (1973). Stimulation of chlorophyll synthesis in cucumber cotyledons by benzyladenine. *Can. J. Bot.* **51**, 937–939.

Giertych, M. M. (1964). Endogenous growth regulators in trees. *Bot. Rev.* **30**, 292–311.

Glenn, J. L. (1973). "Investigations concerning the physiology of gibberellin-induced growth and sexual differentiation phenomena in Arizone cypress (*Cupressus arizonica*, Greene) seedlings". Ph.D. Thesis, University of Calgary, Canada.

Glenn, J. L., Kuo, C. C., Durley, R. C. and Pharis, R. P. (1972). Use of insoluble polyvinylpyrrolidone for purification of plant hormones. *Phytochemistry* **11**, 345–351.

Hashizume, H. (1969). Auxins and gibberellin-like substances existing in the shoots of conifers and their roles in flower bud formation and flower sex differentiation. *Bull. Tottori Univ. For.* **4**, 1–46.

Hejnowicz, A. and Tomaszewski, M. (1969). Growth regulators and wood formation in *Pinus silvestris*. *Physiol. Plant.* **22**, 984–992.

Hiron, R. W. P. and Wright, S. T. C. (1973). The role of endogenous abscisic acid in the response of plants to stress. *J. exp. Bot.* **24**, 769–781.

Jenkins, P. A. (1973). The influence of applied indole acetic acid and abscisic acid on xylem cell diameters of *Pinus radiata* D. Don. *In* "Mechanisms of the Regulation of Plant Growth" (R. C. Beleski and M. M. Cresswell, eds) *Bull. R. Soc. N.Z.* **12**, 737–742.

Jenkins, P. A. and Shepherd, K. R. (1972). Identification of abscisic acid in young stems of *Pinus radiata* D. Don. *New Phytol.* 71, 501–511.
Jones, R. L. (1973). Gibberellins: their physiological role. *A. Rev. Pl. Physiol.* 27, 571–598.
Knegt, E. and Bruinsma, J. (1973). A rapid, sensitive and accurate determination of indol-3-acetic acid. *Phytochemistry* 12, 753–756.
Kozlowski, T. T. (1971). "Growth and Development of Trees", Vols. I and II. Academic Press, New York and London.
Kozlowski, T. T. and Keller, T. (1966). Food relations of woody plants. *Bot. Rev.* 32, 293–382.
Kuo, C. G. (1973). Growth retardation and nutritional stress in relation to vegetative growth and reproductive differentiation of seedlings of *Cupressus arizonica* Greene. Ph.D. Thesis, University of Calgary, Canada.
Kuo, C. G. and Pharis, R. P. (1975). Effects of AMO-1618 and B-995 on growth and endogenous gibberellin content of *Cupressus arizonica* seedlings. *Physiol. Plant.* 34, 288–292.
Lang, A. (1970). Gibberellins: structure and metabolism. *A. Rev. Pl. Physiol.* 21, 537–570.
Lavender, D. P., Sweet, G. B., Zaerr, J. B. and Hermann, R. K. (1973). Spring shoot growth in Douglas-fir may be initiated by gibberellins exported from the roots. *Science, N.Y.* 182, 838–839.
Lenton, J. R., Perry, V. M. and Saunders, P. F. (1971). The identifications and quantitative analysis of abscisic acid in plant extracts by gas–liquid chromatography. *Planta* 96, 271–280.
Little, C. H. A. (1975). Inhibition of cambial activity in *Abies balsamea* by internal water stress: role of abscisic acid. *Can. J. Bot.* 53, 1805–1810.
Little, C. H. A. and Bonga, J. M. (1974). Rest in the cambium of *Abies balsamea. Can. J. Bot.* 52, 1723–1730.
Little, C. H. A. and Eidt, D. C. (1968). Effect of abscisic acid on budbreak and transpiration in woody species. *Nature, Lond.* 220, 498–499.
Little, C. H. A. and Eidt, D. C. (1970). Relationship between transpiration and cambial activity in *Abies balsamea. Can. J. Bot.* 48, 1027–1028.
Little, C. H. A., Strunz, G. M., LaFrance, R. and Bonga, J. M. (1972). Identification of abscisic acid in *Abies balsamea. Phytochem.* 11, 3535–3536.
McGraw, D. (1973). "An investigation into the role of gibberellins in apical dominance in the Cupressaceae". M.Sc. Thesis. University of Calgary, Canada.
Milborrow, B. V. (1969). Abscisic acid in plants. *Sci. Prog. Oxf.* 57, 533–538.
Milborrow, B. V. (1974). The chemistry and physiology of abscisic acid. *A. Rev. Pl. Physiol.* 25, 259–307.
Miller, C. O. (1965). Evidence for the natural occurrence of zeatin and derivatives: compounds from maize which promote cell division. *Proc. natn. Acad. Sci. U.S.A.* 54, 1052–1058.

Murakami, Y. (1966). Bioassay of gibberellins using rice endosperm and problems of its application. *Bot. Mag. (Tokyo)* 79, 315—327.

Nix, L. E. and Wodzicki, T. J. (1974). The radial distribution and metabolism of IAA-^{14}C in *Pinus echinata* stems in relation to wood formation. *Can. J. Bot.* 52, 1349—1355.

Owens, J. N. and Pharis, R. P. (1967). Initiation and ontogeny of the microsoporagiate cone in *Cupressus arizonica* in response to gibberellin. *Am. J. Bot.* 54, 1260—1272.

Owens, J. N. and Pharis, R. P. (1971). Initiation and development of western red cedar cones in response to gibberellin induction and under natural conditions. *Can. J. Bot.* 491, 1165—1175.

Pharis, R. P. and Kuo, C. G. (1976). Physiology of gibberellins in conifers. *Can. J. For. Res.* 7, (in press).

Pharis, R. P., Kuo, C. C. and Glenn, J. L. (1972). Gibberellin, a primary determinant in the expression of apical dominance, apical control and geotropic movement of conifer shoots. *In* "Plant Growth Substances, 1970". (D. J. Carr, ed.), pp. 441—448. Springer-Verlag, Berlin, Heidelberg, New York.

Phillips, I. D. J. (1969). Apical dominance. *In* "The Physiology of Plant Growth and Development" (M. B. Wilkins, ed.), pp. 165—202. McGraw-Hill, London.

Phillips, I. D. J. (1975). Apical dominance. *A. Rev. Pl. Physiol.* 26, 341—367.

Reid, D. M. and Burrows, W. J. (1968). Cytokinin and gibberellin-like activity in spring sap of trees. *Experientia* 24, 189—190.

Roberts, L. W. (1969). The initiation of xylem differentiation. *Bot. Rev.* 35, 201—250.

Romberger, J. A. (1963). "Meristems, Growth, and Development in Woody Plants". Tech. Bull. No. 1293, U.S. Dept. of Agriculture, U.S. Forest Service, Washington, D.C.

Sachs, R. M. (1965). Stem elongation. *A. Rev. Pl. Physiol.* 16, 73—96.

Sembdner, G., Weiland, J., Aurich, O. and Schreiber, K. (1968). Isolation, structure and metabolism of a gibberellin glucoside. *Plant Growth Regulation*, S.C.I. Mono. 31, pp. 70—86. London.

Shepherd, K. R. and Rowan, K. S. (1967). Indole-acetic acid in cambial tissue of radiata pine. *Aust. J. biol. Sci.* 20, 637—646.

Sirois, J. C. (1966). Studies on growth regulators. I. Improved *Avena* coleoptile elongation test for auxin. *Plant Physiol.* 41, 1308—1312.

Thomas, T. H., Carroll, J. E., Isenberg, F. M. R., Pendergrass, A. and Howell, L. (1975). Separation of cytokinins from Danish cabbage by column chromatography on insoluble polyvinylpyrrolidone. *Physiol. Plant* 33, 83—86.

Torrey, J. G., Fosket, D. E. and Hepler, P. K. (1971). Xylem formation: a paradigm of cytodifferentiation in higher plants. *Am. Sci.* 59, 338—352.

Wample, R. L., Durley, R. C. and Pharis, R. P. (1975). Metabolism of gibberellin

A_4 by vegetative shoots of Douglas fir at 3 stages of ontogeny. *Physiol. Plant.* 35, 273–278.

Wardrop, A. B. and Davies, G. W. (1964). Nature of reaction wood. VIII. Structure and differentiation of compression wood. *Aust. J. Bot.* 12, 24–38.

Wareing, P. F., Hanney, C. E. and Digby, J. (1964). The role of endogenous hormones in cambial activity and xylem differentiation. *In* "The Formation of Wood in Forest Trees" (M. H. Zimmermann, ed.), pp. 323–344. Academic Press, New York and London.

Westing, A. H. (1965). Formation and function of compression wood in gymnosperms. *Bot. Rev.* 31, 381–480.

Westing, A. H. (1968). Formation and function of compression wood in gymnosperms II. *Bot. Rev.* 34, 51–78.

Weston, G. C. (1957). Exotic forest trees in New Zealand. *In* "Proc. Seventh British Commonwealth Forestry Conf." pp. 69–74.

Wilson, B. F., Wodzicki, T. and Zahner, R. (1966). Differentiation of cambial derivatives: proposed terminology. *Forest Sci.* 12, 438–440.

Wodzicki, T. J. (1968). On the question of occurrence of indole-3-acetic acid in *Pinus silvestris*. *Am. J. Bot.* 55, 564–571.

Wodzicki, T. J. and Wodzicki, A. B. (1973). Auxin stimulation of cambial activity in *Pinus silvestris*. II. Dependence upon basipetal transport. *Physiol. Plant.* 29, 288–292.

Wort, D. J. (1975). Mechanism of plant growth stimulation by napthenates: some effects on photosynthesis and respiration. *Proc. Annual Meeting Can. Soc. Plant Physiol.* 15, p. 27 (abstr.).

Zaerr, J. B., Lavender, D. P. and Sweet, G. B. (1973). Growth regulators in Douglas-fir during dormancy. *In* "I.U.F.R.O. Int. Sym. on Dormancy in Trees" (M. Giertych, ed.), Inst. of Dendrology and Kornik Arboretum, Pol. Acad. Sci.

Zahner, R. (1968). Water deficits and growth of trees. *In* "Water Deficits and Plant Growth". (T. T. Kozlowski, ed.), pp. 191–254. Academic Press, New York and London.

Zajaczkowski, S. (1973). Auxin stimulation of cambial activity in *Pinus silvestris*. The differential cambial response. *Physiol. Plant.* 29, 281–287.

Zimmerman, M. H. and Brown, C. L. (1971). "Trees: Structure and Function". Springer-Verlag, Berlin, Heidelberg, New York.

WATER STRESS AND WATERLOGGING

18
Water Relations and Tree Improvement

T. T. KOZLOWSKI
Department of Forestry, University of Wisconsin, Madison, U.S.A.

I.	Introduction	307
II.	Development of Water Deficits	308
III.	Characterization of Water Deficits	309
IV.	Drought Resistance	310
	A. Drought Avoidance	310
	B. Drought Tolerance	321
V.	Summary	322
References		323

I. INTRODUCTION

Because of their extensive leaf surfaces perforated with stomata, trees are admirably constructed to lose large amounts of water. An acre of trees with a stem basal area of about 60 ft^2 may expose more than 5 acres of leaf surface. Hence it is not surprising that forests in the southern United States may lose as much as 8000 gallons of water per acre per day (*c.* 9 x 10^4 l ha^{-1} day^{-1}; Kozlowski, 1968).

Rates of water loss on a whole plant basis and per unit of transpiring tissue vary widely among and within species. For example, transpiration on a leaf dry weight basis was 68—159% higher in *Fraxinus americana* than in *Acer saccharum* seedlings (Kozlowski *et al.*, 1974). Water loss of *Populus* clones also varied significantly (Siwecki and Kozlowski, 1973).

Variations among plants in water use efficiency or "transpiration ratio", generally measured as the ratio of transpiration to net photosynthesis (T/P), probably are more important than variations in transpiration alone. For plant communities water use efficiency is often calculated as the ratio of evapotranspiration to total dry matter produced. This ratio usually varies from <200 to >2000. Whereas high production plants have low values, those of arid regions have high ones. The T/P values in arid regions are high (often >2000) for two main reasons: (1) leaf area index (area of leaf surface per unit area of ground surface) usually is between 1 and 2, so the rate of evapotranspiration is high in comparison to the rate of growth; and (2) infrequent precipitation and extensive root systems account for long periods of consumption of reserve foods in respiration without high rates of photosynthesis. When rains do occur, the plants must synthesize foods with limited leaf surfaces. Also the steep vapour pressure gradient from leaves to air tends to increase T/P (Slatyer, 1964).

According to Gardner (1963) it is more important to breed for water use efficiency within a species than among species. If genotypic variation for water use efficiency is appreciable, advances might be made by a simple breeding method such as mass selection. However, if genotypic variation is small, progeny tests may be necessary.

II. DEVELOPMENT OF WATER DEFICITS

If we consider the soil—tree—atmosphere system as a physical continuum through which water moves along a path of decreasing potential energy, then the driving force for movement of water is its decreasing energy gradient. There are several resistances along the flow path. Resistance is greater in the soil than in the tree, and is greatest in the transition from the leaves to the atmosphere where water changes from liquid to vapour (Hillel, 1971). In stems water transport occurs in the sapwood and resistance in the heartwood is very great.

Since transpiration is controlled largely by atmospheric factors (e.g. light intensity, temperature, humidity, wind), influencing the vapour pressure gradient between the leaf and air, water loss tends to exceed absorption by roots, because of resistance to water transport. Hence trees tend to become dehydrated during the day as absorption lags behind transpiration. Recurrent temporary midday water deficits in leaves because of excessive transpiration are not serious when the soil is well watered because leaves usually rehydrate and regain turgor during the night when stomata are closed. However, as droughts intensify, leaves are less likely to regain turgor during the night and progressively greater internal water deficits develop until the soil becomes recharged with water. Thus water deficits are recurrent and inevitable, and they probably are more important than any other factor in growth loss, injury, and killing of trees (Kozlowski, 1968, 1971).

III. CHARACTERIZATION OF WATER DEFICITS

The degree of water deficit in trees has been variously characterized in terms of moisture content, relative water content (RWC), saturation deficit (SD), and water potential (ψ). Although moisture content of tissues is easily determined, this measure sometimes is not very useful. Moisture content generally is expressed as percent of oven dry weight of tissues. But, because dry weight changes with time often are not proportional to changes in the actual amount of water in tissues, variations in moisture content do not necessarily reflect changes in protoplasmic hydration or physiological activity. Such processes as photosynthesis, respiration, and translocation often produce large changes in dry weight of tissues. Over long periods progressive cell wall thickening usually is responsible for dry weight increase of tissues such as rapidly growing leaves. Thus changes in moisture content (as percentage of dry weight) may reflect changes in amount of water, in dry weight increment, or both. Kozlowski and Clausen (1965) showed that seasonal changes in moisture content of leaves were traceable more to increases in leaf dry weight than to changes in hydration of physiologically active cells. Sometimes translocation of solutes into and out of leaves may change the dry weight base appreciably within hours, thereby altering % moisture content without appreciable change in actual water content (Kozlowski, 1968).

Saturation deficit (SD) and relative water content (RWC) compare the moisture content of plant tissue at a given time with moisture content of the same tissue when it is fully turgid. Unfortunately full turgor may occur in leaves of one species at a moisture content that is found in wilted leaves in another species, or in leaves of different ages on the same plant.

Most plant physiologists characterize soil and plant water deficits in terms of water potential (ψ), the difference in chemical potential of water in a system and of pure free water at the same temperature. The chemical potential of water expresses the capability of a unit mass of water to do work in comparison to the work an equal mass of pure free water at the same location would do. Some of the forces affecting ψ in soils and plant tissues are:

1. Solute potential (ψ_s). Solutes lessen the free energy of a solution.
2. Matric potential (ψ_m). The matric component decreases ψ as a function of capillary or colloidal forces by soil colloids, cell colloids, and cell walls. The force of adsorption between the matrix surface and the water molecules decreases ψ below that of pure free water.
3. Pressure potential (ψ_p). The pressure potential may increase or decrease ψ according to whether the molecules are subjected to pressures above or below atmospheric pressure. Under atmospheric pressure, the effect in an open system, such as soil, is zero. Turgor pressure in plants adds free energy to the system, and ψ is increased. At wilting the pressure component approaches or reaches zero and does not appreciably influence ψ.

In cells ψ, the sum of ψ_s, ψ_m, and ψ_p, has a negative value except in fully turgid cells when it is zero. As water deficits in plant tissues increase during droughts, their ψ values become increasingly more negative.

IV. DROUGHT RESISTANCE

Many different features contribute to drought resistance of trees. Survival is accomplished by various degrees of drought avoidance and drought tolerance (Kozlowski, 1964, 1972b). These will be discussed separately.

A. Drought Avoidance

Adaptations for avoiding drought may be found in leaves, stems and roots. Leaf adaptations include: leaf shedding; microphylly; small, few, and sunken stomata; rapid stomatal closure; abundant leaf waxes; strong development of palisade mesophyll and weak development of spongy mesophyll. Although the volume of intercellular spaces often is less in xeromorphic than in mesomorphic leaves, the ratio between the internal exposed surface area of the leaf and its external surface area tends to be higher in xeromorphic leaves (Fahn, 1964). Drought avoidance in a plant often is traceable to more than one adaptation. In Brazil, for example, transpiration of drought-avoiding plants was greatly decreased by stomatal closure during early phases of drought, but as drought severity increased, leaf abscission followed (Ferri, 1953). Adaptations for drought avoidance often vary for plants growing in the same area. The speed at which stomata close at a critical water deficit contributes materially to drought avoidance of *Rhagodia baccata* of western Australia (Hellmuth, 1968). By comparison, *Acacia craspedocarpa* of the same area exhibits poor control of stomatal aperture. In this species, control of water loss is related to anatomical characteristics and development of water-retaining substances in cells of phyllodes (Hellmuth, 1969).

1. Leaf Adaptations

(a) *Leaf shedding and microphylly.* If certain plants do not obtain enough water to balance unavoidable losses they shed their leaves. Generally shedding begins with the oldest leaves and progresses towards the apical meristems. During dry summers shedding of old leaves occurs commonly even in the temperate zone.

Encelia farinosa and *Fouquieria splendens* of the south-western United States drop their leaves during a drought and grow a new crop after rain. In the latter species four or five annual crops may be produced. In the arid "caatinga" and "cerrado" of Brazil the leaves of many species also are shed during the height of the dry season (Ferri, 1961; Eiten, 1972).

According to Orshansky (1954) the most important factor in the survival of desert plants in the Near East is their small leaf surface. Their leaves are often replaced by brachyblasts, reduced to stipules, or are shed during the dormant season. Examples are *Artemisia, Noea, Haloxylon, Anabasis,* and *Zygophyllum.* Seasonal dimorphism is attained by shedding and growth of various types of branches and leaves at different seasons. Large winter leaves are often replaced by small summer leaves.

(b) *Stomatal size and frequency.* Both the size and number of stomata vary greatly among species and genotypes. Among 28 species of trees in Madison, Wisconsin, stomatal length varied from 17 to 56 μm and stomatal frequency from about 100 to 600 stomata per mm^2. In a general way stomatal size and frequency were negatively correlated. Thus a species with few stomata per unit of leaf area tended to have large stomata. For example, *Acer saccharum, Acer saccharinum* and *Rhus typhina* had many small stomata, whereas *Betula papyrifera, Fraxinus pennsylvanica, Ginkgo giloba, Gleditsia triacanthos, Salix fragilis*, and *Vitis vinifera*, had few large stomata. The *Quercus* species were an exception, having both large and numerous stomata. Stomatal size and frequency varied markedly within a genus, as in *Crataegus, Fraxinus* and *Quercus* (Davies *et al.*, 1973). Siwecki and Kozlowski (1973) found wide differences in stomatal size and distribution among species and clones of *Populus*. *Pinus resinosa* seedlings with larger and more numerous stomata than *P. strobus* lost water faster than the latter species (Davies *et al.*, 1974a).

(c) *Control of stomatal aperture.* There is considerable evidence for a genetic component in control of stomatal aperture. Rapid wilting, tip scorching and premature leaf fall in abnormal diploid *Solanum tuberosum* plants were associated with an inability to close the stomata (Waggoner and Simmonds, 1966). Wilting of *Flacca*, a wilty mutant of *Lycopersicon esculentum*, resulted from very high transpirational loss. The wilting tendency of the mutants reflected high stomatal frequency, wide stomatal opening, and resistance to stomatal closure even in the dark (Tal, 1966). The differences in stomatal responses of normal plants and wilty mutants were associated with a deficiency of abscisic acid (ABA) in the mutant (Tal and Imber, 1970). When the wilty mutants were sprayed with ABA, stomatal closure was readily induced (Tal and Imber, 1972).

The capacity for early stomatal closure during drought varies markedly both between and within species (Kozlowski, 1972a, 1972b). Stomata of plants of the Mediterranean maquis close rapidly when soil dries, whereas those of *Phillyrea media* do not (Oppenheimer, 1953). Stomata closed sooner and at a smaller water deficit in *Ilex cornuta* than in *Rhododendron poukhanensis* when subjected to drought. *Ilex* appeared to be more drought resistant than *Rhododendron* because it controlled transpiration more efficiently and had a higher resistance to cuticular transpiration (Kaul and Kramer, 1965). *Eucalyptus rostrata* seedlings were injured more during drought than were seedlings of *E. polyanthemos* or *E. sideroxylon*. Transpiration decline curves demonstrated that *E. rostrata* closed its stomata much later than *E. sideroxylon*, and *E. polyanthemos* was intermediate (Quraishi and Kramer, 1970).

Siwecki and Kozlowski (1973) examined the relation of internal leaf anatomy, stomatal size, stomatal frequency, and control of stomatal aperture on transpiration rates of six *Populus* clones (two clones of *P. maximowiczii*; one clone from each of the species *P. deltoides*, *P. nigra*, and *P. trichocarpa*; and one hybrid, *P. maximowiczii* x *P. nigra*). Water loss of excised leaves varied widely among clones, as did internal leaf anatomy, stomatal size, stomatal frequency and control of stomatal aperture. Transpiration rates were more closely related to stomatal size, frequency and control than to internal leaf anatomy. No consistent pattern was shown over all clones in the correlation of transpiration rate with any individual feature of internal leaf anatomy examined (leaf thickness, epidermal thickness, amount of palisade parenchyma, amount of spongy parenchyma, % of combined thickness of spongy parenchyma plus epidermal layers, or % of palisade parenchyma). Very high transpiration capacity of *P. trichocarpa* was correlated with low stomatal resistance associated with large stomata (but low stomatal frequency). The high rate of water loss of *P. maximowiczii* x *P. nigra* was correlated with high stomatal frequency. In both of these clones the capacity to keep stomata open for long periods also contributed to their high transpiration rates. Although *P. deltoides* and *P. nigra* leaves had relatively large stomata, their low rates of transpiration were attributed to early stomatal closure (Siwecki and Kozlowski, 1973).

Species differ with respect to the soil and plant water potentials, ψ, at which stomatal closure occurs. In drying soil there was a much greater reduction in transpiration of *Pinus ponderosa* and *P. contorta* than in *Pseudotsuga menziesii* or *Abies grandis*. At a soil ψ of −10 bar, transpiration rates of the pines were only about 12% of their maximum rate; those of *Pseudotsuga* and *Abies* were 27−37% of maximum. The greater capacity of the pines to reduce water loss helped explain their occurrence on drier sites (Lopushinsky and Klock, 1974).

A small decrease in leaf ψ resulted in a rapid decrease in transpiration of *Populus tremula*, indicating stomatal closure. Transpiration and control of stomatal aperture were less sensitive in *Betula verrucosa* and *Pinus sylvestris*; and transpiration of *Picea abies* remained high over a wide range of water potentials (Jarvis and Jarvis, 1963). Stomatal closure in *Pinus ponderosa* and *P. contorta* seedlings occurred at leaf ψ values in the range of −14 to −17 bars; in *Pseudotsuga menziesii* in the range of −19 to −22 bars; and in *Abies grandis* at −25 bars (Lopushinsky, 1969).

We have found considerable difference among species in stomatal responses to such factors as light intensity, temperature, wind and humidity, all of which are involved in inducing plant water deficits.

(*i*) *Light intensity*. Experiments were conducted on stomatal aperture (diffusion resistance) of *Fraxinus americana, Acer saccharum, Quercus macrocarpa, Citrus mitis*, and *Cercis canadensis* seedlings over a range of light intensities of 800 lux and 32 000 lux.

Equilibrium leaf resistances varied with light intensity and species. *Acer* stomata began to close when light intensity decreased below 16 000 lux. By comparison, *Fraxinus* and *Quercus* stomata did not begin to close until light intensity fell below 6500 lux. *Cercis* stomata remained open until light intensity dropped below 5000 lux, and *Citrus* stomata began to close at 3200 lux.

When light intensity was rapidly changed from 0 to 32 000 lux and the reverse, stomata opened faster than they closed in *Fraxinus* and *Quercus* whereas in *Citrus* they closed faster than they opened. Opening and closing times were not different from each other in *Acer* and *Cercis*. In each species both opening and closing responses of stomata to light were much faster in green than in chlorotic plants (Davies and Kozlowski, 1974a).

(*ii*) *Temperature*. Experiments were conducted on seedlings of five species of trees which occur along an ecological gradient from xeric to mesic: *Quercus macrocarpa, Q. velutina, Q. alba, Q. rubra*, and *Acer saccharum*. Leaf resistance and rates of transpiration and photosynthesis as well as water use efficiency were calculated for each species over a temperature range of 20−40°C (Wuenscher and Kozlowski, 1971a, 1971b).

In all species transpiration resistance increased linearly as the temperature was raised, with the rate of increase in resistance greatest for *Q. velutina*, intermediate for *Q. macrocarpa* and *Q. rubra*, and least for *Q. alba* and *A. saccharum*. Water use efficiency increased with increasing leaf temperature up to 35°C and decreased at higher temperatures. Efficiency of water use varied in the following order: *Q. velutina* > *Q. macrocarpa* > *Q. rubra* = *Q. alba* > *A. saccharum*. Under most conditions the water use efficiency of *Q. velutina* was more than four times greater than in *A. saccharum*. Thus *Q. velutina* was most drought resistant. It fixed CO_2 rapidly while losing little water.

(*iii*) *Wind*. Experiments were conducted on effects of wind on transpiration and stomatal resistance of *Fraxinus americana*, *Acer saccharum* and *Pinus resinosa* seedlings. Wind was alternately on and off for 24 h periods. Transpiration rates in both wind and still air varied as follows: *Fraxinus* > *Acer* > *Pinus*. Transpiration over a 24 h period was increased by wind in *Fraxinus*, decreased in *Acer*, and was unaffected in *Pinus*. These differences reflected species variation in stomatal size, structure, and control of stomatal aperture. Whereas *Acer* stomata closed rapidly when exposed to wind, those of *Fraxinus* did not. Rapid stomatal closure resulted in high turgor and reduction in water loss despite an increase in the plant—air vapour pressure gradient. In *Fraxinus*, with relatively insensitive stomata, wind had less effect on stomatal closure, and plant turgor was not maintained. In *Fraxinus*, despite eventual stomatal closure, seedlings in wind lost more water than those in still air. In *Pinus*, with sunken stomata which were occluded with cuticular waxes, wind had little effect on transpirational water loss (Davies et al., 1974b).

(*iv*) *Humidity*. Increase in humidity caused stomatal opening of *Fraxinus americana* and *Acer saccharum*; decrease in humidity caused stomatal closure. A change in humidity at high light intensity affected stomatal responses less than humidity changes at low light intensity. Both opening and closing of stomata in response to humidity change occurred much faster in *Acer* than in *Fraxinus* (Davies and Kozlowski, 1974a).

(d) *Epicuticular waxes.* Once the stomata close it becomes critical for the cuticle to prevent further water loss. Transpiration rates of drought-evading plants with closed stomata may vary from 1/5 to

1/50 of the rates when their stomata are open. By comparison, mesophytes lose from 1/2 to 1/5 as much water with stomata closed as they do when their stomata are open (Levitt, 1972).

Large differences in composition and structure of leaf cuticles occur between and within species and between plants of different ages. The genotype, as well as the environment, controls the deposition of epicuticular waxes. "Normal" or glaucous forms are dominant and the "glossy" or glabrous forms are recessive. The differences in appearance are associated with the amount, composition, and structure of leaf wax. "Glossy" forms of several species of plants always had wax deposits consisting of smooth films or of platelets lying flat on the leaf surface. In "glaucous" variants, the wax consisted predominantly of rods or filaments growing outward from the leaf surface (Hall *et al.*, 1965).

Leaf wax composition is a useful taxonomic character to separate species of *Eucalyptus* (Hallam and Chambers, 1970), *Pinus* (Herbin and Sharma, 1969) and *Cupressus* (Herbin and Robbins, 1969; Dyson and Herbin, 1970). Within the eucalypts can be found a very heterogeneous range of leaf waxes. These occur as plates, tubes, or a mixed wax of both plates and tubes. However, most species have simple unornamented waxes.

In some species adaptive changes in waxes occur in response to selection by some ecological variable (Harland, 1947; Barber, 1955). Clinal analysis of glaucousness in species of *Eucalyptus* in Tasmania indicated high selective coefficients for genes controlling this characteristic. The non-glaucous (green) phenotypes were characteristic of the more sheltered ecological habitats, with the glaucous phenotypes becoming increasingly frequent the more exposed the environment. Thus in such species as *E. gigantea, E. pauciflora, E. salicifolia, E. gunnii* and *E. urnigera*, the populations at low altitudinal ranges were uniformly green whereas those at upper altitudinal limits were all glaucous (Barber and Jackson, 1957).

Our scanning electron microscope studies showed wide variations among species in cutinization of angiosperm leaves (Fig. 1). *Fraxinus americana* leaves had cuticular lips overlapping large guard cells. On leaves of *Acer saccharum* and *Cercis canadensis* many stomatal pores were partially or totally occluded by waxy deposits (Kozlowski *et al.*, 1974; Davies and Kozlowski, 1974a).

Some plants show marked resistance to desiccation, partly because

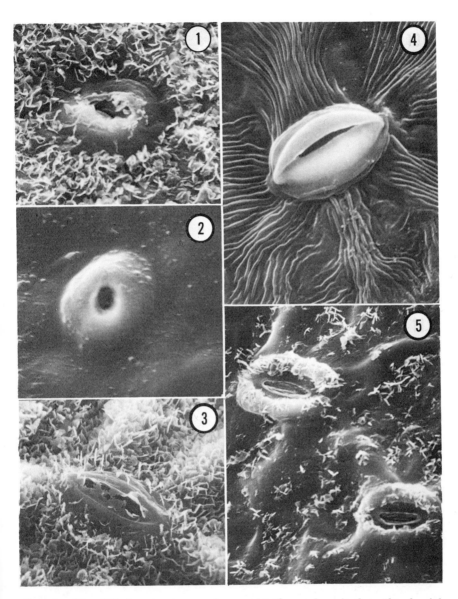

FIG. 1 The differing structure and amounts of wax deposited on the abaxial leaf surfaces of (1) *Acer saccharum*; (2) *Citrus mitis*; (3) *Cercis canadensis*; (4) *Fraxinus americana*; (5) *Quercus macrocarpa*. (From Davies and Kozlowski, 1974a). x1400.

of their thick needle waxes (Hall and Jones, 1961; Horrocks, 1964). In a number of gymnosperms considerable wax is present in stomatal antechambers as well as on the needle surface (Jeffree et al., 1971; Hanover and Reicosky, 1971; Lehala et al., 1972; Davies et al., 1974a), and these antechamber waxes may well be more important than those on epidermal surfaces in controlling water loss. The presence of an antechamber in pines increases the length of the diffusion path for water and CO_2. Wax in the antechamber makes the pathway more tortuous and decreases the cross-sectional area available for diffusion. It has been estimated that wax in antechambers of stomatal pores of *Picea sitchensis* reduced transpiration by approximately two-thirds when the stomata were fully open (Jeffree et al., 1971).

There are marked differences among gymnosperms in amounts and importance of leaf waxes (Oppenheimer, 1970). *Pinus sylvestris* needles have more wax than do *Pinus radiata* needles, and the former species has much better control of cuticular water loss (Leyton and Juniper, 1963; Leyton and Armitage, 1968). Occlusion of stomatal antechambers with wax occurs in *Pinus resinosa* and *P. strobus* needles (Davies and Kozlowski, 1974b; Davies et al., 1974a). In excised needles of *Pinus pinea* and *P. halepensis*, stomatal transpiration during the first 27 h depleted about one-third of the water in each species. Thereafter, transpiration, which was almost entirely cuticular, was much faster in *Pinus pinea*, emphasizing much more effective control of cuticular transpiration in *P. halepensis* (Oppenheimer and Shomer-Ilan, 1963).

2. Stem Adaptations

Photosynthesis of branches and bark appears to be an important adaptation to arid environments. In the Sonoran desert, for example, several plants have rigid thorny branches and functional leaves are almost completely absent. Species include *Canotia holocantha*, *Holocantha emoryi*, *Koeberlinia spinosa* and *Dalea spinosa*. In these plants photosynthesis occurs in the persistent stems and branches (Jaeger, 1955). In Israel the young green branches of *Retama raetam* and *Calligonum comosum* carry on photosynthesis (Fahn, 1964). Bark photosynthesis is very important in *Cercidium floridum*, which grows in the deserts of the south-western United States. This species, which has many small-diameter chlorophyllous branches, is leafless during much of the year. Adams and Strain (1969) showed that stems of leafless trees contributed more than 40% of the total annual photosynthetic production of leafy trees. After a prolonged drought, leafless plants had low rates of photosynthesis but these increased after autumn and winter rains. By mid-winter, photosynthesis in a leafless plant was 86% as great as that in a plant with leaves.

Fahn (1964) suggested that the presence of living wood fibres containing reserve materials was another stem adaptation for survival during severe drought in Jerusalem. Living fibres were found in almost 75% of the plants examined. Other adaptations include (a) the development of a wide cortex that protects vascular tissues from desiccation in early ontogeny, before they develop periderm tissues, and (b) the development of interxylary cork. In *Artemisia cana*, interxylary cork forms early in the growing season as a sleeve over the previous year's xylem increment. The interxylary cork retards desiccation, especially near wounds, and restricts upward water transport to a narrow zone, thereby conserving water (Moss, 1940). Interxylary periderm in *Epilobium angustifolium* also affords the plant protection against desiccation (Moss, 1936).

3. Root Adaptations

Major drought-avoiding adaptations of roots are capacity for extensive root growth and production of adventitious roots near the soil surface.

(a) *Root—shoot ratios.* Because water in the soil which is not penetrated by roots is largely unavailable to plants, trees able to produce deeply penetrating and branching root systems use water most efficiently and prevent or postpone drought injury (Kozlowski, 1972b). For example, on dry sites *Eucalyptus socialis* outcompetes *E. incrassata*, due at least in part to higher root—shoot ratios and slower growth of aerial parts, thus preventing water depletion (Parsons, 1969). Deep rooting also is an important factor in drought avoidance of several species of western Australian sclerophylls (Grieve and Hellmuth, 1970).

In studying genetic variations in root—shoot ratios, it is important to consider that these change with plant age and weight according to the law of allometric growth. Ledig and Perry (1965) emphasized that the differences they found in root—shoot ratios among *Pinus taeda* progenies were due primarily to differences in seedling weights. A useful research approach may involve grafting scion stem pieces on clonal rootstocks. It is known that rootstocks influence meristematic activity and tissue development as well as the morphogenetic potential of apical and lateral meristems in many genera of forest trees (Tubbs, 1967, 1973). In forestry the emphasis in use of rootstocks has been on wood production or early seed production for breeding. We need to extend such work toward achieving drought avoidance through altered root—shoot balance. It seems highly relevant that dwarfing rootstocks or scions induce short laterals and spur shoots in some species, and in others they maintain an effective leaf area in the compound tree over a shorter season than do vigorous rootstocks or scions (Tubbs, 1967, 1973).

(b) *Adventitious roots.* When sands shift and dunes develop, some desert shrubs produce adventitious roots close to the surface where some soil moisture is available. For example, dune stabilization in New Mexico was aided in this way by *Rhus trilobata, Poliomintha incana, Atriplex canescens, Populus wislizenii* and *Yucca elata* (Shields, 1961).

4. Allelopathy

There is a growing body of evidence that some plants can avoid drought by releasing chemicals, which inhibit seed germination and growth of adjacent plants. Such allelopathic chemicals may be released from plants by leaching, volatilization, excretion, exudation, and by decay, either directly or by activity of micro-organisms. Among the naturally-occurring compounds which have inhibitory effects on growth of neighbouring plants are organic acids, lactones, fatty acids, quinones, terpenoids, steroids, phenols, benzoic acids, cinnamic acid, coumarins, flavonoids, tannins, amino acids, polypeptides, alkaloids, cyanohydrins, sulfides, mustard oil glycosides, purines and nucleosides (Rice, 1974).

There are many examples of allelopathic effects. Perhaps the best known allelopathic chemical is juglone in *Juglans*. It is washed into the soil from leaves and fruits and inhibits growth of adjacent plants. Went and Westergaard (1949) observed, in California deserts, that *Larrea* seedlings died in the vicinity of adult *Larrea* plants because of the toxic action of chemicals excreted by roots of adult plants. Inhibition of seed germination by existing shrubs was shown in the case of *Salvia mellifera*, which almost completely prevented establishment of *Adenostema* seedlings under it (Went, 1952). Naturally-occurring plant growth inhibitors are widely distributed in tropical and subtropical vegetation. For example, inhibitors of fruits of *Ilex vomitoria* inhibited germination of seeds of *Prosopis juliflora* (Bovey and Diaz-Colon, 1969). In India *Prosopis juliflora* forms dense thickets of small shrubs to large trees. Very few plants come up within the community of these trees and shrubs and the ground is covered with a thick layer of leaf litter. Inhibitors in the leaves inhibit seed germination and growth of shoots and roots of seedlings (Lahiri and Gaur, 1969).

B. Drought Tolerance

In addition to drought avoidance, the capacity of some woody plants to tolerate severe water deficits over long periods undoubtedly contributes to their drought resistance. Pronounced resistance to desiccation was shown by Slatyer (1961) for *Acacia aneura* of central Australia. The individual plants of this species are low and shrubby and its "leaves" are phyllodes about 1–3 mm wide, 1 mm thick, and 3–10 cm long. The plants persist under extreme aridity, often with no rain for 3 consecutive months. *Acacia aneura* survived extreme desiccation (plant ψ values of -130×10^6 erg g^{-1}). By comparison, when ψ values of tomato (*Lycopersicon esculentum*) dropped to only -5×10^6 erg g^{-1} the plants were killed.

1. Mechanisms of Drought Tolerance

According to Levitt (1972), desiccation tolerance involves variation in indirect metabolic strains (starvation, protein losses) and direct plastic strains. Mechanisms of desiccation tolerance in the case of indirect injury are generally not well understood. On the other hand considerable attention has been given to explain direct drought injury on the basis of mechanical stress and protein aggregation.

Iljin (1930, 1931) believed that it was not desiccation *per se* that killed plants, but rather the mechanical injuries associated with dehydration and rehydration. When plant tissues dry, the cells collapse. Rigid cell walls, when present, oppose the collapse and subject protoplasm to tension which may cause injury. If the walls are soft and thin they are pulled in together with the protoplast. Cells that survive desiccation are subjected to additional stresses on rehydration, and these may cause death. Iljin's theory is supported by various lines of evidence, as summarized by Levitt (1972): (1) cell contraction and pulling in of walls occur on hydration; (2) high tensions have been measured in plant tissues; (3) torn protoplasm sometimes adheres to the cell wall; (4) plasmolysing solutions release walls and prevent tensions; and (5) factors associated with drought tolerance can be explained by this theory.

Since drought injury involves cell dehydration, it has been explained by protein aggregation because of intermolecular bonding of the dehydrated, more closely packed protein molecules. Levitt (1972) suggested that mechanical stress subjects the proteins to a strain, causing incipient denaturation. Concurrently, desiccation brings molecules close together so as to allow for formation of intermolecular bonds between sensitive groups exposed by the incipient denaturation. The plasma membrane proteins, which are subject to mechanical stress, are most likely to undergo these changes.

V. SUMMARY

Various measures of plant water deficits (tissue moisture content, relative water content, saturation deficit and plant water potential) are discussed. Drought resistance in plants is the result of various degrees of drought avoidance and desiccation tolerance. Adaptations for avoiding drought may be found in leaves, stems and roots. Leaf adaptations include leaf shedding; microphylly; small, few and sunken stomata; rapid stomatal closure; abundant leaf waxes; strong development of palisade mesophyll and weak development of spongy mesophyll. Important stem adaptations include bark photosynthesis, living wood fibres, wide cortex and interxylary cork. The major drought-avoiding adaptations of roots are capacity for extensive root growth and production of adventitious roots near the soil surface. Drought avoidance may also involve decreasing plant competition by release of allelopathic chemicals which inhibit seed germination and growth of adjacent plants. Variations among plants in desiccation tolerance have been attributed to indirect metabolic strains (e.g. starvation, protein loss) and direct plastic strains. The indirect strains are not well understood. Direct drought injury has been related to mechanical stress and protein aggregation.

REFERENCES

Adams, M. S. and Strain, B. R. (1969). Seasonal photosynthetic rates in stems of *Cercidium floridum* Benth. *Photosynthetica* 3, 55–62.

Barber, H. N. (1955). Adaptive gene substitution in Tasmanian eucalypts. I. Genes controlling the development of glaucousness. *Evolution* 9, 1–14.

Barber, H. N. and Jackson, W. D. (1957). Natural selection in action in *Eucalyptus. Nature, Lond.* 179, 1267–1269.

Bovey, R. W. and Diaz-Colon, J. D. (1969). Occurrence of plant growth inhibitors in tropical and subtropical vegetation. *Physiol. Plant.* 22, 253–259.

Davies, W. J. and Kozlowski, T. T. (1974a). Stomatal responses of five woody angiosperms to light intensity and humidity. *Can. J. Bot.* 52, 1525–1534.

Davies, W. J. and Kozlowski, T. T. (1974b). Short- and long-term effects of antitranspirants on water relations and photosynthesis of woody plants. *J. Am. Soc. Hort. Sci.* 99, 297–304.

Davies, W. J., Kozlowski, T. T., Chaney, W. R. and Lee, K. J. (1973). Effects of transplanting on physiological responses and growth of shade trees. *Proc. 48th Int. Shade Tree Conf.* (1972), 22–30.

Davies, W. J., Kozlowski, T. T. and Lee, K. J. (1974a). Stomatal characteristics of *Pinus resinosa* and *Pinus strobus* in relation to transpiration and antitranspirant efficiency. *Can. J. Forest Res.* 4, 571–574.

Davies, W. J., Kozlowski, T. T. and Pereira, J. (1974b). Effect of wind on transpiration and stomatal aperture of woody plants. *In* "Mechanisms of Regulation of Plant Growth". (R. L. Bieleski, A. R. Ferguson, and M. M. Creswell, eds). Bull. 12, Royal Society of New Zealand.

Dyson, W. G. and G. A. Herbin. (1970). Variation in leaf wax alkanes in cypress trees grown in Kenya. *Phytochemistry* 9, 585–589.

Eiten, G. (1972). The cerrado vegetation of Brazil. *Bot. Rev.* **38**, 201—341.
Fahn, A. (1964). Some anatomical adaptations of desert plants. *Phytomorphology* **14**, 93—102.
Ferri, M. G. (1953). Water balance of plants from the "Caatinga". *Revta bras. Biol.* **13**, 237—244.
Ferri, M. G. (1961). Problems of water relations of some Brazilian vegetation types with special consideration of the concepts of xeromorphy and xerophytism. *In* "Plant-water Relationships in Arid and Semi-arid Conditions", pp. 191—197. UNESCO, Madrid Symposium.
Gardner, J. L. (1963). Aridity and agriculture. *In* "Aridity and Man". (C. Hodge and P. C. Duisberg, eds), pp. 239—276. Publ. 74. Am. Assoc. for the Advancement of Science. Washington, D.C.
Grieve, B. J. and Hellmuth, E. O. (1970). Eco-physiology of western Australian plants. *Oecol. Plant.* **5**, 33—68.
Hall, D. M. and Jones, R. L. (1961). Physiological significance of surface wax on leaves. *Nature, Lond.* **191**, 95—96.
Hall, D. M., Matus, A. I., Lamberton, J. A. and Barber, H. N. (1965). Intraspecific variation in wax on leaf surfaces. *Aust. J. biol. Sci.* **18**, 323—332.
Hallam, N. D. and Chambers, T. C. (1970). The leaf waxes of the genus *Eucalyptus* L. Heritier. *Aust. J. Bot.* **18**, 335—386.
Hanover, J. W. and Reicosky, D. A. (1971). Surface wax deposits on foliage of *Picea pungens* and other conifers. *Am. J. Bot.* **58**, 681—687.
Harland, S. C. (1947). An alteration in gene frequency in *Ricinus communis* L. due to climatic conditions. *Heredity* **1**, 121—125.
Hellmuth, E. O. (1968). Eco-physiological studies on plants in arid and semi-arid regions in western Australia. I. Autoecology of *Rhagodia baccata* (Labill.) Moq. *J. Ecol.* **56**, 319—344.
Hellmuth, E. O. (1969). Ecophysiological studies on plants in arid and semi-arid regions in western Australia. II. Field physiology of *Acacia craspedocarpa* F. Muell. *J. Ecol.* **57**, 613—634.
Herbin, G. A. and Robins, P. A. (1969). Patterns of variation and development in leaf wax alkanes. *Phytochemistry* **8**, 1985—1998.
Herbin, G. A. and Sharma, K. (1969). Studies on plant cuticular waxes. V. The wax coatings of pine needles: a taxonomic survey. *Phytochemistry* **8**, 151—160.
Hillel, D. (1971). "Soil and Water". Academic Press, New York and London.
Horrocks, R. L. (1964). Wax and the water vapor permeability of apple cuticle. *Nature, Lond.* **203**, 547.
Iljin, W. S. (1930). Die Ursache der Resistenz von Pflanzenzellen gegen Austrocknung. *Protoplasma* **10**, 379—414.

Iljin, W. S. (1931). Austrocknungsresistenz des Farnes *Notochlaena Marantae* R. Br. *Protoplasma* 13, 322—330.
Jaeger, E. C. (1955). "The California Deserts". Stanford Univ. Press, Stanford, California.
Jarvis, J. G. and Jarvis, M. S. (1963). The water relations of tree seedlings. II. Transpiration in relation to soil water potential. *Physiol. Plant.* 16, 236—253.
Jeffree, C. E., Johnson, R. P. C. and Jarvis, P. G. (1971). Epicuticular wax in the stomatal antechamber of Sitka spruce and its effects on the diffusion of water vapor and carbon dioxide. *Planta* 98, 1—10.
Kaul, R. N. and Kramer, P. J. (1965). Comparative drought tolerance of two woody species. *Indian Forester* 91, 462—469.
Kozlowski, T. T. (1964). "Water Metabolism in Plants". Harper and Row, New York.
Kozlowski, T. T. (1968). "Water Deficits and Plant Growth", Vol. I. Academic Press, New York and London.
Kozlowski, T. T. (1971). "Growth and Development of Trees", Vols I and II. Academic Press, New York and London.
Kozlowski, T. T. (1972a). Shrinking and swelling of plant tissues. *In* "Water Deficits and Plant Growth". (T. T. Kozlowski, ed.), chapter 1. Vol. III. Academic Press, New York and London.
Kozlowski, T. T. (1972b). Physiology of water stress. *In* "Wildland Shrubs — Their Biology and Utilization" pp. 229—244. U.S.D.A. Forest Service General Tech. Rept. INT-1. Ogden, Utah.
Kozlowski, T. T. and Clausen, J. J. (1965). Changes in moisture contents and dry weights of buds and leaves of forest trees. *Bot. Gaz.* 126, 20—26.
Kozlowski, T. T., Davies, W. J. and Carlson, S. D. (1974). Transpiration rates of *Fraxinus americana* and *Acer saccharum* leaves. *Can. J. Forest Res.* 4, 259—267.
Lahiri, A. N. and Gaur, Y. D. (1969). Germination studies on arid zone plants. V. The nature and role of germination inhibitors present in leaves of *Prosopis juliflora. Nat. Inst. Sci. of India Proc.* 35B, 60—71.
Ledig, F. T. and Perry, T. O. (1965). Physiological genetics of the shoot—root ratio. *Proc. Soc. Am. Foresters*, 39—43.
Lehala, A., Day, R. J. and Koran, Z. (1972). A close-up of the stomatal region of white spruce and jack pine. *Forest Chron.* 48, 32—35.
Levitt, J. (1972). "Responses of Plants to Environmental Stresses". Academic Press, New York and London.
Leyton, L. and Armitage, I. P. (1968). Cuticle structure and water relations of the needles of *Pinus radiata* (D. Don.). *New Phytol.* 67, 31—38.
Leyton, L. and Juniper, B. E. (1963). Cuticle structure and water relations of pine needles. *Nature, Lond.* 198, 770—771.

Lopushinsky, W. (1969). Stomatal closure in conifer seedlings in response to leaf moisture stress. *Bot. Gaz.* 130, 258–263.

Lopushinsky, W. and Klock, G. O. (1974). Transpiration of conifer seedlings in relation to soil water potential. *Forest Sci.* 20, 181–186.

Moss, H. E. (1936). The ecology of *Epilobium angustifolium* with particular reference to rings of periderm in the wood. *Am. J. Bot.* 23, 114–120.

Moss, H. E. (1940). Interxylary cork in *Artemisia* with a reference to its taxonomic significance. *Am. J. Bot.* 27, 762–768.

Oppenheimer, H. R. (1953). An experimental study on ecological relationships and water expenses of Mediterranean forest vegetation. *Palestine J. Bot. Rehovot Ser.* 8, 103–124.

Oppenheimer, H. R. (1970). Drought resistance of cypress and thuya branches. *Israel J. Bot.* 19, 418–428.

Oppenheimer, H. R. and Shomer-Ilan, A. (1963). A contribution to the knowledge of drought resistance of Mediterranean pine trees. *Mitt. Flor.-soz. Arbeitsg. Stolz.* (Weser) N.S. 10, 42–55.

Orshansky, G. (1954). Surface reduction and its significance as a hydroecological factor. *J. Ecol.* 42, 442–444.

Parsons, R. F. (1969). Physiological and ecological tolerances of *Eucalyptus incrassata* and *E. socialis* to edaphic factors. *Ecology* 50, 386–390.

Quraishi, M. A. and Kramer, P. J. (1970). Water stress in three species of *Eucalyptus*. *Forest Sci.* 16, 74–78.

Rice, E. L. (1974). "Allelopathy". Academic Press, New York and London.

Shields, L. M. (1961). Morphology in relation to xerophytism. In "Bioecology of the Arid and Semi-arid Lands of the Southwest" (L. M. Shields and L. J. Gardner, eds), pp. 15–22. New Mexico Highlands Univ., Las Vegas, New Mexico.

Siwecki, R. and Kozlowski, T. T. (1973). Leaf anatomy and water relations of excised leaves of six *Populus* clones. *Arboretum kórn.* 18, 83–105.

Slatyer, R. O. (1961). Internal water balance of *Acacia aneura* F. Muell. in relation to environmental conditions. In "Plant–water Relationships in Arid and Semi-Arid Conditions", pp. 137–146. UNESCO, Madrid Symposium.

Slatyer, R. O. (1964). Efficiency of water utilization by arid zone vegetation. *Ann. Arid Zone* 3, 1–12.

Tal, M. (1966). Abnormal stomatal behavior in wilty mutants of tomato. *Plant Physiol.* 41, 1387–1391.

Tal, M. and Imber, D. (1970). Abnormal stomatal behavior and hormonal imbalance in flacca, a wilty mutant of tomato. II. Auxin and abscisic acid-like activity. *Plant Physiol.* 46, 373–376.

Tal, M. and Imber, D. (1972). The effect of abscisic acid on stomatal behavior in *flacca*, a wilty mutant of tomato, in darkness. *New Phytol.* 71, 81–84.

Tubbs, F. R. (1967). Tree size control through dwarfing rootstocks. *Proc. VII Int. Hort. Congr.* (1966) 3, 43–56.

Tubbs, F. R. (1973). Research fields in the interaction of rootstocks and scions in woody perennials. *Hort. Abstr.* 43, 247–253; 325–334.

Waggoner, P. E. and Simmonds, N. W. (1966). Stomata and transpiration of droopy potatoes. *Plant Physiol.* 41, 1268–1271.

Went, F. W. (1952). Fire and biotic factors affecting germination. *Ecology* 33, 351–364.

Went, F. W. and Westergaard, M. (1949). Ecology of desert plants. III. Development of plants in the Death Valley National Monument, California. *Ecology* 30, 26–38.

Wuenscher, J. E. and Kozlowski, T. T. (1971a). The response of transpiration resistance to leaf temperature as a desiccation resistance mechanism in tree seedlings. *Physiol. Plant.* 24, 254–259.

Wuenscher, J. E. and Kozlowski, T. T. (1971b). Relationship of gas-exchange resistance to tree seedling ecology. *Ecology* 52, 1016–1023.

19
Physical Parameters of the Soil–Plant–Atmosphere System: Breeding for Drought Resistance Characteristics That Might Improve Wood Yield

M. T. TYREE
Department of Botany, University of Toronto, Canada

I.	Introduction	329
II.	The Ability of Plants to Draw on Soil Water Reserves	330
	A. Water Potential and Water Movement	330
	B. Soil Water Potential Isotherms and Drought Tolerance	331
	C. Resistances to Water Flow in the Soil–Plant System	332
III.	The Measurement of Physical Water Relations Parameters in Leaves and Shoots	334
	A. Analysis of "Pressure–Volume Curves"	334
	B. Sources of Error	338
IV.	Drought-Resistance Characteristics That Might Improve Wood Yield	339
	A. Osmotic Pressure	340
	B. The Elastic Modulus of Cell Walls	340
	C. Negative Turgor Pressures?	343
	D. Important Characteristics of Drought Avoidance	345
V.	Summary	346
References		346

I. INTRODUCTION

This contribution has two objectives. The first is to define the physical water relations parameters in the soil–plant–atmosphere system; this is relatively easy to do. The other objective is to indicate which water relation parameters "might be worth characterizing by physiological geneticists"; this second objective is not so easy to reach, and I have personally found it to be a very challenging question which, it seems to me, is bound to invoke an invidious response however I attempt to answer it.

Kozlowski (Chapter 18) gives an excellent compendium of the morphological shoot and root characteristics that appear to confer on woody plants a kind of drought resistance that can be classified as drought *avoidance*. It is clear that characteristics of drought avoidance have allowed certain species to invade many arid and semi-arid regions. However, some characteristics of drought avoidance are correlated with low wood yield. I have decided to confine my deliberations to the water relations parameters that might confer marginally greater drought *tolerance*. I believe a greater drought tolerance will permit species to maintain longer periods of net CO_2 assimilation (and therefore yield more wood) whether the tree grows in a semi-arid climate or in a temperate climate with only occasional periods of drought.

II. THE ABILITY OF PLANTS TO DRAW ON SOIL WATER RESERVES

Because most of the net CO_2 assimilation in commercial trees occurs in leaves, it seems logical to focus our initial attention on leaves. Leaves must maintain a pathway for the diffusion of CO_2 from the atmosphere, because of this leaves unavoidably experience evaporative water loss. Net water loss must occur if the leaf water potential is to fall to a low enough level to extract water from the soil and partly replace evaporative loss.

A. Water Potential and Water Movement

Two factors determine the rate at which water flows from the soil to the leaf: one is the water potential (Ψ) difference between the soil and leaf and the other is the overall resistance to water movement from the soil to the leaf. Water naturally flows from a region of higher water potential to one of lower water potential. Thus, in order for water to flow from the soil to the leaf, the Ψ of the bulk soil must exceed the Ψ of the root surface which must exceed the Ψ of the leaf (in symbols $\Psi_{bulk\ soil} > \Psi_{root\ surface} > \Psi_{leaf}$).

B. Soil Water Potential Isotherms and Drought Tolerance

I shall equate drought tolerance with the ability of a leaf to maintain a 24 h averaged positive net CO_2 assimilation, at low Ψ_{leaf}. Since stomata must remain open to sustain net CO_2 assimilation, our success at increasing yield by increasing drought tolerance turns very strongly on the soil characteristics of the region in which we propose to exercise silviculture. This point is illustrated by Fig. 1 which shows the water potential isotherms of three soil types. Let us imagine that strain A can maintain a positive net CO_2 assimilation

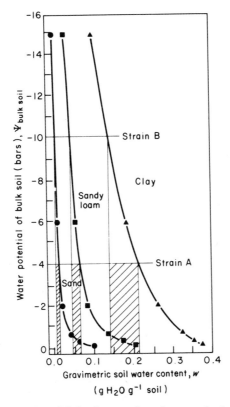

FIG. 1 Soil water potential isotherms for clay, sandy loam and sand soils adapted from Slatyer (1967, Fig. 3.3). Strain A and strain B represent hypothetical limits of soil water potential from which plants can extract soil water while maintaining a relatively high net CO_2 assimilation rate. Strain B has the advantage of being able to withdraw a larger fraction of the gravimetric soil water content than strain A (cf. widths of hatched areas).

(and a concomitantly high daytime evapotranspiration rate) down to $\Psi_{bulk\ soil} = -4$ bars whereas strain B can accomplish the feat down to $\Psi_{bulk\ soil} = -10$ bars. (Remember that Ψ_{leaf} in both strains is probably considerably more negative than $\Psi_{bulk\ soil}$). In a clay soil strain B would be able to draw on a much larger fraction of the gravimetric soil water content, w, than strain A (the difference is a $\Delta w = 0.073$), but in a sandy soil the advantage is negligible (the difference is $\Delta w = 0.007$). Without additional soil water reserves to draw upon, the stomata will close early in the day during a drought period.

C. Resistances to Water Flow in the Soil–Plant System

When net CO_2 assimilation is near the optimum level during daylight hours, the water flow rate q (cm^3 s^{-1} per plant) is high and therefore the difference in water potential between the leaf and bulk soil ($\Delta\Psi = \Psi_{leaf} - \Psi_{bulk\ soil}$) is large. The flow rate, q, and $\Delta\Psi$ are related through $R_{plant} + R_{soil}$, the sum of the resistances to water flow in the plant and soil,

$$q = \frac{\Delta\Psi}{R_{soil} + R_{plant}}. \qquad (1)$$

Growth and net CO_2 assimilation studies on tree seedlings indicate that a marked decline in net CO_2 assimilation and q occurs when Ψ_{leaf} falls low enough for stomatal closure to begin ($\Psi_{leaf} = -15$ to -30 bar Jarvis and Jarvis, 1963c, 1963d; Brix, 1962), but this decline in q, Ψ_{leaf}, and net CO_2 assimilation occurs at a relatively high $\Psi_{bulk\ soil}$ equal to -2 to -5 bars (Jarvis and Jarvis, 1963a, 1963b; Babalola et al., 1968). Since $\Delta\Psi (= \Psi_{bulk\ soil} - \Psi_{leaf})$ is large under these conditions $R_{soil} + R_{plant}$ must be substantial.

It is not certain which resistance (R_{soil} or R_{plant}) is larger. Part of the uncertainty arises because R_{soil} is very strongly dependent on the relative water content of the soil; as soils dry the mode of water transport shifts over from the relatively efficient bulk fluid flow to the relatively inefficient vapour transport. As the $\Psi_{bulk\ soil}$ drops from -0.1 to -10 bar the specific soil resistance can change by six orders of magnitude (see Fig. 2). Plants, unlike soils, maintain continuous columns of water and consequently maintain bulk fluid flow at quite low water potentials; therefore I would expect R_{plant} to be relatively constant. Theoretical calculations by Gardner (1960) and Cowan (1965) suggest that in dry soils R_{soil} exceeds R_{plant} so that $\Psi_{bulk\ soil} - \Psi_{root\ surface}$ is quite large compared to $\Psi_{root\ surface} - \Psi_{leaf}$. These calculations have been questioned by Williams (1974) who suggests that rooting densities were underestimated and soil water fluxes were overestimated by Gardner (1960) and Cowan (1965). Experiments by Lawlor (1972) on pasture grass also suggest that $R_{plant} > R_{soil}$ (see Newman, 1969, for a review of the experimental evidence that bears on this problem with regard to herbaceous plants). The relative resistance, R_{plant} compared to R_{soil}, in trees is unknown. Hellkvist et al. (1974) have compiled evidence that shows R_{plant} of large trees is one to three orders of magnitude *less* than R_{plant} of herbaceous species! Therefore the situation in trees needs to be studied in more detail.

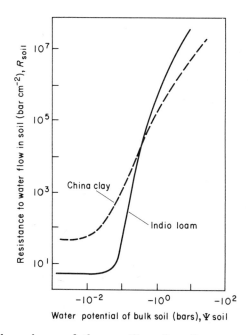

FIG. 2 The dependence of the specific soil resistance to water transport, R^*_{soil}, on the bulk soil water potential, Ψ_{soil}, adapted from Slatyer (1967, Fig. 4.6). The specific soil resistance is defined by $R^*_{soil} = (d\Psi_{soil}/dx)/J$, where J is the flux of soil water cm^{3}H$_2$O s^{-1} cm^{-2} driven by a water potential gradient of $d\Psi_{soil}/dx$ bar cm^{-1}.

The importance of knowing the dependence of R_{soil} on $\Psi_{bulk\ soil}$ became evident when I carried out some (unpublished) theoretical calculations. Under any given extraction rate (q) and rooting density a theoretical limit is reached at which $\Psi_{root\ surface}$ approaches minus infinity at a finite $\Psi_{bulk\ soil}$. Whether or not this theoretical limit is reached at bulk soil water potentials above the permanent wilt point for trees depends on several parameters which can only be guessed given our present knowledge of soil water extraction by trees. But we clearly do not want to breed for characteristics of drought tolerance that would be of no value to increasing yield if said characteristics (metaphorically speaking) brought the tree up against theoretical limitations imposed by soil physics characteristics.

III. THE MEASUREMENT OF PHYSICAL WATER RELATIONS PARAMETERS IN LEAVES AND SHOOTS

Before proposing that leaf water relations parameters might confer greater drought tolerance, I shall explain what can be measured in shoots and leaves by the Scholander-Hammel pressure bomb.

A. Analysis of "Pressure—Volume Curves"

The pressure-bomb technique, as developed by Scholander *et al.* (1964, 1965) has been very useful in studying the water relations of higher plants. The technique can be used to construct an entire Höfler diagram relating the osmotic pressure, the turgor pressure, and the water potential of leaves and shoots to the relative water content. The pressure-bomb technique involves the measurement of all or part of what is called a "pressure—volume curve". A leaf or shoot is completely enclosed inside a pressure chamber except for the cut end of the stem or petiole, which protrudes through an air-tight seal into the open air (see Fig. 3). At the time of collection the shoot had an original water potential,

$$\text{original } \Psi = P_{VAT} - \pi \qquad (2)$$

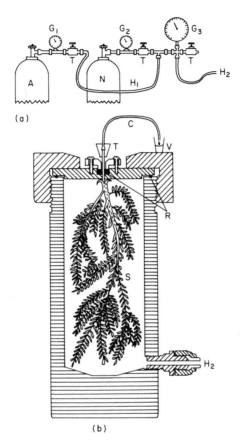

FIG. 3 The Scholander-Hammel pressure bomb apparatus. Part A. The arrangement of gas tanks, pressure gauges and gas taps. (A = compressed air tank; G = pressure gauges monitoring the air tank pressure (G_1), the nitrogen tank pressure (G_2), and bomb pressure (G_3); H = flexible hose connecting air and nitrogen sources (H_1) and gas sources to the bomb (H_2); N = compressed air tank; T = gas taps). Part B. A sectional view of the pressure bomb containing a hemlock shoot being subjected to a gas pressure greater than the balance pressure. (C = flexible capillary tube (microcatheter); H_2 = flexible hose connecting pressure bomb to gas source; R = rubber gaskets; S = shoot of hemlock; T = translucent silicone-rubber stopper; V = vial collecting expressed xylem sap.)

where P_{VAT} = the volume averaged turgor (VAT) pressure (= Ψ_P in Kozlowski's terminology) and π = the bulk osmotic pressure (= $-\Psi_S$ in Kozlowski's terminology). In the bomb the gas pressure, P, can be controlled. The gas pressure increases Ψ in an additive fashion:

$$\text{in the bomb } \Psi = P_{VAT} - \pi + P. \tag{3}$$

At some balance pressure P^* the water potential of the shoot just reaches zero and nearly pure xylem water appears at the cut end of the shoot; thus at the balance pressure,

$$\Psi = 0 = P_{VAT} - \pi + P^*,$$

therefore

$$P^* = \pi - P_{VAT} = - \text{original } \Psi. \tag{4}$$

The shoot can be artifically dehydrated in the bomb by applying pressures above P^* and the volume of fluid expressed, V_e, can be measured. One can deduce the data for a Höfler diagram by analysing the dependence of other P^*s on V_e. The analysis turns on the verifiable assumption that when enough water is expressed, i.e. when V_e is large enough, the VAT pressure, P_{VAT}, falls to, and remains at, zero for greater values of V_e. In this region

$$P^* = \pi = \frac{RTN_S}{V} = \frac{RTN_S}{V_0 - V_e} \tag{5}$$

where N_S = the total number of osmoles of solute in the living cells or symplasm, V = the total volume of water remaining in the living cells or symplasm, and $V = V_0 - V_e$ where V_0 is the symplasmic water volume at the time the shoot was first placed in the bomb. A plot of $1/P^*$ versus V_e should give a straight line when P_{VAT} is zero (whereas a curved line would result if P_{VAT} were nonzero); this can be seen by taking the inverse of eqn (5)

$$\frac{1}{P^*} = \frac{1}{\pi} = \frac{V_0}{RTN_S} - \frac{1}{RTN_S} V_e \tag{6}$$

The intercept, V_0/RTN_S, is the inverse of the original bulk osmotic pressure $\pi_0 = RTN_S/V_0$; the slope is $-1/RTN_S$; and from the intercept and slope V_0 can be calculated (see Tyree and Hammel, 1972; and Tyree et al., 1973).

The pressure-bomb technique has been refined to the point that "pressure–volume curves" can be obtained on single leaves (of broad leaf species) weighing at least 0.15 g in fresh weight provided the leaf has a petiole at least 1 cm long (Cheung et al., 1975a). A typical result on a sugar maple leaf (*Acer saccharum*) is shown in Fig. 4. In this sample the "pressure–volume curve" is nonlinear for V_e up to 0.20 cm^3 because both P_{VAT} and π contribute to P^*; beyond $V_e = 0.20$ cm^3 $1/P^* = 1/\pi$. The straight line extrapolated back to point A gives $1/\pi_0$; extrapolating the other direction to B′ gives $V_0 = 0.65$ cm^3 (which can also be computed from the slope and intercept point A). The point E at which the curve first becomes linear is difficult to determine accurately but reading across gives $1/\pi_p$ where π_p is the osmotic pressure at "incipient plasmolysis", i.e.

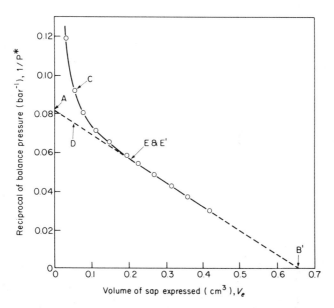

FIG. 4 Pressure–volume curve obtained for leaves of sugar maple (*Acer saccharum*) by plotting volumes of sap expressed, V_e, against the reciprocals of balance pressures $1/P^*$. Points labelled with letters on the pressure volume curve are read from the ordinate unless they are primed (e.g. E') in which case they are to be read from the abscissa. A is the inverse of the bulk osmotic pressure, π_0, B' is the original volume of osmotic water, V_0. $1/D - 1/C$ is the volume averaged turgor (VAT) pressure at V_e equal to 0.05 cm^3 on this graph. E is the inverse of the bulk osmotic pressure at incipient plasmolysis, π_p, $B' - E'$ is the osmotic water volume at incipient plasmolysis, V_p.

where P_{VAT} first becomes zero. For V_es between 0.0 and 0.2 cm^3 P_{VAT} can be determined. For example, drawing a vertical line from $V_e = 0.05$ cm^3 two intercepts are obtained, D and C. The inverse of D read off the vertical axis gives π(at $V_e = 0.05$ cm^3) and the inverse of C read off the vertical axis gives π(at $V_e = 0.05$ cm^3) $- P_{VAT}$ (at $V_e = 0.05$ cm^3) from which P_{VAT} can be determined by difference.

B. Sources of Error

The pressure–volume curve can be used to deduce the two main components of leaf water potential (π and P_{VAT}) by an analysis of the leaf water potential isotherm (Ψ_{leaf} versus some measure of leaf water content). It is important to note that the entire analysis can be done with acceptable accuracy without ever explicitly invoking a dependence of the leaf water potential on matric potential (Ψ_m). The matric potential is generally a poorly understood and frequently misused concept. In leaf water potential isotherms the matric potential (and surface tension effects) need to be invoked only to understand why apoplastic (cell wall and xylem) water is in equilibrium with symplasmic (cell) water. It is in the cell wall that the matric potentials are most important, but in the cell wall the matric potential changes very rapidly with very small cell wall water volume changes; for this reason Ψ_m does not enter significantly into the Ψ_{leaf} isotherm analysis and can probably be safely disregarded.

It is important to maintain leaves and shoots in an aerobic environment during pressure bomb measurements; leaf cells of many species die and rupture under anaerobiosis. This can be prevented by mixing N_2 with compressed air to maintain the partial pressure of O_2 between 1 and 5 times the atmospheric level (Tyree *et al.*, 1973; Cheung *et al.*, 1975a). Oxygen concentrations exceeding 10–15 times the atmospheric level also affect pressure bomb measurements.

A very legitimate concern can be raised about the pressure bomb evaluation of leaf water potential. If the measurement of Ψ_{leaf} by the bomb is in error then the analysis of the isotherm will be in error. Kaufmann (1968) reported quite substantial deviations between Ψ_{leaf} measured by the pressure bomb and Ψ_{leaf} measured by a thermocouple psychrometer technique. Talbot *et al.* (1975) refer to 32 woody and herbaceous species on which comparisons have been made between Ψ_{leaf} measurements by the pressure bomb and some other (usually psychrometer) method. In 14 species agreement is found between techniques within experimental error. In 18 species agreement is not found, but in 17 of these, cut leaves were used and we suggest that in some of these cases the errors arose because leaf tissue was cut before being put into the psychrometer. The leaves of *Tsuga canadensis* and *Picea abies* are much more sensitive to cutting than sunflower and pepper leaves (Barrs and Kramer, 1969). Careful work still needs to be done to resolve these questions. However, if anyone seriously doubts the accuracy of the pressure bomb on any one species, an identical, albeit more tedious, analysis will yield values of P_{VAT} and π by use of water potential isotherms obtained from the psychrometer technique. The analysis involves a plot of $-1/\Psi_{leaf}$ versus relative water content (RWC) as defined by Weatherley (see Fig. 7).

Serious errors can arise when "pressure–volume curves" are generated from large shoots of broad-leafed species bearing large vessels because gas pressures in the bomb can drive symplasmic water into originally empty vessels; this affects the accurate determination of the balance pressure, P^*. Alternatively high gas pressures can expel water from originally filled vessels or pithy tissue giving a false balance pressure. These problems can be minimized by doing pressure bomb analyses on single leaves of broad leaf species because this eliminates the possibility of symplasmic water exchange with large volumes of woody stem. I have not attempted pressure bomb analyses on single conifer needles, but can foresee some technical problems. But symplasmic water exchange with the woody stems of conifers is much less likely because the boardered pits of tracheids in conifers prevent the filling or emptying of tracheids even when high pressure differentials exist between the air outside the tracheid and water within the tracheid.

IV. DROUGHT RESISTANCE CHARACTERISTICS THAT MIGHT IMPROVE WOOD YIELD

Stocker (1956, p. 732) very aptly pointed out that the survival of a plant in dry habitats has to do with its ability "to muddle through between thirst and hunger". Breeding for greater drought resistance combined with greater yield must entail making the tree more tolerant to "thirst" so that it can avoid "hunger" further into the drought period between rains. At the stomatal level this means reaching a better compromise between photoactive opening and hydroactive closure; we want to postpone closure of stomata until lower water potentials are reached, and we want the leaves to tolerate these low water potentials. The advantage to be gained is that the tree can further draw on soil water reserves between rains (see Section II B).

A. Osmotic Pressure

A high osmotic pressure (= a low solute potential, Ψ_S) ought to confer a large measure of drought tolerance. I suggest that we should breed for large values of π_o and π_p. The osmotic pressure at "incipient plasmolysis" π_p sets the lower limit to which the leaf water potential can fall when the VAT pressure first reaches zero ($\Psi_{leaf}(P_{VAT} = 0) = -\pi_p$). Although hydroactive stomatal closure may reach its completion in a leaf before $\Psi_{leaf} = -\pi_p$ there may well be a correlation between Ψ_{leaf} at complete closure and π_p. Although my general textbook knowledge of the competitive photoactive opening and hydroactive closing of stomata seems to support my prediction, I am aware that the situation may not be as simple as I suggest. I refer you to the iconoclastic observations of Schultze et al., (1972); in the Negev desert these authors observed the stomatal response of apricot (*Prunus armeniaca*) and two desert shrubs (*Hammada scoparia* and *Zygophyllum dumosum*). Although the mechanism of stomatal action in these three species was not elucidated, they produced clear evidence that the degree of stomatal opening responded directly to the absolute humidity of the air and not at all to the RWC of the photosynthetic organs. These and other species do also appear to respond directly to leaf temperature and Ψ_{leaf} (Schultze et al., 1973). Clearly it would be prudent to screen forest trees for the dependence of stomatal aperture on absolute humidity before π_p is used as a parameter for genetic screening.

B. The Elastic Modulus of Cell Walls

During the active photosynthetic period leaves generally undergo a net water loss, i.e. the RWC declines. A leaf must undergo a net water loss in order to reduce its water potential because this increases

19. Drought Resistance: Physical Parameters

the rate of water extraction from the soil (see eqn 1). A decline in the RWC tends to increase the concentration of all the leaf cell solutes proportionately. However, the reaction rates at different steps in the metabolic pathway respond in a differential and disproportionate way to substrate concentration changes; these complicated events could lead to rather large changes in substrate concentrations and could cause a reduction in net CO_2 assimilation or in the rate of wood formation.

Plant cells reduce the magnitude of variation of cell sap concentration through the regulation of turgor pressure. The factor that determines the rate of change of turgor pressure with cell volume is the elastic modulus of the cell wall. If the turgor pressure falls rapidly with the decline in RWC then $\Psi_{leaf}(= P_{VAT} - \pi)$ will decline rapidly with only a small increase in π. For a leaf the parameter that measures the rapidity of change of P_{VAT} with water content is the bulk modulus of elasticity, ϵ, defined by

$$\epsilon = \frac{dP_{VAT}}{dF} \tag{7}$$

where
$$F = \frac{V - V_P}{V_P}. \tag{8}$$

The quantity V_p is the volume of symplasmic water in the leaf when P_{VAT} first reaches zero, and V is the symplasmic volume when P_{VAT} is greater than zero. The bulk elastic modulus, ϵ, is equal to the slope of a plot of P_{VAT} versus F as determined from a "pressure–volume curve". We want to breed for a large value of ϵ.

There is no reason to believe that ϵ would be a constant, and the value of ϵ determined from a "pressure–volume curve" is not related in any simple way to the elastic properties of the individual cells. This makes comparison of ϵ between phenotypes complicated. Hellkvist *et al.* (1974) report that ϵ in Sitka spruce increases linearly with P_{VAT}, i.e. ϵ is not constant. Cheung *et al.* (1975b) have shown, however, that the inconstancy of ϵ in Sitka spruce does not necessarily reflect a true inconstancy of the cell wall properties of the individual cells. A theoretical analysis has shown that a hypothetical shoot, in which each cell has the same constant value of elastic modulus but in which π_0 of the individual cells is normally distributed about a mean, will generate a "pressure–volume curve" from which ϵ appears to be linearly dependent on P_{VAT}. Cheung *et al.* (1975b) computed values of ϵ for several species of trees (Fig. 5) and found that ϵ generally assumes a constant value above a certain value of P_{VAT}; the value of ϵ at high P_{VAT} more nearly reflects the elastic properties of the plant cell walls. Because of this, I suggest that we should breed for large values of ϵ computed for samples near maximum turgor pressure.

Figure 6 illustrates how osmotic pressure and VAT pressure combine to confer drought tolerance. It is a modified Höfler diagram for two very different trees growing 2000 miles apart in two different habitats, but this need not detract from the usefulness of the comparison. The π and Ψ_{leaf} are plotted against relative symplasmic water loss ($= (V_0 - V)/V_0$) for *Salix lasiandra* (growing in river sand bars in a semi-arid region of Alberta) and *Ginkgo biloba*

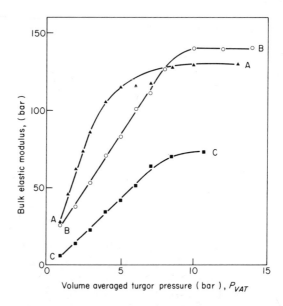

FIG. 5 The relation between the bulk modulus of elasticity, ϵ, and the volume averaged turgor pressure, P_{VAT}. Curve AA is for *Acer saccharum*; curve BB is for *Populus balsamifera*, and curve CC is for a hypothetical leaf, i.e. a model in which every cell has the same constant elastic modulus (= 100 bar and independent of P_{VAT}) but in which the cell osmotic pressures are normally distributed about a mean. The pressure–volume analysis underestimates ϵ even for high values of P_{VAT}.

19. Drought Resistance: Physical Parameters

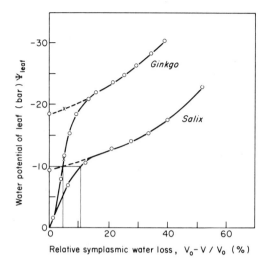

FIG. 6 A modified Höfler diagram for leaves of *Ginkgo biloba* and *Salix lasiandra*, showing how leaf water potential and osmotic pressure depend on the relative loss of symplasmic water, given by $(V_0 - V)/V_0$ where V_0 = symplasmic water volume before tissues were pressurized in Scholander-Hammel bomb, and V = volume of water remaining in symplasm after pressurization.

(cultivated on the campus of the University of Toronto). *G. biloba* has a much larger π_p than *S. lasiandra* (22.2 bar versus 11.8 bar), therefore *G. biloba* leaves can probably sustain much lower leaf water potentials before stomatal closure is complete. Turgor pressure regulation in both species accounts for most of the decline of Ψ_{leaf} as relative osmotic water loss increases up to about 16% when P_{VAT} becomes zero for both species. But ϵ for *G. biloba* is about twice the ϵ of *S. lasiandra*; consequently when both species have reached a leaf water potential of (say) -10 bar *G. biloba* has lost only 4.8% of its osmotic water whereas *S. lasiandra* has lost 10.8% of its osmotic water.

C. Negative Turgor Pressures?

Since it can be argued that the regulation of positive turgor pressures may be an important characteristic of drought tolerance, it is natural to ask whether this mode of regulation can extend into the region of negative turgor pressures. Negative turgors in the sense of water tensions routinely exist in the xylem; can sclerophyllic leaves of desert species maintain negative pressures (tensions) in the protoplasm? From the literature the answer would appear to be yes. Slatyer's (1960) data on *Acacia aneura* is frequently quoted as evidence for negative turgor pressure down to −60 bar. Substantial negative turgor pressures have been reported in *Pennisetum typhoides* (Begg et al., 1964), *Ceratonia siliqua* (Noy-Meir and Ginzburg, 1969), *Larrea divaricata* (Odening, 1971) and *Artemisia herba-alba* (Kappen et al., 1972).

In all cases the negative turgor pressures were deduced from two measurements. First the water potential of the leaf tissue is measured in a reliable way. Second the leaf tissue is either mashed or frozen and thawed, and the water potential of the sample (or sap extracted from the sample) so treated is taken as a measure of the cell osmotic pressure. By difference the turgor pressure is deduced. The disadvantage of this technique is that the magnitude of the osmotic pressure and the turgor pressure is underestimated because after mashing or freezing the cell sap is diluted by apoplastic water. In my opinion all of these reports of negative turgor pressure are wrong (Tyree, 1976). A reliable way of detecting the existence of negative turgor pressure is by the analysis of the leaf water potential isotherm, i.e. by analysis of the relation between Ψ_{leaf} and RWC. The relative water content, RWC, in pressure bomb terminology equals $(V_0 + K - V_e)/(V_0 + K)$, where K is the apoplastic water volume, V_0 is the symplasmic volume at full hydration and V_e is the net symplasmic water loss. A plot of $-1/\Psi_{leaf}$ versus RWC from 1 to 0 should yield a straight line below a certain RWC if $\Psi_{leaf} = \pi$, and a curved line if negative turgor pressures arise. I have plotted $-1/\Psi_{leaf}$ versus RWC for every leaf water potential isotherm I have been able to find in the literature including isotherms for plants that are supposed to have negative turgor pressures; I have always found a distinct linear region. Figure 7 shows the isotherms for *A. aneura* and *C. siliqua*. I would expect some distortion of the plots in Fig. 7 due to drainage of water from xylem vessels at high water potentials, but since there is no distortion in the linear regions there is no evidence for negative turgor pressures or of matric potentials entering into the isotherm. (My guess is that negative turgor pressures of just −2 or −3 bars would slip by unnoticed in this kind of analysis.) If we accept Iljin's (1930, 1931) hypothesized mechanism of drought-induced damage to protoplasm, it is not surprising that sclerophyllic leaves fail to develop negative turgors.

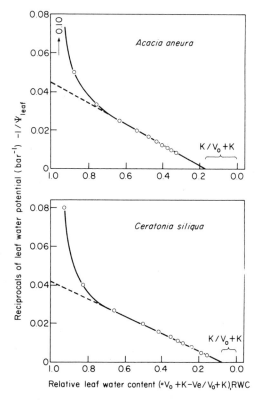

FIG. 7 Leaf water potential isotherms for *Acacia aneura* and *Ceratonia siliqua* obtained by plotting relative water contents (RWC) against reciprocals of leaf water potential. V_0 = symplasmic water volume before tissues were pressurized in Scholander-Hammel bomb and K = apoplastic water volume. These two species are reported to have negative turgor pressures, but the water potential isotherms show no indication of negative turgor pressure. The extrapolation of the linear part of $-1/\Psi_{leaf}$ versus RWC should intersect the abscissa at the relative apoplastic water content = $K/(V_0 + K)$.

D. Important Characteristics of Drought Avoidance

We must bear in mind that breeding for characters of drought tolerance (high π_p and ϵ) may bring the species into peril. A high π_p will allow the leaf to fall to very low leaf water potentials; consequently the tree can deplete soil water reserves to a dangerously low level. During an unusually dry growing season that may occur once every 20 or 50 years the entire crop may be lost. The only way

I can see to minimize this danger is to breed for two important characteristics of drought avoidance: thick cuticles and effective hydroactive stomatal closure. These characteristics will allow the tree to "ride-out" periods of drought by minimizing stomatal and cuticular transpiration. This could also be accomplished by drought deciduous characteristics; although I suspect this would divert dry matter incorporation from the wood to the leaves if more than one leaf fall occurred per season. Finally it should be pointed out that breeding for high ϵ *may* correlate with rigid cell walls which could subject the protoplasts to lethal negative turgor pressures.

V. SUMMARY

A method is described for measuring the physical water relations parameters of shoots and leaves that might correlate with drought resistance. These measurements can be done with acceptable accuracy by the Scholander-Hammel pressure bomb technique. It is suggested that wood yield might be marginally increased by breeding for two characteristics of drought tolerance and two characteristics of drought avoidance. The drought tolerance characteristics are a large bulk osmotic pressure at "incipient plasmolysis" (π_p) and a high bulk modulus of elasticity (ϵ). The drought avoidance characteristics are thick cuticles and effective hydroactive stomatal closure at very low leaf water potential. The strategy proposed is to breed leaves to tolerate lower water potential while maintaining a high net CO_2 assimilation. However, it is important to be cognizant of the physical limitations imposed by soil physics and hydrology in the silviculture area.

REFERENCES

Babalola, O., Boersma, L. and Youngberg, C. T. (1968). Photosynthesis and transpiration of monterey pine seedlings as a function of soil water suction and soil temperature. *Plant Physiol.* 43, 515–521.

Barrs, H. D. and Kramer, P. J. (1969). Water potential increase in sliced leaf tissue as a cause of error in vapour phase determinations of water potential. *Plant Physiol.* 44, 959–964.

Begg, J. E., Bierhuizen, J. F., Lemon, E. R., Misra, D. K., Slatyer, R. O. and Stern, W. R. (1964). Diurnal energy and water exchanges in bulrush millet in an area of high solar radiation. *Agric. Met.* 1, 294–312.

Brix, H. (1962). The effect of water stress on the rates of photosynthesis and respiration in tomato plants and loblolly pine seedlings. *Physiol. Plant.* 15, 10–20.

Cheung, Y. N. S., Tyree, M. T. and Dainty, J. (1975a). Water relations parameters on single leaves obtained in a pressure bomb and some ecological interpretations. *Can. J. Bot.* 53, 1342–1346.

Cheung, Y. N. S., Tyree, M. T. and Dainty, J. (1975b). Some possible sources of error in determining bulk elastic modulii and other parameters from pressure volume curves of shoots and leaves. *Can. J. Bot.* 54, 758–765.

Cowan, I. R. (1965). Transport of water in the soil–plant–atmosphere system. *J. appl. Ecol.* 2, 221–239.

Gardner, W. R. (1960). Dynamic aspects of water availability to plants. *Soil Sci.* 87, 63–73.

Hellkvist, J., Richards, G. P. and Jarvis, P. G. (1974). Vertical gradients of water potential and tissue water relations in sitka spruce trees measured with the pressure chamber. *J. appl. Ecol.* 11, 637–668.

Iljin, W. S. (1930). Die Ursache der Resistenz von Pflanzenzellen gegen Austrocknung. *Protoplasma* 10, 379–414.

Iljin, W. S. (1931). Austrocknungsresistenz des Farnes *Notochlaena marantae* R. Br. *Protoplasma* 13, 322–330.

Jarvis, P. G. and Jarvis, M. S. (1963a). The water relations of tree seedlings. I. Growth and water use in relation to soil water potential. *Physiol. Plant.* 16, 215–235.

Jarvis, P. G. and Jarvis, M. S. (1963b). The water relations of tree seedlings. II. Transpiration in relation to soil water potential. *Physiol. Plant.* 16, 236–253.

Jarvis, P. G. and Jarvis, M. S. (1963c). The water relations of tree seedlings. III. Transpiration in relation to osmotic potential of the rooting medium. *Physiol. Plant.* 16, 269–275.

Jarvis, P. G. and Jarvis, M. S. (1963d). The water relations of tree seedlings. IV. Some aspects of the tissue water relations and drought resistance. *Physiol. Plant.* 16, 501–516.

Kappen, L., Lange, O. L., Schulze, E.-D., Evenari, M. and Buschbom, U. (1972). Extreme water stress and photosynthetic activity of the desert plant *Artemisia herba-alba* Asso. *Oecologia* (Berl.) 10, 177–182.

Kaufmann, M. R. (1968). Evaluation of the pressure chamber technique for estimating plant water potential of forest tree species. *Forest Sci.* 14, 369–374.

Lawlor, D. W. (1972). Growth and water use of *Lolium perenne*. I. Water transport. *J. appl. Ecol.* 9, 78–98.

Newman, E. I. (1969). Resistance to water flow in soil and plant. II. A review of experimental evidence on the rhizosphere resistance. *J. appl. Ecol.* 6, 261–273.

Noy-Meir, I. and Ginzburg, B-Z. (1969). An analysis of the water potential isotherm in plant tissue. II. Comparative studies of leaves of different types. *Aust. J. biol. Sci.* 22, 35–52.

Odening, W. R. (1971). "The effect of decreasing water potential on net CO_2 exchange of intact woody desert shrubs". Ph.D. Thesis, Duke University.

Scholander, P. F., Hammel, H. T., Hemmingsen, E. A. and Bradstreet, E. D. (1964). Hydrostatic pressure and osmotic potentials in leaves of mangroves and some other plants. *Proc. natn. Acad. Sci. U.S.A.* 51, 119–125.

Scholander, P. F., Hammel, H. T., Bradstreet, E. D. and Hemmingsen, E. A. (1965). Sap pressure in vascular plants. *Science, N.Y.* 148, 339–346.

Schultze, E-D., Lange, O. L., Buschbom, U., Kappen, L. and Evenari, M. (1972). Stomatal responses of intact growing plants to changes in humidity. *Planta* 108, 259–270.

Schultze, E-D., Lange, O. L., Kappen, L., Buschbomb, U. and Evenari, M. (1973). Stomatal responses to changes in temperature and increasing water stress. *Planta* 110, 20–42.

Slatyer, R. O. (1960). Aspects of the tissue water relationships of an important arid zone species (*Acacia aneura* F. Muell.) in comparison with two mesophytes. *Bull. Res. Coun. Israel* 8D, 159–168.

Slatyer, R. O. (1967). "Plant-Water Relationships". Academic Press, London and New York.

Stocker, O. (1956). *In* "Ruhland's Hdb. d Pfl.-Phys". 3, 696–741, Springer-Verlag, Berlin, Goettingen and Heidelberg.

Talbot, A. J. B., Tyree, M. T. and Dainty, J. (1975). Some notes concerning the measurement of water potentials of leaf tissue with specific reference to *Tsuga canadensis* and *Picea abies. Can. J. Bot.* 53, 784–788.

Tyree, M. T. (1976). Negative turgor pressure in plant cells: Fact or Fallacy? *Can. J. Bot.* (in press).

Tyree, M. T. and Hammel, H. T. (1972). The measurement of the turgor pressure and the water relations of plants by the pressure-bomb technique. *J. exp. Bot.* 23, 267–282.

Tyree, M. T., Dainty, J. and Benis, M. (1973). The water relations of hemlock (*Tsuga canadensis*). I. Some equilibrium water relations as measured by the pressure-bomb technique. *Can. J. Bot.* 51, 1471–1480.

Williams, J. (1974). Root density and water potential gradients near the plant root. *J. exp. Bot.* 25, 669–674.

20
Morpho-Physiological Characteristics Related to Drought Resistance in *Pinus taeda*

J. P. VAN BUIJTENEN
Texas Forest Service, College Station, Texas, U.S.A.
M. VICTOR BILAN
Stephen F. Austin State University, Nacogdoches, Texas, U.S.A.
and
RICHARD H. ZIMMERMAN
Plant Genetics and Germplasm Institute, Beltsville, Maryland, U.S.A.

I.	Introduction	349
II.	Factors Related to Water Uptake	351
	A. Root Growth and Morphology	351
	B. Osmotic Stress Studies	353
III.	Factors Related to Water Use	355
	A. Needle Characteristics of Drought-Hardy and Drought-Susceptible Provenances	355
	B. Stomatal Response and Transpiration	356
IV.	Discussion and Conclusions	358
References		358

I. INTRODUCTION

The U.S. Southwide Geographic Seed Source Study, a cooperative study started in 1952 and coordinated by the Southern Forest Experiment Station (Wells, 1969), has shown that geographic races of *Pinus taeda* (loblolly pine) differ greatly in their ability to survive drought. Generally speaking, the northern and western sources show better survival throughout the region. A study by the Texas Forest

Service (Zobel and Goddard, 1955) indicated that seedlings from the Lost Pines area are especially drought resistant. These Lost Pines occur more than 160 km west of the continuous range of that species, in the Texas counties of Bastrop, Fayette and Caldwell. They generally receive 25—50 cm less annual rainfall than trees in the pine belt of East Texas. Climatic data for Bastrop County (West Texas) and Polk (East Texas) show that in addition to less total annual rainfall, trees in the Lost Pines, Bastrop, area receive the lowest monthly rainfall in August, while trees in the Polk area have the lowest monthly rainfall in September and October. The Texas Forest Service tests showed that the drought-conditioned Lost Pine seedlings had better survival than seedlings of eastern provenances (Zobel and Goddard, 1955; Goddard and Brown, 1959; van Buijtenen, 1966). Apparently natural selection has produced a more drought-resistant strain of loblolly pine in the Lost Pine area.

The studies described in this contribution were carried out at Stephen F. Austin State University, Nacogdoches, Texas and by the Texas Forest Service, at various locations throughout East Texas.

Those carried out at Stephen F. Austin State University were based on two provenances: Bastrop County (west, dry) and Polk County (east, wet), Texas. Three studies were involved: one on needle morphology as related to drought resistance, another on stomatal opening, transpiration, and needle water content, and a third on root development and needle water content under favourable as well as unfavourable moisture conditions.

The Texas Forest Service studies were all outgrowths from a selection programme for drought resistance. A series of field tests were done comparing drought resistance of different seed sources. Each seed source generally was a mixture of seed from three parents growing in close proximity to each other. These studies were followed by tests of open-pollinated families, grown outdoors in specially created beds, covered by a plastic shelter during the summer months to create artificial drought. These studies provided basic data on the drought resistance of seed sources and individual families used. The Texas Forest Service tests reported here involve one set of 8 and another of 16 families of known drought resistance, and include one study on root growth, two on growth under osmotic stress, and one study on transpiration rates.

II. FACTORS RELATED TO WATER UPTAKE

A. Root Growth and Morphology

1. Glasshouse Experiments

Davies (1973) studied root elongation and development of paired seedlings of provenances from West Texas (Bastrop) and East Texas, grown in a glasshouse, with one seedling of each provenance in each of twelve 1.2 m-long acrylic resin tubes filled with soil. Root elongation of individual seedlings was determined from weekly root tracings made on sheets of transparent acetate as described by Bilan (1964). Six of the twelve tubes were washed out after 12 weeks, the rest after 24 weeks.

Growth of both the main and the lateral roots was greater in the Bastrop source than in the East Texas source, although the differences were not statistically significant. At 12 weeks of age the root dry weight and total dry weight of East Texas seedlings, were respectively 27.9 and 98.6 mg as compared to 36.9 and 110.5 mg (significant $P = 0.05$) for the Bastrop seedlings. Differences at 24 weeks were not significant. Thus, only small differences existed between the growth and development of the two ecotypes during the first 24 weeks.

In a second, similar experiment, Davies droughted paired Bastrop and East Texas loblolly pine seedlings when they were dormant at the end of their second growing season. Needle water content, as percent of dry weight, was used as an indicator of water stress, and was determined weekly for both seedlings grown in each tube. As the water content in needles of one or both seedlings in a tube dropped to 60–80%, the soil in that tube was rewatered and maintained moist until growth was evident. Root elongation began in the upper half of each tube at an average depth of 40 ± 16 cm. Roots in the lower third of each tube began growth 7–14 days later than those in the upper third. In all cases the new growth originated from first order laterals attached to the root collar (cf. Bilan, 1961).

In about two-thirds of all plant pairs the East Texas plants (wet-zone provenance) reached the assigned needle water content sooner than Bastrop plants. A good early indication of plant survival was the needle water content 24 h after rewatering: plants did not survive when their needle water contents did not rise to at least 110% within 24 h. Following Stransky's (1963) example, a critical range of needle water contents was established at which an individual seedling might either survive or die. The critical range for the East Texas provenance was 71—85%, and for the Bastrop provenance 69—77%.

A third glasshouse experiment, done at Texas A&M University, compared the amount of root growth produced by four drought-hardy and four drought-susceptible loblolly pine families grown in large, free-draining metal drums (c. 30 gal or 140 l capacity, similar to Brix, 1959), filled with 7.5 cm of gravel, 7.5 cm of perlite and c. 50 cm of peat : perlite (1 : 1 v/v). After 4 months highly significant differences occurred among families in seedling dry weights (205—302 mg per plant) and root lengths (85—132 cm per plant), but they bore no clear-cut relation to the family differences in drought resistance determined in the drought test beds.

2. Studies in Nursery Beds

Seven 1-year-old seedlings of Bastrop (dry-zone) and Polk (wet-zone) County provenances were grown in a nursery bed in alternating rows, 36 cm apart, with the seedlings 28 cm apart within the rows. After one growing season seedlings were carefully excavated and the positions were mapped of (a) the actively elongating roots throughout the soil profile (Fig. 1) and (b) the horizontal position and the extent of the first-order lateral roots in the top 3 cm of soil (Fig. 2).

Average Polk seedlings had a total of 45 root tips, whereas average Bastrop seedlings had 62 root tips. The deepest root-tip in the Polk provenance was encountered at 25 cm, while the corresponding value for Bastrop was 36 cm. Each provenance had about 40 tips located above 16 cm depth, but below this level, Polk seedlings had an average of only 5 root tips while Bastrop seedlings had 22 (significant by χ^2; see Fig. 1).

The Bastrop seedlings had surprisingly few first-order laterals in the top soil layers, while the East Texas plants had numerous shallow roots (see Fig. 2). Many of the first-order lateral roots of the Bastrop seedlings extended for several centimetres in the top 3 cm of soil, but then began to go into deeper horizons. A further investigation showed that the average number of root tips per plant originating

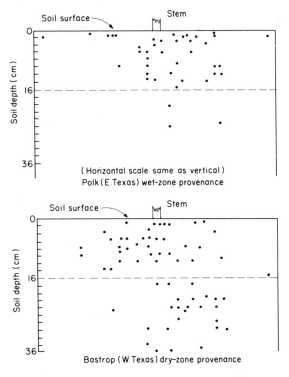

FIG. 1 Schematic diagram depicting root tip distribution with soil depth of two 1-year-old *Pinus taeda* provenances growing in a nursery bed. Data based on seven seedlings per provenance, with each dot representing the position of one root tip. (After Davies, 1973.)

from the upper 3 cm of the main root was 122 in Bastrop provenance and 440 in the Polk provenance (significant $P = 0.05$) while the corresponding values for root tips originating deeper than 3 cm were 72 and 45.

B. Osmotic Stress Studies

1. Unchanging Osmotic Stress

The same four drought-tolerant and four drought-susceptible families used in the study of growth in large metal drums were grown in solution culture with various amounts of polyethyleneglycol (mol.

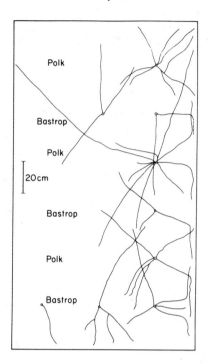

FIG. 2 Plan view of first-order lateral-root distribution in the upper 3 cm layer of soil for 1-year-old seedlings of *Pinus taeda*. Seedlings of Polk (E. Texas) wet-zone provenance alternate with seedlings of Bastrop (W. Texas) dry-zone provenance. (After Davies, 1973.)

wt. = 400) to give osmotic pressures of 0.8, 1.4, 2.8, 5.5 and 11.5 atm as determined by freezing point depression. Meaning all treatments, significant differences among families existed in dry weight of the shoots (465—693 mg per plant) and roots (139—221 mg per plant), numbers of lateral roots, total lateral root lengths and shoot : root ratios. Treatment differences were also highly significant, but there were no significant family × treatment interactions. In spite of the variation which occurred, drought-hardy families could not be distinguished from drought-susceptible ones on plant weights or shoot and root lengths.

2. Gradually Increasing Osmotic Stress

Sixteen seedlots were germinated, consisting of 13 families of potentially drought-resistant trees, and three controls. Twenty seedlings of each seedlot were grown in nutrient solution, and over a

20. Drought Resistance: Morpho-Physiological Characteristics

period of 2 months the osmotic value of the solution was gradually increased to 40 atm by small daily increments of polyethyleneglycol (mol. wt. = 400). Each container had 16 seedlings, 4 each of 4 families. The osmotic pressures at which 50% of the trees died varied from 18.0 to 26.5 atm. However, there was no relationship between this measurement and drought resistance as measured by drought survival in test beds. This may be due to the fact that the drought-resistance tests were carried out in coarse sand, where the ability to take up water under greater osmotic stress would not make much additional water available to the plants, or, alternatively, polyethyleneglycol of 400 mol. wt. was not suitable for the drought study.

III. FACTORS RELATED TO WATER USE

A. Needle Characteristics of Drought-Hardy and Drought-Susceptible Provenances

Loblolly pine seedlings from Bastrop (dry-zone) and Polk (wet-zone) Counties were grown in the School of Forestry nursery at Stephen F. Austin State University (Knauf and Bilan, 1974). One mature three-needle fascicle from the first year's growth was taken from each of 10 randomly selected trees from each seed source when the seedlings were in their second growing season. A 5 mm section was cut from the middle of each needle for measurement. In addition, a batch of five fascicles from each seed source was gathered for counts of stomata, measurements of stomatal craters and distances between stomatal rows. The surface area of the cross section of needles was significantly smaller in Bastrop than in the Polk County seed source. No significant differences were found in needle length and volume. The distance between the stomatal rows was significantly greater in the Bastrop seedlings, as were the depths of stomatal craters on adaxial surfaces. The latter is probably associated with the greater thickness of the combined epidermis plus hypodermis layer on Bastrop plants.

Similar measurements were made on some 15—17-year-old trees of Bastrop County as compared to an East Texas (wet zone) seed source, both of which were grown at the Arthur Temple Research Area, Cherokee County, Texas. The surface area of Bastrop needles was, again, smaller, but the distances between rows of stomata were not different, and the stomatal craters on these mature Bastrop trees were relatively shallow.

It was concluded that needle characteristics conferring resistance to drought were present only at the seedling stages. This was consistent with field observations which showed that differences in drought hardiness during the first 2 years were most critical for survival.

B. Stomatal Response and Transpiration

1. Stomatal Opening, Needle Water Content and Transpiration of Bastrop and Polk County Loblolly Pine.

Hogan (1974) studied the effect of various soil watering regimes on percentages of open stomata, transpiration rates and needle water contents in 4—12-month-old seedlings from Bastrop and Polk Counties. Methyl-violet dye infiltration was used to determine percentages of open stomata. Transpiration rates were measured using a modified hygrosensor apparatus developed by Bierhuizen and Slatyer (1964). Hogan concluded that: (1) the percentage of open stomata was positively related to the rate of transpiration of seedlings at all ages; (2) stomatal transpiration ceased at lower needle water contents in younger than in older seedlings, but long before the lethal range was reached; (3) under favourable conditions Bastrop seedlings transpired more and had more stomata open than Polk County seedlings; (4) under water stress Bastrop seedlings had fewer open stomata and transpired less; (5) Bastrop seedlings conserved water better than Polk County seedlings; and (6) considerable variability existed within each seed source.

2. Transpiration Rates of Eight Loblolly Pine Families Tested for Drought Resistance

Twenty seedlings of each of the eight loblolly pine families used in previous studies were grown in a 1 : 1 mixture of peat and perlite in cartons. When the seedings were 3 months old the cartons were no longer watered, were sealed and weighed daily thereafter for 2 months.

The drought-hardy families had somewhat greater dry weights than the drought-susceptible ones and transpired significantly less ($P = 0.05$) during days 21 to 40 of the experiment. This was the period during which water stress first developed (Table I).

One progeny (BA1-1) transpired profusely when water was abundant, but stopped transpiring almost entirely as stress increased. This progeny also gained most dry weight and has considerable value for further breeding work.

TABLE I. Mean dry weights of 5-month-old pot-grown seedlings of drought-hardy and drought-susceptible families of *Pinus taeda* following 2 months of imposed drought, and the water lost per seedling during the drought period when water stress developed (days 21—40).

	Dry weight shoots (mg)[2]	Dry weight roots (mg)[2]	Water used days 21—40 (g)[2]
Drought-hardy families[1]			
BA3-F10-20	761 abc	398 a	97.4 a
BA1-1	870 a	386 a	92.1 a
BA1-2	692 bc	345 bc	106.9 b
GR1-2	660 bc	326 bc	95.8 a
Means	746	364	98.1[3]
Drought-susceptible families			
FA1-1	644 c	339 bc	108.8 b
WA1-5	696 bc	371 bc	105.9 b
BA3-R13-41b	751 abc	347 bc	104.9 b
GR1-4	798 ab	325 c	106.3 b
Means	722	345	106.5[3]

[1] Drought hardiness indicated by survival in field tests
[2] Means not followed by the same letter within a column are significantly different at the 5% level by Duncan's Multiple Range Test
[3] Water used during days 21—40 is significantly different ($P = 0.05$) between drought-hardy and drought-susceptible families

IV. DISCUSSION AND CONCLUSIONS

The evidence indicates that drought-hardy loblolly pines owe their drought hardiness largely to various *avoidance* mechanisms. The only experiment indicating that drought-*tolerance* mechanisms were involved was the one showing that the critical water content in needles of drought-hardy (Bastrop) seedlings was a few percent lower than in the East Texas (Polk) seedlings (*c.* 73% cf. 78%). Another striking feature was the variability among families in growth characteristics which showed no significant relation to their drought hardiness as measured in the test beds (see Table I).

Survival during the first 2 years is the most critical. After that, the root systems are sufficiently deep for the trees to be able to withstand all but the most severe droughts. The various drought-avoidance mechanisms work in conjunction with each other and not necessarily at the same time, and the large variability may indicate an adaptation to variable conditions. The various mechanisms that operate, roughly in order of their apparent importance, are as follows. (1) Stomatal control: drought-hardy seedlings appeared to transpire rapidly when water was available but conserved water under stress. (2) Root morphology: drought-hardy seedlings seemed to have deeper root systems and wider ranging laterals. (3) Needle morphology: the needles of drought-hardy seedlings were somewhat smaller, with deeper stomatal pits than needles on drought-susceptible seedlings. (4) Numbers of stomata per mm^2: drought-hardy seedlings had fewer stomata per mm^2 because the rows of stomata were somewhat further apart.

REFERENCES

Bierhuizen, J. F. and Slatyer, R. O. (1964). An apparatus for the continuous and simultaneous measurement of photosynthesis and transpiration under controlled environmental conditions. *CSIRO Australian Division of Land Research. Technical Paper* No. 24.

Bilan, M. V. (1961). Effect of planting date on regeneration and development of roots of loblolly pine seedlings. *Proc. XIII Congr. of International Union of Forest Research Organizations.* Vienna, Austria. Part 2, Vol. 1, Sec. 22-15.

Bilan, M. V. (1964). Acrylic resin tubes for studying root growth in tree seedlings. *Forest Sci.* 10, 461–462.

Brix, M. (1959). "Some Aspects of Drought Resistance in Loblolly Pine Seedlings". Ph.D. Dissertation, Texas A&M University, College Station, Texas.

Davies, G. (1973). "Response of Loblolly Pine (*Pinus taeda* L.) Seedlings from two Seed Sources to Favorable and Unfavorable Moisture Regimes". M.S. Thesis, Stephen F. Austin State University, Nacogdoches, Texas.

Goddard, R. E. and Brown, C. L. (1959). Growth of drought resistant loblolly pines. *Texas Forest Service. Res. Note* No. 23.

Hogan, C. T. (1974). "Stomatal Opening, Transpiration, and Needle Moisture Content in Loblolly Pine Seedlings from two seed Sources". M.S. Thesis, Stephen F. Austin State University, Nacogdoches, Texas.

Knauf, T. A. and Bilan, M. V. (1974). Needle variation in loblolly pine from mesic and xeric seed sources. *Forest Sci.* 20, 88–90.

Stransky, J. J. (1963). Needle moisture as mortality index for Southern pine seedlings. *Bot. Gaz.* 124, 176–179.

van Buijtenen, J. P. (1966). Testing loblolly pine for drought resistance. *Texas Forest Service. Tech. Rep.* No. 13.

Wells, O. O. (1969). Results of the Southwide pine seed source study through 1968–69. *Proc. 10th Southern Forest Tree Improvement Conference*, Houston, Texas. pp. 117–129.

Zobel, B. J. and Goddard, R. E. (1955). Preliminary results on tests of drought-hardy strains of loblolly pine (*Pinus taeda* L.). *Texas Forest Service. Res. Note* No. 14.

21
Role of Oxygen Transport in the Tolerance of Trees to Waterlogging

M. P. COUTTS
Forestry Commission Northern Research Station, Roslin, Scotland
and
W. ARMSTRONG
Department of Plant Biology, University of Hull, England

I.	Introduction	361
II.	Flooding Damage	363
III.	Adaptations in Trees and Herbaceous Plants	364
	A. Production of Adventitious and Surface Roots	364
	B. The Shaving-brush Pattern	365
	C. Aerenchyma Production	365
	D. Rhizosphere Oxidation and Oxygenation	366
	E. Metabolic Adaptations	366
IV.	Factors Affecting the Oxygen Balance in Roots	368
V.	Gas Pathways in Trees	370
	A. Evidence for Gas Movement	370
	B. Entry of Gas	370
	C. Composition of Gas in Trees	372
	D. Movement in Secondary Tissues outside the Xylem	372
	E. Pathways in the Xylem	374
	F. The Primary Root	376
	G. Mechanism of Gas Transport	377
	H. Blockage of Gas Pathways	378
VI.	Prospects for Tree Improvement	379
References		380

I. INTRODUCTION

Effects of excess soil water on tree physiology have received relatively little attention, compared with effects of water stress, despite the prevalence of wetland conditions in many parts of the world. In Britain some 50% of Forestry Commission land is subject to periodic or permanent waterlogging, and in current planting areas the proportion may approach 100% (Toleman and Pyatt, 1974). In such soils there is often an abrupt decrease in oxygen flux with increasing soil depth (Fig. 1).

The tolerance of trees to waterlogging, and especially to inundation of the entire root system, has been reviewed by Gill (1970). The emphasis of our contribution is on the more usual condition where only the lower portion of the root system is subject to permanent or periodic waterlogging.

It is frequently impossible to establish forest trees without some form of drainage, as many tree species grow poorly in wet soil (e.g. *Picea sitchensis*, Lees, 1972; *Pinus contorta*, Boggie, 1972; *Pinus serotina*, Graham and Rebuck, 1958; *Pinus radiata* and *Pinus pinaster*, Poutsma and Simpfendorfer, 1963; fruit trees, Boynton and Compton, 1943). Also, trees tend to be shallow-rooted in waterlogged soil and so become unstable in high winds (Fraser and

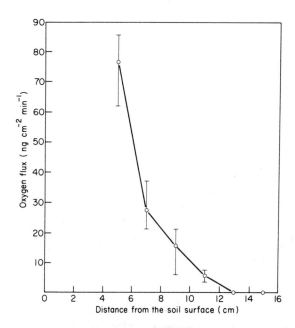

FIG. 1 Decrease in oxygen flux with soil depth in a peaty-gley soil in the U.K. in winter. Data are means of four replicates; the vertical bars indicate the maxima and minima of the replicates (Armstrong and Read, unpublished).

Gardiner, 1967; Day, 1950). Soil aeration can be increased by drainage, but this is costly and often of limited value on the impermeable soils characteristic of many forest areas in Britain. Another approach is to select and breed tree species which are tolerant to waterlogging. Clearly, it would be helpful to understand the stresses imposed by waterlogging and the adaptive features developed by trees to counter those stresses. In this contribution we review the nature of waterlogged soil environments and their effect on plants, discuss the scanty literature on adaptive processes in trees and indicate promising areas for future research.

II. FLOODING DAMAGE

Regardless of the soil type there are certain properties which are common to all saturated soil horizons (Ponnamperuma, 1972) other than those subject to flushing influences (Armstrong and Boatman, 1967). Foremost among these is oxygen deficiency; complete anaerobiosis may develop within a few hours of flooding owing to gas displacement from the soil pore space, and uptake by micro-organisms of the remaining dissolved oxygen (Scott and Evans, 1955; Turner and Patrick, 1968). The diffusion coefficient of oxygen in waterlogged soil is only $c.\ 1 \times 10^{-5}\ cm^2 s^{-1}$ whereas in air it is $c.\ 0.205\ cm^2 s^{-1}$. Crumb-structured horizons can become essentially anaerobic without being completely saturated because water fills the crumb pore spaces (Currie, 1962; Greenwood and Goodman, 1967). Oxygen deficiency *per se* can limit the exploration of wet horizons by roots of non-wetland plants (Yu *et al.*, 1969).

The nature of the damage done to plant roots as a result of excess soil water varies with species, plant age, dormancy status, and with the type of soil.

Toxic compounds are produced by the roots of many species under anaerobic stress. Such substances include ethanol and acetaldehyde (Fulton and Erickson, 1964), ethylene (Kawasi, 1972), and cyanogenic compounds (Rowe, 1966; Salesses and Juste, 1971). Harmful substances are also produced by waterlogged soils themselves, and may have greater effects on plant roots than the lack of oxygen. Hence sulphides (Culbert and Ford, 1972), high concentrations of carbon dioxide (Hook *et al.*, 1971), soluble iron and manganese, and lower-fatty-acid accumulation have all been implicated as agents causing root death in a wide range of species (Jones, 1972; Wang et al., 1967). More recently, interest has been shown in the potentially harmful levels of ethylene in wet soil (Smith and Jackson, 1973; Smith and Cook, 1974).

III. ADAPTATIONS IN TREES AND HERBACEOUS PLANTS

Most of the adaptive features that are well documented for herbaceous plants also occur in trees, although the literature on trees is much less extensive (Gill, 1970; Hook et al., 1972). The principal distinguishing features of tree roots are their greater size, the development of secondary tissues with ensheathing cambia, and the consequent loss of primary cortex.

A. Production of Adventitious and Surface Roots

When the roots of herbaceous species, which have developed under aerobic conditions, are submerged, they stop growing, and may eventually die. However, a great many species eventually respond by producing new roots which seem better adapted to the anaerobic environment (Yu et al., 1969).

Numerous herbaceous species survive on wet sites by exploiting only the surface aerated soil horizons (Yu et al., 1969), and in some instances adventitious roots can be produced which grow horizontally within saturated horizons along a tolerable redox plane, with ascending lateral roots, e.g. *Molinia caerulea* and various sedges (Armstrong and Boatman, 1967). Forest trees can also make good growth on wet sites by surface rooting and, like most flood-tolerant plants, can produce new roots immediately below the water table or in the upper layers of aerated soil when the lower parts of their root systems are inundated (Gill, 1970; Hook et al., 1970; Braun, 1973). It is noteworthy that *Salix* species which readily produce adventitious roots in waterlogged soils, are easy to propagate vegetatively (Wilkinson, 1946), therefore propagation programmes may select for this mechanism of flood-tolerance. *Salix* has also been observed with upwardly growing primary roots (Makarevič, 1956), as has *Pinus contorta* (unpublished). The vertical knee roots of *Taxodium* and the pneumatophores of mangroves consist of secondary tissues, which grow out from woody root systems.

B. The Shaving-brush Pattern

In wet soils cycles of death and regrowth occur at the ends of major branches of the root system and give rise to the brush-like formations known as shavers. This growth pattern is frequently observed in both herbaceous and woody species, and is common in forest tree species such as *Picea sitchensis* and *Pinus contorta*. Death may be restricted to the apical regions, or may extend back to include smaller woody roots. In our view this localized dieback should not be overlooked by those seeking to improve the ability of trees to withstand waterlogging.

C. Aerenchyma Production

An enlargement of the air spaces within plants lessens resistance to gas flow, and is frequently complemented by relatively low oxygen demands. Almost all herbaceous wetland species have voluminous gas-filled pore space in their tissues, and there is ample evidence that their internal aeration is adequately provided for by aerenchyma (Coult, 1964; Armstrong, 1975). Species that do not readily respond to anaerobic situations by enlarging their internal air spaces suffer anoxia in their roots (Luxmore *et al.*, 1972; Healy and Armstrong, 1972). Air spaces enlarge principally by cell separation (schizogenous spaces) and cell disintegration (lysigenous spaces) and in both instances may be further enlarged by cell collapse (Sifton, 1945; 1957; Esau, 1965). Aerenchyma occurs in the secondary cortex of submerged stems of *Populus* spp. and in *Salix viminalis* growing in water or on marshy soil (Arber, 1920) and is highly developed in secondary tissues of mangroves (Troll and Dragendorff, 1931; see section V). It may occur within many tree species but observations are lacking. The anatomy of timber has been more intensively studied, and it is more certain that xylem aerenchyma is rare, although it is known to occur in the secondary xylem of *Aeschynomene aspera*, the leguminous shrub used for making pith helmets (Arber, 1920).

D. Rhizosphere Oxidation and Oxygenation

Toxic compounds produced in waterlogged soils may be oxidized to harmless substances by oxygen which is transported internally from the aerated parts of plants and diffuses out of the roots. Such rhizosphere oxidation may occur enzymatically, by microbes associated with roots, and directly by molecular oxygen (Armstrong, 1975).

The extent of the oxygenated rhizosphere zone depends on the root's internal porosity and its respiratory demand. This zone is believed to buffer the roots against the reducing action of the soil. Reduced dye solutions (e.g. alpha-naphthylamine) applied to herbaceous plants are re-oxidized to coloured forms by the roots: Fukui (1953) relates this capacity of roots to their ability to penetrate reduced paddy soil. Similarly, Bartlett (1961) showed that high oxidizing capacities enabled roots to limit iron uptake from reducing environments. Oxidation of reduced dyes has been demonstrated in roots of a few tree species, viz. *Betula pubescens* (Huikari, 1954), *Salix atrocinerea* (Leyton and Rousseau, 1958), *Nyssa aquatica* and *Fraxinus pennsylvanica* (Hook and Brown, 1973). The latter workers were unable to detect rhizosphere oxidation in the non-swamp species *Liquidambar styraciflua*, *Liriodendron tulipifera* and *Platanus occidentalis*. Armstrong (1968) demonstrated radial oxygen loss from rooted cuttings of *Salix* spp. and from *Myrica gale*.

A feature deserving more study is the influence of waterlogging on mycorrhizas. Mycorrhizal fungi cannot grow anaerobically, but experiments on *Pinus contorta* roots in deoxygenated agar showed that the fungal symbiont continued to grow on parts of the roots where internally transported oxygen was diffusing outwards (Read and Armstrong, 1972).

E. Metabolic Adaptations

There seems little doubt that those herbaceous species which survive long periods of anoxia possess various forms of anaerobic metabolism. Species which accumulate metabolic end products such as malic and shikimic acid to high concentrations are regarded by some as better adapted than species which accumulate ethanol (Boulter *et al.*, 1963; Tyler and Crawford, 1970; McMannon and Crawford, 1971). However, submerged organs do not appear to grow unless there is a readily available source of oxygen.

It has yet to be demonstrated that tree species which tolerate flooding possess biochemical pathways which prevent alcohol accumulating under anaerobic conditions. Hook and Brown (1973) measured ethanol accumulation in a range of species and found that the most flood-tolerant one, *Nyssa aquatica*, accumulated more ethanol in its roots under flooded conditions than any of the other species tested. Dubinina (1961) measured changes in organic acids in flooded roots of *Salix cinerea*, pumpkin and tomato, but no useful differences were detected which would separate the tolerant from the intolerant species. Rowe and Beardsell (1973) point out that ethanol is produced in almost identical amounts in roots of pear and peach under anaerobic stress, yet pear is tolerant to flooding and peach extremely sensitive. Garcia-Novo and Crawford (1973) have shown a relationship between flooding tolerance and the amount of nitrate reductase in flooded roots of *Pinus contorta, Alnus incarna, Larix laricina, Picea mariana, P. sitchensis* and *Pinus sylvestris*. Nitrate-reducing capacity has also been demonstrated in the roots of *Alnus rubra* (Li *et al.*, 1972), and even the leaves of this species contain nitrate-reducing enzymes (Li *et al.*, 1967), whereas no nitrate-reducing system could be detected in *Pseudotsuga menziesii*, a species intolerant to flooding.

Biochemical studies are usually made on roots that have been waterlogged, without considering the effects on root metabolism of oxygen transported internally from the shoot system. Hook and Brown's (1973) experiments were unusual in that they took the precaution of subjecting detached portions of flooded roots to periods of anoxia before analysis. Genetically controlled biochemical differences must not be discounted, for they probably enhance survival (Crawford, Chapter 22), but it is doubtful whether the existence of specialized anaerobic pathways alone enables tree roots to grow in waterlogged conditions. Whenever root growth has been observed in waterlogged soils, and attention has been paid to oxygen movement, improved internal gas pathways have been found.

IV. FACTORS AFFECTING THE OXYGEN BALANCE IN ROOTS

It is our contention that an increase in the internal aeration of tree roots should be sought as the major step towards tree improvement for wetlands, giving better establishment and stability on wet sites.

The oxygen demand of wetland soils and the respiratory and diffusion characteristics of roots can now be simulated and their interplay studied electrically (Armstrong, 1975; Armstrong and Wright, unpublished; Gaynard et al., unpublished) or mathematically (Luxmoore et al., 1970, 1972). Thus, we may express the root's longitudinal resistance to oxygen diffusion by the expression

$$R_p = \frac{L}{P\tau D_O^t A} \text{ (s cm}^{-3})$$

where R_p is the non-metabolic diffusional resistance, L the length of root, P the gas-filled porosity, τ is a tortuosity factor, D_O^t is the diffusion coefficient for oxygen in air at temperature t, and A is the root cross-sectional area. The effective porosity of the root, ϵ, is therefore the product, $P\tau$.

Figure 2 illustrates how the oxygen status in the sub-apical regions of a root may be expected to vary with different effective root porosities and internal respiratory demands. It was assumed that the roots were of constant radius (0.05 cm), that the soil oxygen demand was relatively high, at 4×10^{-5} cm^3 (O$_2$) cm^{-3} (soil) s^{-1}, and that the root respiration will be unhindered at internal concentrations of oxygen above 1% (Greenwood, 1968). Root wall permeability was considered to be at a maximum of 100% at the apex (i.e. negligible radial resistance) falling to a minimum of 60% at 6 cm from the apex (Luxmoore et al., 1970; Luxmoore and Stolzy, 1972). The rate of oxygen consumption in root tissues varies markedly with temperature and from plant to plant, so two levels of potential respiration rate were used; the higher of which is in the range frequently reported for root tissues at laboratory temperatures (20–25°C).

The root porosities of dryland herbaceous species, growing in aerated soils, ranges from about 0.5% to 10%, 3–4% being most frequent. Assuming a tortuosity factor of 0.433 (Jensen et al., 1967), it seems reasonable to assume that many dryland roots have an

21. Oxygen Transport and Tolerance to Waterlogging

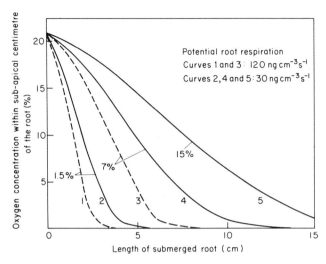

FIG. 2 Calculated oxygen concentrations within the sub-apical centimetre of roots penetrating up to 15 cm into a saturated anaerobic soil. Values are presented for roots with three effective internal porosities (% effective gas-filled volumes) and two potential respiration rates. (See text.)

effective porosity of 1.5% (Fig. 2, curves 1 and 2). If root elongation is halted when internal oxygen levels approach zero, and root survival is threatened by inadequate rhizosphere oxygenation, it would appear that roots with 1.5% porosity are particularly unsuited to a wetland environment. Internal aeration would only fully support potential respiratory activity in roots shorter than 3 cm at 20–25°C (curve 1), and the roots would extend a maximum of 4 cm into anaerobic soil (intercept on x axis). Some improvement accompanies the substantial decrease in respiratory activity to 30 ng cm^{-3} s^{-1} (curve 2), but even then the distal parts of roots exceeding 6 cm would probably suffer anaerobiosis.

Curves 3 and 4 (Fig. 2) suggest that roots would penetrate substantially deeper into saturated soils if their effective root porosities were increased from 1.5 to 7%. If greater porosity is associated with lower root respiration rates we may predict that anaerobiosis will not appear until the roots have penetrated to a depth of c. 12–13 cm. If tree roots extended this far below the water table each year, made possible each season by permanent oxygen pathways in secondary tissues, they might thereby be able to root deeply and dry out the soil.

Maize roots which grow out after the plants have been flooded can have porosities of 15—17% and penetrate saturated soil to a depth of 16—18 cm (Yu et al., 1969). Similar treatment given to sunflower increased root porosities to 12%, and soil penetration to c. 12 cm, which agrees well with our predictions (Fig. 2).

Gas paths become less tortuous as root porosities increase, and when lacunae begin to form the tortuosity factor may increase to near unity. In rice the first lacunae are evident at porosities above 12% (Fig. 2, curve 5). Substantial gas-filled voids may also appear in the pericycle of *Pinus contorta* roots growing in wet soil (Coutts, unpublished).

V. GAS PATHWAYS IN TREES

A. Evidence for Gas Movement

The existence of gas pathways through the tissues of tree species may be inferred from experiments on the movement of gas under applied pressure (Troll and Dragendorff, 1931; Scholander et al., 1955; Hook and Brown, 1972; MacDougal, 1936; MacDougal et al., 1929), and from the presence of comparatively high levels of oxygen in the xylem despite respiratory consumption. Further evidence is provided by oxidation of dyes in the vicinity of tree roots in anaerobic media (see section III D). The most direct evidence has been obtained by measuring oxygen diffusion from roots by a polarographic method, which demonstrates that oxygen flux from roots decreases rapidly when oxygen is prevented from reaching the shoots (Armstrong, 1968 on *Salix atrocinerea, S. fragilis, S. repens, Myrica gale*; Armstrong and Read, 1972 on *Pinus contorta, P. nigra, P. sylvestris, Picea sitchensis, P. abies*).

B. Entry of Gas

MacDougal *et al.* (1929) used a vacuum technique to show that air could enter the xylem of the stem via the leaves in *Quercus agrifolia*. Oxygen can enter the stem and roots from the leaves in

Populus petrowskiana and *Salix alba* cuttings (Chirkova, 1968), but experiments with *Nyssa sylvatica* seedlings demonstrated that significant amounts of oxygen did not enter the root system from the leaves in this species (Hook *et al.*, 1971), and Armstrong (1968) reached the same conclusion with *Salix fragilis, S. atrocinerea* and *Myrica gale* cuttings. Chirkova may have used cuttings with leaves close to the base of the stem. Although gaseous diffusion rates of up to 36 m h^{-1} have been reported in herbaceous species (Barber *et al.*, 1962), it is unlikely that the roots of trees, other than small seedlings or cuttings, will be supplied by gas entering the leaves because of resistance to gas movement and the consumption of oxygen by respiration during its long journey down the stem.

In woody species lenticels at the stem base are more important entry points. Armstrong (1968) found that obstruction of the lenticels on the basal 3 cm of *Salix* spp. cuttings was sufficient to drastically decrease oxygen diffusion from the root into an anaerobic medium, and Hook *et al.* (1971) showed that blocking the lenticels on *Nyssa sylvatica* seedlings prevented oxidation within the rhizosphere. On forest sites in Britain, it is the new roots and those parts containing their first secondary tissues, together with the lower portions of the woody roots, that are subject to periodic waterlogging. The nearest entry point for air on sites where the lower portion of the root system becomes submerged is via lenticels on roots in the upper, aerated soil horizons. Conspicuous lenticels develop on many species, and become especially prominent on roots growing in wet soils. They proliferate on parts of the root system above the water table, a feature which has been demonstrated experimentally on *Pinus contorta* (Coutts, unpublished). The pine lenticels are hydrophobic, so even if submerged by periodic changes in soil water level, they will become functional as soon as the water recedes. Specialized knee roots, or pneumatophores, may serve as entry points for oxygen in certain swamp species but these organs need not necessarily be for aeration. The pneumatophores of *Xylopia staudtii* have no air spaces (Jenik, 1967), and Kramer *et al.* (1952) concluded that the well known knee roots of *Taxodium distichum* did not function for aeration because no consistent change in oxygen consumption was detected when the knees were severed from their parent roots. However, somewhat clumsy techniques were employed and it is worth noting that Troll and Dragendorff (1931) reached similar conclusions on the function of the pneumatophores of mangroves, yet Scholander *et al.* (1955) subsequently demonstrated that these do indeed function for aeration. When pneumatophores were submerged by the tides, the respiratory consumption of oxygen by roots in anaerobic mud and the concomitant absorption of the more soluble carbon dioxide produced, caused a decrease in internal gas pressure so that air was drawn in when the tides receded.

C. Composition of Gas in Trees

Gas in tree tissues always contains less oxygen and more carbon dioxide than the atmosphere, but gas containing 10–18% oxygen has been extracted from bore-holes extending to the centre of the stem in *Quercus macrocarpa* (Chase, 1934) and from the sapwood of *Acer saccharinum* (Thacker and Good, 1952) and *Abies grandis* (Coutts, 1971), implying that the cambium is relatively permeable. *Abies grandis* frequently forms a wet and highly impermeable type of heartwood susceptible to colonization by micro-organisms which almost entirely deplete it of oxygen. The healthy sapwood is freely permeable to gas, but changes comparable to those leading to heartwood formation gradually occur in the vicinity of injuries to the sapwood, so that it soon becomes difficult to withdraw any gas from sample boreholes. Moreover, the gas which is extracted contains only traces of oxygen. Features of this type probably account for many of the very low oxygen concentrations in trees reported in the literature (Chase, 1934; Carter, 1945; Hartley *et al.*, 1961; Jensen, 1969).

D. Movement in Secondary Tissues outside the Xylem

Having passed through the loose material in the lenticel, oxygen must penetrate the associated underlying phellogen from which the lenticel tissues arise. The phellogen has been shown to contain intercellular spaces in *Nyssa sylvatica* (Hook *et al.*, 1970) and these presumably exist in other species where gas has been found to enter through lenticels. The secondary cortex of roots and stems is generally well endowed with intercellular spaces as seen in transverse section (Hyland, 1974). A *Myrica pennsylvanica* stem figured by Esau (1964) had longitudinally and tangentially connected air spaces. MacDougal *et al.* (1929) demonstrated the longitudinal permeability of the inner bark of oak by drawing air through strips of bark under reduced pressure.

Although the cambium is believed to present a barrier to gas movement into the xylem (Kramer and Kozlowski, 1960), it is permeable to some degree. MacDougal *et al.* (1929) sealed the lower end of an oak stem in a vessel partially filled with water, greased various surfaces on the portion projecting from the vessel, and reduced the gas pressure above the water in the vessel. Permeability of the cambium was shown by the fact that gas bubbled out of the submerged portion. Hook and Brown (1972) used a similar technique to demonstrate the movement of air radially across the cambium into the xylem of the two flood-tolerant species *Nyssa aquatica* and *Fraxinus pennsylvanica*. Air moved within the xylem both radially and longitudinally. Species intolerant to flooding were also permeable, but less so than the tolerant species, especially in the radial direction. In the two flood-tolerant species intercellular spaces were evident on microscopic examination of sections of cambial ray initials. They were not visible in the cambia of the intolerant species, although they were present in the lenticels and extended through the rays of the non-functional and functional phloem practically up to the ray initials of the cambium. A further experiment on *Nyssa aquatica* which had been growing in waterlogged soil showed that rhizosphere oxidation could proceed even when the stem base was girdled, therefore oxygen must have been able to pass the girdle by diffusing down in the xylem. However, girdling appeared to reduce the rate of rhizosphere oxidation suggesting that some of the oxygen supplied to the roots normally moved longitudinally in tissues external to the xylem.

Even where movement across the cambium can be demonstrated by applying gas pressure to the stem it is sometimes impossible to detect open pores microscopically (MacDougal, 1936). Intercellular spaces were detected in the stem cambium of the flood-tolerant *Nyssa aquatica* and *Fraxinus pennsylvanica*, but not in that of the intolerant *Liquidambar styraciflua*, *Liriodendron tulipifera*, *Platanus occidentalis*, *Populus occidentalis*, or *Populus deltoides* (Hook and Brown, 1972).

E. Pathways in the Xylem

Gas in the xylem was considered by Hook and Brown (1972) to move radially through intercellular spaces in the medullary ray parenchyma. Ziegler (1964) observed gas-filled intercellular spaces in the medullary rays of *Picea abies*, and these could provide radial pathways for gas movement, but no spaces were visible in the cambium. Back (1969) has shown that the intercellular spaces in medullary rays in xylem of *Pinus sylvestris* and *Picea abies* open to the ray parenchyma cells through specialized pits believed to serve for the aeration of the parenchyma cells. The volume of the gas-filled intercellular spaces was estimated to represent 0.2–0.3% of the xylem volume, but this represents only a small proportion of the total pore space of the sapwood.

In some species substantial pathways for the longitudinal transport of gas may become available in the secondary xylem through the normally water-filled vessels or tracheids. Physiologists have been debating the importance of gas in xylem elements since the eighteenth century because bubbles can effectively prevent the upward movement of water. Conversely, the presence of water in a vessel or tracheid greatly retards the movement of gas; the diffusion rate of oxygen in the gaseous state is 10 000 times faster than in water, therefore if water in the lumina of tracheids or vessels is replaced by gas, a new pathway for oxygen movement will appear. Hook *et al.* (1972) appear to overlook this pathway because of the difficulty with which gas can be made to pass from an embolized xylem element into one containing water. However, extensive zones of gas-filled elements undoubtedly occur in both angiosperms and gymnosperms, and can be shown experimentally to be capable of conducting gas.

In pines and certain other conifers, water in the late-wood tracheids in each growth ring becomes replaced by gas soon after differentiation (Vintila, 1939; MacDougal *et al.*, 1929), possibly because bordered pits in the late wood do not close the pit aperture as they do in the early wood (Harris, 1961; Petty, 1970). Gas can therefore pass from tracheid to tracheid, permitting the withdrawal of water under the normal hydrostatic tension in the xylem. A permanent longitudinal and tangential pathway for gaseous diffusion is thereby formed.

Concentric gas-filled zones have also been observed in the sapwood of angiosperms (Overton, 1930), where the replacement of water by gas in conducting elements appears to be partly reversible. In *Alnus incana* and *Salix* spp. the xylem contained ample gas in summer and autumn but much of the space was taken up by water again during winter and spring, although some vessels and groups of vessels remained gas-filled. So much water is replaced by gas that the porosity of the secondary xylem of trees is much larger than that of primary tissues discussed earlier. MacDougal *et al.* (1929) forced air into a borehole in a *Salix* stem and almost immediately recorded a pressure increase in a borehole 4.75 m above. The authors concluded that there was a continuous gaseous connection between the two boreholes, mainly in the vessels, but care is obviously needed in such experiments because of the possibility of displacing water from vessels and thereby transmitting the pressure or even creating new pathways.

Broad zones of dry wood, which include the early-wood portion of the growth rings, are sometimes present in roots of *P. contorta*, a relatively flood-tolerant species, but are absent from *P. sitchensis*, a species with roots intolerant to waterlogging. Air can readily be passed through root xylem of the former species but not through that of the latter. The mechanism by which gas displaces water in the early wood is obscure because closure of the bordered pits prevents the movement of gas from tracheid to tracheid.

Gas replaces water in conducting elements of the inner sapwood of most species prior to heartwood formation, and embolism can be induced in the outer sapwood of conifers by injecting solutions of mercuric chloride or dinitrophenol, or by introducing various micro-organisms; it is always accompanied by gradual death of some of the medullary ray parenchyma (Coutts, 1971). It is tempting to suggest that in some tree species, products of anoxia in the roots induce embolism in the xylem, leading to enhanced gas pathways such as those detected in *P. contorta*.

It is not always clear how the radial and longitudinal gas pathways in the xylem inter-connect. Intercellular spaces in the rays may connect directly with similar longitudinal spaces in the xylem, but the latter are not always present. Ray tracheids and longitudinal tracheids are connected by bordered pits, but ray tracheids only occur in certain species, and in the outer xylem of *Pinus contorta*, for example, they are normally filled with liquid. Although gas molecules doubtless diffuse through cell walls, perhaps via the microcapillaries demonstrated in longitudinal tracheids by Bailey and Preston (1969), the hydrated walls must present considerable resistance to movement.

F. The Primary Root

In the primary root the main pathway of gas transport is usually assumed to lie in the cortex, because intercellular spaces are visible. In many herbaceous species decreased oxygen tensions result in enlargement of the cortical gas spaces and increased root porosity (Bryant, 1934; Luxmoore et al., 1972). Apart from mangroves, there are few recorded instances of aerenchyma in tree roots. Measurements on adventitious roots of *Salix viminalis* showed a porosity of 20—30% in stagnant water culture (Armstrong, unpublished). Makarevič (1956) described cortical aerenchyma in roots of flooded *S. pentandra, S. fragilis* and *S. nigricans*. The roots protruded above ground level and may have served to aerate the submerged part of the root systems. Primary roots of *Laguncularia racemosa*, having a loose aerenchymatous cortex, also grew upwards out of wet soil and are presumed to function in the same way (Jenik, 1970). Cortical aerenchyma is most extensive in roots under conditions of poor aeration. However, Hook et al. (1971) could detect no visible difference in intercellular spaces between roots of *Nyssa sylvatica* grown in oxygenated or deoxygenated culture solutions.

The enlargement of gas pathways in the primary root is not necessarily confined to the cortex. Roots of *Pinus contorta* cuttings which had grown down into waterlogged soil developed gas-filled spaces in the stele (Fig. 3) (Coutts, unpublished). Much smaller spaces were also present in roots growing in non-waterlogged soil but did not form continuous longitudinal cavities like those in roots which had become adapted to waterlogging. Gas could not be passed through the former whereas the roots grown in waterlogged soil were permeable to air under a pressure of 20 mmHg. The cavities communicated with the lenticels which developed above the water table.

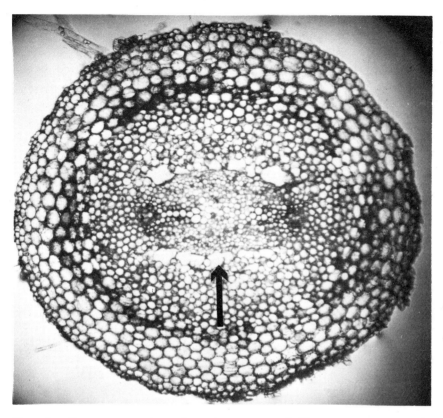

FIG. 3 Transverse section of a primary root of *Pinus contorta* grown in waterlogged soil, showing spaces in the stele adjacent to the vascular tissue (arrow). Much larger spaces sometimes occur in roots adapted to waterlogged conditions.

G. Mechanism of Gas Transport

In herbaceous species the rate of movement of $^{15}O_2$ from shoot to root, determined by laboratory measurement, is of the magnitude expected when gas moves by diffusion along a tortuous pathway (Evans and Ebert, 1960; Barber *et al.*, 1962), and oxygen movement in trees probably proceeds mainly by the same mechanism. However, laboratory measurements are always made under still conditions, whereas tensions and pressures set up in the tissues when a tree sways in the wind could be communicated to internal gas pathways (Hook *et al.*, 1972). Stem expansion and contraction with changes in internal water potential (e.g. Kozlowski, 1971) may also help to aspirate the tissues (Hook *et al.*, 1972), but such changes are small and occur mainly in tissues external to the differentiated xylem.

Sifton (1945) was the first to suggest that submerged parts of plants are aerated by mass flow. Air may be drawn in as respiratory carbon dioxide is absorbed. This mechanism occurs in mangroves and may occur more generally.

Hook *et al.* (1972) consider the inner tissues of most broad-leaved tree species to be so poorly aerated that living cells in the xylem must rely on oxygen dissolved in the transpiration stream. However, our calculations, based on measured rates of respiration (Goodwin and Goddard, 1940) and water movement (Kramer and Kozlowski, 1960), suggest that the oxygen content of water which was initially saturated with air, would be reduced to zero after the water had traversed *c.* 1–3 m of xylem, if no oxygen could enter from the outside.

Cytoplasmic streaming has been observed in the xylem of tree stems, in medullary rays situated across the width of the sapwood and heartwood transition zone (Coutts, 1971). This feature may enhance the radial movement of dissolved gases.

H. Blockage of Gas Pathways

Grable (1966) discussed the blockage of the intercellular spaces in submerged roots of herbaceous species. If pores became water-filled the rate of oxygen diffusion would be drastically reduced, but most of the evidence for blockage has been obtained from measurement of the respiration of excised roots. In the intact plant hydrostatic tension in the xylem may keep intercellular spaces free from water. Pathways in gas-filled xylem elements of certain angiosperms must presumably become partially blocked by the seasonal influx of water reported by Overton (1930). The lacunae in the pericycle of *Pinus contorta* roots eventually become closed by growth of the secondary xylem (Coutts, unpublished).

VI. PROSPECTS FOR TREE IMPROVEMENT

Efforts have been made to select inherently flood-tolerant root stocks in apples and peaches (Rowe and Beardsell, 1973), but relatively little has been done on forest trees. Most of the work on flooding tolerance of trees has revealed differences between species and, more rarely, between provenances. The only inheritance study known to us is Wilkinson's (1946); he observed that tolerance was inherited by the hybrids of natural crosses between *Salix* species of differing flood tolerance.

There are many criteria that could be used when selecting for flood tolerance. Selection could be simply based on growth and survival on wet sites, or on the performance of experimentally flooded plants under more controlled conditions. Alternatively, measurements could be made of the roots' ability to penetrate flooded soil, or their ability to tolerate applied chemicals which are known to be produced under waterlogged conditions. Additionally, biochemical tests could be done on plant tissues, or measurements made on the oxygen transport system.

The breeder still needs to know which are the most significant traits to preserve and enhance. At present he is restricted to selection on the basis of generalized plant responses, with all the pitfalls which that implies. The theme of this review has been that features which enhance oxygen transport to the root will be those most likely to repay further study, but further work should aim to evaluate the contribution made by all of the various adaptive mechanisms; only then will it be possible for the breeder to develop sound practical tests for tolerance. If we are to progress beyond hit-or-miss selection procedures, the commonly employed techniques of suddenly flooding entire seedling root systems, and observing effects on the shoot, must give way to selection and experimentation which consider the entry points of oxygen into the plant, and the likely pathways of gas in the tissues. In general, the possession of porous tissues and the capacity to increase porosity under waterlogged conditions would be expected to correlate well with tolerance to both sudden and permanent waterlogging. Such a correlation has already been demonstrated in certain herbaceous species (Fukui, 1953), and it is important to test this relationship in trees. We have seen that porosity may increase by the development of lysigenous or schizogenous lacunae or by the replacement of water in the xylem by gas. In no instance is the physiology of these processes understood, and it is not known to what extent these characteristics are under genetic control in trees.

From the present state of knowledge it is difficult to distinguish characteristics that enable tree roots to tolerate sudden waterlogging from those that are developed to a varying degree by certain species in response to a normally waterlogged environment. For example, a high degree of porosity in the primary root, even when growing in well aerated conditions, is a desirable feature, but the capacity of a root to increase porosity on encountering waterlogged conditions may be equally important.

A further area for research is to establish the relative importance of the various primary and secondary tissues for oxygen transport. Only when this has been done will it become possible to assess the value of using seedlings or rooted cuttings in testing for tolerance.

Simple screening experiments are still feasible, in spite of our rudimentary knowledge of the interactions between the plant root and waterlogged soil. It is time for breeders to look seriously at the advantages to be gained by developing flood-tolerant varieties but such work will have to be accompanied by physiological investigations in order to provide the basic information on which the breeder can act.

REFERENCES

Arber, A. (1920). "Water Plants". Wheldon and Wesley (Reprinted 1963), London.

Armstrong, W. (1968). Oxygen diffusion from the roots of woody species. *Physiol. Plant.* 21, 539–543.

Armstrong, W. (1975). Waterlogged soils. In "Environment and Plant Ecology" (J. R. Etherington, ed.), pp 184–216. John Wiley, London.

Armstrong, W. and Boatman, D. J. (1967). Some field observations relating the growth of bog plants to conditions of soil aeration. *J. Ecol.* 55, 101–110.

Armstrong, W. and Read, D. J. (1972). Some observations on oxygen transport in conifer seedlings. *New Phytol.* 71, 55–62.

Back, E. L. (1969). Intercellular spaces along the ray parenchyma — the gas canal system of living wood? *Wood Sci.* 2, 31—34.

Bailey, P. J. and Preston, R. D. (1969). Some aspects of softwood permeability. I. Structural studies with Douglas fir sapwood and heartwood. *Holzforschung* 23, 113—120.

Barber, D. A., Ebert, M. and Evans, N. T. S. (1962). The movement of ^{15}O through barley and rice plants. *J. exp. Bot.* 13, 397—403.

Bartlett, R. J. (1961). Iron oxidation proximate to plant roots. *Soil Sci.* 92, 372—379.

Boggie, R. (1972). Effect of water-table height on root development of *Pinus contorta* on deep peat in Scotland. *Oikos* 23, 304—312.

Boulter, D., Coult, D. A. and Henshaw, G. G. (1963). Some effects of gas concentrations on metabolism of the rhizome of *Iris pseudacorus. Physiol. Plant.* 16, 541—548.

Boynton, D. and Compton, O. C. (1943). Effect of oxygen pressures in aerated nutrient solutions on production of new roots and on growth of roots and top by fruit trees. *Proc. Am. Soc. hort. Sci.* 37, 19—26.

Braun, von H. J. (1973). Zum Wachsverhalten von Pappeln bei wechselndem und langandauerndem Hochwasser. *Alg. Forst.-u. J.-Ztg.* 144, 89—91.

Bryant, A. E. (1934). Comparison of anatomical and histological differences between barley grown in aerated and non-aerated culture solutions. *Plant Physiol.* 9, 389—391.

Carter, J. C. (1945). Wetwood of elms. *Illinois Nat. History Survey* 23, 401—448.

Chase, W. W. (1934). The composition, quantity and physiological significance of gases in tree stems. *Minnesota Agr. Expt. Sta. Tech. Bull.* 99.

Chirkova, T. V. (1968). [Features of the O_2 supply of roots of certain woody plants in anaerobic conditions]. *Fiziologiya Rast.* 15 (*For. Abstr.* 30, No. 146).

Coult, D. A. (1964). Observations on gas movement in the rhizome of *Menyanthes trifoliata* L, with comments on the role of the endodermis. *J. exp. Bot.* 15, 205—218.

Coutts, M. P. (1971). "Reaction of Grand Fir to Infection by *Fomes annosus*". Ph.D. Thesis, University of Cambridge.

Culbert, D. L. and Ford, H. W. (1972). The use of a multi-celled apparatus for anaerobic studies of flooded root systems. *Hort. Science* 7, 29—31.

Currie, J. A. (1962). The importance of aeration in providing the right conditions for plant growth. *J. Sci. Fd Agric.* 13, 380—385.

Day, W. R. (1950). Soil conditions which determine wind-throw in forests. *Forestry* 23, 90—95.

Dubinina, I. M. (1961). [Metabolism of roots under various levels of aeration]. *Fiziologiya Rast.* 8, 395—406.

Esau, K. (1964). Structure and development of bark in dicotyledons. *In* "The Formation of Wood in Forest Trees" (M. H. Zimmermann, ed.), pp 37–50. Academic Press, New York and London.

Esau, K. (1965). "Plant Anatomy". John Wiley, New York.

Evans, N. T. S. and Ebert, H. (1960). Radioactive oxygen in the study of gas transport down the root of *Vicia faba*. *J. exp. Bot.* **11**, 246–257.

Fraser, A. I. and Gardiner, J. B. H. (1967). Rooting and stability in Sitka spruce. *For. Comm. Bull.* No. 40. H.M.S.O., London.

Fukui, J. (1953). Studies on the adaptability of green manure and forage crops to paddy-field conditions. *Proc. Crop Sci. Soc. Japan.* **22**, 110–112.

Fulton, J. M. and Erickson, A. E. (1964). Relation between soil aeration and ethyl alcohol accumulation in xylem exudate of tomatoes. *Proc. Soil Sci. Soc. Am.* **28**, 610–614.

Garcia-Novo, F. and Crawford, R. M. M. (1973). Soil aeration, nitrate reduction and flooding tolerance in higher plants. *New Phytol.* **72**, 1031–1039.

Gill, C. J. (1970). The flooding tolerance of woody species – a review. *For. Abstr.* **31**, 671–688.

Goodwin, R. H. and Goddard, D. R. (1940). The oxygen consumption of isolated woody tissues. *Am. J. Bot.* **27**, 234–237.

Grable, A. R. (1966). Soil aeration and plant growth. *Adv. Agron.* **18**, 57–106.

Graham, B. F. and Rebuck, A. L. (1958). The effect of drainage on the establishment and growth of pond pine (*Pinus serotina*). *Ecology* **39**, 33–36.

Greenwood, D. J. (1968). Root growth and oxygen distribution in soil. *Trans. Ninth Int. Congress Soil Sci.* **1**, 823–832.

Greenwood, D. J. and Goodman, D. (1967). Direct measurement of the distribution of oxygen in soil aggregates and in columns of fine soil crumbs. *J. Soil Sci.* **18**, 182–196.

Harris, J. M. (1961). Water-conduction in the stems of certain conifers. *Nature, Lond.* **189**, 678.

Hartley, C., Davidson, R. W. and Crandall, B. S. (1961). Wetwood, bacteria and increased pH in trees. *Forest Products Lab. Report* No. 2215, Princes Risborough, England.

Healy, M. T. and Armstrong, W. (1972). The effectiveness of internal oxygen transport in a mesophyte (*Pisum sativum* L.). *Planta* **103**, 302–309.

Hook, D. D. and Brown, C. L. (1972). Permeability of the cambium to air in trees adapted to wet habitats. *Bot. Gaz.* **133**, 304–310.

Hook, D. D. and Brown, C. L. (1973). Root adaptations and relative flood tolerance of five hardwood species. *Forest Sci.* **19**, 225–229.

Hook, D. D., Brown, C. L. and Kormanik, P. P. (1970). Lenticel and water root development of swamp tupelo under various flooding conditions. *Bot. Gaz.* **131**, 217–224.

Hook, D. D., Brown, C. L. and Kormanik, P. P. (1971). Inductive flood tolerance in swamp tupelo (*Nyssa sylvatica* var. *biflora* (Walt.) Sarg.). *J. exp. Bot.* 22, 78–89.

Hook, D. D., Brown, C. L. and Wetmore, R. H. (1972). Aeration in trees. *Bot. Gaz.* 133, 443–454.

Huikari, O. (1954). Experiments on the effect of anaerobic media upon birch, pine and spruce seedlings. *Commun. Inst. Forest. Fenn.* 42, 1–13.

Hyland, F. (1974). Fiber analysis and distribution in the leaves, juvenile stems and roots of ten Maine trees and shrubs. *Life Sci. Agric. Exp. Sta. Tech. Bull. No. 71*, University of Maine, U.S.A.

Jenik, J. (1967). Root adaptations in West African trees. *J. Linn. Soc. (Bot.)* 60, 25–29.

Jenik, J. (1970). Root system of tropical trees. 5. The peg-roots and the pneumathodes of *Laguncularia racemosa* Gaertn. *Preslia (Praha)* 42, 105–113.

Jensen, K. F. (1969). Oxygen and carbon dioxide concentrations in sound and decaying red oak trees. *Forest Sci.* 15, 246–251.

Jensen, C. R., Stolzy, L. H. and Letey, J. (1967). Tracer studies of oxygen diffusion through roots of barley, corn and rice. *Soil. Sci.* 103, 23–29.

Jones, R. (1972). Comparative studies of plant growth and distribution in relation to waterlogging. V. The uptake of iron and manganese by dune and slack plants. *J. Ecol.* 60, 131–140.

Kawasi, M. (1972). Effect of flooding on ethylene concentration in horticultural plants. *J. Am. Soc. hort. Sci.* 97, 584–588.

Kozlowski, T. T. (1971). "Growth and Development of Trees", Vol. 2. Academic Press, New York and London.

Kramer, P. J. and Kozlowski, T. T. (1960). "Physiology of Trees". McGraw-Hill, New York.

Kramer, P. J., Riley, W. S. and Bannister, T. T. (1952). Gas exchange of cypress knees. *Ecology* 33, 117–121.

Lees, J. C. (1972). Soil aeration and Sitka spruce seedling growth in peat. *J. Ecol.* 60, 343–349.

Leyton, L. and Rousseau, L. Z. (1958). Root growth of tree seedlings in relation to aeration. *In* "The Physiology of Forest Trees" (K. V. Thimann, ed.), pp. 467–475. Ronald Press, New York.

Li, C. Y., Lu, K. C., Trappe, J. M. and Boilen, W. B. (1967). Enzyme systems of red alder and Douglas-fir in relation to infection by *Poria weiri*. *In* "Biology of Alder" (J. M. Trappe, J. F. Franklin, R. F. Tarrant and G. M. Hansen, eds.), pp 241–250. *Proc. Northwest Sci. Ass. Fortieth Ann. Meet.* 1967. U.S. Dept. Agr., Portland, Oregon.

Li, C. Y., Lu, K. C., Trappe, J. M. and Boilen, W. B. (1972). Nitrate-reducing capacity of roots and nodules of *Alnus rubra* and roots of *Pseudotsuga menziesii*. *Pl. Soil* 37, 409–414.

Luxmoore, R. J. and Stolzy, L. H. (1972). Oxygen diffusion in the soil–plant system V and VI. *Agron. J.* 64, 720–729.

Luxmoore, R. J., Stolzy, L. H. and Letey, J. (1970). Oxygen diffusion in the soil–plant system. *Agron, J.* 62, 317–332.

Luxmoore, R. J., Sojka, R. E. and Stolzy, L. H. (1972). Root porosity and growth responses of wheat to aeration and light intensity. *Soil Sci.* 113, 354–357.

MacDougal, D. T. (1936). The communication of the pneumatic systems of trees with the atmosphere. *Proc. Am. phil. Soc.* 76, 823–845.

MacDougal, D. T., Overton, J. B. and Smith, G. M. (1929). "The Hydrostatic-Pneumatic System of Certain Trees: Movements of Liquids and Gases". Carnegie Inst., Washington, Publ. No. 397.

MacMannon, M. and Crawford, R. M. M. (1971). A metabolic theory of flooding tolerance: the significance of enzyme distribution and behaviour. *New. Phytol.* 70, 299–306.

Makarevič, V. N. (1956). [Some observations on land plants in the waterlogged or periodically-flooded zone of the Rybinsk reservoir]. *Bot. Z.* 41, 1647–1652. (*For. Abstr.* No. 2599).

Overton, J. B. (1930). Seasonal variations in the hydrostatic-pneumatic system of certain trees. *Am. J. Bot.* 17, 1403–1423.

Petty, J. A. (1970). Permeability and structure of the wood of Sitka spruce. *Proc. R. Soc.* B 175, 149–166.

Ponnamperuma, F. N. (1972). The chemistry of submerged soils. *Adv. Agron.* 24, 29–95.

Poutsma, T. and Simpfendorfer, K. J. (1963). Soil moisture conditions and pine failure at Waarre, near Port Campbell, Victoria. *Aust. J. agric. Res.* 13, 426–433.

Read, D. J. and Armstrong, W. (1972). A relationship between oxygen transport and the formation of the ectotrophic mycorrhizal sheath in conifer seedlings. *New. Phytol* 71, 49–53.

Rowe, R. N. (1966). "Anaerobic Metabolism and Cyanogenic Glycoside Hydrolysis in Differential Sensitivity of Peach, Plum and Pear Roots to Water-saturated Conditions". Ph.D. Thesis, Davis, California.

Rowe, R. N. and Beardsell, D. V. (1973). Waterlogging of fruit trees. *Hort. Abstr.* 43, 533–548.

Salesses, G. and Juste, C. (1971). Recherches sur l'asphyxie radiculaire des arbres fruitiers a noyan. II. Comportement des porte-greffes de type pêcher et prunier: Étude de leur teneur en amygdaline et des facteurs intervenant dans l'hydrolyse de celle-ci. *Ann. Amél. Plantes* 21, 265–280.

Scholander, P. F., van Dam, L. and Scholander, S. I. (1955). Gas exchange in the roots of mangroves. *Am. J. Bot.* 42, 92–98.

Scott, A. D. and Evans, D. D. (1955). Dissolved oxygen in saturated soil. *Proc. Soil Sci. Soc. Am.* 19, 7–16.

Sifton, H. B. (1945). Air-space tissue in plants. *Bot. Rev.* 11, 108–143.

Sifton, H. B. (1957). Air-space tissues in plants II. *Bot. Rev.* 23, 303–312.

Smith, A. M. and Cook, R. J. (1974). Implications of ethylene production by bacteria for biological balance of soil. *Nature, Lond.* 252, 703–705.

Smith, K. A. and Jackson, M. B. (1973). Ethylene, waterlogging and plant growth. *A. Rep. Letcombe Laboratory* 60–75, Wantage, Berks., England.

Thacker, D. G. and Good, H. M. (1952). The composition of air in trunks of sugar maple in relation to decay. *Can. J. Bot.* 30, 475–485.

Toleman, R. D. L. and Pyatt, D. G. (1974). Site classification as an aid to silviculture in the Forestry Commission of Great Britain. *Proc. Tenth Commw. For. Conf.* U.K. H.M.S.O. London.

Troll, W. and Dragendorff, O. (1931). Über die Luftwurzeln von *Sonneratia* Linn. f. und ihre biologische Bedeutung. Mit einem rechnerischen Anhang von Hans Fromherz. *Planta* 13, 311–473.

Turner, F. T. and Patrick, W. H. (1968). Chemical changes in waterlogged soils as a result of oxygen depletion. *Trans. Ninth Int. Congr. Soil Sci.* 4, 53–65.

Tyler, P. D. and Crawford, R. M. M. (1970). The role of shikimic acid in waterlogged roots and rhizomes of *Iris pseudacorus* L. *J. exp. Bot.* 21, 677–682.

Vintila, E. (1939). Untersuchungen über Raumgewicht und Schwindmass von Früh- und Spätholz bei Nadelhölzern. *Holz Roh-u. Werkstoff.* 2, 345–357.

Wang, T. S. C., Cheng, S. Y. and Tung, H. (1967). Dynamics of soil organic acids. *Soil Sci.* 104, 138–144.

Wilkinson, J. (1946). Some factors affecting the distribution of the Capreae group of *Salix* in Gower. *J. Ecol.* 33, 214–221.

Yu, P. T., Stolzy, L. H. and Letey, J. (1969). Survival of plants under prolonged flooded conditions. *Agron. J.* 61, 844–847.

Ziegler, H. (1964). Storage, mobilization and distribution of reserve material in trees. *In* "The Formation of Wood in Forest Trees" (M. H. Zimmermann, ed.), pp 303–320. Academic Press, New York and London.

22
Tolerance of Anoxia and the Regulation of Glycolysis in Tree Roots

R. M. M. CRAWFORD
Department of Botany, The University, St. Andrews, Scotland

I.	Low Oxygen Stress in Tree Roots	387
II.	Metabolic Adaptations to Anoxia	392
III.	Conclusions	399
	Acknowledgements	399
	References	400

I. LOW OXYGEN STRESS IN TREE ROOTS

Anoxic (i.e. oxygen-deficient) conditions are extremely limiting to the growth of most higher plants. Roots that penetrate below the level of the water table, and produce new growth in areas that are perpetually flooded, all possess some means of aerating their tissues. This is achieved either by the downward diffusion of oxygen — alone or coupled with the metabolic production of oxygen as suggested by Vámos and Köves (1972). The downward diffusion of oxygen takes place readily over short distances. In grasses and herbs where the photosynthetic regions of the shoots are in immediate juxtaposition to the roots, numerous authors have demonstrated an outward movement of oxygen from the submerged roots (van Raalte, 1941; Greenwood, 1967; Armstrong, 1964; Vartapetian, 1964). This radial loss of oxygen has usually been interpreted as an indication that the downward diffusion of oxygen from shoot to root is in excess of the respiratory needs of the root. However, as shown below, this assumption may not be completely justified.

In the tree the development of the trunk separates the oxygen-producing shoots from the roots and thus effectively prevents the leaves from serving as a source of molecular oxygen. In addition, the interior of large woody tissues has been known to be at least partially anaerobic ever since the detection of ethanol in the branches of trees by Devaux (1899). The more recent and sophisticated examination of the gas composition of woody stems by mass spectroscopy (Carrodus and Triffett, 1975) provides ample support for the view that there is an acute shortage of oxygen in woody tissues. In flood-tolerant species of *Salix* a limited amount of aeration is provided for roots by the marked development of lenticel tissues that takes place at the base of the trunk. The lenticels allow the upward diffusion and removal of toxic substances from the roots (Chirkova, 1968) as well as the downward diffusion of oxygen (Armstrong, 1968; Coutts and Armstrong, Chapter 21). The distance over which this diffusion mechanism is active is however relatively short. The maximum length of conifer root from which a radial loss of oxygen has been reported is 6 cm (Armstrong and Read, 1972), and in the flood-tolerant species of *Salix* root aeration ceases when the lenticels at the base of the tree are covered with water to a depth of 3 cm (Armstrong, 1968). Oxygen diffusion may therefore be of significance in aerating adventitious roots in flooded trees but can have no effect in aiding the survival of the deeper anchoring roots of flood-tolerant trees. The flood-plain forest of the upper Amazon provides an excellent example where many tree species are able to survive a complete inundation of their trunks up to a depth of 5 m without any aeration mechanism to alleviate the effects of anoxia (Gessner, 1968).

As already stated above, oxygen is essential for new root growth. However it does not follow that in all species oxygen has to be available continuously to the roots. There are many species of plants, both herbs and trees, where once the root system is established it can endure prolonged periods of anoxia. Lodgepole pine (*Pinus contorta*) can survive prolonged periods of inundation and resume active growth when the water table is lowered (Fig. 2), yet when grown on a soil with a permanently high water table root development is largely restricted to the aerated soil above the water table (Boggie, 1972). Although no tree species is able to penetrate a permanently anaerobic water table, the established roots of some trees can tolerate flooding. The only observed exceptions to this rule are the adventitious roots which are produced by some flood-tolerant tree species (Gill, 1975).

When different tree species are compared, the length of time for

22. Glycolysis Regulation and Tolerance to Waterlogging

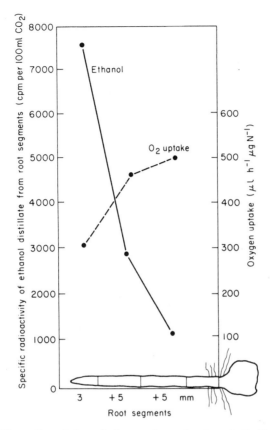

FIG. 1 Specific radioactivity of the combusted ethanol distillate from root segments of maize that were incubated with 0.05 M 3^{14}C-pyruvate for 4 h under air, together with the oxygen consumption of the same segments.

which they can withstand flooding varies from 24 h to 2 years. On the one hand there are those species that die suddenly when flooded, such as peach and apricot, where waterlogging is associated with the liberation of hydrogen cyanide from glycosides found in the root bark (Rowe and Catlin, 1971). On the other hand some species are able to withstand flooding for up to 2 years (Gill, 1970). This upper limit appears to be set by the need to renew the younger absorptive regions of the root. The active salt- and water-absorbing areas of the root need to be renewed constantly and, as active growth is inhibited under anaerobic conditions, very few trees are able to survive if their root systems are flooded for more than two consecutive growing seasons. It must be concluded therefore that trees able to survive prolonged periods of flooding will have an inbuilt tolerance to the potentially auto-toxic effects of prolonged anaerobiosis. Other externally produced toxicities will also impinge on the metabolism of flooded roots (sulphide, ferrous iron and manganese poisoning: see Armstrong, 1975). However it has been possible to show in herbaceous plants that flooding tolerance is always accompanied by a control of metabolic rate (Crawford, 1975). Thus resistance to or avoidance of internally produced toxins appears to be fundamental to survival under anoxia and may also be linked with the ability of the root to detoxicate its external environment.

Some controversy has taken place in the past as to the level of anoxia that is necessary before oxygen becomes limiting for root growth. Because of the low K_m values for cytochrome oxidase (4×10^{-6} to 2.4×10^{-8} M) it has been argued that the rate of oxygen uptake is limited only by availability at concentrations of 1% or less (Greenwood, 1968). It is dangerous to draw general conclusions concerning the whole root system in this way as an oxygen supply that is adequate for fully elongated young roots may not be sufficient for woody tissues or for actively growing meristems. Figure 1 shows the specific radioactivity of ethanol that has been distilled from portions of maize roots which have been incubated with radioactive pyruvate. The marked increase in ethanol radioactivity in the extracts from the distal portions of the roots is ample evidence of the conditions of partial anoxia that exist in the meristems of well aerated roots. Admittedly this zone of anaerobiosis is confined to the apical 3–8 mm of the root but it serves to indicate that any decrease in the supply of oxygen to the root will inevitably increase the quantity of tissue that is exposed to anaerobic conditions. Similar anaerobic phenomena have been observed in onion root meristems from manometric measurements (Berry, 1949) and by an enzymatic analysis of ethanol production in pea meristems (Betz, 1958).

Figure 2 extends these observations of ethanol production in the root meristems of crop species to the level of ethanol production in whole root systems of trees. The graph illustrates the effect of discontinuing the aeration of a water culture on the ethanol content of roots of *Pinus contorta*. When the aeration pumps were switched off there was an immediate decline in oxygen content of the culture

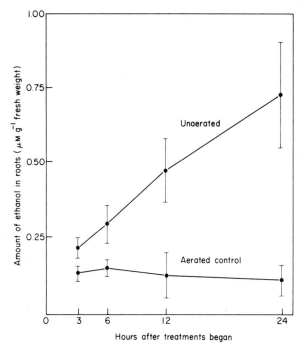

FIG. 2 Accumulation of ethanol in roots of *Pinus contorta* grown in water culture after aeration ceased, as compared with continuously aerated controls. The vertical bars indicate the maximum and minimum values from estimates on four different trees.

solution accompanied by an increase in the ethanol content of the root. All solutions in these experiments were sterilized and the tree roots were pre-treated with a 0.02% solution of mercuric chloride in order to decrease microbial activity. In all cases ethanol was determined enzymatically from an extract made from the roots. Any rhizosphere production of ethanol would not be assayed as this would pass into the culture solution. When similar experiments were done with *Picea sitchensis* the increase in ethanol level in flooded roots was approximately 10 times greater than with *Pinus contorta* rising in spruce to 25 μM g^{-1} fresh weight after 24 h without aeration of the water culture. In both species of tree stopping aeration caused an increase in ethanol production. Therefore, in spite of the radial loss of oxygen reported in *Pinus contorta* grown in water culture (Armstrong and Read, 1972), large portions of the root system must suffer an oxygen deficit which is increased when the medium around the roots becomes anaerobic.

In earlier experiments with herbaceous plants (Crawford, 1966, 1967) it was found that flood-tolerant species could be distinguished from intolerant species by the greater degree of control of glycolytic rate that was found in the flood-tolerant plants. The above experiments with *Picea sitchensis* and *Pinus contorta* suggest that a similar situation may exist in trees. Because of the possible downward diffusion of oxygen the results are not conclusive on this point, as it is always possible that the physical presence of oxygen in portions of the roots of *Pinus contorta* may be controlling the rate of glycolysis. The experiments reported below overcome this problem by using excised roots incubated under nitrogen, and suggest that trees conform to the general pattern of a homeostatic control of glycolysis already observed in species tolerant of anoxia. A similar situation occurs in animals where, in species that are capable of surviving temporary periods of anaerobiosis, lowering the concentration of oxygen also lowers the metabolic rate (Crawford, 1975). This is in marked contrast with most aerobic organisms where a decrease in oxygen availability accelerates glycolysis.

II. METABOLIC ADAPTATIONS TO ANOXIA

Previous studies on herbaceous plants have suggested certain aspects of metabolic adaptation that enable plants to withstand periods of low oxygen supply.

1. *A control of metabolic rate.* Species that are flood-tolerant have a minimal Pasteur effect and show little induction of glycolytic activity on exposure to anoxia. The reverse is true for intolerant species.

2. *Diversification of the end-products of glycolysis.* Flood-tolerant species produce a range of metabolic products under anoxia which facilitates proton disposal without the production of more ethanol.

22. Glycolysis Regulation and Tolerance to Waterlogging

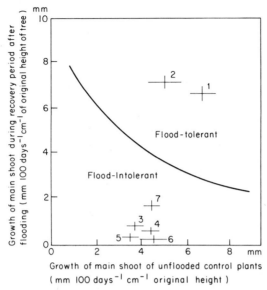

FIG. 3 Effects of a 4-month period of flooding on subsequent shoot growth during a 2-month recovery period, of seven tree species (see Table I, column 1). Four replicates were used in each test and the bars represent the 95% confidence limits.

If we examine each of these features in turn it is possible to demonstrate certain basic differences in the reactions of tree roots to anoxia which can be related to their differential tolerance of flooding. The tree species used in these experiments together with their provenances are listed in Table I. When the growth rates of the coniferous trees are examined it can be seen (Fig. 3) that two species, *Pinus contorta* and *Larix laricina*, distinguish themselves by being able to increase their growth rates after flooding over that of the unflooded controls, whereas the growth of all other species was decreased. When sample roots were excised from flooded and unflooded plants and placed under nitrogen, rates of carbon dioxide production increased in the species that were intolerant of flooding (Table II). Not only was there the greatest increase in Q^{N_2} CO_2 in the flood-intolerant species but they also had the highest absolute rates of anaerobic carbon dioxide evolution. By contrast, little change occurred in the tolerant species which thus behaved in the same way as herbaceous plants; i.e. anaerobiosis caused an increased glycolytic rate only in the intolerant species (Crawford and McMannon, 1968).

TABLE I. Tree species and provenances used in flooding tolerance tests together with details of habitat and location.

Species and number in Fig. 3	Provenance number[a]	Latitude	Longitude	Habitat
Pinus contorta (1)	65(797)	47°20'N	124°W	Coastal site
Pinus sylvestris (6)	67(4122)11	57°40'N	3°45'W	Sandy loam, Culbin, Moray
Picea mariana (3)	67063	45°53'N	77°25'W	Wet organic soil, Quebec
Picea mariana	5114	48°53'N	80°54'W	Wet organic soil, Quebec
Picea mariana (4)	5088	48°31'N	80°50'W	Sandy dry plain Ontario
Picea sitchensis (5)	66(7111)1	54°0'N	132°10'W	Queen Charlotte Islands, British Columbia
Larix europea (7)	62(4371)	49°44'N	15°55'E	Vlasim, Bohemia Czechoslovakia
Larix laricina (2)	67021	45°58'N	77°26'W	Wet organic soil, Ontario
Larix laricina	68018	48°50'N	54°44'W	Cut over *Picea mariana* stand Newfoundland
Alnus incana	66(4127)500	56°30'N	4°10'W	Brown earth Perthshire, Scotland

[a] Numbers refer to the British Forestry Commission's provenance register.

From a number of experiments with herbaceous plants Crawford and Tyler (1969) suggested that dark fixation of carbon dioxide with the production of malic acid acted as a detoxicating mechanism whereby flood-tolerant species avoided the accumulation of potentially toxic concentrations of ethanol. When excised portions of tree roots were kept in anaerobic conditions in Warburg flasks it was observed that there was marked variation in their ability to continue dark fixation of carbon dioxide. The ratios of carbon dioxide fixation in roots under air and under nitrogen are compared in Table III. In

TABLE II. Anaerobic respiration rates of tree roots that were kept continuously in unflooded conditions, compared with those that were flooded for 4 months ($Q^{N_2} CO_2 = \mu l\ CO_2$/g fresh weight).

Species and provenance number	Unflooded	Flooded	Flooded/unflooded
Pinus contorta	82.5	74.4	0.9
Alnus incana	62.8	75.4	1.2
Larix laricina (67021)	53.7	78.7	1.5
Picea mariana (67063)	62.7	172.2	2.7
Picea mariana (5088)	51.2	128.0	2.5
Picea sitchensis	48.4	105.8	2.2
Pinus sylvestris	81.9	162.4	2.0

addition, the quantities fixed under these two conditions into the malic, glutamic and aspartic acid pools were also compared. The flood-tolerant species *Pinus contorta* and *Larix laricina* continued their rates of dark fixation of carbon dioxide with little alteration under nitrogen, whereas the process was markedly depressed in the intolerant species. As these experiments were carried out with

TABLE III. Amounts of ^{14}C fixed by roots of tree species into various metabolites in the dark expressed as ratios between amount fixed in nitrogen and air. Roots were unflooded and were exposed to $^{14}CO_2$ for 1 h.

Species and provenance numbers	N_2 : air fixation ratios			
	Total ethanol extract	Malic acid	Glutamic acid	Aspartic acid
Pinus contorta	0.99	1.05	0.65	0.98
Alnus incana	0.86	0.85	0.54	0.66
Larix laricina (67021)	0.82	0.76	0.70	0.52
Picea mariana (67063)	0.71	0.60	0.47	0.16
P. mariana (5088)	0.55	0.69	0.16	0.40
P. sitchensis	0.52	0.61	0.15	0.11
Larix europea	0.40	0.50	0.11	0.41
Pinus sylvestris	0.48	0.45	0.26	0.17

excised roots in an atmosphere of nitrogen there was no confusion with the effects of aeration on the metabolism of the plants. The relative distribution of the products of dark fixation in these roots (Table IV) shows that in all species malic acid is the principal product. Intolerant trees are, therefore, directly comparable with intolerant herbaceous plants in that the effect of anoxia is to decrease the production of malic acid in roots and accelerate glycolysis.

In spring, much greater concentrations of malic acid were found in the ascending xylem sap in birch trees (*Betula pubescens*) growing on wet soils, compared with trees growing on similar adjacent soils which were better drained (Crawford, 1972). The ratio of malate to ethanol in these extracts was in the region of 100 : 1 and it would appear that, in these trees, it was the malic acid that was the main carrier of the oxygen debt from root to shoot. In the leaf, malate can be assimilated to carbohydrate via pyruvate with a minimal loss of fixed carbon to the plant. The malic acid emanating from roots,

TABLE IV. Percentage of ^{14}C fixed into various metabolites in the dark by unflooded roots of two flood-tolerant and two flood-intolerant tree species. Roots were allowed to fix $^{14}CO_2$ in air for 1 h.

Substance	Flood-tolerant		Flood-intolerant	
	Pinus contorta	*Larix laricina*	*Pinus sylvestris*	*Larix europea*
Malic acid	60.1	59.1	62.7	38.7
Citric acid	8.7	29.5	13.3	17.6
Fumaric acid	0.0	0.0	2.0	3.3
Aspartic acid	2.2	0.4	3.0	3.6
Glutamic acid	14.6	5.0	8.5	19.6
Alanine	3.3	0.0	1.2	3.1
Tyrosine	3.4	4.5	3.9	8.1
Uridine monophosphate	2.7	0.0	2.5	1.9
Sugar-phosphate esters	0.0	0.0	0.0	1.9
Sucrose	3.0	1.4	2.9	2.2
Glucose	2.0	0.0	0.0	0.0

together with the high concentrations of carbon dioxide found in the gases of woody tissues, may provide considerable sources of carbon dioxide for leaf photosynthesis.

It is pertinent to note that, in early spring, the deep anchoring roots of trees begin to mobilize the considerable quantities of starch stored in the xylem parenchyma. Where these roots grow in wet soils they will suffer from at least partial anoxia for the reasons outlined above. The use of their carbohydrate reserves at this time of the year can be likened to the mobilization of glycogen that takes place in the liver of the human infant before parturition, which enables the infant to tolerate up to 30 min of anoxia at birth. Similar glycogen stores are found in the muscles of oysters and other intertidal molluscs, and are again related to the ability of the tissues to endure anoxia. It is possible to argue therefore that the carbohydrate stores of the tree root allow this organ to respire anaerobically in spring, when increased metabolic work is required and the high winter water tables have not yet subsided. The utilization of starch and the production of malate allow the root to respire anaerobically, produce a non-toxic end-product for glycolysis and transfer the oxygen debt of the root to the well aerated shoot (Crawford, 1975).

Plants that are tolerant of anoxia characteristically produce a number of metabolites when subjected to low oxygen stress. This behaviour is analogous to the mixed fermentations discussed by Krebs (1972). Most biochemical syntheses involve an excess of oxidations over reductions, and therefore if a tissue is to perform any anabolic activities under anoxia, additional forms of proton disposal are necessary. Thus, in the oyster there is simultaneous production of succinate, alanine and propionic acid (Hochachka and Somero, 1973), which not only facilitates proton disposal but, in this case, increases the yield of ATP to 8 mol/mol of glucose. In eutrophic marsh plants flooding increases nitrate reductase activity, suggesting that the uptake of nitrate and the subsequent synthesis of amino acids is providing an efficient method of proton disposal in flooded roots. In trees there are several reports of increased nitrate utilization by flood-tolerant species which is coupled with an increased amino acid content on flooding. As with animal species where this pathway is well known, the coupling of carbohydrate metabolism with amino acid production will facilitate proton disposal. However, in the case of plants the reduction of nitrate to ammonium makes this pathway very much more advantageous in terms of available hydrogen acceptors (Chirkova, 1971; Garcia-Novo and Crawford, 1973).

Glycerol production has also been detected as accumulating under flooded conditions in the roots of *Alnus incana* (Crawford, 1972). The detection of this metabolite is of interest because it occurs in the silk worm at diapause as well as in certain blood-inhabiting trypanosomes. The production of this metabolite does not give a net gain in ATP but may be of significance in that it deflects pyruvate from reduction to ethanol and this together with the production of a potentially readily usable metabolite may make this metabolic variation of considerable ecological advantage to the species concerned. Figure 4 summarizes the various end-products of glycolysis that can be detected in plants. This scheme is a modified form of the original outline given by McMannon and Crawford (1971). The original version of this pathway postulated that the deletion of malic enzyme was responsible for the accumulation of malic acid in flood-tolerant species. Further experimentation by Davies *et al.* (1974) has shown that this enzyme is universally present in both

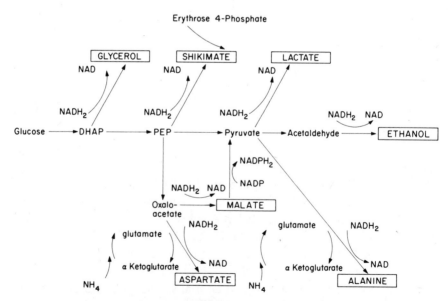

FIG. 4 Summary diagram illustrating the various means of proton disposal and the range of end-products of glycolysis found in plants capable of enduring prolonged periods of partial or total anoxia.

flood-tolerant and -intolerant species. However in the experiments of Chirkova *et al.* (1973) anoxic conditions were found partially to inhibit this enzyme in flood-tolerant plants and induce additional activity in flood-intolerant species. It would now appear (Crawford, 1975) that qualitative changes in the properties of a number of enzymes are responsible for the alterations in the end-products of glycolysis that are associated with the possession or lack of flooding tolerance in higher plants.

III. CONCLUSIONS

1. Anoxia (i.e. a deficiency of oxygen) is a general condition in root meristems and in non-adventitious roots of trees.

2. In *Pinus contorta* (a species which can resume growth after flooding) and in *Picea sitchensis* (a species in which flooding retards growth after flooding), restricting the aeration of the water culture increases the production of ethanol by the roots. Thus, a flood-tolerant species and an intolerant species of tree both suffer from anoxia when the rooting medium is oxygen deficient.

3. Rates of glycolytic activity in anoxic conditions are greater in flood-intolerant trees than in flood-tolerant species.

4. Detoxification of the end-products of glycolysis by the dark fixation of carbon dioxide is restricted by anoxia in flood-intolerant species. Flood-tolerant trees continue to fix carbon dioxide and produce malic, glutamic and aspartic acids as rapidly in nitrogen as in air.

5. Nitrate reduction and assimilation into amino acids can provide flood-tolerant trees with an additional means of proton disposal during periods of anaerobiosis.

ACKNOWLEDGEMENTS

This research was supported by financial assistance from the Natural Environment Research Council. The Northern Research Branch of the Forestry Commission gave unstinted help with materials and advice which is gratefully acknowledged together with the careful study of the trees under investigation in St Andrews by Fräulein Dr Gertrud Vester and Miss Margaret Baines.

REFERENCES

Armstrong, W. (1964). Oxygen diffusion from the roots of some British bog plants. *Nature, Lond.* **204**, 801–802.

Armstrong, W. (1968). Oxygen diffusion from the roots of woody species. *Physiol. Plant.* **21**, 539–543.

Armstrong, W. (1975). Waterlogged soils. In "Environment and Plant Ecology". (J. R. Etherington, ed.), pp. 184–216. John Wiley, London.

Armstrong, W. and Read, D. J. (1972). Some observations on oxygen transport in conifer seedlings. *New Phytol.* **71**, 55–62.

Berry, L. J. (1949). The influence of oxygen tension on the respiratory rate in different segments of onion roots. *J. cell. comp. Physiol.* **33**, 41–66.

Betz, A. (1958). Aerobe Gärung und Äthanol-Umsatz. *Naturwissenschaften* **45**, 88–89.

Boggie, R. (1972). Effects of water table height on root development of *Pinus contorta* on deep peat in Scotland. *Oikos* **23**, 304–312.

Carrodus, B. B. and Triffett, A. C. K. (1975). Analysis of composition of respiratory gases in woody stems by mass spectrometry. *New Phytol.* **74**, 243–246.

Chirkova, T. V. (1968). [Features of the oxygen supply of roots of certain woody plants in anaerobic conditions]. *Fiziologiya Rast.* **15**, 565–568.

Chirkova, T. V. (1971). Role nitratnovo dichaniya kornei v zhiznediyatelnosti nekotori drevesnich rastenii v usloviach vremnovov anaerobioza. *Vestnik Leningr. gos. Univ.* **21**, 118–124.

Chirkova, T. V., Khazova, I. V. and Astafurova, T. P. (1973). Metabolic regulation of plant adaptation to conditions of temporary anaerobiosis. *Fiziolgiya Rast.* **21**, 102–107.

Crawford, R. M. M. (1966). The control of anaerobic respiration as a determining factor in the distribution of the genus *Senecio*. *J. Ecol.* **54**, 403–413.

Crawford, R. M. M. (1967). Alcohol dehydrogenase activity in relation to flooding tolerance in roots. *J. exp. Bot.* **18**, 458–464.

Crawford, R. M. M. (1972). Physiologische Ökologie: Ein Vergleich der Anpassung von Pflanzen und Tieren an sauerstoffarme Umgebung. *Flora* **161**, 209–223.

Crawford, R. M. M. (1975). Metabolic adaptations to anoxia in plants and animals. *Proc. 12th Internat. Bot. Congress, Leningrad.*

Crawford, R. M. M. and McMannon, M. (1968). Inductive responses of alcohol and malic dehydrogenase to flooding tolerance in roots. *J. exp. Bot.* **19**, 435–441.

Crawford, R. M. M. and Tyler, P. D. (1969). Organic acid metabolism in relation to flooding tolerance in roots. *J. Ecol.* **57**, 237–246.

Davies, D. D., Nascimiento, K. H. and Patil, K. D. (1974). The distribution and properties of NADP malic enzyme in flowering plants. *Phytochemistry* **31**, 2417–2425.

Devaux, H. (1899). Asphyxie spontanée et production d'alcool dans les tissus profonds des tiges ligneuses poussant dans les conditions naturelles. *C.r. hebd. Séanc. Acad. Sci. Paris* **128**, 1346–1349.

Garcia-Novo, F. and Crawford, R. M. M. (1973). Soil aeration, nitrate reduction and flooding tolerance in higher plants. *New Phytol.* **72**, 1031–1039.

Gessner, F. (1968). Zur ökologischen Problematik der Überschwemmungswälder des Amazonas. *Int. Rev. ges. Hydrobiol.* **53**, 525–547.

Gill, C. J. (1970). The flooding tolerance of woody species – a review. *For. Abstr.* **31**, 671–688.

Gill, C. J. (1975). The ecological significance of adventitious rooting as a response to flooding in woody species with special reference to *Alnus glutinosa* (L.) Gaertn. *Flora* **164**, 85–97.

Greenwood, D. J. (1967). Studies on the transport of oxygen through the stems and roots of vegetable seedlings. *New Phytol.* **66**, 337–347.

Greenwood, D. J. (1968). Effect of oxygen distribution in the soil on plant growth. *Proc 15th Easter School, Agric Science Univ. of Nottingham*, 202–221.

Hochachka, P. W. and Somero, G. N. (1973). "Strategies of Biochemical Adaptation". Saunders, London.

Krebs, H. (1972). The Pasteur effect and the relations between fermentation and respiration. *In* "Essays in Biochemistry" (P. N. Campbell and F. Dickens, eds), Volume 8, pp. 1–34. Academic Press, London and New York.

McMannon, M. and Crawford, R. M. M. (1971). A metabolic theory of flooding tolerance: the significance of enzyme distribution and behaviour. *New Phytol.* **70**, 299–306.

van Raalte, M. H. (1941). On the oxygen supply of rice roots. *Annls Jard. bot. Buitenz.* **51**, 43–57.

Rowe, R. N. and Catlin, P. B. (1971). Differential sensitivity to waterlogging and cyanogenesis by peach, apricot and plum roots. *J. Am. Soc. hort. Sci.* **96**, 305–308.

Vámos, R. and Köves, E. (1972). Role of light in the prevention of the poisoning action of hydrogen sulphide in the rice plant. *J. appl. Ecol.* **9**, 519–525.

Vartapetian, B. B. (1964). [Poloragraphic study of oxygen transport in plants.] *Fiziologiya Rast.* **11**, 774.

FROST HARDINESS

23
Frost Hardiness of Forest Trees

C. GLERUM

Ministry of Natural Resources, Forest Research, Maple, Ontario, Canada

I.	Introduction	403
II.	The Various Aspects of Frost Hardiness	404
III.	Frost Hardiness and Dormancy	408
IV.	Carbon Assimilation and Distribution in Autumn	411
V.	Some Observations on Frost Hardiness and Tree Breeding	414
VI.	Summary	417
	References	418

I. INTRODUCTION

Frost hardiness is vital for the survival of plants growing in climates where freezing occurs. As a result, frost hardiness has received considerable attention in most of the temperate climatic regions of the world as indicated by the voluminous, but often contradictory, literature on the subject. In spite of all this effort little progress has been made toward elucidating the mechanism of frost hardiness, although recently some promising leads have appeared. This reflects the complexity of the subject and leads one to conclude that it is highly unlikely that one comprehensive, all-embracing theory can be established.

It seems clear that frost hardiness is an expression of many physiological processes. Consequently the geneticist should have an understanding of these processes when attempting to determine the genetic mechanisms involved.

It is not my intention to give a comprehensive survey of the literature on frost hardiness because several reviews are available (Levitt, 1956; Vasil'yev, 1956; Parker, 1963; Alden and Hermann,

1971), including the most recent one by Levitt (1972). However, I shall touch on those aspects which I consider important and which, as yet, have received insufficient attention in the literature. Furthermore, I shall confine my observations strictly to trees.

II. THE VARIOUS ASPECTS OF FROST HARDINESS

The frost hardiness of a tree can be defined in general terms as the lowest temperature below the freezing point to which a tree can be subjected without being damaged. The phenomenon which enables a tree to increase or decrease its cold resistance is called the frost-hardiness process.

The ability of protoplasm to withstand dehydration caused by ice formation is one of the major features of frost hardiness. This was clearly demonstrated in the 1930s by Scarth, Levitt and Siminovitch at McGill University. They showed that two types of ice formation should be recognized in plants. They are: extra-cellular ice formation, where the ice crystals form outside the cells in the intercellular spaces; and intracellular ice formation, where ice crystals form inside cells. The first type of ice formation predominates in nature and is non-fatal in hardy tissues, while the second type generally disrupts the cytoplasm and is nearly always fatal, but this type occurs infrequently in nature (Scarth, 1944).

When the frost-hardiness process is activated, and the tree goes from its summer condition of minimum frost hardiness to its winter condition of maximum frost hardiness, major changes take place within living cells of the tree. The flow of water in and out of cells is facilitated by the increased permeability of the cell membranes, and of the cytoplasm, which changes from a translucent sol to an opaque gel state. These changes reflect the numerous biochemical reactions that occur in the cells and which have consequently attracted considerable attention. Numerous conflicting observations are available on the subject. In general many good correlations have been found between increases in frost hardiness and increases in chemical substances such as sugars, proteins, lipids and nucleic acids. However, these correlations rarely apply to these substances simultaneously and frequently they are considerably poorer during the dehardening period. It can be expected that these increases in substances should result in a general protoplasmic augmentation during hardening, which has been clearly demonstrated by Siminovitch et al. (1967a, 1967b, 1968) with bark parenchyma cells of black locust (*Robinia pseudoacacia*) and more recently by Pomeroy et al. (1970) for red pine (*Pinus resinosa*). The increases in the various substances are frequently due to the accumulation of reserves for the winter, but they also appear to serve as protective agents against freezing injury. This has led to the observation (Levitt, 1972) that if the factors that control frost hardiness are to be elucidated, all the changes associated with the build-up of reserves for any process not directly related to frost hardiness must be eliminated. Some of my own findings, which will be discussed later, support this observation.

In recent years increasing attention has been focused on the cell membrane and its chemical composition, which is lipoidal-proteinaceous in nature. In general there is an increase in phospholipids and also in the degree of unsaturation of the fatty acids during hardening. However, there remains some doubt as to the exact extent of the changes in unsaturation. Since unsaturated fatty acids have lower melting points than saturated fatty acids, the increase in unsaturation should assist in keeping the membrane more pliable at lower temperatures and consequently less prone to mechanical damage. Although it is not known to what extent these changes in phospholipids and fatty acids represent changes in the cellular membrane, the quantity of phospholipids has been used as an indication of the amount of membrane material in cells.

In the well known ringing experiment on *Robinia pseudoacacia* (Siminovitch and Briggs, 1953) where phloem continuity was disrupted by girdling at various times during late summer (11 August–23 September), bark segments were created which only differed in their time of isolation from the crown. They found that the segments isolated earliest from the crown were the least hardy, while the segments which remained in contact with the crown the latest, developed an almost normal hardiness with the expected accompanying chemical changes. However, in the early isolated sections they found a noticeable increase in hardiness without any accompanying chemical changes, which they suggested might be due to a subsidiary hardening mechanism. Subsequent work with isolated (encircled) bark segments (Siminovitch et al., 1967a, 1968) confirmed that these isolated, starved segments were still able to achieve considerable hardiness (−30 to −45°C). This occurred without protoplasmic augmentation and without increases in water-soluble protein and ribonucleic acids. However, there was a continued synthesis of phospholipids without any change in total lipid content. It seems feasible that this is not a subsidiary mechanism, but a major one, which requires a certain level of substances for optimum function. This level is not reached in the starved sections. It also suggests that the frost-hardiness mechanism is located in each cell or at least in each type of tissue, which would in part explain the differential hardiness that is known to occur among the different tissues of trees. This localized nature of the frost-hardiness mechanism was clearly demonstrated by Timmis and Worrall (1974) with climatically "split" 2-year-old Douglas fir (*Pseudotsuga menziesii*). They induced a high degree of frost hardiness in one branch and not in the other of the same seedling by exposing the branches to different environmental conditions.

Several observers believe that frost hardiness occurs in two to three stages in woody plants native to temperate zones (Tumanov and Krasavtsev, 1959; Krasavtsev, 1968; Weiser, 1970; Glerum, 1973b). According to Levitt (1972) Tumanov lists three periods or stages in the preparation of plants for winter: (a) the onset of dormancy, (b) the first stage of hardening at about 0°C and (c) the second stage of hardening during a gradual lowering of the temperature below 0°C. This is in agreement with Weiser (1970) who believes that the first stage of hardening (b) is induced by short days. The second stage is apparently induced by temperatures just below 0°C and a third stage by temperatures of −30 to −50°C. Only the extremely hardy species would be able to attain this third stage and as Weiser suggests, this kind of hardiness is quickly lost. Levitt is not in agreement with this three-stage concept because he suggests that one can obtain as many stages of hardening as desired by means of a graded series of hardening treatments. However, he based his argument on cabbage which has only a maximum hardiness of −20°C. Furthermore, the fact that these stages, particularly the first two, were observed under natural to near natural hardening conditions was apparently not taken into consideration. In my frost hardiness study on seven coniferous species I observed two stages of hardening with the transition from one stage to the next occurring at approximately −18°C. This is illustrated for example by the white pine (*Pinus strobus*) frost-hardiness trend in Fig. 1. I also suspect that there are two stages of dehardening but since dehardening in the spring occurs so rapidly it is difficult to identify separate dehardening stages (Glerum, 1973b).

Some investigators appear to favour the concept that a translocatable factor is involved in the frost-hardiness process (Weiser, 1970; Timmis and Worrall, 1974). A major question according to Weiser is whether the translocatable hardiness—promoting factor is (a) a growth inhibitor which indirectly influences hardiness by stopping growth, (b) a simple sugar, or (c) a regulating substance like a hormone which plays a direct role in mobilizing the metabolic machinery responsible for the first stage of cold acclimation. I tend to believe that the hardiness-promoting factor plays an indirect role in

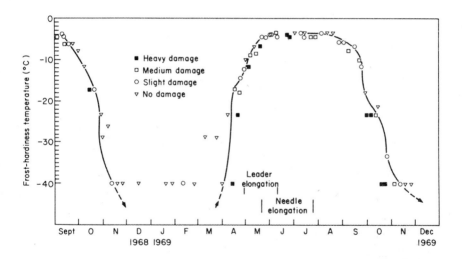

FIG. 1 Frost-hardiness trend of *Pinus strobus* (white pine) where the frost-hardiness temperature delineated is the lowest temperature to which trees can be exposed without being damaged.

In general the frost hardiness increased from late August until late November, when it had reached its maximum hardiness below −40°C. In April the dehardening rate was rapid to about −15°C after which it became more gradual to the minimum frost hardiness of about −4°C by June. The transition between the first and second stage of hardening was not detectable in October 1968 but was in October 1969. The transition is not too distinct because the time scale is calibrated in months, whereas the transition generally occurs over a period of a few days. (After Glerum, 1973b.)

cold acclimation by simply slowing or stopping growth, because it does not appear to behave as a hormone.

In the field of frost hardiness it is generally easier to find points of disagreement than points of agreement. However, most investigators will agree that temperature and light are the two main environmental factors controlling the development of frost hardiness. These factors play a significant role in the hardening stages as already mentioned but they also play a vital role in dormancy. This leads one to the question of what sort of relationship exists between frost hardiness and dormancy. This particular aspect has received insufficient attention by both frost hardiness and dormancy investigators alike.

III. FROST HARDINESS AND DORMANCY

The type of relationship between frost hardiness and dormancy is unknown but since both processes are complex one can expect the relationship between them to be complex as well. It is difficult to give a general definition of dormancy because it will be subject to many exceptions. However, the most acceptable definition appears to be that of Doorenbos (1953) who applied the term dormancy to all cases where a living tissue predisposed to elongate does not do so. Romberger (1963) has summarized clearly the various proposed definitions on dormancy, and suggests that the term dormancy should be restricted to behavioural attributes of meristematic tissues only. This means that only meristems can be dormant and consequently dormancy will be confined to the various meristematic zones, which make up a very small tissue volume of the entire tree. Frost hardiness, on the other hand, is a behavioural attribute of all living tree tissue, which has a vastly greater volume than the meristematic zones. Consequently any relationship between dormancy and frost hardiness would depend on some sort of signal (translocatable factor?) which is transmitted either directly or indirectly from the meristematic zones to the living non-meristematic tissues or vice versa. For example, a decrease in secondary meristematic activity may cause an increase in reserve accumulation, which in turn, may assist the cells in their hardening process (protoplasmic augmentation).

It has been suggested by several investigators, including Weiser

(1970), that the key factor in the induction of hardening appears to be growth cessation rather than the onset of dormancy (rest period) because low temperatures can stop growth and bring about substantial hardening without inducing dormancy. However, it appears that only dormant trees can develop a high degree of frost hardiness. Just because some trees become dormant but not frost hardy (e.g. southern pines), this does not mean that there is no relationship between these two processes. The available evidence suggests that with certain species dormancy will be directly influenced by external factors and so influence frost hardiness indirectly, while with other species the frost hardiness will be more directly influenced by external factors. Consequently, as I have suggested previously (Glerum, 1973a), it is useful to classify those species which have to go into winter dormancy (rest period) in order to attain their maximum frost hardiness and those which do not. This is by no means an easy task since there will be considerable variation even within species, due to several factors, including geographic differences.

Most tree species in the northern temperate zone, particularly the coniferous species, go into winter dormancy before they attain their maximum winter hardiness. For example, under natural conditions *Pinus strobus* is in a state of winter dormancy by September (Glerum, 1973a), and the only way to break this dormancy is to satisfy the chilling requirements of the trees. This is in accordance with Perry (1971) who suggested that dormancy should be defined in a highly restricted sense, namely that it possesses a chilling requirement. From September to December the chilling requirements were increasingly satisfied so that some time during December they had been totally satisfied and the trees went from a condition of winter dormancy to one of imposed dormancy (quiescence). Similar results have been obtained by Mergen (1963), also with *P. strobus* and by Little and Bonga (1974) with *Abies balsamea*.

During this period, when the chilling requirements were being satisfied, trees increased in hardiness until they reached their maximum frost hardiness (winter hardiness) near the beginning of December (Fig. 1). This coincides approximately with the time that the trees go from their winter dormancy to their imposed dormancy. Consequently, it is possible to suggest that when maximum frost hardiness has been attained, the chilling requirements have been satisfied. This term "imposed" could also be applied to winter hardiness, since it persists until conditions are favourable for growth.

During the dehardening period the first visible signs of growth become evident, well before the trees have fully dehardened (Figs 1 and 2). This is particularly true for a conifer such as larch (*Larix laricina*) where the buds had flushed while the young needles were still considerably hardy, between −17 and −11°C (Fig. 2). From these and other observations in the literature it is evident that the frost-hardiness process lags behind the dormancy process, both in the autumn when the trees enter dormancy and increase in frost hardiness and in the spring, when the trees break dormancy and decrease in frost hardiness. This could be interpreted as suggesting that, in trees, dormancy is a prerequisite for frost hardiness. This may be true for some species but not necessarily for all.

The early attainment of imposed dormancy and winter hardiness in the autumn may well be responsible for the rapid dehardening in

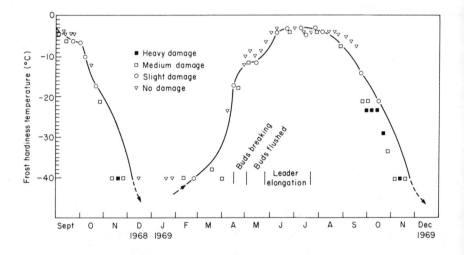

FIG. 2 Frost-hardiness trend of *Larix laricina* (American larch) where the frost-hardiness temperature delineated is the lowest temperature to which trees can be exposed without being damaged.

In general the frost hardiness increased from late August until December, when it had reached its maximum hardiness below −40°C. Towards the end of February there was a gradual dehardening rate, which increased in April until May when it levelled off at about −11°C for several days. The minimum frost hardiness of about −3°C was reached in June. The hardening and dehardening rates were more gradual in larch than in pine. (After Glerum, 1973b.)

the spring when conditions suitable for growth are restored (Fig. 1). The rapidity of the dehardening is undoubtedly responsible in part for the fact that in the spring the correlations between frost hardiness and the various chemical and physical factors have invariably been poorer than the same correlations in the autumn, when the hardening rate is considerably more gradual.

In their reviews on dormancy Doorenbos (1953) and Vegis (1964) mentioned several researchers who observed increases in respiration during chilling. These observations have never received much attention in dormancy studies. Recently, Siminovitch and his co-workers have noted that hardy cells had a greater respiratory capacity than non-hardy cells. This has been attributed to the greater number of mitochondria in hardy cells during autumn and winter than in non-hardy cells during spring and summer (Pomeroy and Siminovitch, 1971). This suggests that hardy trees have a greater metabolic capacity and possibly a higher rate of respiration, at least during the early stages of the hardening process. Levitt (1972) suggests that metabolism will occur at a decreased rate, at hardening temperatures. This observation is not necessarily contradictory, because the data of Siminovitch and Pomeroy only imply that at relatively warm autumn temperatures the rate of respiration might be greater than during the summer, and that the later autumn rates would not decrease as rapidly as would be the case if the respiratory capacity had not increased. The observation by Levitt appears to be based on work on cabbage and it suggests that some major differences probably exist between the frost hardiness of woody plants and that of herbaceous plants.

I feel that the relationship between frost hardiness and dormancy is not only of great physiological importance but also of immense practical application, particularly in forest nursery practice, and merits intensive research. Another aspect which has received little attention is carbon assimilation during the frost-hardening period and its influence on food reserves.

IV. CARBON ASSIMILATION AND DISTRIBUTION IN AUTUMN

To determine what happens to the carbon that is assimilated during the frost-hardening period and its subsequent distribution, carbon-14, in the form of $^{14}CO_2$ (50 μCi per tree) was photosynthetically introduced into 3-year-old potted *Pinus banksiana, Pinus strobus* and *Picea glauca* (10 trees per treatment) at 2-weekly intervals from August to December, 1972. Some preliminary results on *Pinus banksiana* have been presented previously (Glerum, 1974). The trees were grown under natural conditions and brought indoors only for the carbon-14 treatment. Radioactivity was monitored by a liquid scintillation technique. Approximately 90% of the administered radioactive $^{14}CO_2$ was incorporated during the first seven tests (9 August to 31 October) while approximately 75% was incorporated during the last three tests (15 November to 12 December). This suggests that hardened trees have a lower photosynthetic rate than unhardened trees, possibly due to a decrease in chlorophyll content which occurs during winter, because chlorophyll degradation exceeds its synthesis at low temperatures (Kramer and Kozlowski, 1960). Although some decrease in chlorophyll content has been noted, El Aouni and Mousseau (1974) attributed the winter drop in photosynthetic activity to chloroplast degradation due to cold, which was independent of the total chlorophyll content. These findings are in general agreement with those of Perry and Baldwin (1966).

The trees were sampled from January to June 1973 for microscopic and chemical examination. Hand sections from the trees were placed on X-ray plates to give a quick although rough indication of the carbon-14 concentration and distribution. When developed, the X-ray plates indicated that good incorporation and distribution was obtained particularly during the first 3 months of carbon-14 feeding treatments (August, September, October). The incorporation during November and December was considerably lower and more irregular. Paraffin embedded sections (10 μm) covered with autoradiographic stripping film indicated the location of the radioactivity more accurately.

The autoradiograph in Fig. 3 shows a band of radioactive tracheids in the 1972 annual growth layer, close to the boundary between the 1972 and 1973 annual layers. This band of tracheids was being formed at the time (9 August) of the carbon-14 feeding and it demonstrates the rapid incorporation of carbon in the cell walls of the xylem during active xylem formation. The subsequently formed tracheids in this annual layer show an ever decreasing amount of radioactive carbon incorporation and the new 1973 annual growth layer shows no radioactivity. However, the X-ray plates showed that

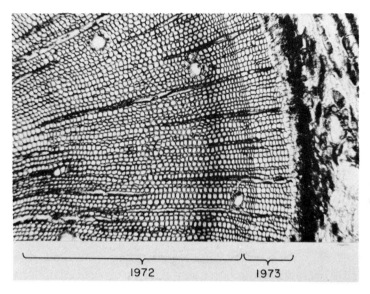

FIG. 3 Autoradiograph of a 10 μm thick, stem cross-section of *Pinus banksiana* (Jack pine). Labelled $^{14}CO_2$ was photosynthetically introduced on August 9, 1972 and the tree was harvested on June 14, 1973. The dark band of tracheids near the end of the 1972 annual growth layer indicates appreciable ^{14}C incorporation, which is missing from tracheids produced later in 1972 and in 1973. × 70.

radioactive compounds had been translocated into the new 1973 terminal leader. Therefore, it appears that the radioactivity occurs in compounds which are not used in wood formation and it suggests that the link between the photosynthetic pathway and wood formation is a direct one. It also suggests that the importance of reserves in relation to subsequent wood formation has been over-emphasized and should be re-evaluated. It is possible that these reserves play a more significant role in other processes, such as the frost hardening—dehardening process, than has previously been supposed. It is hoped that these initial findings will be corroborated by the chemical analyses which are now in progress. Considerable research efforts to elucidate the role of the reserves in relation to the processes mentioned are highly desirable.

It has been mentioned from the outset that frost hardiness is an expression of several physiological processes. The mechanisms involved in the frost-hardiness process as proposed by Siminovitch *et al.* (1967a, 1967b, 1968) and subsequently strengthened by Pomeroy *et al.* (1970) and Pomeroy and Siminovitch (1971) seem to me at present to be the most applicable in regard to trees. The proposal is based on the augmentation of total cellular protoplasm, including augmentation of cell organelles and other membranous components of the cell. Siminovitch suggests that frost hardiness is a combined expression of three processes consisting of (1) phospholipid synthesis which probably means some form of special membrane replication; (2) phospholipid synthesis in conjunction with total protoplasmic augmentation including replication of membrane and membrane structures throughout the cell and including synthesis of proteins, nucleic acids and cell organelles and (3) the starch–sugar transformation. These processes are capable of acting independently. However, little is known about the synergistic effect when these processes are working concurrently. Consequently, we should be careful in deciding the relative importance of each of these various processes.

V. SOME OBSERVATIONS ON FROST HARDINESS AND TREE BREEDING

Frost hardiness has received some attention by tree breeders, particularly in Europe (Dietrichson, 1961; Eiche, 1966). However, little work has been done to select and breed specifically for frost resistance and the literature on the subject tends to be badly scattered. Most of the commercially important native tree species in the northeastern part of the North American continent are frost hardy except for the southern provenances. Frost hardiness becomes a problem when one works with fast-growing non-native species and races. For example, in eastern Canada it is an important problem in poplar propagation.

Considerable work has been done on producing late-flushing varieties of spruce. In Europe spring frost injury frequently causes serious losses to spruce. This damage is avoided or greatly reduced in trees with late bud break (Holzer, 1969). In North America spring frost injury frequently causes serious losses to *Picea glauca*. Therefore, late-flushing is an important selection criterion for varieties in central and north-eastern United States and Canada, as it is in Europe, particularly since flushing time is a highly hereditary characteristic (Nienstaedt and King, 1969; Nienstaedt and Teich, 1972; Yeatman and Venkatesh, 1974; Thielges and Beck, Chapter 14).

It is obvious that frost hardiness is a multigenic property, which makes breeding specifically for frost hardiness difficult and frequently impractical. However, frost hardiness should be part of any breeding programme for developing fast-growing varieties, particularly in view of the ever increasing demand for these varieties. Since cessation of growth is essential for the development of frost hardiness, the time when growth ceases or resumes and the time of the first or the last frost are important. However, one should keep in mind that fast-growing varieties do not have to be more susceptible to frost than slow-growing varieties during the critical periods of the year, provided growth ceases on time.

In addition to the fact that the tree should have the inherent frost-hardiness capability, there are three aspects of frost hardiness, which should always be taken into consideration in a breeding programme:

(1) the time when frost hardening begins and the rate of hardening;

(2) the degree of maximum winter hardiness and whether or not it is high enough for the climate; and

(3) the time when the dehardening process begins and the rate of dehardening.

Sakai and Weiser (1973) have found that considerable variation can exist between species in these aspects. Indeed it must exist within a species if selection and breeding is to be effective.

The prospects for producing fast-growing, hardy varieties are promising because of the tremendous variation that exists within species. For example, Mergen (1963) found that in general *Pinus strobus* from southern latitudes broke dormancy more slowly, was less hardy, but produced more shoot growth than *P. strobus* from northern latitudes. Sakai (1970) found that even tropical and subtropical willows have a genetic potential to withstand freezing, although their capacity to resist intensive dehydration in winter was much lower than in willows native to temperate climates. With such a tremendous source of genetic variability at the tree breeder's disposal one can only be optimistic about the future.

The development of methods for testing trees for frost hardiness has received considerable attention. The ultimate test is the survival and growth of the tree in the field and as a result the favoured method by tree breeders has been to outplant the trees under as many different climatic conditions as possible. This type of testing is labour intensive and time consuming because test winters occur on an average only once every 10 years. Consequently, since the need for a quicker and more accurate method of testing is desirable, laboratory techniques have been developed, which are being used with good results. All these techniques are similar in that they are based on taking a part of the tree, usually twigs or needles, and testing them under a range of freezing temperatures followed by damage evaluation. Scheumann (1965, 1968) has developed a bulk test using a growth chamber for artificial hardening and dehardening while simultaneously determining repeatedly critical freezing temperatures in a freezing chamber, using twigs or 1-year-old seedlings. This technique is suitable for mass investigations and has been successfully used in tree breeding, according to Scheumann. Scheumann and Schmiedel (1972) have subsequently corroborated the results of the bulk tests with field trials, using 2-year-old *Pseudotsuga menziesii*. Another technique, which uses the natural hardening and dehardening cycles, is to cut twigs or needles from trees and subject them to a range of freezing temperatures in the laboratory. This technique has been used extensively both in physiological studies (e.g. Sakai and Weiser, 1973) and in genetic studies (e.g. Mergen, 1963). Recently Bialobok *et al.* (1974) used it to determine the winter hardiness of most trees and shrubs that grow in Poland.

At present no direct method to determine frost hardiness is as yet available. After the tree has been subjected to frost, the extent of injury is determined, which enables us to ascertain the frost hardiness of the tree. Consequently, one of the critical aspects of determining frost hardiness is the method used for evaluating the injury or viability of the tree (or twig) after it has been subjected to freezing temperatures. There are many methods of evaluating viability at our disposal. Some are based on the loss of the selective permeability of the plasma membrane such as the plasmolysis test, vital staining test, leaching test and electrical impedance test. Others are based on the extend of browning, subsequent growth or regenerative capacity. The method selected will depend on a variety of circumstances, but it is advisable to use at least two methods simultaneously for subsequent corroboration.

In conclusion, I would like to stress that all these techniques give relative frost-hardiness ratings only, because a twig or needle that is severed from the tree always appears to be hardier than one that is still attached to the tree.

VI. SUMMARY

Various aspects of frost hardiness, such as ice formation, cytoplasmic changes and accompanying biochemical changes have been discussed, with emphasis on the plasma membranes. The concepts that hardening and dehardening occur in stages and that a translocatable factor is involved were examined. This was followed by an examination of relationships between frost hardiness and dormancy, an aspect that is of fundamental and practical importance, but is poorly understood. The hardening period seems to coincide approximately with the period when most food reserves accumulate. Preliminary results of an experiment suggest that reserves play a more important role in the frost hardening—dehardening process than in subsequent wood formation; evidently relationships between reserves and subsequent wood formation need to be re-evaluated.

Thus, frost hardiness is an expression of several physiological processes. The mechanism proposed that can be applied most widely is one based on the augmentation of total cellular protoplasm, including augmentation of cell organelles and other membraneous components.

There are several aspects of frost hardiness, such as the time and rate of hardening and dehardening, and maximum winter hardiness, which tree breeders should consider. Some of the available methods for testing frost hardiness were mentioned. Considerable variation in growth and frost hardiness exists within species, suggesting that breeding for fast-growing, frost-hardy varieties can be successful.

REFERENCES

Alden, J. and Hermann, R. K. (1971). Aspects of the cold-hardiness mechanism in plants. *Bot. Rev.* 37, 37–142.

Bialobok, S., Pukacki, P., Wnuk, B. and Chylarecki, H. (1974). Variation of cold hardiness of woody plants. *Polish Acad. Sci., Inst. Dendrology and Kornik Arboretum* Final Report 1–96.

Dietrichson, J. (1961). Breeding for frost resistance. *Silvae Genet.* 10, 172–179.

Doorenbos, J. (1953). Review of the literature on dormancy in buds of woody plants. *Meded. LandbHoogesch. Wageningen* 53, 1–23.

Eiche, V. (1966). Cold damage and plant mortality in experimental provenance plantations with Scots pine in northern Sweden. *Studia Forestalia Suecica* 36, 1–216.

El Aouni, M. H. and Mousseau, M. (1974). Relation d'échange de CO_2 chez les aiguilles du Pin noir d'Autriche (*Pinus nigra* Arn.) avec l'âge, la teneur en chlorophylle et réassimilation. *Photosynthetica* 8, 78–86.

Glerum, C. (1973a). The relationship between frost hardiness and dormancy in trees. *IUFRO Symposium on Dormancy in Trees.* Kornik, Poland. Sept. 1973. pp. 9.

Glerum, C. (1973b). Annual trends in frost hardiness and electrical impedance for seven coniferous species. *Can. J. Plant Sci.* 53, 881–889.

Glerum, C. (1974). Carbon assimilation during the frost hardening period and its influence on reserves and wood formation in jack pine. *Proc. 3rd North Am. Forest Biol. Workshop* C.S.U. Fort Collins, U.S.A. p. 354, Abstr. 11.

Holzer, K. (1969). Cold resistance in spruce. *Proc. 2nd World Consult. Forest Tree Breeding.* FAO-IUFRO. Washington D.C. pp. 13.

Kramer, P. J. and Kozlowski, T. T. (1960). "Physiology of Trees". McGraw-Hill, New York.

Krasavtsev, O. A. (1968). Über die Gefriervorgänge bei pflanzlichen Geweben. *In* "Klimaresistenz, Photosynthese und Stoffproduktion". Tagungsber. d. D. A. L. 100, 23–34.

Levitt, J. (1956). "The Hardiness of Plants". Academic Press, New York and London.

Levitt, J. (1972). "Responses of Plants to Environmental Stresses". Academic Press, New York and London.

Little, C. H. A. and Bonga, J. M. (1974). Rest in the cambium of *Abies balsamea. Can. J. Bot.* 52, 1723–1730.

Mergen, F. (1963). Ecotypic variation in *Pinus strobus* L. *Ecology* 44, 716–727.

Nienstaedt, H. and King, J. P. (1969). Breeding for delayed bud break in *Picea glauca* (Moench) Voss. *Proc. 2nd World Consult. Forest Tree Breeding* FAO-IUFRO Washington, D.C. 1, 61–80.

Nienstaedt, H. and Teich, A. H. (1972). Genetics of white spruce. *U.S.D.A. For. Serv. Res. Paper* WO-15, pp. 24.
Parker, J. (1963). Cold resistance in woody plants. *Bot. Rev.* **29**, 123–201.
Perry, T. O. (1971). Dormancy of trees in winter. *Science* **171**, 29–36.
Perry, T. O. and Baldwin, G. W. (1966). Winter breakdown of the photosynthetic apparatus of evergreen species. *Forest Sci.* **12**, 298–300.
Pomeroy, M. K. and Siminovitch, D. (1971). Seasonal cytological changes in secondary phloem parenchyma cells in *Robinia pseudoacacia* in relation to cold hardiness. *Can. J. Bot.* **49**, 787–795.
Pomeroy, M. K., Siminovitch, D. and Wightman, F. (1970). Seasonal biochemical changes in the living bark and needles of red pine (*Pinus resinosa*) in relation to adaptation to freezing. *Can. J. Bot.* **48**, 953–967.
Romberger, J. A. (1963). Meristems, growth and development in woody plants. *U.S.D.A. For. Serv. Tech. Bull.* 1293.
Sakai, A. (1970). Freezing resistance in willows from different climates. *Ecology* **51**, 485–491.
Sakai, A. and Weiser, C. J. (1973). Freezing resistance of trees in North America with reference to tree regions. *Ecology* **54**, 118–126.
Scarth, G. W. (1944). Cell physiological studies of frost resistance: A review. *New Phytol.* **43**, 1–12.
Scheumann, W. (1965). Möglichkeiten und Ergebnisse der Frostresistenz-prüfung in der Douglasien- und Lärchen-züchtung. *Tagungsber. d. D.A.L.* **69**, 189–199.
Scheumann, W. (1968). Die Dynamik der Frostresistenz und ihre Bestimmung an Gehölzen im Massentest. *In* "Klimaresistenz, Photosynthese und Stoffproduktion". *Tagungsber. d. D.A.L.* **100**, 45–54.
Scheumann, W. and Schmiedel, H. (1972). Die Prüfung der Frostresistenz von Kreuzungsnachkommenschaften der Douglasie (*Pseudotsuga menziesii* [Mirb.] Franco). im Labortest und die Bestätigung der Ergebnisse im Anbauversuch. *Biol. Zbl.* **91**, 707–713.
Siminovitch, D. and Briggs, D. R. (1953). Studies on the chemistry of the living bark of the black locust tree in relation to frost hardiness. IV. Effects of ringing on translocation, protein synthesis and the development of hardiness. *Plant Physiol.* **28**, 177–200.
Siminovitch, D., Rheaume, B. and Sachar, R. (1967a). Seasonal increase in protoplasm and metabolic capacity in tree cells during adaptation to freezing. *In* "Molecular mechanisms of temperature adaptation" (C. L. Prosser, ed.), pp. 3–40. Pub. No. 84. Amer. Assoc. Adv. Sci. Washington, D.C.
Siminovitch, D., Gfeller, F. and Rheaume, B. (1967b). The multiple character of the biochemical mechanism of freezing resistance of plant cells. *In* "Cellular injury and resistance in freezing organisms" (E. Asahina, ed.), pp. 93–117. Inst. Low Temp. Sci. Hokkaido Univ., Sapporo, Japan.

Siminovitch, D., Rheaume, B., Pomeroy, K. and Lepage, M. (1968). Phospholipid, protein and nucleic acid increases in protoplasm and membrane structures associated with development of extreme freezing resistance in black locust tree cells. *Cryobiology* 5, 202–225.

Timmis, R. and Worrall, J. (1974). Translocation of dehardening and bud-break promoters in climatically "split" Douglas fir. *Can. J. Forest Res.* 4, 229–237.

Tumanov, I. I. and Krasavtsev, O. A. (1959). Hardening of northern woody plants in temperatures below zero. *Sov. Plant Physiol.* 6, 654–667.

Vasil'yev, I. M. (1956). "Wintering of Plants". Translated from Russian 1961. Amer. Inst. Biol. Sci. Washington, D.C.

Vegis, A. (1964). Dormancy in higher plants. *A. Rev. Plant Physiol.* 15, 185–224.

Weiser, C. J. (1970). Cold resistance and injury in woody plants. *Science* 169, 1269–1278.

Yeatman, C. W. and Venkatesh, C. S. (1974). Parent-progeny correlation of bud break in white spruce at Petawawa, Ontario. *Proc. 21st. Northeastern For. Tree Improvement Conf.*, U.N.B. Fredericton, N.B. Aug. 1973. pp. 58–65.

24

Methods of Screening Tree Seedlings for Frost Hardiness

ROGER TIMMIS
Forestry Research Center, Weyerhaeuser Company, Centralia, Washington, U.S.A.

I.	Introduction	421
II.	Selected Screening Methods	424
III.	Evaluation	427
IV.	Relative Sensitivity of Tissues	429
V	Discussion	430
VI	Summary	432
References		433

I. INTRODUCTION

Relative differences in the frost hardiness of trees have been satisfactorily predicted by testing the hardiness of seedlings, commonly by freezing them, and then determining whether or not the tissues have been killed (Bellman and Schönbach, 1964; Scheumann and Schmiedel, 1972). Numerous viability tests are available which determine whether (a) enzyme and metabolic functions have been impaired, or (b) cell membranes have been damaged or destroyed. Many of the potentially rapid viability tests available are summarized in Table I. Few have been evaluated as practical methods of screening seedlings on a commercial scale. This contribution describes a comparative study of five selected viability tests that showed promise as means of screening *Pseudotsuga menziesii* seedling progenies for frost hardiness within one day. Preliminary studies had confirmed that these selected methods were effective in discriminating live from dead tree seedlings, and were likely to present fewest technical problems when applied on a large scale (Table I).

TABLE I. Methods of testing whether plant tissues are alive or dead, which can give an answer within a day.

Name and class of method	Theory for injured tissue	How measured	Literature references
METHODS BASED ON INACTIVATION OF ENZYME AND METABOLIC FUNCTION			
1. *Morphological*			
Bud tissue browning	Phenol-amine group reactions and subsequent oxidations?	Visual scoring of macroscopic section	Alden, 1971
2. *Physiological*			
[a]Photosynthesis	Chloroplasts breakdown. Mesophyll diffusion resistance increases	Infrared gas analysis	Neilson et al., 1972; Weise and Polster, 1962
[a]Leaf segment flotation	Photosynthesis inhibition lowers O_2 in intercellular space	Rate of sinking of leaf segments	Truelove et al., 1974
3. *Chemical*			
Chlorophyll stability	Chlorophyll breaks down	Optical absorbance of acetone extract	Kaloyareas, 1958
Nitrite accumulation	Photosynthesis inhibition blocks normal reduction of nitrite in the light	Colour test for nitrite, and optical absorbance after brief incubation in light	Klepper, 1975
[a]Tri-phenyl tetrazolium chloride	Inactivated dehydrogenases cannot reduce these vacuum-infiltrated substances	Overnight incubation and absorbance of red alcohol extract	Stepomkus and Lanphear, 1967
[b]Nitro-blue tetrazolium		Absorbance of green, dimethyl-formamide product extracted	Altmann, 1969
Fluorescein di-acetate	As above, but also based on membrane permeability	Ultraviolet fluorescence of sectioned tissue	Heslop-Harrison, 1970
[b]Tissue fluorescence	Undetermined reactions occur in the cytoplasm	1) Intensity of yellowish green u.v. fluorescence at freezing temperature. 2) Intensity of flash of ultraweak fluorescence on thawing tissue in water.	Krasavtsev, 1962

METHODS BASED ON LOSS OF SEMI-PERMEABILITY OR DESTRUCTION OF CELL MEMBRANE

Method	Basis	Detection	Reference
Odour	Volatile substance released to intercellular space?	Presence of characteristic odour	—
Ninhydrin	Amino acids leak from cell	Optical absorbance of ninhydrin reagent in bathing solution	Siminovitch et al., 1964
4. *Electrical*			
Electrolytic method	Ions leak from cell	Conductivity of bathing solution	Dexter et al., 1932; Wilner, 1960
[a]Impedance	Ionic conductance of membrane increases	Inserted electrodes and impedance bridge circuit	Greenham and Daday, 1957; van den Driessche, 1973
[b]Corona discharge	}	Blue discharge of applied high voltage from plant surfaces photographed in dark	Kirlian and Kirlian, 1961
[b]Electrical refreeze pattern	Essential compartmentalization of water in tissue is irreversibly disrupted	Attached electrodes. Absence of double decline in conducted current during re-freeze	Olien, 1961; Timmis, 1973
5. *Physical*			
[b]Exotherm analysis	}	Thermocouples. Absence of multiple peaks in re-freeze exotherm	McLeester et al., 1969
[b]Plasmolysis		Number of cells that plasmolyse and deplasmolyse in thin sections in solutions	Siminovitch and Briggs, 1953
[a]Plant water stress	Water potential gradient across tissue is reduced	Pressure chamber technique	Bixby and Brown, 1974
[b]Drying rate		Weight loss after brief vacuum drying	Paton, 1972
[b]Infrared reflectance		Increase in foliar I.R. reflectance after drying	—

[a] Methods evaluated in detail and reported here.
[b] Methods shown to be ineffective on *Pseudotsuga menziesii*, or which presented technical problems when applied on a large scale.

Tree breeders are not concerned with the maximum frost hardiness possible, but with the speed and time at which the tissues become hardy in response to changing climatic conditions. Frost hardiness develops in 2–3 phases determined by photoperiod, chilling and non-injurious frost (Weiser, 1970; Timmis and Worrall, 1974, 1975; Glerum, 1973). Successive hardiness phases may be interdependent, or linked with shoot growth phenology, but are thought to involve different underlying mechanisms (Weiser, 1970) which can differ among organs and tissues (Olien, 1974; Alden, 1971). Various lines of evidence indicate that factors governing the onset, speed and extent of frost hardening and dehardening may be inherited independently (e.g. Stushnoff, 1972; Scheumann and Schmiedel, 1972; Olien, 1971). When screening for frost hardiness, it is necessary to know the critical stage at which damage is likely to occur. Thus, short-day induced hardiness may be a good way of predicting whether glasshouse-grown seedlings will survive when planted in autumn at high elevations (Tanaka, 1974), but will be of little value for sites where frost damage occurs in late winter owing to premature dehardening (Scheumann and Schmiedel, 1972).

II. SELECTED SCREENING METHODS

Because frost hardiness is such a complex trait, screening necessarily involves three procedures: (a) conditioning the seedlings to varying stages of hardening or dehardening, (b) freezing them at different temperatures, and (c) testing whether the seedlings are alive or dead and estimating the temperatures which kill 50% of them (lethal temperature LT_{50}) (see also Scheumann and Hoffman, 1967; Young and Hearn, 1972).

Tests for frost hardiness were done on 2-year-old potted seedlings of *Pseudotsuga menziesii* of a coastal provenance from 300 m altitude, near Aberdeen, Washington, U.S.A. They were placed in controlled environments and given successive changes in photoperiod, temperature and night frost during 28 weeks which induced a sequence of 2–3 hardiness stages and 2 stages of dehardening (Fig. 1). Seedlings were removed every 2–3 weeks, and groups of 15 were subjected to one of several freezing temperatures 3°C apart, estimated to cover the lethal range at the particular stage of hardiness

(Fig. 1). Freezing was done in a programmable chamber in which the plant pots and root systems were maintained at 2–10°C, while the shoots were cooled at the rate of 5°C h^{-1}, maintained at their minimum temperature for 2 h, and brought back to laboratory temperature at 20°C h^{-1}.

After treatment, the following five viability tests were applied to each group of 15 seedlings:

1. *Photosynthesis.* Direct measurements of photosynthesis were made using whole seedlings, preconditioned at 20°C, and placed in a closed chamber at 20°C linked to an infrared gas analyser.

2. *Leaf segment flotation.* Truelove *et al.* (1974) indicated that leaf discs remain afloat on phosphate buffer solutions for relatively long periods of time only if photosynthesis is unimpaired. They proposed that buoyancy is maintained by gaseous oxygen produced during photosynthesis and accumulated in the intercellular spaces. Flotation time may therefore be an indirect measure of frost damage to the photosynthetic apparatus. This method was tested using 3 mm lengths of needles, recording the time taken for 50% of the needle pieces to sink (after Truelove *et al.*, 1974).

3. *Enzyme activity.* Dehydrogenase activity can be detected by the appearance of a pink colour when living cells are incubated in the presence of triphenyl tetrazolium chloride (TTC) and a suitable oxidizable substrate. The TTC is reduced to insoluble red formazan which, after extraction by ethanol, can be measured at 530 nm in a colorimeter. Samples, weighing 100 mg, of both needle and stem segments, were vacuum-infiltrated with 0.6% aqueous TTC and tested following the procedures of Steponkus and Lanphear (1967).

4. *Electrical impedance.* There is an increase in ionic permeability of frost-injured cell membranes which can be measured by a decrease in electrical impedance (Blazich *et al.*, 1974; van den Driessche, 1973; Glerum, 1970; Pukacki, 1973). Impedance of an a.c. circuit

has both resistive and capacitive components, each of which may provide separate information about injury. Both were recorded in upper and basal parts of the seedling stems at 20 Hz, 1 kHz and 1 MHz frequencies, after 40 min at room temperature. This was done by pushing into the stem two stainless steel pins embedded 1.5 cm apart in a plastic block from which they protruded 3 mm. A small voltage from a tunable oscillator was applied across the pins and impedance was measured using a standard bridge circuit (Glerum and Krenciglowa, 1970). Several variables were derived from the impedance measurements. These included the 20 Hz : MHz and kHz : MHz resistive impedance ratios which eliminate some of the variation not due to injury (Greenham and Daday, 1957; Glerum and Krenciglowa, 1970).

5. *Plant water stress.* When tissues are frost-injured there can be a loss of cell membrane semi-permeability which results in a measurable increase in (i.e. less negative) xylem water potential. Water

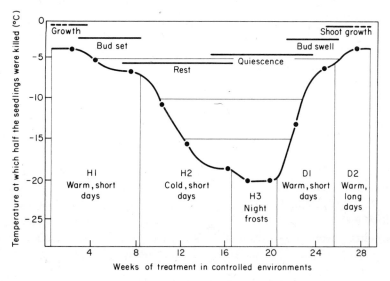

FIG.1 Phases of frost hardening (H1, H2, H3) and dehardening (D1, D2) induced in seedlings of *Pseudotsuga menziesii* by placing them in successive environmental regimes. The idealized curve is the temperature which killed half the seedlings, derived from data in Fig. 3. Points on this curve represent sampling dates. H1 = c. 15°C 8 h photoperiods; H2 = c. 3°C 8 h; H3 = c. 3°C with c. −4°C night frost for 2 h, 8 h photoperiods; D1 = 15°C, 8 h; D2 = 23°C, 12 h.

potentials were measured in detached branches after 2 h light-preconditioning, using a pressure chamber (Wareing and Cleary, 1967).

The results of the five viability tests were compared with the presence or absence of frost browning which became evident several weeks after harmful treatments. Injury was recorded as absent, partial or present in the bark, pith, needles and buds, taken from four age zones on each seedling.

Multiple discriminant function analyses were done to compare the ability of the different viability tests to distinguish healthy plants from those which were subsequently judged as injured or killed. Analyses were done for each frost-hardiness phase (Fig. 1), and discriminant functions were calculated, taking the differing tests singly and together in various combinations.

III. EVALUATION

All five viability tests reported here correctly classified at least 71% of the seedlings as alive or dead in one or more of the hardiness phases (Table II).

Measurements of electrical impedance were consistently reliable. Between 88% and 92% of the seedlings were correctly classified as alive or dead, and estimates of the temperature which gave a 50% kill were made within 1°C at any phase of hardening or dehardening. Resistive impedance was better than capacitance or phase angle, and measurements in the upper parts of the stems were more reliable than those taken in the lower parts. Discriminant functions for electrical impedance were improved by including the reciprocal of stem diameter as a second variable, but expressing the impedance as a fraction of the pre-freezing impedance gave no marked improvement. The best single impedance parameter was the kHz : MHz resistive impedance ratio which enabled 87% of the seedlings to be correctly classified. Discriminant functions derived using this variable could not be significantly improved by adding any other electrical variable. The critical value for this ratio varied significantly between hardiness phases, from 2.4 in phase D1 to 5.3 in phase H3.

All the non-electrical measurements of viability tended to be unsatisfactory when the seedlings were most frost hardy, in phases H2 and H3 (Fig. 1). Thus, measurements of dehydrogenase activity in the needles, which were the second most effective means of classifying seedlings as alive or dead, were ineffective when the plants were in phase H3 (Table II). The same tests carried out on stem tissues were unreliable measures of frost damage at all hardiness phases. Rates of photosynthesis, expressed as fractions of photosynthetic rates before seedlings were given the freezing treatments, were as effective as impedance ratios in predicting frost damage when the seedlings were least frost hardy (phase H1 and D1/D2), but again, were ineffective during phases H2 and H3. Needle flotation half-times and changes in water potential were the least reliable measures of frost damage (Table II).

Frost damage could be detected only slightly better by using two "unrelated" variables together; for instance, ones measuring enzyme and cell membrane damage or the viability of different plant organs. Thus, combining measurements of 20 Hz impedance with ones of photosynthesis, water stress or needle flotation increased the percentage of correctly classified seedlings by a maximum of 8%.

Taking all hardiness phases together, the best "survival function" was obtained by combining impedance ratios with the accumulated day-degrees below 12°C during the 30 days in the controlled

TABLE II. Percentage of seedlings of *Pseudotsuga menziesii* which were correctly classified as alive or dead using five methods, after the seedlings had received near-lethal freezing temperatures for 2 h. Seedlings were tested at five frost hardiness stages (see Fig. 1).

Stages of frost hardiness (see Fig. 1)	Needle dehydrogenase activity (TTC)	Impedance ratio	Photosynthesis	Plant water stress	Leaf segment flotation
H1 (short-day)	81	89	92	72	71
H2 (cold)	69	92	74	n.s.	61
H3 (frost)	n.s.	88	65	n.s.	62
D1/D2 (dehardening)	79	89	83	75	n.s.
Means	76	87	72	71	61

n.s. denotes no significant difference between live and dead groups as determined by multiple discriminant analysis.

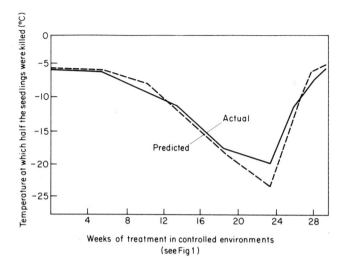

FIG. 2 Temperatures which killed half the seedlings of *Pseudotsuga menziesii* during the phases of hardening and dehardening shown in Fig. 1, as assessed by their abilities to regrow following freezing treatments. The predicted values are those estimated using a survival function based on electrical impedance ratios of the stems immediately after freezing, plus "accumulated day-degrees below $12°C$" in the 30 days before freezing.

environments prior to the freezing tests. The latter parameter corrected somewhat for an increase in impedance ratio with increasing hardness. When this complex function was used, it was possible to estimate the temperature which had killed half the seedlings within $±2°C$ (see Fig. 2).

IV. RELATIVE SENSITIVITY OF TISSUES

The temperatures that proved lethal during the differing stages of hardening varied for the stems, buds and needles (Fig. 3). Thus, during the "winter" rest and early quiescent periods the buds were the least frost-hardy organs, whereas, during the "autumn" period of bud-set and "spring" bud swell, the bark tissues were least hardy, particularly near the root collar. The needles were seldom the most frost-susceptible organs, although young needles suffered damage.

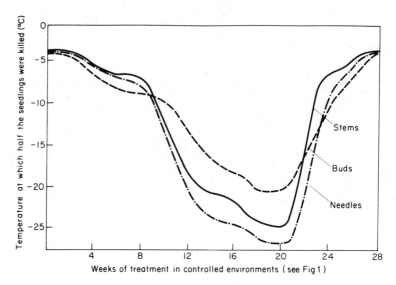

FIG. 3 Temperatures which killed the stem, bud and leaf tissues on half the seedlings of *Pseudotsuga menziesii* during the phases of frost hardening and dehardening shown in Fig. 1. The curves are based on 11 sample dates.

V. DISCUSSION

The main finding of this study was that "impedance ratio" was the most accurate and convenient measure of whether seedlings had or had not suffered frost damage, and was reliable at all phases of hardening and dehardening. This agrees with van den Driessche (1973) and Blazich *et al.* (1974). Frost damage could not be detected any better by using capacitance or phase angle because capacitance and resistive impedance were highly inversely correlated ($r = 0.80$), as found by Tattar *et al.* (1974) working with healthy and decaying wood. Measurements of dehydrogenase activity were relatively less reliable than previously reported, particularly using stem segments (Stergios and Howell, 1973; Pukacki, 1973). The periods of needle flotation were closely linked with the ability of the needles to photosynthesize, but even photosynthesis proved to be an unexpectedly poor measure of frost injury. Plant water stress gave poorer predictions of survival than indicated by Bixby and Brown (1974). Indeed, frost damage to short-day hardened seedlings, in phase H1, could be detected more reliably by their "mown-grass" odour than by using either of the last two methods.

It was generally more difficult to assess whether or not frosted seedlings had been damaged when they were in the "winter" periods of rest and quiescence (H2 and H3, Table II). During these stages, the critical values used to discriminate between living and dead plants were very variable among individuals, although the difference between the means of live and dead groups remained about the same. Also, in these "winter" phases plant survival was determined principally by the frost susceptibility of the buds (Fig. 3). An ability to photosynthesize rapidly was a poor discriminator because rates were low before and after the freezing treatment, since night frosts were given as part of the conditioning.

Even using several methods to determine frost damage, it is clearly not possible to predict the growth potential of 100% of the seedlings in one day (Edgren, 1970). Some plants can be severely damaged, yet grow again from few buds.

The lethal temperatures for 50% kill (LT_{50}) may be predicted directly from measurements of single rather than several freezing tests, if LT_{50} values can be shown to be linearly related with some measure of survival after frost. Such a relationship did occur in this study between (a) impedance ratio measured after a $-15°C$ frost, and (b) LT_{50} in the range -9 to $-20°C$ (Fig. 4). That is, the impedance ratio after a single $-15°C$ freezing treatment could be used to predict the likely LT_{50}. It would seem profitable to seek further such relationships using several variables.

Although this account has been confined to measurements taken on whole seedlings placed in freezing environments, techniques have also been tested in which excised buds, needles and branch segments were frozen to different degrees in vials placed along a bar with a linear temperature gradient (Hodges et al., 1970). This apparatus has obvious advantages of size and throughput. However, excised tissues which were frosted in this way, and then tested as alive or dead, did not allow such accurate predictions to be made about the survival of the whole plants. About 17% more seedlings were incorrectly classified, compared with the freezing chamber tests, which could mean an error of $1.5-4°C$ in the estimated LT_{50}, depending on the hardiness phase. Nevertheless, it is expected that this freezing method, combined with the methods described here for detecting

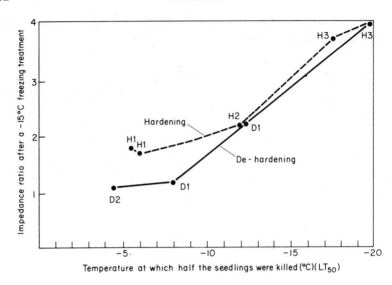

FIG. 4 Relationship between the "impedance ratio" of stems of *Pseudotsuga menziesii* following a single freezing treatment at −15°C, and the freezing temperature that was found to have killed half the seedlings. H1, H2, H3, D1, D2, refer to the different stages of hardiness shown in Fig. 1.

frost damage, will eventually provide economical and reliable "same-day" frost hardiness estimates which can be applied on a meaningful scale to progeny.

VI. SUMMARY

Methods of testing for frost hardiness rapidly and on a commercial scale were evaluated by: (a) pre-conditioning seedlings to different "phases" of hardening and dehardening, corresponding to natural changes during autumn, winter and spring; (b) freezing whole plants to different degrees during the different phases; and (c) then evaluating different methods of testing whether the seedlings had or had not survived the freezing tests. Differences in kHz : MHz electrical impedance ratio in the upper stem predicted frost survival with an 87% accuracy and enabled lethal temperatures for 50% kill to be estimated within 2°C at all phases of hardening or dehardening. Measures of enzyme activity, photosynthesis and change in water potential were progressively less effective, especially during "winter" phases of hardiness. Only slight improvement resulted from combining variables by multiple discriminant analysis. Having evaluated methods for testing viability, the hardiness screening procedure may be further speeded by testing the frost hardiness of excised organs rather than whole seedlings, using a temperature gradient bar.

REFERENCES

Alden, J. N. (1971). "Freezing Resistance of Tissues in the Twig of Douglas-fir". Ph.D. Thesis, Oregon State University, Corvallis, Oregon.
Altmann, F. P. (1969). The quantitative elution of nitro-blue formazan from tissue sections. *Histochemie* 17, 319–326.
Bellman, E. and Schönbach, H. (1964). Erfolgsaussichten der Auslesezüchtung auf Frostresistenz bei der grünen Douglasie (*Pseudotsuga menziesii* (Mirb.) Franco). *Arch. Forstw.* 13, 307–331.
Bixby, J. A. and Brown, G. N. (1974). Rapid determination of cold hardiness in black locust seedlings using a pressure chamber. North American Forest Biology Workshop, Boulder, Colorado, Abstr. No. 12.
Blazich, F. A., Evert, D. R., and Bee, D. E. (1974). Comparison of three methods of measuring winter hardiness of internodal stem sections of *Forsythia intermedia* "Lynwood". *J. Am. Soc. hort. Sci.* 99, 211–214.
Dexter, S. T., Tottingham, W. E. and Graber, L. F. (1932). Investigations on the hardiness of plants by measurement of electrical conductivity. *Plant Physiol.* 7, 63–78.
Driessche, R. van den (1973). Prediction of frost hardiness in Douglas-fir seedlings by measuring electrical impedance in stems at different frequencies. *Can. J. Forest Res.* 3, 256–264.
Edgren, J. W. (1970). Growth of frost-damaged Douglas-fir seedlings. *U.S. Dept. Agric. For. Serv. Res. Note* PNW-121.
Glerum, C. (1970). Vitality determinations of tree tissue with kilocycle and megacycle electrical impedance. *For. Chron.* 46, 63–64.
Glerum, C. (1973). Annual trends in frost hardiness and electrical impedance for seven coniferous species. *Can. J. Plant Sci.* 53, 881–890.
Glerum, C. and Krenciglowa, E. M. (1970). The dependence of electrical impedance of woody stems on various frequencies and tissues. *Can. J. Bot.* 48, 2187–2192.
Greenham, C. G. and Daday, H. (1957). Electrical determination of cold hardiness in *Trifolium repens* L. and *Medicago sativa* L. *Nature, Lond.* 180, 541–543.

Heslop-Harrison, J. and Heslop-Harrison, Y. (1970). Evaluation of pollen viability by enzymatically induced fluorescence: intracellular hydrolysis of fluorescein diacetate. *Stain Technol.* **45**, 115–120.

Hodges, H. F., Svec, L. V., and Barta, A. L. (1970). A technique to determine cold hardiness in plants. *Crop Sci.* **10**, 318–319.

Kaloyereas, S. A. (1958). A new method of determining drought resistance. *Plant Physiol.* **33**, 232–233.

Kirlian, S. D. and Kirlian, V. K. (1961). Photography and visual observation by means of high-frequency currents. *Zhurnal Nauchnoy i Prikladnoy Fotografii i Kinematografii.* **6**, 397–403.

Klepper, L. A. (1975). Inhibition of nitrite reduction by photosynthetic inhibitors. *Weed Sci.* **23**, 188–190.

Krasavtsev, O. A. (1962). Fluorescence of cells of woody plants in the frozen state. *Fiziologiya Rast.* **9**, 359–367.

McLeester, R. C., Weiser, C. J. and Hall, T. C. (1969). Multiple freezing points as a test for viability of plant stems in the determination of frost hardiness. *Plant Physiol.* **44**, 37–44.

Neilson, R. E., Ludlow, M. M. and Jarvis, P. G. (1972). Photosynthesis in Sitka spruce (*Picea sitchensis* (Bong.) Carr.). II. Response to temperature. *J. appl. Ecol.* **9**, 721–745.

Olien, C. R. (1961). A method of studying stresses occurring in plant tissue during freezing. *Crop Sci.* **1**, 26–34.

Olien, C. R. (1971). Freezing stresses in barley. *In* "International Barley Genetics Symp., 1969 Proc". 2D, pp. 356–363.

Olien, C. R. (1974). Winter hardiness in barley. Stress analysis. *Michigan State Univ. Agric. Expt. Sta., East Lansing, Res. Rep.* **247**, 1–10.

Paton, D. M. (1972). Frost resistance in Eucalyptus: A new method for assessment of frost injury in altitudinal provenances of *E. viminalis. Aust. J. Bot.* **20**, 127–139.

Pukacki, P. (1973). Laboratory methods of evaluating the resistance of woody plants to low temperatures. *Arboretum Kornickie* **18**, 188–198.

Scheumann, W. and Hoffman, K. (1967). Die serienmässige Prüfung der Frostresistenz einjähriger Fichtensämlinge. *Arch. Forstw.* **16**, 701–705.

Scheumann, W. and Schmiedel, H. (1972). Die Prüfung der Frostresistenz von Kreuzungsnachkommenschaften der Douglasie (*Pseudotsuga menziesii* (Mirb.) Franco) im Labortest und die Bestätigung der Ergebnisse im Anbauversuch. *Biol. Zbl.* **91**, 707–713.

Siminovitch, D. and Briggs, D. R. (1953). Studies on the chemistry of the living bark of black locust and its relation to frost hardiness. III. The validity of plasmolysis and desiccation tests for determining the frost hardiness of bark tissue. *Plant Physiol.* **28**, 15–34.

Siminovitch, D., Therrien, H., Gfeller, F. and Rheaume, B. (1964). The quantitative estimation of frost injury and resistance in black locust, alfalfa, and wheat tissues by determination of amino acids and other ninhydrin-reacting substances released after thawing. *Can. J. Bot.* 42, 637–649.

Steponkus, P. L. and Lanphear, F. O. (1967). Refinement of the triphenyl tetrazolium choloride method of determining cold injury. *Plant Physiol.* 42, 1423–1427.

Stergios, B. G. and Howell, G. S., Jr. (1973). Evaluation of viability tests for cold stressed plants. *J. Am. Soc. hort. Sci.* 98, 325–330.

Stushnoff, C. (1972). Breeding and selection methods for cold hardiness in deciduous fruit crops. *Hort. Sci.* 7, 10–13.

Tanaka, Y. (1974). Increasing cold hardiness of container grown Douglas-fir seedlings. *J. For.* 72, 349–352.

Tattar, T. A., Blanchard, R. O. and Saufley, G. C. (1974). Relationship between electrical resistance and capacitance of wood in progressive stages of discoloration and decay. *J. exp. Bot.* 25, 658–662.

Timmis, R. (1973). "Cold Acclimation and Freezing in Douglas-fir Seedlings". Ph. D. Thesis, University of British Columbia, Canada.

Timmis, R. and Worrall, J. G. (1974). Translocation of dehardening and bud-break promoters in climatically split Douglas-fir. *Can. J. For. Res.* 4, 229–237.

Timmis, R. and Worrall, J. G. (1975). Environmental control of cold acclimation in Douglas-fir during germination, active growth and rest. *Can. J. For. Res.* 5, 464–477.

Truelove, B., Davis, D. E. and Jones, L. R. (1974). A new method for detecting photosynthesis inhibitors. *Weed Sci.* 22, 15–17.

Wareing, R. H. and Cleary, B. D. (1967). Plant moisture stress: evaluation by pressure bomb. *Science N. Y.* 155, 1248–1254.

Weise, G. and Polster, H. (1962). Untersuchungen über den Einfluss von Kältebelastungen auf die physiologische Aktivität von Forstgewächsen. II. Stoffwechselphysiologische Untersuchungen zur Frage der Frostresistenz von Fichten und Douglasienherkünften (*Picea abies* (L.) Karst. und *Pseudotsuga taxifolia* (Poir.) Britton). *Biol. Zbl.* 81, 129–143.

Weiser, C. J. (1970). Cold resistance and injury in woody plants. *Science N.Y.* 169, 1269–1278.

Wilner, J. (1960). Relative and absolute electrolytic conductance tests for frost hardiness of apple varieties. *Can. J. Plant Science* 40, 630–637.

Young, R. and Hearn, C. J. (1972). Screening citrus hybrids for cold hardiness. *Hort. Sci.* 7, 14–18.

MINERAL NUTRITION

25
Genetic Factors Affecting the Response of Trees to Mineral Nutrients

P. A. MASON and J. PELHAM
Institute of Terrestrial Ecology, Bush Estate, Penicuik, Scotland.

I.	Introduction and Review	437
II.	Nutritional Studies in the Absence of Mycorrhizas	439
III.	Studies on Host—Fungus Interactions	441
IV.	Discussion	445
V.	Conclusions	446
References		447

I. INTRODUCTION AND REVIEW

The growth of natural populations of many plants differs on contrasting soils, suggesting that they vary inherently in their abilities to absorb and/or utilize mineral nutrients. Numerous studies have demonstrated inherent differences in nutrient response between species and among cultivars of field crops.

Bradshaw *et al.* (1960) using sand culture, studied the reaction of eight grass species to different phosphate concentrations: *Lolium perenne, Agrostis stolonifera* and *A. tenuis* grew appreciably better with increasing amounts of phosphate, whereas *Cynosurus cristatus, Festuca ovina* and *Nardus stricta* showed little or no response. Field observations on the growth of these species led them to conclude that soil phosphate level is an important factor in determining their natural distributions.

Subsequent work demonstrated variation in nutrient response within species. Thus, natural populations of *Trifolium repens* differed markedly in their responses to a wide range of phosphate levels, including concentrations within the range tolerated in their native habitat (Snaydon and Bradshaw, 1962). The same authors

(1961) showed that populations of *Festuca ovina* from acid, low-calcium soils did not respond to added calcium, whereas the growth of those from calcareous soils was increased. Although experimental evidence is scanty, it would seem that such edaphic races also exist within forest tree species. Stern and Roche (1974) found that the concentration of manganese within individuals of *Pinus sylvestris* taken from the same stand varied greatly, apparently because of unequal uptake of this element by the different genotypes within the population. Also, Steinbeck (1971) showed that four clones of *Platanus occidentalis* responded differently to increasing concentrations of nutrients in sand culture.

Relatively few foresters or tree breeders consider the presence of edaphic ecotypes, much less genetical factors that may regulate differences in nutrient responses. This is surprising, bearing in mind that one tree species is often expected to grow on a diverse array of sites. But tree nutrition should not be considered without recognizing the involvement of fungi in potentially beneficial mycorrhizal associations.

Some conifers, such as *Pinus taeda*, must develop sheathing mycorrhizas (ectomycorrhizas) in order to develop normally after their year of growth. When introducing *Pinus occidentalis, P. attenuata, P. caribaea* and *P. radiata* to Puerto Rico, Hacskaylo (1962) found that it was essential to introduce simultaneously soil contaminated with the appropriate compatible mycorrhizal fungi. With these fungi, mycorrhizas developed and the pines grew well; without them mycorrhizas failed to develop and the saplings died. In a further study, seedlings of *Pinus caribaea* were inoculated in a nursery with pure cultures of four mycorrhizal fungi (Hacskaylo and Vozzo, 1967). In all cases, seedlings with mycorrhizas were more vigorous than non-mycorrhizal seedlings, irrespective of whether fertilizers were added. In contrast, McComb and Griffith (1946), working with *Pinus strobus* found that mycorrhiza formation was dependent upon a threshold concentration of available phosphate, which apparently activated the mycorrhiza-forming fungi. However, the best mycorrhizal development and seedling growth was obtained by introducing new fungal species which established mycorrhizas without additional phosphate. Foresters now recognize the need to ensure that appropriate fungal species are present in soils (Bowen, 1965). Trappe and Strand (1969) found that severely stunted, phosphorus-deficient seedlings of *Pseudotsuga menziesii* from a newly developed nursery responded to added phosphate only after the site had been inoculated with soil containing suitable mycorrhizal fungi.

However, responses to inoculation by appropriate fungal species are not assured, or consistent. Thus, Shemakhanova (1967) found that the fungus *Cenococcum graniforme* promoted the growth of tree seedlings more effectively than other fungi, whereas Lundeberg (1970) found it was positively harmful. These conflicting observations can be reconciled if one recognizes that there might be differing strains of a single symbiotic fungus (Levisohn, 1961) and inherent differences between and within host species which might influence the development of the mycorrhizas (Marx and Bryan, 1971). Great differences have already been noticed among strains of *Boletus scaber* (Levisohn, 1959) and *Rhizopogon luteolus* (Theodoru and Bowen, 1970). In fact, some species appear to occur in nature as both "mycorrhizal" and "non-mycorrhizal" races, for instance *Xerocomus subtomentosus* (Lundeberg, 1970) and *Paxillus involutus* (Laiho, 1970).

Studies on tree nutrition *per se* have rarely been integrated with those concerned with the inherent variation both in mycorrhizal fungi and tree hosts. This contribution describes ongoing studies with this approach, to identify and characterize the ability of different populations of given "host" tree species to use nutrients both in the absence and presence of mycorrhizas, working in laboratory-controlled, aseptic conditions (see also Mason, 1975). The host species used are *Betula verrucosa* and *B. pubescens*, and the mycorrhizal fungus *Amanita muscaria*.

II. NUTRITIONAL STUDIES IN THE ABSENCE OF MYCORRHIZAS

Species of *Betula* were chosen for studies on the genetics of nutrition because: (a) they have not been artificially selected to any great extent in Britain; (b) they can be induced to flower within 18 months of sowing; and (c) they usually produce abundant seed.

Seeds were collected from *Betula* growing on a wide range of sites, and selected seedlings which responded very differently to added phosphate in preliminary experiments were subsequently cloned. Single-node cuttings were first cultured in a general rooting medium, and then transplanted into test media contained in sterile plastic tubes. The tubes had translucent tops which were pierced and plugged with cotton wool to allow gaseous exchange. The tubes were small enough to enable up to 1000 to be used per m^2 of controlled environment, but large enough to allow the cuttings to grow unrestricted to about 5 cm height during the 7 weeks of each experiment. The test media were nutrient solutions gelled with agar. They were allowed to solidify as slopes so that the maximum surface was obtained over which roots could spread. "High" and "low" levels of P were chosen to be 26 ppm and 0.4 ppm, being those which normally gave satisfactory and reduced growth, respectively. Trojan-square experimental designs were usually used, and there were normally eight replicates of 16 treatments in each experiment. Observations were made through the sides of the tubes during the course of each experiment, and weights and leaf area measurements were taken after 7 weeks.

A series of experiments confirmed that clones of both *Betula verrucosa* and *B. pubescens* responded very differently when amounts of P were increased from 0.4 ppm to 26 ppm. Thus, in one experiment with *B. verrucosa* the leaves of clones A and H were increased in length by 116% and 318% respectively, whereas the leaves of clone C were increased in length by only 7%. That is, there were genotype—P interactions. Similar "high" and "low" responders were found among clones of *B. pubescens* (see Table I).

In a further experiment, other facets of growth were recorded (Table II). The "high" and "low" responders behaved as before, but it was apparent that final differences in total dry weights, leaf areas and stem lengths, attributable to increased P, were usually greater among the "high" than "low" responders. Also, it is apparent from Table II that the term "responders" is somewhat misleading, as the "low" responders were clones that used small amounts of P efficiently, whereas "high" responders used P inefficiently — both groups grew equally well when supplied with generous amounts of P, irrespective of species.

Interestingly, increasing the supply of P from 0.4 ppm to 26 ppm decreased the numbers of roots that the cuttings produced. This effect was particularly noticeable on clones that used small amounts

TABLE I. Percentage increase in mean leaf lengths of selected clones of *Betula* spp. in response to two levels of phosphorus, in the absence of mycorrhizas. Single node cuttings were grown in translucent tubes in sterilized media under controlled laboratory conditions.

Species	Designation of clone	Percentage increase of high P (26 ppm) over low P (0.4 ppm)	Assessment of response
B. verrucosa	A	116	high
	H	318	high
	C	7	low
	D	52	low
B. pubescens	D	123	high
	E	123	high
	A	17	low
	H	61	low

of P efficiently (the "low" responders). Thus, adding P to these clones decreased their mean numbers of roots 3.2-fold, from 54 to 15, whereas root numbers on the less efficient "high" responders were decreased from 29 to 22 (Table II).

In summary, these experiments highlighted the varying abilities of *Betula* genotypes to exploit available phosphorus when it is in short supply, in particular by producing different numbers of roots in deficient conditions.

III. STUDIES ON HOST–FUNGUS INTERACTIONS

The fly agaric, *Amanita muscaria*, forms mycorrhizas with a wide range of hardwood and coniferous trees (Trappe, 1962). Toadstools were collected from a range of pure stands of *Betula* and *Pinus* both in Britain and the U.S.A., and 80 cultures obtained by growing mycelium, derived from small pieces of cap tissue, on agar media.

Young microbe-free *Betula* seedlings, growing on agar slants in clear plastic tubes as described above, were inoculated with blocks of agar culture containing selected isolates of *A. muscaria*, and then allowed to incubate in the laboratory for 6–8 weeks. Mycorrhizas

TABLE II. Effects of increasing the supply of phosphorus on the growth of "low" and "high" responding clones of *Betula* spp., in the absence of mycorrhizas. Single node cuttings were grown in translucent tubes in sterile media under controlled laboratory conditions.

	Species	"High" responders		"Low" responders	
		0.4 ppm P	26.0 ppm P	0.4 ppm P	26.0 ppm P
Top dry weight (mg)	B. verrucosa	12	21	15	21
	B. pubescens	14	22	24	15
Leaf area (cm²)	B. verrucosa	4	8[a]	8	9
	B. pubescens	5	15[a]	6	10
Stem length (mm)	B. verrucosa	14	39[a]	34	47
	B. pubescens	14	41[a]	30	41
Root number	B. verrucosa	19	14	56	20[a]
	B. pubescens	39	30	52	9[a]

[a] Significant difference between 0.4 and 26 ppm phosphorus at $P < 0.001$.

were observed through the sides of the tubes 3—4 weeks after inoculation — a shorter time than hitherto recorded. Furthermore, cytological examinations confirmed that the mycorrhizas were like those formed in nature.

Experiments, primarily aimed at developing media for screening numerous isolates of *A. muscaria* against selected seed lots of *Betula* gave the following results.

(a) Some strains of *A. muscaria* formed mycorrhizas on the test seed lot of *Betula verrucosa*, whereas others, like isolate 4, did not (see Table III).

(b) Of the strains which formed mycorrhizas, isolate 2 formed them at 3.25, 6.5 and 26 ppm P, whereas isolates 1, 3 and 5 did so only at the two higher concentrations. By contrast, isolate 6 formed mycorrhizas at the intermediate concentration but failed to do so at the highest concentration. Clearly, media with any one P concentration would select particular isolates as "effective".

(c) Isolates of *A. muscaria* from *Betula* stands did not behave differently from those from *Pinus* forests.

Subsequent experiments showed that more roots became mycorrhizal per seedling when grown in media with 6.5 ppm phosphorus than in 26 ppm (3.7 cf. 1.4 per seedling, Table IV). The former concentration was therefore adopted for screening isolates of the fungus.

TABLE III. Effects of three phosphorus concentrations on the development of mycorrhizas by six isolates of *Amanita muscaria* with seedlings of *Betula verrucosa* (ex Scotland). Seedlings grown in translucent tubes in controlled laboratory conditions.

Cultures of *A. muscaria*		Presence (+) or absence (−) of mycorrhizas with *B. verrucosa* growing in media with different P concentrations		
Isolate	Origin	3.25 ppm	6.5 ppm	26 ppm
1	Britain (*Betula*)	−	+	+
2	Britain (*Betula*)	+	+	+
3	Britain (*Pinus*)	−	+	+
4	U.S.A. (*Pinus*)	−	−	−
5	U.S.A. (*Pinus*)	−	+	+
6	U.S.A. (*Pinus*)	Not tested	+	−

TABLE IV. Effects of two phosphorus concentrations on the numbers of mycorrhizal roots formed per seedling between *Betula verrucosa* (ex Scotland) and different isolates of *Amanita muscaria*.

Cultures of *A. muscaria* (see Table III)	Numbers of mycorrhizas per seedling in media with different P concentrations	
	6.5 ppm	26 ppm
2	5.8	2.6
4	3.2	1.6
6	2.2	0.0
Means	3.7	1.4

These effects of P concentration on mycorrhizal formation were not readily explained, because they were not correlated with differences in fungal growth. The fungus grew equally well over the roots of *Betula* seedlings at 6.5 and 26 ppm P. It may be that mycorrhizal formation depends upon such factors as the concentration of P within the plant roots, or the production and balance of growth regulators produced by both the plant and fungus.

Mycorrhizal formation can depend not only on the isolate and P concentration, but also on the host genotype. Using the same test procedures, it was shown that, at 6.5 ppm P, isolate 1 of *A. muscaria* formed a mean of 28 mycorrhizas per seedling on a *B. verrucosa* provenance from Latvia, but only 13 mycorrhizas per seedling on a provenance from Scotland (significant $P = 0.05$). However, when grown at 26 ppm P there was no difference (13–14 mycorrhizas per seedling; Mason, 1975). Could the Latvian seed lot be more suited to sites poor in available phosphorus?

Some isolates of *A. muscaria* produced small numbers of highly branched mycorrhizas, whereas others stimulated large numbers of sparsely branched ones. Thus, isolate 2 produced only 4.5 mycorrhizas per seedling on *Betula verrucosa* of one Scottish provenance, but 44.4% of them were branched, some with as many as 10–20 branches. Isolate 1, on the other hand, produced 13.5 mycorrhizas per seedling, but only 18.5% of them were branched, and none had more than 5 branches. Selection from a cross between

25. Genetic Factors and Response to Mineral Nutrients 445

these two types of isolates might result in a strain which could produce large numbers of repeatedly branched mycorrhizas.

The fungus not only controlled mycorrhizal development, but also influenced the growth of all types of roots. Both root numbers and lengths were significantly stimulated, but to different extents, depending on which isolate of *A. muscaria* was placed in association with seedlings of *Betula verrucosa*. Thus, isolates 2 and 3 induced the seedlings to produce longer individual roots than the uninoculated controls, and 1.4–2.0 times as many roots, whereas isolate 1 did not increase the length of individual root very much, but stimulated over 2.4 times as many to be produced (Table V). Thus, the fungal isolate appeared able to influence both the development and form of its host's root system.

IV. DISCUSSION

Evidently, seedlings from natural populations of *Betula verrucosa* and *B. pubescens* have inherently different abilities to respond to P, and in particular to grow on P-deficient media. These observations

TABLE V. Effects of three isolates of *Amanita muscaria* on the development of roots of seedlings of *Betula verrucosa* (ex Scotland). Data are presented as log $(n + 1)$ of original data with detransformed values in parenthesis.

		Growth of roots of *B. verrucosa*		
	Inoculation treatments	Number of roots per seedling	Total root length per seedling (mm)	Mean length per root (mm)
Cultures of *A. muscaria* (see Table III)	1	4.70 (111)	5.48 (254)	1.11 (2.23)
	2	4.48 (91)	5.43 (241)	1.28 (2.61)
	3	3.89 (65)	4.79 (178)	1.27 (2.57)
	Uninoculated control	3.60 (46)	4.29 (105)	1.13 (2.11)
Least significant difference ($P = 0.05$)		0.145	0.122	0.061

parallel those made by Bradshaw and co-workers on grasses. The laboratory technique developed enables large numbers of plants to be screened for these attributes within 7 weeks. Hence, clones which show extreme or desirable responses to mineral nutrients can now be selected within a much shorter time than traditional methods allow, although they will, clearly, have then to be further studied in more natural conditions.

The studies with combinations of *B. verrucosa* and *Amanita muscaria* suggest that the genetical control of mycorrhiza formation has features in common with the well studied symbiotic associations between legumes and *Rhizobium* bacteria. In the latter, it is known that genetic factors in both the host and the bacterium control the formation of root nodules; we found that host and fungal genotypes affected the formation of sheathing mycorrhizas, as originally suggested by Marx and Bryan (1971). After formation, the ability of nodules to fix nitrogen depends on a second set of factors in the bacterial *Rhizobium*; we found that different isolates of *A. muscaria* that successfully formed mycorrhizas had different effects on the growth of *Betula* seedlings, especially the growth of roots.

At present, studies on nutrient responses in the absence of fungi, and on the development of mycorrhizas, need to be brought together. Thus, can *A. muscaria* benefit the "low" responding *Betula* clones that grew relatively well on poor media? If so, how big is the response compared with "high" responders both in relative and absolute terms? Such information, when tested in the field, might markedly affect the detail of future programmes of tree improvement.

V. CONCLUSIONS

Our experiments confirm that different populations of *Betula verrucosa* and *B. pubescens* respond differently to added mineral nutrients, with some being noticeably tolerant of small amounts. Studies on the plant—symbiont system have indicated at least four factors that can be genetically controlled by either the host or the fungus: (a) mycorrhiza formation; (b) the extent of mycorrhizal development; (c) the pattern of mycorrhizal branching; and (d) the shape and size of the host root system. Evidently, the performance of tree genotypes on given sites might be greatly increased by exploiting *all* the genetical factors controlling tree nutrition, in particular those that affect the formation of "effective" mycorrhizas and the "efficient" use of nutrients in poor supply.

REFERENCES

Bowen, G. D. (1965). Mycorrhiza inoculation in forestry practice. *Aust. Forest.* 29, 231–237.

Bradshaw, A. D., Chadwick, M. J., Jowett, D., Lodge, R. W. and Snaydon, R. W. (1960). Experimental investigations into the mineral nutrition of several grass species. III. Phosphate level. *J. Ecol.* 48, 631–637.

Hacskaylo, E. (1962). Research on mycorrhizae in the United States. *Proc. 13th I.U.F.R.O. Congress, Vienna 1961.* 2 : 1; Sections 24–26. 7 pp.

Hacskaylo, E. and Vozzo, J. A. (1967). Inoculation of *Pinus caribaea* with pure cultures of mycorrhizal fungi in Puerto Rico. *Proc. 14th I.U.F.R.O. Congress* 5, 139–148.

Laiho, O. (1970). *Paxillus involutus* as a mycorrhizal symbiont of forest trees. *Acta forest. fenn.* 106, 1–72.

Levisohn, I. (1959). Strain differentiation in a root-infecting fungus. *Nature, Lond.* 183, 1065–1066.

Levisohn, I. (1961). "Researches in Mycorrhiza". Forestry Commission Report of Forest Research, March 1960. H.M.S.O., London.

Lundeberg, G. (1970). Utilisation of various nitrogen sources, in particular bound soil nitrogen, by mycorrhizal fungi. *Stud. for. suec.* 79, 1–95.

Marx, D. H. and Bryan, W. C. (1971). Formation of ectomycorrhizae on half-sib progenies of slash pine in aseptic culture. *Forest Sci.* 17, 488–492.

Mason, P. A. (1975). The genetics of mycorrhizal associations between *Amanita muscaria* and *Betula verrucosa*. *In* "The Development and Functions of Roots". (J. G. Torrey and D. T. Clarkson, eds) Academic Press, New York and London.

McComb, A. L. and Griffith, J. E. (1946). Growth stimulation and phosphorus absorption of mycorrhizal and non-mycorrhizal Northern white pine and Douglas fir seedlings in relation to fertilizer treatment. *Plant Physiol.* 21, 11–17.

Shemakhanova, N. M. (1967). "Mycotrophy of Woody Plants". Israel Program for Scientific Translations, Jerusalem.

Snaydon, R. W. and Bradshaw, A. D. (1961). Differential response to calcium within the species *Festuca ovina* L. *New Phytol.* 60, 219–234.

Snaydon, R. W. and Bradshaw, A. D. (1962). Differences between natural populations of *Trifolium repens* L. in response to mineral nutrients. 1. Phosphate. *J. exp. Bot.* 13; 422–434.

Steinbeck, K. (1971). Growth responses of clonal lines of American sycamore grown under different intensities of nutrition. *Can. J. Bot.* 49, 353–358.

Stern, K. and Roche, L. (1974). "Genetics of Forest Ecosystems", Springer-Verlag, Berlin, Heidelberg and New York.

Theodoru, C. and Bowen, G. D. (1970). Mycorrhizal responses of radiata pine in experiments with different fungi. *Aust. Forest.* 34, 183–191.

Trappe, J. M. (1962). Fungus associates of ectotrophic mycorrhizae. *Bot. Rev.* 28, 538–606.

Trappe, J. M. and Strand, R. F. (1969). Mycorrhizal deficiency in a Douglas fir region nursery. *Forest Sci.* 15, 381–389.

26
Responses of *Pinus taeda* and *Pinus elliottii* to Varied Nutrition

R. E. GODDARD[1], B. J. ZOBEL[2] and C. A. HOLLIS[1]

1. *School of Forest Resources and Conservation, University of Florida, Gainsville, U.S.A.*
2. *School of Forest Resources, North Carolina State University, Raleigh, U.S.A.*

I.	Introduction	449
II.	Loblolly Pine Field Studies (*Pinus taeda*)	450
III.	Slash Pine Field Studies (*Pinus elliottii*)	453
IV.	Glasshouse Experiments (*Pinus elliottii*)	455
V.	Discussion	460
VI.	Conclusions	461
References		462

I. INTRODUCTION

The soils of the lower coastal plains of the south-eastern United States are generally sandy and inherently infertile. Their use is primarily restricted to forest production, principally southern pines (especially *Pinus taeda* and *Pinus elliottii*). Since 1950 there has been a dramatic increase in the intensity of forest management, including site preparation and the establishment of even-aged, planted stands. A more recent, but increasing, trend is the application of fertilizers to these forests. Over the past 5 years, approximately 100 000 ha of forest land in the lower coastal plain have received routine fertilizer applications (Pritchett and Smith, 1975). Tree improvement activities have also been intensive in the region, and most seedlings currently being planted originate from seed orchards.

With wide-scale planting of improved trees on relatively infertile soil, needing substantial fertilizer inputs, an obvious development has been to seek genetic lines which respond well to added nutrients. Long-standing university—industry tree improvement cooperatives involving North Carolina State University and the University of Florida have, as their objectives, genetic improvement of southern tree species, primarily *P. taeda* and *P. elliottii*. More recently, comparable cooperative programmes were organized on the use of fertilizers, and tests of select tree progenies were established, both to determine their general response to added nutrients and to observe inherent variation in response.

II. LOBLOLLY PINE FIELD STUDIES (*Pinus taeda*)

The North Carolina Tree Improvement Cooperative has over 800 ha of control-pollinated progeny tests established, of which over 30% have been included in fertilizer tests. The trials were not designed to test different fertilizers, but to assess genetic reaction to the "best" fertilizer for each site, as determined by soil analyses carried out at North Carolina State University. The type of fertilizer prescribed ranged from phosphates alone, to balanced NPK fertilizers which are used more generally in the *P. taeda* range. The objectives of the fertilizer phase of the progeny tests were to determine (a) whether there was an overall fertilizer response, and (b) whether families or progenies responded differently, that is, were there genotype × fertilizer interactions.

All progeny plantings were repeated on three differing sites in three successive years, so that each progeny in each test was subjected to differing seasons. At each planting, three out of six 10-tree rows of each progeny received fertilizers, and three were left untreated. Fertilizers were usually applied at the beginning of the second and sixth year's growth after planting, but on the best sites the first fertilizer applications were postponed until the beginning of the fifth year. In all instances, responses were assessed 8 years after planting.

In general, the progeny plantations grew better with fertilizers than without, although the magnitude of the responses varied greatly from site to site. However, the fertilizer × family interactions were usually insignificant. Thus, the F-ratios for interactions in five trials done on differing sites with balanced fertilizers ranged from only 0.5 to 1.0, and the rankings of the different crosses were remarkably similar with and without fertilizers. Responses to fertilizers were most variable on wet, phosphate-deficient sites in the coastal plain of North Carolina. For instance, at one site, which had minimal preparation and care, P and K, applied at planting and at the start of the third growing season, doubled the average stem volume attained 6 years after planting (Table I). Progenies which grew poorly without fertilizers usually responded most to their application.

Although *P. taeda* progeny tests frequently showed less change in rank than illustrated in Table I, particular families did change rank substantially when they received fertilizers. Thus, progenies of all three crosses involving tree 8-31 were higher in rank with fertilizers than without, suggesting that tree 8-31 was a responsive genotype. More frequently, only specific crosses shifted substantially in rank, rather than all crosses involving an individual genotype. That is, it was usually easier to identify good, specific combinations than to find generally responsive genotypes which gave responsive progenies in all crosses.

Dramatic increases in site quality were achieved by adding fertilizers to wet, phosphate-deficient sites, as illustrated in Table II. Eight years after planting the trees that received fertilizers were, on average, almost 50% taller than those that received none. Again, many crosses did not change rank (i.e. there was no interaction), although a few crosses responded to the fertilizer addition much more strongly than the average (e.g. 8-33 and 8-21, Table II).

The results of numerous 8-year-old *P. taeda* tests can be summarized as follows:

1. Every family responded to fertilizer application when there was a significant overall response on the site.

2. Usually, families which grew rapidly without fertilizers, also grew most rapidly with fertilizers. There were occasional major exceptions suggesting genotype x fertilizer interactions, but F-tests did not show any statistically significant interaction when all tests were combined.

3. Nonetheless, the fertilizer responses of "faster-growing" families were generally less than the responses of "slower-growing" families.

TABLE I. Responses of 22 *Pinus taeda* families to P and K fertilizers, grown for 6 years on a P-deficient site in North Carolina, U.S.A.

	stem volumes		Percentage increase in stem volume produced by fertilizer application
Progenies	Rank without fertilizers	Rank with fertilizers	
8-76 x 8-68	1	3	31.7
8-46 x 8-141	2	4	44.3
8-68 x 8-73	3	12	48.0
8-33 x 8-21	4	9	71.9
8-74 x 8-68	5	2	97.8
8-102 x 8-76	6	1	121.4
8-68 x 8-53	7	20	54.3
Seed Prod. Area	8	22	37.6
8-31 x 8-64	9	7	97.6
8-33 x 8-142	10	17	73.1
8-46 x 8-21	11	21	50.8
8-102 x 8-68	12	5	112.9
8-65 x 8-33	13	14	96.2
8-33 x 8-31	14	8	121.4
Comm. Check	15	10	118.3
8-33 x 8-64	16	18	114.9
8-78 x 8-68	17	11	148.8
8-102 x 8-63	18	15	145.4
8-33 x 8-53	19	16	162.3
8-31 x 8-53	20	6	162.3
8-103 x 8-33	21	13	196.5
8-33 x 8-73	22	19	178.9
Mean			104.2

The planting had minimal site preparation. The fertilizers applied per tree at planting were 135 g triple superphosphate, 135 g muriate of potash, and 1300 g dolomitic lime, and after two growing seasons, 400 g triple superphosphate and 400 g muriate of potash.

TABLE II. Mean heights of *Pinus taeda* progenies 8 years after planting on a P-deficient North Carolina coastal plain site, illustrating varied responses to fertilizer treatment.[a]

	Without fertilizers		With fertilizers	
Progenies	Mean heights (m)	Rank	Mean heights (m)	Rank
No progeny × fertilizer interaction:				
8-33 × 8-103	3.3	8	4.7	8
8-65 × 8-33	3.1	17	4.5	18
8-31 × 8-64	3.0	20	4.4	21
8-46 × 8-141	2.9	23	4.4	23
Progeny × fertilizer interaction:				
8-76 × 8-31	3.6	4	4.4	24
8-64 × 8-76	3.1	19	4.9	4
8-33 × 8-21	2.9	25	4.9	3
8-33 × 8-73	2.8	30	4.7	9
Commercial check	2.8	27	4.2	27
Plantation average	3.1		4.5	

[a] Fertilizers were applied at planting and at the end of the second growing season, as for Table I.

III. SLASH PINE FIELD STUDIES (*Pinus elliottii*)

In the Florida Cooperative Program, special tests were established to determine early responses to fertilizers applied at the time of planting. A split-plot design was used, the major plots being fertilizer treatments (usually no treatment, N, P and NP, or no treatment, P, NP and NPK), and sub-plots being open-pollinated *P. elliottii* families. Tree heights were observed 3 years after planting.

These tests were established on poorly drained "flatwoods" soils which are mainly acidic sands and sandy loams, characteristically having a weakly cemented organic pan 30–100 cm below the surface. Significant fertilizer responses were obtained in about 80% of these tests. Similarly, significant responses to fertilizer treatments were observed in almost all of the progeny tests of the cooperative genetics programme. Less frequently, but in several tests, significant differences in response were observed between families. Examples of such varied responses, observed in one 3-year-old test in north Florida, are given in Table III. Note particularly the desirable

TABLE III. Effects of fertilizer treatments applied at planting, on the heights of 3-year-old open-pollinated *Pinus elliottii* progenies grown in Florida.

Response type and family	Fertilizer treatment			
	none	P	NP	NPK
	Mean tree height (m)			
Typical response				
107	1.69	2.12	1.84	2.35
112	1.74	1.93	2.19	1.99
Inferior growth, untreated Strong response				
Commercial control	1.58	1.91	2.21	2.42
604	1.53	2.00	2.17	2.47
Superior growth, untreated Weak response				
1205	1.89	1.97	2.28	2.31
115	1.83	1.85	2.33	1.86
Superior growth, untreated Strong response				
901	1.81	2.19	2.57	2.57
121	1.87	2.24	2.30	2.53
Mean (33 lots)	1.63	1.97	2.16	2.29

The fertilizer treatments were as follows: P, 290 kg per ha triple superphosphate; NP, 290 kg per ha diammonium phosphate; NPK, 1130 kg per ha of 4 : 12 : 12.

combination in the lower group, of good growth without fertilizers and a strong response to fertilizers. It appears that these trees responded favourably to nitrogen once the phosphorus deficiency was corrected.

IV. GLASSHOUSE EXPERIMENTS

A detailed investigation of inherent differences in the growth of *P. elliottii* seedlings in relation to added soil nutrients was conducted by Jahromi (1971). One-year-old control-pollinated seedlings of 20 full-sib families were established in large pots containing severely P-deficient soil. Four combinations of N and P were applied as follows: $N_0 P_0$, $N_0 P_{50}$, $N_{100} P_0$, $N_{100} P_{50}$ (in ppm). The seedlings were measured, harvested, and root and shoot tissues analysed, 16 months after nutrients were applied.

Fertilizer treatments greatly affected the growth of all families. Nitrogen alone decreased shoot dry weights by an average of 18%, whereas phosphorus alone, and in combination with nitrogen, increased the shoot dry weights by 46% and 69% respectively (Table IV).

Genetic effects were also apparent. The balanced design enabled variances to be partitioned, and mean squares for primary effects and interactions to be estimated separately (Table V). Using the method of Dickerson (1961), Jahromi calculated heritabilities for "general

TABLE IV. Mean effects of fertilizer treatments on the growth of young *Pinus elliottii* trees grown in pots of P-deficient soil. Plants were treated about 16 months after germination, and assessed when about 2½ years old. (From Jahromi, 1971.)

Variable measured	Treatments (ppm)			
	$N_0 P_0$	$N_{100} P_0$	$N_0 P_{50}$	$N_{100} P_{50}$
Height growth (cm)	65	56	80	74
Basal diameter growth (cm)	0.9	0.9	1.3	1.4
Shoot fresh wt (g)	246	224	366	438
Shoot dry wt (g)	83	68	121	141

adaptability" and for "local adaptability". The former indicates heritability in the usual sense, calculated by the formula:

$$h^2 = \frac{2(\sigma_F^2 + \sigma_M^2)}{\text{phenotypic variance}}$$

where σ_F^2 and σ_M^2 are variances due to female and male effects, respectively. Heritability for "local adaptability", calculated by the formula:

$$h^2 = \frac{2(\sigma_F^2 + \sigma_M^2) + \sigma_{G \times T}^2}{\text{phenotypic variance}}$$

can be used to calculate genetic gain obtainable by using genotypes adapted to the specific environmental conditions in which the variances were observed. In this case, including the genotype x treatment interaction in the numerator ($G \times T$) gives the heritabilities of growth traits when nutrients were added to a deficient soil.

Table V shows that dominance and other non-additive genetic effects (attributed to particular crosses), reflected in the large female x male component, made up a much greater proportion of the total phenotypic variance than did the male or female components which estimated additive genetic effects. This was consistent with the

TABLE V. Estimates of components of variance and heritabilities for growth and nutrient uptake of 20 full-sib families of *Pinus elliottii* seedlings given the differing nutrient treatments shown in Table IV. (From Jahromi, 1971).

Sources of variance	Height growth	Shoot dry wt	N-uptake	P-uptake
Females (F)	8.2	36.5	2,942	9.7
Males (M)	6.4	−3.6	1,183	9.9
F x M	21.1	81.2	4,416	40.1
Genotype x Fertilizer treatments	39.7	97.1	22,449	75.9
Phenotypic variance	93.1	265.5	41,804	170.1
h^2 general adaptability	0.31	0.23	0.20	0.23
h^2 local adaptability	0.74	0.61	0.73	0.68

frequent observation in the *P. taeda* field trials that good, specific combinations occurred more frequently than generally responsive genotypes (like 8-31). On the other hand, the high heritabilities for "local adaptability" indicated that progenies of some genotypes may be particularly suited to certain soils and fertilizer treatments. Thus, this study added weight to the existing evidence that there are *P. elliottii* genotypes which are able to respond particularly well to P and N added to severely P-deficient soils (see Table VI).

We recently conducted a study to gain some insight into the mechanisms by which some families take greater advantage of added P than others. Six open-pollinated families of *P. elliottii* were selected, all of which grew rapidly without added nutrients. Three families (A, B and C), had shown strong growth responses to added P in earlier field tests, while the other three (D, E and F) had not. The

TABLE VI. Relative rankings of *Pinus elliottii* seedling progenies (average over 4 crosses each) for shoot dry weights and N and P uptake, indicating a general combining ability among parents and correlation among traits. (From Jahromi, 1971).

	Shoot dry wt	N-uptake	P-uptake
Female parents			
17	2	2	2
268	5	5	4
254	1	1	1
262	3	3	5
87	4	4	3
Male parents			
1	2	2	2
5	4	3	4
12	5	5	5
98	1	1	1
196	3	4	3

N and P uptake were measured as the total amount of these elements in plant tissues 16 months after treatment.

experiment was done in two phases, first testing the effects of different amounts of P on growth in sand culture ("pre-treatment" phase) and then examining ^{33}P-uptake when the differently treated seedlings were put in water culture.

Seedlings were taken from nursery beds in November, and were planted in pots of acid-washed quartz sand. They were grown for 98 days in a glasshouse at 18–25°C, in 16 h photoperiods, without added nutrients, so that the concentration of P in their foliage dropped to c. 900 ppm. Five seedlings of each family were harvested to determine the P contents of roots and shoots. The remaining seedlings of each family were then given two nutrient regimes, a balanced nutrient solution, with respect to N, K and minor nutrients, supplemented by either high (25 ppm) or low (5 ppm) levels of P. After 2 months seedlings were removed from the sand and their roots were placed for 2 days in a continuously aerated aqueous nutrient solution having the same composition as before. Subsequently, the seedlings were placed in two changes of a phosphorus-free nutrient solution for 12 h per change, before being transferred to a nutrient solution to which was added carrier-free ^{33}P with an initial activity of 4 μCi l^{-1}. After 30 min in the ^{33}P solution the roots were washed and placed in a nutrient solution without P for 2 h. The seedlings were subsequently measured, roots and tops were separated, tissues dried, weighed, ground, ashed and analysed by liquid scintillation spectrometry. There were five replicate seedlings per family and per P pre-treatment combination. An additional five seedlings of each treatment combination were harvested without ^{33}P application in order to determine the nutrient status of their roots and shoots.

Seedlings of all six families responded well to the initial P treatments in sand culture. The average gain over the 2-month pre-treatment period was 10.3 cm in height and 3.1 g in shoot dry weight. However, the most pronounced effects of increased P were changes in root fresh weights and shoot and root P contents. The fresh weights of roots of families A, B and C, which had been previously classed as responsive to P-fertilizers, became on average 4.1 g heavier per tree when amounts of P were increased from 5 to 25 ppm. In contrast, the unresponsive families D, E and F had 0.2 g less average root fresh weight with the higher P-application, that is, they had equally good root growth under both P-levels (Table VII). Regarding P content, during the two months growth in sand with

TABLE VII. Effects of two levels of P on fresh root weights and subsequent ^{33}P-uptake of seedlings of 6 open-pollinated *Pinus elliottii* families grown in sand culture (see text).

			Difference between seedlings receiving 25 and 5 ppm P		
			Fresh wt of roots per seedling after P-treatment[a] (g)	Counts per minute of ^{33}P taken up into the	
				shoots (cpm)	roots (cpm × 10^{-3})
Families responsive to P-application in field trials		A	1.3	482	−189
		B	7.3	360	− 58
		C	3.7	166	4
		mean	4.1		
Unresponsive families		D	1.4	118	−116
		E	0.4	− 35	−167
		F	−2.3	583	189
		mean	−0.2		

[a] Fresh weights of roots per plant ranged from 16 to 26 g.

5 ppm, roots and shoots lost on average 170 and 205 mg P per g dry weight, respectively, whereas with 25 ppm P they gained 481 and 510 mg P per g dry weight in roots and shoots, respectively.

The subsequent effects of these two pre-treatments on the uptake of ^{33}P from nutrient solution were not consistent, and families did not differ substantially. However, families A, B and C took up greater amounts of ^{33}P into their shoots than families D, E and F (Table VII), although family F was more like the responsive families than families D and E.

Rates of P uptake could not be related to obvious differences in root morphology, nor to differences in mycorrhizal development. However, fresh and dry weights of roots were positively related to the amount of radioactive P in the seedling tissues. Fresh weights of roots were more closely related to the total amounts of radioactive P absorbed than dry weights, and are perhaps a better measure of the potential absorptive capacity of roots than dry weights, as mycorrhizas and other absorptive structures of seedling roots are relatively succulent. Correlation coefficients were 0.65 and 0.72 between root fresh weights and total counts per minute in roots and shoots, respectively. Radioactivity counts on a per gram dry weight basis were not significantly related to root weights, probably because of dilution effects in the tissues of the larger plants.

The use of radioactive P was expected to indicate genotypic differences in P-uptake capability *per se*, but none were detected. Instead, families appeared to differ in the ways their root systems grew in response to different P levels and in the proportion of P absorbed to P translocated to aerial portions of the plant (Table VII).

The glasshouse experiments, considered together with the large effects of specific genetic combinations in field trials, suggest that there are inherent factors that influence the transport and utilization of P. It is suggested that such factors, along with ones controlling the size and extent of absorptive root surfaces, could be responsible for the differing responses to P fertilizers in the field.

V. DISCUSSION

There is no question about the need for fertilizer applications on certain P-deficient sites in the coastal southern pine region of the U.S.A., and they are already being applied there on a large scale. In other areas, especially the Piedmont, where a lack of N is of key importance, some large-scale fertilization is being practised, but widespread use of fertilizers is not yet generally accepted.

Many field tests suggest that for both *P. taeda* and *P. elliottii* genotype × fertilizer interactions can be ignored on all but P-deficient sites. In most areas, the fastest-growing trees also grow well when fertilized, but there are major implications that they are often the least responsive per unit of fertilizer applied. However, this implication is still speculative and is less applicable to *P. taeda* and *P. elliottii* planted on P-deficient soils. As improved first- and second-generation seed orchards are developed, the relative responses of families to added fertilizers must be closely monitored. Clearly, on sites where fertilizers are applied routinely, most wood will be produced by the most responsive families, whether or not they have superior growth rates without fertilizers.

There has been concern that improved nutrition may adversely affect wood qualities. Although numerous tests have shown that P-fertilizers have little effect, N may lower wood specific gravity to some extent. But on sites where growth responses to fertilizers are good, the added wood volume from fertilizers compensates many times over for the loss in wood weight caused by a drop in specific gravity. As a generalization, one can reasonably ignore such negative effects of fertilizers, although tracheids may be considerably shorter in fertilized trees.

VI. CONCLUSIONS

The many field trials and controlled environment experiments done to date lead us to the following conclusions:

1. Until proven otherwise, one must always test for a fertilizer x genotype interaction on forest land that will receive routine applications of fertilizers.

2. A fertilizer x genotype interaction does not appear to be strong enough in *Pinus taeda* to justify developing separate seed orchards of especially responsive clones except, perhaps, for planting on P-deficient sites. Such special seed orchards may be more beneficial for *Pinus elliottii* in which P-responsive strains can be developed. The development of strains for N-response does not now appear to be promising.

3. Wood quality is not harmed so much by the use of fertilizers to outweigh the gain in yield.

4. Progeny assessments may be speeded up, and certain characteristics may be better expressed, by growing trees at optimum nutrient levels.

5. There appear to be certain specific crosses in *Pinus taeda* and *Pinus elliottii* that respond particularly well to fertilizers. These could be exploited by creating specialty orchards, containing only those parent genotypes which will produce the desired crosses.

Only the surface has been scratched in obtaining information on genotype x fertilizer interactions, and for the present, such interactions may be ignored on the majority of *P. taeda* and *P. elliottii* sites. However, in order to maximize production on deficient sites, much more research is needed to understand both interactions between genotypes and nutrients, and interactions among nutrient elements in relation to genotypes.

REFERENCES

Dickerson, G. E. (1961). *In* "Statistical Genetics and Plant Breeding" (W. D. Hanson and H. F. Robinson, eds), pp. 95–106. 1963. National Academy of Sciences, 982.

Jahromi, S. (1971). "Genetic Variation in Nutrient Absorption in Slash Pine". Ph.D. Dissertation, University of Florida, Gainesville.

Pritchett, W. L. and Smith, W. H. (1975). *In* "Forest Soils and Forest Land Management" (B. Bernier and C. H. Wingest, eds), 1975. pp. 467–476. Proc. 4th North American Forest Soils Conf. Laval Univ., Quebec.

PROBLEMS CONCERNING THE USE OF PHYSIOLOGICAL SELECTION CRITERIA

27
Competition, Genetic Systems and Improvement of Forest Yield

E. D. FORD
Institute of Terrestrial Ecology, Bush Estate, Penicuik, Scotland

I.	Introduction	463
II.	Competition Between Plants	464
III.	The Genetic System and the Development of Plant Monocultures	467
IV.	Some Consequences for Tree Breeders	469
References		471

I. INTRODUCTION

The assumption that "competition" operates in closed stands of trees is universal, yet there is no universally agreed definition of the process. When plants are grown close together mean plant size is smaller than when they are well separated, and large size differentials occur in the population. The term "competition" implies that some plants have been "successful" i.e. become large, whilst others remain small. Clements defined competition with reference to the environment: "When the immediate supply of a single factor necessary (for growth) falls below the combined demands of the individual plants, competition begins" (Donald, 1963). Thus, finite resources for growth are distributed between plants, although Clements did not pre-judge on what basis this distribution of resources was determined. Sakai (1961) accepted that there would be a decrease in environmental resources available per plant at close spacings but did not consider this as competition. He considered that "within a plant population where different genotypes are grown mixed, an intergenotypic competition could take place". That is, at least some of the difference in size amongst individuals could be the consequence of genotypic differences.

In addition to differences of definition, there are contrasting attitudes about the role which competition plays in population genetics of forest trees.. Assmann (1970) developed the argument that the growth of individuals in a stand produces a selection procedure for superior genotypes: "The number of trees per hectare in beech stands on good sites reduces from over 500 000 at the age of 10 to approximately 400 at the age of 100 years. This conveys an idea of how often the eliminating situations — one might say, the selective 'trials' — must have repeated themselves for the final remaining trees." On the other hand Stern (1969) has stated that "... intra-specific competition for a variety of reasons does not lead to the establishment of a genetically uniform population, but, in most of the cases, most probably leads to balanced systems with high genetic variation...".

However complex it may seem, the process of inter-plant competition cannot be ignored by tree breeders. In the majority of cases their objective is a closed stand with improved yield and the initial tree selections are themselves made from closed stands. The tree breeder must examine the relationship between inter-plant competition and the genetic structure of his population and answer the following questions:

(1) is it possible to select high-yielding individuals from a forest and expect to obtain high-yielding stands from their progenies;

(2) should the tree breeder work with populations of trees rather than with individuals;

(3) should he attempt to recognize and select for "components of yield" rather than select high-yielding individuals?

II. COMPETITION BETWEEN PLANTS

I should say at this point that my comments on plant competition are made with reference to even-aged monocultures and particularly to plantations of forest trees raised from seed. In such forests we can expect that some individuals will be larger than others. In itself this does not say much about inter-plant competition since size differentials are also present in collections of open-grown trees. However, the relative proportion of large and small individuals differs markedly with plant density (Fig. 1a). At the high density the proportion of trees in the small size categories is larger, and maximum plant size is smaller. The processes which give rise to these characteristic frequency distributions may be judged by examining the relative growth rates (RGR, cm girth cm^{-1} $year^{-1}$) and mortality rates of trees of different sizes. If large plants are shading small ones, or otherwise obtain a greater proportion of the finite resources for growth, then they would be expected to have greater RGRs than small plants. Figure 1b illustrates that this is so. Note that at the low density, even plants in the smallest size category had positive values of RGR but at the higher densities positive values are increasingly confined to the largest plants. The overall mortality rate was greatest at the high plant density and extended further into the intermediate size categories.

Figure 1 illustrates an important feature of inter-plant competition: as plant density increases, mortality increases, and the larger plants become responsible for an increasing proportion of the

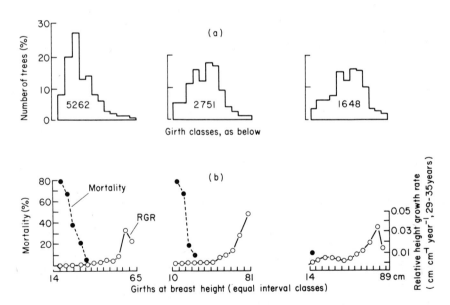

FIG. 1 (a) Size frequency histograms for 29-year-old *Picea sitchensis* growing in Wales, initially planted at 0.91 m (left), 1.83 m (centre) and 2.44 m (right) spacings. The numbers of surviving trees per hectare are written within each histogram.
(b) Mean relative growth rates (○) and mortalities (●) occurring between 29 and 35 years after planting, for each girth class. (After Ford, 1975.)

total growth. This is in agreement with the thesis that competition is for resources in finite supply. But is the outcome of competition the result of genotypic differences? The most effective way to judge this is to examine the spatial distribution of large and small plants. Two rather different patterns might be anticipated. If competition is the outcome of selection between genotypes, then the distribution of successful plants would be expected to follow the initial distribution of "large plant" genotypes and, unless forest workers are able to recognize this during planting, we would expect this to be random. Alternatively, if competition occurs in response to the finite concentration of resources which exists at any point, then the spatial distribution of large plants can be expected to be even. Investigation of the spatial distribution of plants in a wide range of monocultures (Ford, 1975) showed that (a) where competition is intensive enough to produce a skewed distribution of plant size the large plants are evenly distributed, and (b) an even distribution is maintained although there is mortality and a decrease in number of plants which can be deemed large. There is a constant re-adjustment which maintains an even distribution. These results are similar to those of Cooper (1961) and Laessel (1965), both of whom investigated naturally regenerated stands of pines and found that the initially very clumped, or at least random, distributions of seedlings progressed towards an even distribution of survivors as growth took place.

The growth of young plants is characteristically a positive feedback system; the larger a plant becomes the more growth it makes (Burdon and Sweet, Chapter 29). But this situation cannot be maintained when plants grow together in stands, because environmental resources of light, water and nutrients are insufficient to maintain all plants at ever increasing absolute rates of growth. There is, consequently, a negative feedback on growth, which is greatest where plants are closest together. This type of control system, where a positive feedback inner loop is stabilized by a negative feedback outer loop is a common one in biological systems (Maruyama, 1963) and produces homeostasis.

There are two situations in which this system of control may be modified. One in conditions of severe environmental stress such as low air temperature, when crops may "stagnate" (Smithers, 1961). Such a stress must apply to all plants equally, and the distinction may be difficult to draw between "stagnation" and a slowing down of the competition process due, say, to low nutrient levels (Yoda *et al.*, 1963). The other situation arises where a mixture of pure line genotypes is cultivated. For instance, tall and short varieties of wheat perform very differently when they are grown in mixture compared with monoculture (Khalifa and Qualset, 1974). Under these circumstances the operation of a spatial homeostasis will be dependent on the sowing density of the major component. When a pure line variety is grown at close spacing there *is* a competition effect (Ford, 1975).

III. THE GENETIC SYSTEM AND THE DEVELOPMENT OF PLANT MONOCULTURES

Under normal forest conditions interplant competition does not result in "natural selection" for large size. Competition occurs as a direct result of differences between genotypes only if the genetic structure of the population is artificially manipulated, and it is noticeable that attempts to select individuals from a stand on the basis of their "high competitive abilities" (Sakai, 1961) have been unsuccessful; "competitive ability" has a poor heritability.

One of the most interesting observations to be made about the individual plants in a densely planted stand is that they show distinct "plasticity", that is there are not only differences in size, but also associated differences in morphology. These differences have been frequently studied in annual plants (Donald, 1963) and the general pattern is that, as the stand develops and a few plants become large, the smaller plants appear to "adjust" to their new environment and become morphologically different. Differences in morphology between large and small trees in closely grown monocultures have also been found, particularly in relation to the structure of individual crowns (Assmann, 1970; Ford, 1976). What sort phenomenon is plasticity? Bradshaw (1965) has presented evidence that plasticity must be considered for each character separately, not as an overall feature of plant growth. It is (a) specific for that character, (b) specific in relation to particular environmental influences, (c) specific in direction, i.e. particular combinations of environment and character produce the same response, (d) under genetic control, not necessarily related to heterozygosity and (e) can be altered by selection. Bradshaw also reviewed evidence that characters associated with long periods of meristematic activity, such as plant size and leaf number, are more likely to have an associated plasticity than characters which develop rapidly, like floral characters. For the breeder the genetical control of plasticity is of great importance when devising strategies of tree improvement.

Another problem is well known to all involved in breeding, but must be restated, as it is an essential component of the breeder's world. "Most characters of economic value to plant and animal breeders are metric characters, and most of the changes concerned in micro-evolution are changes in metric characters" (Falconer, 1964). A metric character is one in which there is simultaneous segregation of many genes affecting the character, i.e. polygenic inheritance. As Cannell *et al.* (Chapter 10) have indicated, shoot growth of trees is polygenically controlled.

In summary, tree breeders are concerned with, and must accommodate, three interconnected mechanisms which affect yield characteristics.

(1) Polygenic inheritance can be anticipated. The polygene is one member of a system whose parts are generally interchangeable, and although the individual effects of one polygene are not large, the members of a system may act together to produce large differences. Necessarily, then, methods of plant breeding in polygenic systems are focussed on work with populations (Mather and Jinks, 1971). The exploitation of polygenic systems has depended upon the "bulking" of multiple genotypes.

(2) The significance of "plasticity" in arable crop breeding has been discussed by Allard and Bradshaw (1964). It may be of even greater importance in tree breeding, when the individual tree, rather than the total crop yield, is the item to be produced for sale. Characters are "plastic" not only within single stands, but also between populations, when different planting environments are used. At this level, plasticity, or rather the lack of it, is frequently referred to as "stability".

(3) The developmental homeostasis which is a consequence of growing plants closely together is also of particular importance for the tree breeder. His crops are long-term, and frequently initial spacings are much closer than final spacings, in Britain the ratio is typically 10 : 1.

IV. SOME CONSEQUENCES FOR TREE BREEDERS

The answer to the first question that I posed on the breeders' behalf is "no". The competition mechanism and the associated plant plasticity suggest that there is no reason why large trees selected from within a monoculture where there is inter-plant competition should be expected to be of a "large plant" genotype. Even if it were so, the pattern of polygenic inheritance is likely to make it difficult to obtain superior progeny. And should a population of "large plant" types be obtained, then the spatial homeostasis which is the automatic outcome of growing plants close together over the normal forest rotation, makes it doubtful whether yield would be increased on a per unit area basis.

The answer to the second question is "yes". Tree breeders should work with populations. The growth of a forest is a population process from the aspects of both genotype structure and stand development. The genetic characteristics of individual trees are likely to be as difficult to discover as those of whole populations and less relevant to the problem of increasing yield, and if work with populations is a practical impossibility then it may be necessary to forego the objective of breeding directly for increased yield.

The answer to the third question is "yes". Where yield is under polygenic control its components must be recognized if progress is to be made. In tree improvement there is considerable pressure to work with clones, which provides an alternative to breeding for long-generation outbreeding species. However, it must be stressed that work with clones or any form of pure line variety produces problems of genetic structure and population development (Simmonds, 1962). With the long-term nature of the forest crop, will it be sufficient to rely on the heterozygosity of one clone? The British rotation length is 50 years in which time management objectives, cultural practice, and the physical environment may change substantially and in this type of situation it could prove important to maintain variability in stands.

One technique for providing controlled variation would be the use of multiline mixtures based on clonal stocks. Multiline mixtures are used in cereal culture to increase overall disease resistance by mixing strains with different types of individual resistance. The technique of multiline planting for high yield has attracted attention and I would like to add a few comments on how the production of such mixtures should be judged for tree crops, since for practical purposes potentially "high-yielding" clones are likely to be the material which first comes to hand.

The essential problem of increasing yield per hectare can be related to the spatial homeostasis which is the outcome of competition. We must increase the growth of the individual plant, the positive feedback loop, but at the same time decrease the competition mechanism which limits growth of the individual, the negative feedback loop. The ideal plant is *not* a strong competitor. To increase yield per hectare two questions need to be resolved:

(1) are there differences in the efficiency with which individuals use space; and

(2) to what extent does a growth increment by one individual automatically mean environmental depletion for another?

There is some evidence for differences in efficiency. Hamilton (1969) found that narrow-crowned trees were more efficient producers of increment per unit of ground area than broad-crowned trees (also Assmann, 1970). But how can we test whether a collection of such trees produces a higher stand yield?

One approach is to examine the relationship between increase in mean tree weight (or volume) and the decrease in stand density due to self-thinning as a stand develops. In Fig. 2 the solid line represents an idealized version of the normal progression of an unthinned stand. In the early stages of stand development, A—B, the plants increase in volume but there is no mortality. Once mortality commences, B—C, it progresses in a regular manner and at a rate which can be thought of as the *unit of material which is added to effect a unit of self-thinning*. During this phase the total weight of the crop is increasing. Eventually a ceiling yield may be reached and the gradient of thinning, C—D, declines. There are a number of possible adjustments to this progression that may be considered. The trees may be of a larger weight before thinning commences, B', the unit of increment in mean weight to effect an increment of self-thinning may increase so that the gradient, $B'-C'$ is steeper than B—C. The ceiling yield of the crop may be higher, $C'-D'$. It is possible that all of these parameters of crop growth can be influenced by the efficiency with which solar radiation is used, so that the standing volume of a crop of a certain density might be increased (cf. White and Harper, 1970; Ford, 1975). In Fig. 2 there is no indication of

27. Competition, Genetic Systems and Yield Improvement

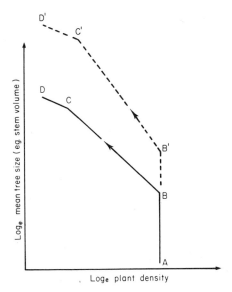

FIG. 2 Hypothetical relationships between the increase in mean tree size (from point A), as a closely grown forest monoculture grows, and the decrease in number of trees resulting from competition (from point B). (See text.)

timescale and the rate of progression along either pathway A → B → C, A → B' → C' could be important in judging productivity.

Experiments to compare the growth rates and spaces occupied by individual trees in stands with different plant forms could be very long term and there is an obvious attraction in modelling the situation. This could be achieved if bulk canopy models (Ledig, Chapter 2) can be amalgamated with models based on the dynamics of the stand population (Fries, 1974), or of the individual tree.

REFERENCES

Allard, R. W. and Bradshaw, A. D. (1964). The implications of genotype-environment interactions in applied plant breeding. *Crop Sci.* 4, 503–508.

Assmann, E. (1970). "The Principles of Forest Yield Study". Pergamon Press, Oxford.

Bradshaw, A. D. (1965). Evolutionary significance of phenotypic plasticity in plants. *Adv. Genetics* 13, 115–155.

Cooper, C. F. (1961). Pattern in ponderosa pine forests. *Ecology* 42, 493–499.

Donald, C. M. (1963). Competition among crop and pasture plants. *Adv. Agron.* **15**, 1–114.

Falconer, D. S. (1964). "Introduction to Quantitative Genetics". Ronald Press, New York.

Ford, E. D. (1975). Competition and stand structure in some even-aged monocultures. *J. Ecol.* **63**, 311–333.

Ford, E. D. (1976). The canopy of a Scots pine forest: description of a surface with complex roughness. *Agric. Meteorol.* (in press).

Fries, J. (1974). Growth models for tree and stand simulation. *Rapporter och Uppsater, Skogshogskolan* **30**, 1–379.

Hamilton, G. J. (1969). The dependence of volume increment of individual trees on dominance, crown dimensions and competition. *Forestry* **42**, 133–144.

Khalifa, M. A. and Qualset, C. O. (1974). Intergenotypic competition between tall and dwarf wheats. I. In mechanical mixtures. *Crop Sci.* **14**, 795–799.

Laessel, A. M. (1965). Spacing and competition in natural stands of sard pine. *Ecology* **46**, 65–72.

Maruyama, M. (1963). The second cybernetics: deviation-amplifying mutual causal processes. *Am. Scientist* **51**, 164–179.

Mather, K. and Jinks, J. L. (1971). "Biometrical Genetics" (2nd edition). Chapman and Hall, London.

Sakai, K. I. (1961). Competitive ability in plants: its inheritance and some related problems. *Symp. Soc. exp. Biol.* **15**, 245–263.

Simmonds, N. W. (1962). Variability in crop plants, its use and conservation. *Biol. Rev. Cambridge Phil. Soc.* **37**, 422–465.

Smithers, L. A. (1961). Lodgepole pine in Alberta, Canada Dept. For. Bull. **127**, 1–53.

Stern, K. (1969). Einige Beitrage genetischer Forschung zum Problem der Konkurrenz in Pflanzenbestanden. *Allg. Forst. und J.-zty.* **140**, 253–262.

White, J. and Harper, J. L. (1970). Correlated changes in plant size and number in plant populations. *J. Ecol.* **58**, 467–485.

Yoda, K., Kira, T., Ogawa, H. and Hozumi, K. (1963). Self-thinning in overcrowded pure stands under cultivated and natural conditions. *J. Inst. Polytech. Osoka Cy Univ.* Ser. D. **14**, 107–129.

28
Maternal Effects on the Early Performance of Tree Progenies

THOMAS O. PERRY
School of Forest Resources, North Carolina State University, Raleigh, U.S.A.

I.	Introduction	473
II.	Maternal Influences on Seed Size	474
III.	Other Maternally Determined Seed Characteristics	476
IV	Variation in Maternal Effects	477
V	Persistence of Maternal Effects	478
VI	Separating Maternal From Genetic Effects	478
VII	Summary	479
References		480

I. INTRODUCTION

Progeny tests with many gymnosperm species reveal that 60—90% of the variation in seedling size is closely correlated with maternal factors. Such correlations decrease with age, although data from mature provenance trials and progeny tests indicate that maternal effects can sometimes account for 45—80% of the variation in tree size up to 10—40 years of age.

This contribution summarizes observations regarding the genetic and environmental circumstances of the seed parent which affect the performance of *Pinus taeda* and *Pinus elliottii* progenies. Selected references are presented to justify the thesis that observations with these species may be generally applicable to the Pinaceae and possibly to other taxons of the Coniferales.

II. MATERNAL INFLUENCES ON SEED SIZE

The importance of seed size in determining progeny performance has been recognized for many years (e.g. Righter, 1965; Nanson, 1965, 1967, 1969). Thus, in 1960, Mergen and Voigt reported that the application of fertilizers to a *Pinus elliottii* seed production area increased the weight per seed by 55% and the weights of the resulting one-year-old seedlings by 40%. However, quantitative data concerning the role of the female parent in determining seed size has only recently become available. Many tree improvement workers, including the author, assumed that cytoplasmic inheritance was of little consequence in tree breeding, and therefore the vigour of two-parent progenies was likely to be similar whichever was the seed parent. Indeed, when seed from a given combination of parents was in short supply, many tree breeders pooled seed from reciprocal crosses and treated them as equivalent.

The data presented here are from the North Carolina State Cooperative Forest Tree Improvement Program, where it has been found that clones of selected plus-tree genotypes produce heavier seeds than unselected trees (Table I). In this programme paternal and maternal effects on seed weight can be separately assessed, as the general combining ability of each select tree is tested by crossing it with four or more pollen parents. Taking data from the Riegel Paper Co. seed orchards of *P. taeda,* it was found that the average weight per seed produced by select female parent trees spanned a range of

TABLE I. Relative weights of open-pollinated seed taken from selected and unselected trees of *Pinus taeda* (Riegel Paper Co., North Carolina, U.S.A.). Commercial check = seed from trees selected at random in natural stands. Rogued clones = seed from mediocre clones which have been rejected in the tree improvement programme. Select clones = seed from outstanding trees.

Classification	Seed weight (mg seed^{-1})	% of commercial check	% of rogued clones
Commercial check	23.2	—	90
Rogued clones	26.8	116	—
Select clones	36.5	157	137

14.7 mg (18.1–32.8 mg per seed), whereas seed obtained from crosses with given pollen parents ranged in weight by only 2.4 mg (26.6–29.6 mg). The total range in seed weight for all crosses was 16.7 mg (18.1–34.8 mg) so it could be said that female or seed parents accounted for 88% (14.7/16.7) of the total variation, whereas the pollen parents accounted for only 14% (2.4/16.7). The differences in weight per seed were closely correlated with the heights of the progenies at 4 years of age (Fig. 1).

The major role of the female parent in determining seed weight is not surprising when one realizes that the gymnosperm seed coat is diploid maternal tissue, and the gametophyte is haploid maternal tissue. Only the embryo contains genes from the pollen parent, and X-ray analyses and seed dissections indicate that the embryo constitutes only 10–20% of the total seed weight.

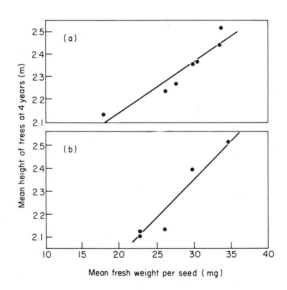

FIG. 1 Relationship between mean weight per seed of female parents of *Pinus taeda*, and the heights of the resulting seedling progenies at 4 years of age.
(a) Each point represents a female parent, averaged for four or more crosses with different pollen parents.
(b) Each point represents a female parent crossed with the same male parent (clone 8–14).
(Data from Riegel Paper Co. progeny tests, U.S.A.).

III. OTHER MATERNALLY DETERMINED SEED CHARACTERISTICS

The female parent also affects the thickness of the seed coats, seed stratification requirements, resistance to seed and cone insects, and polyembryony, all of which are often correlated with progeny vigour, and frequently obscure the more obvious and long recognized effects of seed size.

The importance of these other maternal effects was illustrated in a replicated *Pinus elliottii* progeny testing programme (Union Camp Corp.) involving 12 select trees and 4 pollen parents. The mean heights of 4-year-old progenies from each select tree spanned a range of 43 cm (287—331 cm), whereas the average height range of progenies produced by the four pollen testers was only 10 cm (301—311 cm). However, the correlation between weights per seed of given select tree clones, and heights of their 4-year-old progenies, was poor, principally because of clones 260 and 201 which gave unexpectedly small progenies (Fig. 2).

X-rays revealed that seeds of clone 201 contained many defective embryos, or contained more than one embryo, giving an average of

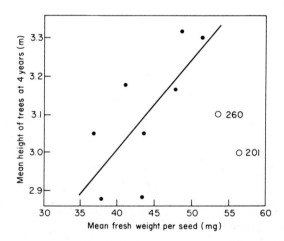

FIG. 2 The relationship between weight per seed of different seed parents of *Pinus elliottii*, and the average height of the resulting seedling progenies at 4 years of age. Seed parent clones 201 and 260 were shown to produce abnormal seed (see text).
(Data from Union Camp. Corp. progeny tests, U.S.A.).

110 embryos per 100 seeds, regardless of the pollen parent. Clone 260, on the other hand, produced seeds with seed coats about twice as heavy as those produced by clones nearer the regression line in Fig. 2 (45 mg cf. 22 mg per seed coat). Observations confirmed that these abnormal seeds rarely produced vigorous seedlings. Embryos within polyembryonic and heavily coated seeds had difficulty in shedding their seed coats, so that the tips of the cotyledons remained trapped and died; this also increased the risk of water droplets persisting, and thus favoured the development of damaging salt accumulations and invasion by pathogens.

Maternal differences in ability to produce sound seed may, in some instances, be a reflection of differences in susceptibility to cone and seed insects (DeBarr and Ebel, 1973). Additionally, it may be noted that the strong maternal influence on the differing seed stratification requirements observed, for instance, among provenances and clones of *Pinus taeda*, may be attributed to the fact that the controlling factors are located in the maternal tissues of the inner and outer seed coats (Barnett, 1972).

IV. VARIATION IN MATERNAL EFFECTS

The weight of individual seeds can vary with position in the cone, size of the cone and position of the cone on the tree. Yearly differences in seed weight of *P. elliottii* can be as high as 14% of the mean (W. Beers, pers. comm.). Also the severity of attack by cone insect pests can vary from one year to the next, and seed stratification requirements are often affected by the climate during the fall when the seeds are maturing. Indeed, any change in seed orchard management such as irrigation, spraying, or fertilizer application can alter the size and quality of the seed produced. These, and many other variations in the environment of the seed parent can alter the degree to which maternally determined seed characteristics regulate progeny performance.

"Commercial check" seed, the unselected seed used to index the progress of tree improvement programmes, is often collected and stored for many years before it is used. Any fluctuation in the quality of the seed collected in a given year, or any alteration in its quality during storage, will result in a change in the magnitude of the differences among select tree progenies and the relative performance of "improved" seedlings when compared with the progenies of unselected trees.

V. PERSISTENCE OF MATERNAL EFFECTS

Inherent differences in pathogen and insect resistance, response to photoperiod, or to other aspects of the physical environment can quickly erode any initial differences in seedling size. Accordingly, correlations between maternal or nursery factors and progeny size decrease with age.

Repeated measurements of progenies from diallel crosses within *Pinus strobus* revealed that maternal effects accounted for 78% of the variation in progeny size at 4 years of age, but only 52% at 6 years (Kriebel *et al.*, 1972a,b). However, data from older progeny trials indicated that the early advantages of large plants did not disappear completely. The combined effects of genetic variation and variation in nursery management practices accounted for 1.5- to 2-fold differences in the size of 7- to 34-year-old plantations of *Pinus taeda* (Wakeley, 1963, 1969). In provenance tests with *Pinus sylvestris* and *Picea abies* similar effects can persist for 35 years or more (Nanson, 1965, 1967, 1969).

VI. SEPARATING MATERNAL FROM GENETIC EFFECTS

The relative importance of (a) maternal factors and (b) factors regulated by the numbers of zygotes and their cytoplasmic constituents can be distinguished using the equation for exponential growth:

$$dw/dt = KW$$

which when integrated gives:

$$\log W_2 - \log W_1 = K(t_2 - t_1) \quad \text{or} \quad W_2 = W_1 \exp K t,$$

where W_1 and W_2 are plant weights at times t_1 and t_2, and K is a constant of "proportionality", which has the units: size time^{-1}.

The thesis of this contribution is that W_1 is often a reflection of maternal factors, and that genetic factors which physiologists, tree breeders and silviculturists need to estimate, determine the values of K.

The value of K is the product of interaction between genotype and environment and can be materially altered by varying the competitive and other environmental circumstances of a plant or a plot of plants of a given genotype. Also, K cannot be measured directly; it can only be estimated by measuring W_1 and W_2 at selected time intervals, either in the laboratory, nursery, or under forest conditions, depending on the purposes of the investigation. Furthermore, K can be calculated on a single tree basis or on a per hectare basis.

Several attempts to correlate genetic differences in dry matter production per plant with various physiological attributes have failed. These include estimates of inherent differences in rates of photosynthesis (per unit of leaf surface or per milligram of chlorophyll), response to fertilizers, root : shoot ratios, and so on. These failures may be attributed partially to the large and continuing effects that initial plant size has on subsequent growth, and our failure to recognize the major role that maternal factors play in determining initial plant size.

In field trials involving "single tree" or "row-plot" designs, progenies with large seedlings tend to suppress progenies with small ones. Since K is materially altered by competitive and other environmental factors, it may be misleading to extrapolate yields per tree derived from single-tree or row plots to yields per hectare. Such designs only exaggerate the maternal and environmental effects. Measurements taken on trees within square or rectilinear plots (as is standard practice in field crop agronomy) may be the only valid bases for estimating differences in dry matter yield per hectare among progenies (see Ford, Chapter 27).

VII. SUMMARY

Progeny tests of *Pinus taeda, P. elliottii* and other gymnosperm species indicate that most of the variation in seedling heights at the end of the first growing season, and as much as 45—80% after 4—8 years' growth, can be attributed to maternally influenced factors. These factors include seed weight, seed coat thickness, stratification requirement and polyembryony, and can be modified by the local and climatic environments in which the seeds develop. Less than 15% of the weight of a conifer seed is in the embryo, which is the only portion with a genetic component from the male. Maternal factors can obscure the true nature of genetically controlled growth processes, and maternal effects may be exaggerated by recording growth differences on widely spaced trees and extrapolating to stand conditions.

REFERENCES

Barnett, J. (1972). Seedcoat influences dormancy of loblolly pine seeds. *Can. J. Forest Res.* 2, 7–10.

DeBarr, G. L. and Ebel, E. H. (1973). How seedbugs reduce the quality and quantity of pine seed. *In* "Proc. Twelfth Southern Forest Tree Improvement Conference", pp. 97–103. Southern Forest Experiment Station, U.S.A.

Kriebel, H. B., Roberds, J. H. and Cox, R. V. (1972a). Genetic variation in vigor in a white pine incomplete diallel cross experiment at age 6. *In* "Proc. Eighth Central States Forest Tree Improvement Conference", pp. 40–42. School of Forestry, Univ. of Missouri, U.S.A.

Kriebel, H. B., Namkoong, G. and Usanis, R. E. (1972b). Analysis of genetic variation in 1, 2, and 3-year-old Eastern white pine in incomplete diallel cross experiments. *Silvae Genet.* 21, 44–47.

Mergen, F. and Voigt, G. K. (1960). Effects of fertilizer applications on two generations of slash pine. *Proc. Soil Sci. Soc. Am.* 407–408.

Nanson, A. (1965). The value of early tests. I. International *Picea abies* provenance trial 1938. *Trav. Stn Rech. Groenendaal* Sér. E1, pp. 60. *For. Abstr.* 27(4), 650–675.

Nanson, A. (1967). Contribution à l'étude de la valeur des tests précoces. II. Expérience internationale sur l'origine des graines de pin sylvestre 1906. Ministère de l'Agriculture, Administration des Eaux et Forêts, Groenendaal-Hoeilaart, Belgique, Trav. Sér. E., no. 2, pp. 7–43.

Nanson, A. (1969). Juvenile and correlated trait selection and its effect on selection programs. *In* "Proc. Second Meeting Working Group on Quantitative Genetics Sec. 22, IUFRO", pp. 19–26. Southern Forest Experiment Station, U.S.A.

Righter, F. I. (1965). *Pinus*: The relationship of seed size and seedling size to inherent vigor. *J. For.* 43, 131–137.

Wakeley, P. C. (1963). Reducing the effects of nursery influences upon provenance tests. *In* "Proc. Southern Forest Tree Improvement Workshop", pp. 28–32. Southern Forest Experiment Station, U.S.A.

Wakeley, P. C. (1969). Results of southern pine planting experiments established in the middle twenties. *J. For.* 67, 237–241.

29

The Problem of Interpreting Inherent Differences in Tree Growth Shortly After Planting

R. D. BURDON and G. B. SWEET
Forest Research Institute, Rotorua, New Zealand.

I.	Introduction	483
II.	Analysis of a Clonal Trial	485
	A. Covariance Adjustment Procedures	486
	B. Extrapolating to Half-sib Progeny Trials	491
III.	Results	493
IV.	Discussion	497
V.	Conclusions	500
VI.	Summary	501
Acknowledgements		502
References		502

I. INTRODUCTION

"Superior" tree genotypes can be identified at an early age only if differences in juvenile seedling growth are strongly associated with differences in eventual productivity, and if genetic factors can be separated from non-genetic ones. The latter can be attempted in three ways:

(1) by decreasing non-genetic influences, by standardizing growing, handling and planting procedures;

(2) by designing trials in which non-genetic influences attributable to "site" can be identified statistically;

(3) by adjusting data to remove non-genetic effects, in particular by using covariance corrections.

Many non-genetic influences have been identified (Perry, Chapter 28). In this contribution we consider the effects of differences in tree size at planting on the subsequent selection of potentially superior genotypes.

The growth of young trees in the nursery is often influenced by seed size and germination rate (Zarger, 1965; Sweet and Wareing, 1966; Terman *et al.*, 1970), or, in the case of cuttings, by their size and initial ability to root. Differences in plant size, created by these effects, which are not necessarily genetically controlled, may subsequently influence and bias the interpretation of experiments designed to study genetic differences in tree growth in the field. Whether or not initial size differences persist depends upon how large they are in relation to genetic differences in subsequent potential rates of growth, always presupposing that the latter and initial size are unrelated (Hatchell *et al.*, 1973; Mullin, 1974). Thus, they may persist for appreciable periods when genetically controlled differences in subsequent growth rates are small. This occurs with members of a local population, individuals within a family or, in an extreme instance, ramets within a clone. The interpretation of observed differences in growth is further complicated by the fact that initial size differences are not always simply related to subsequent rates of growth (Smith and Walters, 1965). Whereas medium-sized seedlings may retain an advantage over small seedlings, large seedlings are sometimes severely checked at planting.

Tree breeders are largely concerned with growth after transplanting, rather than in the nursery, and need a measure of growth which is not affected by the size of the plants at planting. Such a measure should be correlated with later performance. To this end attempts should be made to (a) minimize or isolate unavoidable planting-stock size effects by using appropriate experimental designs and (b) quantify the final "carry-over" of initial differences in plant size, so that the data can be adjusted.

Although seed size and germination behaviour must, to some extent, be affected genetically, albeit through maternal influences, these differences can be largely overcome by such practices as stratifying the seed and grading it by size. Nevertheless, we should not rule out the possibility that genotypic differences that appear within populations of seedlings in the nursery might be relevant after planting. Thus, with clones, some differences in ease of propagation might be genetically related to subsequent potential rates of growth.

29. Interpreting Progeny Differences in the Early Years

TABLE I. Heights (dm), stem basal diameters (mm), and derived statistics for 216 clones of *Pinus radiata* immediately after planting in 1968 (year 0), and after 2 and 4 years' cultivation in Kaingaroa Forest, New Zealand. (After Shelbourne and Thulin, 1974.)

Statistic	Height			Diameter in fourth year
	year 0	year 2	year 4	
Experiment mean ($\bar{X}$)	4.71	15.1	52.1	80.4
Clonal variance (σ^2_C)	0.50	2.97	22.12	79.77
Within-clone variance (σ^2_W)	0.94	9.19	33.40	152.26
Clonal coefficient of variation ($\sigma_C/\bar{X}$)	0.152	0.114	0.090	0.112
Within clone coefficient of variation ($\sigma_W/\bar{X}$)	0.208	0.201	0.111	0.154
Phenotype coefficient of variation $\sqrt{(\sigma^2_C + \sigma^2_W)}/\bar{X}$	0.261	0.231	0.143	0.190

A clonal experiment provided the opportunity to study the magnitude of planting stock size effects (some of which were known to be non-genetic) and the potential implications of such effects.

II. ANALYSIS OF A CLONAL TRIAL

Shelbourne and Thulin (1974) described a trial in which rooted cuttings of 216 selected trees of *Pinus radiata* were planted in Kaingaroa Forest, New Zealand; they were not previously arranged in any replicated or random lay-out in the nursery. The nine planted ramets of each clone were distributed in three randomized blocks; an average of 8.52 survived. Heights were measured immediately after planting in 1968 (H0) and again in 1970 (H2) and 1972 (H4); stem diameters were recorded in 1972 (D4) (Table I). In the event, the minor effects of blocks and of clone x block interactions could be safely ignored. The data have been re-examined with the following objectives: (a) to assess the relationships between heights at planting (H0) and heights 4 years later (H4); (b) to compare different statistical methods of eliminating height associations found in (a);

and (c) explore the potential importance of possible relations between tree "heights at planting" and subsequent heights in half-sib progeny tests. Four approaches were adopted. Analyses were made of (i) actual data; (ii) heights and diameters adjusted by simple covariance on previous heights; (iii) height increments between the second and fourth years after planting (H4 minus H2); and (iv) relative rates of height growth (RHGR). The results of (ii), (iii) and (iv) were compared with one another and with analyses of unadjusted data, giving alternative estimates of real and apparent genetic gains and heritabilities.

A. Covariance Adjustment Procedures

1. Adjusting Heights and Diameters on Heights at Planting

The provisional assumption was made that "heights at planting" was an independent variable with respect to clonal as well as intraclonal variation; but the presence and implications of departures from this assumption are duly covered by later analyses.

Adjustments involved fitting average within-clone regressions on previous heights. Straight-line regressions were used, because there was no reason to believe that supraoptimal-sized ramets were planted. Using an average within-clone regression, it was possible to partition both between- and within-clone variances into components which were, or were not, attributable to initial height effects.

Analyses of variance were made on data with and without covariance adjustment, the adjusted sums of squares being calculated either (i) by direct adjustment of the sums of squares for the sums of squares for the covariate and the sums of cross products (Mather, 1951), or (ii) by adjusting individual values, according to the average within-clone regression, for deviations from the overall covariate mean.

These two methods give slightly different adjusted sums of squares for clones. Although the former is normally preferred for statistical tests, the latter is more relevant considering our practical application of the adjustment.

29. Interpreting Progeny Differences in the Early Years

The analyses of 216 clones, with a mean of 8.52 effective ramets per clone (Snedecor, 1956, 10.16) took the following form:

Source of variation	Degrees of freedom	Expectation of mean squares
Between clones (unadj.)	215	$\sigma^2_W + 8.52\,\sigma^2_C = \widehat{\sigma^2_E} + \widehat{\sigma^2_m} + 8.52\,(\widehat{\sigma^2_G} + \widehat{\sigma^2_M})$
Between clones (adj.)	215	$\widehat{\sigma^2_E} + 8.52\,\widehat{\sigma^2_G}$
Within clones (unadj.)	1624	$\sigma^2_W \quad\quad = \widehat{\sigma^2_E} + \widehat{\sigma^2_m}$
Within clones (adj.)	1623	$\widehat{\sigma^2_E}$

where ^ denotes estimate of superscript parameter

σ^2_C = the variance between clones
σ^2_W = the variance between ramets within clones
σ^2_m = the component of σ^2_W attributable to initial size effects
σ^2_E = the component of σ^2_W attributable to all other non-genetic effects
σ^2_M = the component of σ^2_C attributable to initial size effects
σ^2_G = the component of σ^2_C attributable to genotypic effects, relating to growth after planting.

Note that adjusting with the average within-clone regression removed one degree of freedom from "within clones".

These variances can provide an expected ratio of real genetic gain to apparent genetic gain when selecting from the clonal test using unadjusted measurements. The expectation of real genetic gain (ΔG) given by

$$\Delta G = i\sigma_{\bar{C}} \cdot \frac{\sigma^2_G}{\sigma^2_C} = \frac{i\sigma^2_G}{\sigma_{\bar{C}}} \tag{1}$$

where i = intensity of selection in standard deviations

$\sigma^2_{\bar{C}}$ = variance of unadjusted clonal means
$= \sigma^2_C + (\sigma^2_W)/n$

and   n = effective average number of individuals per clone
and that of apparent genetic gain by

$$\Delta G \text{ (apparent)} = i\sigma_{\bar{C}} \cdot \frac{\sigma^2_C}{\sigma^2_{\bar{C}}} = \frac{i\sigma^2_C}{\sigma_{\bar{C}}} \quad (2)$$

Hence,
$$\frac{\Delta G}{\Delta G \text{ (apparent)}} = \frac{\sigma^2_G}{\sigma^2_C} = \frac{\sigma^2_G}{\sigma^2_G + \sigma^2_M} \quad (3)$$

Whereas true clonal repeatability, or broad-sense heritability (h^2_{bs}), is taken as $\sigma^2_G/(\sigma^2_G + \sigma^2_E)$, the estimate of clonal repeatability, without covariance adjustment, is given by:

$$\widehat{h^2}_{bs} = \frac{\sigma^2_G + \sigma^2_M}{\sigma^2_G + \sigma^2_M + \sigma^2_E + \sigma^2_m} \quad (4)$$

The efficiency of selection (E), or the proportion of the possible genetic gain which would be achieved if the adjustment were not made, is given by

E = ΔG without adjustment/ΔG with adjustment.

$$\Delta G \text{ without adjustment} = i \cdot \frac{\sigma^2_G}{\sqrt{\sigma^2_G + \sigma^2_M + (\sigma^2_E + \sigma^2_m)/n}}$$

$$\Delta G \text{ with adjustment} = i \cdot \frac{\sigma^2_G}{\sqrt{\sigma^2_G + (\sigma^2_E)/n}}$$

Therefore, $E = \sqrt{\dfrac{\sigma^2_G (\sigma^2_E)/n}{\sigma^2_G + \sigma^2_M + (\sigma^2_E + \sigma^2_m)/n}}$ \quad (5)

This calculation assumes that size at planting (M) and genetic value (G) for subsequent growth "capacity" are independent.

There is a second method of estimating the efficiency of gain without adjusting. It has the advantage of not assuming a zero genetic correlation between initial size at planting and subsequent growth "capacity" as measured by adjusted height or diameter. In it, the unadjusted parameter (Y) can be considered as a criterion of indirect selection for the adjusted parameter (Y'). This efficiency

29. Interpreting Progeny Differences in the Early Years

(Falconer, 1961), the relative efficiency of indirect selection (E_I), is given by:

$$E_I = r_G \frac{h^2_{\bar{Y}}}{h^2_{\bar{Y}'}} \quad (6)$$

where $h^2_{\bar{Y}}$ is the repeatability of clonal means for the unadjusted parameter

$$= \frac{\sigma^2_{G_Y} + \sigma^2_{M_Y}}{\sigma^2_{G_Y} + \sigma^2_{M_Y} + (\sigma^2_{E_Y} + \sigma^2_{m_Y})/n}$$

and $h^2_{\bar{Y}'}$ is the repeatability of clonal means for the adjusted parameter

$$= \frac{\sigma^2_{C_{Y'}}}{\sigma^2_{C_{Y'}} + \sigma^2_{W_{Y'}}} = \frac{\sigma^2_{G_Y}}{\sigma^2_{G_Y} + (\sigma^2_{E_Y})/n}$$

and r_G, the genotypic correlation between the two parameters

$$= \frac{\text{Cov}_{C_Y \cdot Y'}}{\sqrt{\sigma^2_{C_{Y'}} \cdot \sigma^2_{C_Y}}} = \frac{\text{Cov}_{C_Y \cdot Y'}}{\sqrt{\sigma^2_{G_Y}(\sigma^2_{G_Y} + \sigma^2_{M_Y})}}$$

$\text{Cov}_{C_Y \cdot Y'}$, being the clonal component of covariance between the two parameters, which will be equal to σ^2_G if there is no genetic correlation between Y' and the covariate.

The equation for E_I can be expanded to give

$$E_I = \frac{\text{Cov}_{C_Y \cdot Y'}}{\sqrt{\sigma^2_{G_Y}(\sigma^2_{G_Y} + \sigma^2_{M_Y})}} \times \sqrt{\frac{\sigma^2_{G_Y} + \sigma^2_{M_Y}}{\sigma^2_{G_Y} + \sigma^2_{M_Y} + (\sigma^2_{E_Y} + \sigma^2_{m_Y})/n}} \times$$

$$\sqrt{\frac{\sigma^2_{G_Y} + (\sigma^2_{E_Y})/n}{\sigma^2_{G_Y}}} \quad (7)$$

Assuming that $\text{Cov}_{C_Y \cdot Y'} = \sigma^2_{G_Y}$ and that Y is the reference parameter, eqn (7) simplifies to eqn (5).

2. Use of Relative Growth Rates

Relative growth rates have been widely used to provide measures of growth which are independent of starting size (Williams, 1946). They can be calculated for height as for dry matter, so were potentially useful in studying the pre-competition phase of this trial. However, preliminary examination showed that relative rates of height growth from planting in year 0 to year 4 (RHGR04), far from being independent, were strongly and negatively correlated with H0 (Fig. 1). The obvious inference was that RHGR declined with increasing plant size, particularly as the relationship clearly applied within clones as well. Hence, RHGR *per se* appears to be unsatisfactory as a means of eliminating initial size effects. However, the strength of the relationship with H0 suggests the possibility of adjusting RHGR for the individual covariances on H0, and using adjusted values to rank the clones.

In this instance heights at year 4 (H4) (or even H2) could be regarded as indirect selection criteria for adjusted RHGR

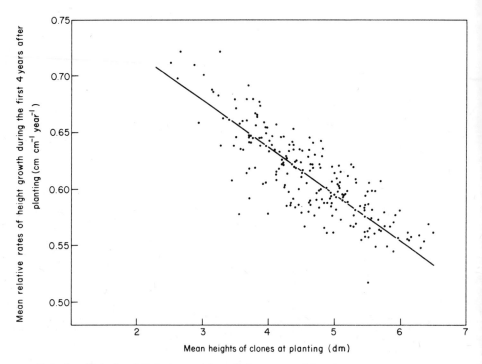

FIG. 1   Relationship between the relative rates of height growth of 216 *Pinus radiata* clones during their first 4 years after planting, and their heights at planting. The data are means of an average of 8.52 ramets per clone in a trial at Kaingaroa, New Zealand.

29. Interpreting Progeny Differences in the Early Years

(RHGRH4/H0). Formula 6 can be adapted to estimate the efficiency of using H4 as the indirect selection criterion by regarding unadjusted H4 and Y and adjusted RHGR as Y'.

3. Use of Height Increments

Increments have sometimes been preferred to cumulative heights as measures of growth capacity (Maugé et al., 1973). Because the increments, H2 minus H0, were virtually independent of planting height ($r^2 = 0.01$ for the average within-clone regression), increments between years 2 and 4 (H4–H2) were tested as indicators of growth capacity. In this instance H4 can be treated as a criterion of indirect selection and eqn 6 can be adapted to estimate the efficiency of using H4 as Y and (H4–H2) as Y'.

B. Extrapolation to Half-sib Progeny Trials

Where covariance adjustments have been used to estimate σ^2_G, σ^2_M and σ^2_m in trials of clones, one can also estimate the potential bias caused by differences in starting size when screening on the basis of seedling progeny tests. We will *assume* that both total genotypic variance and initial size effects are as estimated from the clonal trial, and that planting stock size is not correlated genetically with growth capacity in the field. The example chosen is the half-sib progeny trial which is widely used both for screening parent trees and for estimating heritabilities. Assuming an additivity plus dominance genetic situation, where

$$\sigma^2_G = \sigma^2_A + \sigma^2_D,$$ two cases are considered:

(i) $\sigma^2_D = 0$, hence $\sigma^2_A = \sigma^2_G$

and (ii) $\sigma^2_D = \sigma^2_A = \tfrac{1}{2}\sigma^2_G$ (this is probably a more reasonable assumption than (i) when considering growth rate).

The genetic gain from selecting on the basis of half-sib (hs) families is given by

$$G = \frac{2*i \frac{1}{4}\sigma^2_A}{\sigma_{\overline{hs}}} \qquad (8)$$

In example (i), with $\sigma^2_D = 0$, and where $\sigma^2_{\overline{hs}}$ = variance of half-sib family means

$\sigma^2_{\overline{hs}} = \frac{1}{4}\sigma^2_G + \sigma^2_M + (\frac{3}{4}\sigma^2_G + \sigma^2_E + \sigma^2_m)/n$, not adjusting for planting size, and

$\sigma^2_{\overline{hs}} = \frac{1}{4}\sigma^2_G + (\frac{3}{4}\sigma^2_G + \sigma^2_E)/n$, if an adjustment can be made for planting size.

In example (ii), with $\sigma^2_D = \sigma^2_A = \frac{1}{2}\sigma^2_G$

$\sigma^2_{\overline{hs}} = \frac{1}{8}\sigma^2_G + \sigma^2_M + (\frac{7}{8}\sigma^2_G + \sigma^2_E + \sigma^2_m)/n$, not adjusting for planting size,

and

$\sigma^2_{\overline{hs}} = \frac{1}{8}\sigma^2_G + (\frac{7}{8}\sigma^2_G + \sigma^2_E)/n$, if an adjustment can be made for planting size.

The ratio of real genetic gain to apparent gain, when unadjusted measurements are used, is given by

$$\frac{\Delta G}{\Delta G \text{ (apparent)}} = \frac{\frac{1}{4}\sigma^2_A}{\sigma^2_{\overline{hs}}} \quad (\text{cf. formula (3)})$$

where $\sigma^2_{\overline{hs}} = \frac{1}{4}\sigma^2_A + \sigma^2_M$.

Hence $\dfrac{\Delta G}{\Delta G \text{ (apparent)}} = \dfrac{\frac{1}{4}\sigma^2_G}{\frac{1}{4}\sigma^2_G + \sigma^2_M}$ for example (i) $\qquad (9)$

and $\dfrac{\Delta G}{\Delta G \text{ (apparent)}} = \dfrac{\frac{1}{8}\sigma^2_G}{\frac{1}{8}\sigma^2_G + \sigma^2_M}$ for example (ii) $\qquad (10)$

* The coefficient 2 is inserted because the select parents will be mated with each other instead of with the general population.

Estimates of narrow-sense heritability (h^2), which is equal to $\sigma^2_A/(\sigma^2_G + \sigma^2_E)$, will be as follows:

for example (i)

$$\widehat{h^2} = \frac{\sigma^2_G + 4\sigma^2_M}{\sigma^2_G + \sigma^2_M + \sigma^2_E + \sigma^2_m} \qquad (11)$$

and for example (ii)

$$\widehat{h^2} = \frac{\tfrac{1}{2}\sigma^2_G + 4\sigma^2_M}{\sigma^2_G + \sigma^2_M + \sigma^2_E + \sigma^2_m} \qquad (12)$$

As with the clonal test the expected efficiency (E) of progeny test selection on unadjusted data compared with selection when either an adjustment can be made for initial size, or when initial size effects can be avoided, can be calculated for example (i) as:

$$E = \sqrt{\frac{\tfrac{1}{4}\sigma^2_G + (\tfrac{3}{4}\sigma^2_G + \sigma^2_E)/n}{\tfrac{1}{4}\sigma^2_G + \sigma^2_M + (\tfrac{3}{4}\sigma^2_G + \sigma^2_E + \sigma^2_m)/n}} \qquad (13)$$

and for example (ii) as:

$$E = \sqrt{\frac{\tfrac{1}{8}\sigma^2_G + (\tfrac{7}{8}\sigma^2_G + \sigma^2_E)/n}{\tfrac{1}{8}\sigma^2_G + \sigma^2_M + (\tfrac{7}{8}\sigma^2_G + \sigma^2_E + \sigma^2_m)/n}} \qquad (14)$$

The above illustrates how estimates of planting size effects can be used to predict how they could bias the genetic information from a specified experiment. It is clearly possible to extend the predictions to other types of experiments, and to other types of genetic information, such as estimates of specific combining ability variance.

III. RESULTS

Predictably H4 was better correlated within clones with H2 ($r^2 = 0.55$) than with H0 ($r^2 = 0.12$) (Table II). D4 and RHGR showed appreciable between-clone heterogeneity of regression slopes on previous heights, causing some reservations concerning the covariance adjustment for

TABLE II. Results of covariance analyses on a *Pinus radiata* trial (Table I) adjusting differing growth parameters for previous heights. The adjustments were made using average within-clone regressions. H0, H2 and H4 are heights 0, 2 and 4 years after planting; RHGR04, rate of relative height growth between years 0 and 4; D4, stem basal diameter in year 4.

Dependent variable Covariate	H4 H0	H4 H2	D4 H0	RHGR04 H0	H4−H2 H0	H2 H0	H2−H0 H0
$\widehat{\sigma^2}_G$	15.2 $(14.3)^a$	8.6 $(8.0)^a$	52.2 $(47.8)^a$	0.001 $(0.0004)^a$	9.24^a	1.32 $(1.27)^a$	$1.28^{a,b}$
$\widehat{\sigma^2}_M$	7.0. $(7.8)^a$	13.5 $(14.1)^a$	27.6 $(32.0)^a$	0.0006 $(0.0006)^a$	2.10^a	1.64 $(1.70)^a$	$0.26^{a,b}$
$\widehat{\sigma^2}_E$	29.4	15.1	138.6	0.0008	16.05	7.63	7.32
$\widehat{\sigma^2}_m$	4.0	18.3	13.6	0.0019	0.57	1.56	0.08
$\widehat{\sigma^2}_G/\widehat{\sigma^2}_C$	0.68	0.39	0.65	0.39	0.81	0.45	0.83
Regression heterogeneity $F_{215, 1408}$ d.f.	1.16	1.19	1.37	1.90	1.08	1.14	1.13
Probability of homogeneity	0.07	0.04	0.001	3×10^{-9}	0.22	0.10	0.11
r^2, average within-clone regression	0.12	0.55	0.09	0.70	0.04	0.17	0.01
Average within-clone regression slope ±S.E.	2.1 ± 0.14	1.4 ± 0.03	3.8 ± 0.30	−0.04 ± 0.007	0.8 ± 0.10	1.3 ± 0.07	0.3 ± 0.07
Overall regression slope ±S.E.	2.9 ± 0.13	1.6 ± 0.03	5.6 ± 0.26	−0.04 ± 0.007	1.3 ± 0.10	1.5 ± 0.06	0.5 ± 0.06

[a] Estimates obtained by means of direct adjustment of between-clone sums of squares.
[b] An approximation involved.

these traits. Interestingly, (H4−H2) proved to be more strongly related, within clones, to H0 than was (H2−H0), although the relationship was still weak.

The salient finding here concerns the ratios of $\hat{\sigma}^2_G$ to $\hat{\sigma}^2_C$, which reflect the magnitude σ^2_M relative to σ^2_G. In almost all cases $\hat{\sigma}^2_G$ was substantially less than $\hat{\sigma}^2_C$, and for crude heights and diameters σ^2_M was generally 50% or more of $\hat{\sigma}^2_G$. Estimates of σ^2_G and thence of σ^2_M did differ slightly, depending on the method of calculating the adjusted clonal sums of squares, the adjustment of individual values giving slightly higher estimates of σ^2_G. The overall regressions were consistently steeper (or less strongly negative with RHGR) than the within-clone regressions.

Table III summarizes the ways of estimating some of the (a) clonal correlations between traits, adjusted and unadjusted, and (b) clonal repeatabilities and the repeatabilities of clonal means. Usually, actual heights and diameters at 4 years were closely correlated with the more refined criteria of growth, such as growth increments, despite the large size of σ^2_M. This is clearly related to the fact that H0 was correlated with all the other measures of growth, including those which had been adjusted for H0. From approximate standard errors of clonal correlations (Falconer, 1961, p. 318) it was clear that all the correlations were significant ($P < 0.001$) and that sampling errors were small. Clone diameters were not perfectly correlated with heights, so both must be considered in conjunction to give a true measure of growth capacity. This feature apart, the differing criteria were very closely correlated with each other, although adjusted relative growth rates showed slightly weaker correlations.

In general, clonal repeatabilities were not greatly altered by the adjustments.

The potential importance of initial size effects on estimates of the amount of genetic gain, on heritability estimates, and on the efficiency of selection is illustrated in Table IV. Where heights and diameters were adjusted and where estimates were thereby obtained of σ^2_G, σ^2_M, σ^2_E and σ^2_m, the predictions could be extended to progeny tests. Whereas, in the clonal trial, with unadjusted measurements, the real gain was only 68% of the apparent gain in height growth during the first 4 years after planting, in a progeny test (assuming $\sigma^2_D = \sigma^2_A = \frac{1}{2}\sigma^2_G$) the same initial size effects could give apparent gains five times the real gains with the same trait.

TABLE III. Clonal correlations in a trial with *Pinus radiata* (Table I) between pairs of parameters, together with clonal variances, clonal repeatabilities and repeatabilities of clonal means for individual parameters. Italicized data are variances. H0, H2 and H4, heights 0, 2 and 4 years after planting; RHGR04, rate of relative height growth between years 0 and 4; D4, stem basal diameter in year 4.

	H0	H2	H4	D4	H4/H0	H4–H2	RHGR04/H0	D4/H0	H2/H0	H4/H2
H0	*0.54*									
H2	0.78	*2.97*								
H4	0.65	0.85	*22.12*							
D4	0.71	0.87	0.86	*79.77*						
H4/H0	0.39	0.72	0.95	0.77	*15.15*					
H4–H2	0.51	0.67	0.96	0.76	0.97	*11.34*				
RHGR04/H0	0.32	0.67	0.91	0.74	–	0.93	*0.00038*			
D4/H0	0.48	0.77	0.82	0.96	0.79	0.79	–	*52.22*		
H2/H0	0.34	0.86	0.74	0.72	–	–	–	–	*1.32*	
H4/H2	0.39	0.53	0.90	0.66	–	0.98	–	–	–	*8.63*
Clonal repeatability $\widehat{\sigma}^2_C/(\widehat{\sigma}^2_C + \widehat{\sigma}^2_W)$ [a]	0.36	0.24	0.40	0.34	0.34	0.41	0.61	0.27	0.15	0.36
Repeatability of clonal means $\widehat{\sigma}^2_C/(\widehat{\sigma}^2_C + (\widehat{\sigma}^2_W)/n)$	0.83	0.73	0.85	0.82	0.81	0.85	0.93	0.75	0.60	0.72

N.B. H4/H0 = H4 adjusted for H0.
RHGR04/H0 = RHGR between planting and year 4 adjusted for H0.

[a] σ^2_C for an adjusted parameter is equal to σ^2_G.

Taking growth during the first 2 years, the exaggeration of gain can be even greater.

Estimates of heritability can also be much inflated, though slightly less so than estimates of genetic gain. Least affected is the expected efficiency of selection, but this can still drop to 50% in a half-sib progeny test assessed 4 years after planting, compared with what could be achieved if an appropriate adjustment was possible.

Where correlated responses were used to calculate selection efficiencies instead of "parameter substitution" the loss of efficiency by using actual measurements was small or negligible, excepting comparisons with adjusted RHGR, although the loss there was still less than 20%. When the same efficiencies were calculated using eqn (5), lower values were obtained. The magnitude of this discrepancy evidently reflected the importance of the clonal correlations between H0 and "growth capacity".

IV. DISCUSSION

None of the adjusted or derived measures of growth in the field enabled young *P. radiata* clones to be evaluated any more reliably than the original measurements of heights and diameters. In general they gave more conservative estimates of genetic gains, because size at planting, while strongly influencing size 2–4 years later, was also, in large measure, an expression of clonal "vigour". This relation was reflected in the close correlations between adjusted measurements and the covariate H0 (height at planting), and in the general tendency for the overall regressions on H0 to be steeper than those within clones (e.g. 2.87 ± 0.13, in contrast to 2.07 ± 0.14 when relating H4 and H0, Table II).

Reassuringly, the various refinements all gave efficiencies similar to those for selection on the basis of crude measurements. Of the refinements the simple covariance adjustment of H or D was probably as good as any. Although the figures for efficiency of selection suggested that adjusted RHGR might be the best measure, this was the parameter which showed the most disturbing heterogeneity of within-clone regressions. Moreover, adjusted RHGR is relatively difficult to obtain, even with computer facilities. Height increment (H4–H2), although easily obtained, had the disadvantage of retaining

TABLE IV. Estimated ratios of (i) real to apparent gain, (ii) real to apparent heritabilities and (iii) relative efficiencies of selection, when unadjusted and adjusted measurements were compared in clonal and half-sib progeny trials of *Pinus radiata*.

Criteria of growth	Method of calculating relative efficiency		Type of trial	Genetic model	Real gain/ apparent gain (%)	Real heritability/ apparent heritability (%)	Relative efficiency (%) of selection using unadjusted H or D
H4 adjusted for H0 (compared with unadjusted H4)	Correlated response	(6)[b]	Clonal trial	Any	68	85	96
	Parameter substitution	(5)	Clonal trial	Any			84
	Parameter substitution	(11)	½-sib progenies	$\sigma^2_D = 0$	35	44	63
	Parameter substitution	(12)	½-sib progenies	$\sigma^2_D = \sigma^2_A = \frac{1}{2}\sigma^2_G$	21	27	50
H4 minus H2 (compared with unadjusted H4)	Correlated response	(6)	Clonal trial	Any	—	—	96
RHGR years 0–4 adjusted for H0 (compared with unadjusted H4)	Correlated response	(6)	Clonal trial	Any	—	—	83
D4 adjusted for H0 (compared with unadjusted D4)	Correlated response	(6)	Clonal trial	Any	65	80	105
	Parameter substitution	(5)	Clonal trial	Any			84
	Parameter substitution	(11)	½-sib progenies	$\sigma^2_D = 0$	32	39	61
	Parameter substitution	(12)	½-sib progenies	$\sigma^2_D = \sigma^2_A = \frac{1}{2}\sigma^2_G$	19	23	48
H2 adjusted for H0 (compared with unadjusted H2)	Correlated response	(6)	Clonal trial	Any	45	61	97
	Parameter substitution	(5)	Clonal trial	Any			77
	Parameter substitution	(11)	½-sib progenies	$\sigma^2_D = 0$	17	23	48
	Parameter substitution	(12)	½-sib progenies	$\sigma^2_D = \sigma^2_A = \frac{1}{2}\sigma^2_G$	9	14	36
H4 adjusted for H2[a]	Correlated response	(6)	Clonal trial	Any	39	91[a]	90

Assumed values for n are 8.5 for clonal trial
50 for half-sib progeny test ($\sigma^2_D = 0$)
100 for half-sib progeny test ($\sigma^2_D = \sigma^2_A = \frac{1}{2}\sigma^2_G$)

[a] Parameter substitution not done because it could not reasonably be assumed that H2 was uncorrelated genetically with adjusted H4.
[b] Numbers in brackets refer to formulae in the text.

29. Interpreting Progeny Differences in the Early Years

some within-clone correlation with planting height. Other refinements could undoubtedly be devised, but they seem unlikely to add more information.

Extending the results from clonal trials to progeny tests involved various assumptions. An important question was whether the residual clonal differences (reflected in $\sigma^2{}_G$) represented true genotypic effects, or were inflated by propagation effects such as differences in the physiological age of the cuttings or interactions between maturation state and the genotype (Burdon and Shelbourne, 1974). Propagation effects were suspected, particularly as weak correlations were found for clonal performance between two cycles of repropagation in this trial (Shelbourne and Thulin, 1974). This being so, the progeny test information could potentially be even more distorted than indicated in Table IV. On the other hand, it is unlikely that growth capacity in the field and size at planting are genetically unrelated. Nevertheless, this assumption seems more plausible for seedlings than clones, because so much of the variation in the size of seedling transplants can be accounted for by differences in seed weights and germination behaviour.

Planting size effects pose two problems for progeny tests. First, they are potentially more important compared with clonal tests, leading to a 4—5-fold overestimation of genetic gain. This could lead to considerable inefficiency in identifying parents with superior capacities for growth in the field, and mean that "growth rate" is given too much weight compared with other selection traits which are not so much affected by starting size. This may be particularly true if early differences in tree size are increased after the trees begin to compete with each other.

Secondly, in progeny tests it is difficult to eliminate non-genetic effects by covariance adjustment without also rejecting valid genetic information. This is so because, within progenies, the non-genetic covariance on planting size is confounded with much of the genetic covariance (in fact, three quarters of the additive and almost all the non-additive genetic covariance in the case of half-sib families). If the adjustment, on the basis of the average within-progeny regression, merely damps down progeny differences, one cannot ascertain from within the trial itself to what extent the within-progeny regression reflects genetic covariance. On the other hand one could be confident that the regression is essentially non-genetic, if, despite a substantial adjustment, there remained pronounced progeny differences which were uncorrelated with initial size.

Failing a reliable covariance adjustment it may be possible to use height increment more satisfactorily than was possible in the clonal trial. If extraneous planting size effects are likely to persist until competition begins, it may be desirable to shift the emphasis from diameter to height to circumvent any reinforcement of planting size effects by competition.

Overall, there is a strong argument for taking all reasonable measures to avoid maternal effects and non-genetic variation in planting stock size. But because such variation can never be eliminated, it is prudent to estimate sizes soon after planting in the hope of obtaining a valid adjustment if later required — this procedure is now standard practice at the Forest Research Institute, New Zealand. However, although height immediately after planting is convenient to measure, it is an imprecise measure of tree size, particularly as planting depth is variable. Estimates of stem diameter would be a useful complement. Heights measured after one year in the field could minimize "noise" related to planting check, but this procedure may overlook an important genetical character — the ability to establish well.

V. CONCLUSIONS

Early growth assessments made on trials of clones and/or progenies should be critically analysed so as to separate genetic from non-genetic effects. This needs to be done before trying to extrapolate to older stands. Difficulties need not always arise, but the following questions should always be asked. To what extent can differences between clones or progenies be attributable to size differences at planting? To what extent are size differences at planting genetically controlled? To what extent are size differences at planting indicators of meaningful differences in potential rates of growth in the forest? How important are genetically controlled differences in nursery performance in determining the overall merit of a cultivar or progeny? The interpretation and even the validity of early assessments can depend greatly on the answers to these questions.

VI. SUMMARY

When testing tree progenies and clones in field conditions it is impossible, because of genetic and non-genetic factors, to start with planting stock of uniform size. How do these initial differences affect the interpretation of subsequent growth?

Estimates of height, stem diameter and relative height increments, obtained 4 years after planting clones of *Pinus radiata*, were "adjusted" using average within-clone regressions for covariance on "height at planting". It emerged that neither the "adjusted" values nor absolute height increments were materially more reliable measures of early clonal differences in growth than were the "unadjusted" heights and diameters. This was so because differences in the performance of the clones before planting apparently foreshadowed comparable differences in subsequent rates of growth.

In the case of seedlings, much of the variation in transplant size can often be attributed to seed size and germination behaviour, rather than differences in "vigour". Assuming that "heights at planting" were unrelated to subsequent growth "capacity", it was shown that initial size differences, of the same order as occurred in the clonal trial, could severely bias genetic information from half-sib progeny tests. Failure to "adjust" could halve the screening efficiency of progeny tests and exaggerate genetic gains and heritabilities four or five times. These exaggerated values would mean that some measures of performance used in multi-trait selection would be given too much weight, but could be of value if differences in nursery growth were genetically controlled and important in establishment. Clearly, screening efficiencies would be greatest if performance in the nursery and forest were closely related.

Although the genetic information from seedling progenies appears to be especially subject to bias from size differences at planting, there is no certainty that valid covariance adjustments can be made for these differences, because the within-progeny covariance represents a confounding of non-genetic effects with any genetic ones. Thus size differences at planting should be minimized, by good replication in the nursery, by seed stratification to give uniform germination and by any practicable control over seed weights. Thereby the value of early growth measurements would be much enhanced.

ACKNOWLEDGEMENTS

Special thanks are due to Mr R. H. M. C. Scott for his work in nursing the data through the computer. We also thank Dr C. J. A. Shelbourne and Mr I. J. Thulin for making the data available, and to Dr H. A. I. Madgwick, Dr C. J. A. Shelbourne and Dr M. D. Wilcox for reviewing the draft.

REFERENCES

Burdon, R. D. and Shelbourne, C. J. A. (1974). The use of vegetative propagules for obtaining genetic information. *N. Z. Jl For. Sci.* 4, 418–425.

Falconer, D. S. (1961). "An Introduction to Quantitative Genetics". Oliver and Boyd, London.

Hatchell, G. E., Dorman, K. W. and Langdon, O. G. (1973). Performance of loblolly and slash pine nursery selections. *Forest Sci.* 18, 308–313.

Mather, K. (1951). "Statistical Analysis in Biology" (4th edition). Methuen, London.

Maugé, J. P., Castaing, J. Ph., Arbez, M. and Baradat, Ph. (1973). Première éclaircie génétique sur le verger à graine de Sore. AFOCEL Rapport Annuel (1973), pp. 303–336.

Mullin, R. E. (1974). Some planting effects still significant after 20 years. *For. Chron.* 50, 191–193.

Shelbourne, C. J. A. and Thulin, I. J. (1974). Early results from a clonal selection and testing programme with radiata pine. *N.Z. Jl For. Sci.* 4, 387–398.

Smith, J. H. G. and Walters, J. (1965). Influence of seedling size on growth, survival and cost of growing Douglas fir. University of British Columbia, Faculty of Forestry, Res. Note No. 50.

Snedecor, G. W. (1956). "Statistical Methods" (5th edition). Iowa State University Press, Ames.

Sweet, G. B. and Wareing, P. F. (1966). The relative growth rates of large and small seedlings in forest tree species. Supplement to *Forestry*, pp. 110–117.

Terman, G. L., Bengtson, G. W. and Allen, S. E. (1970). Greenhouse pot experiments with pine seedlings: methods for reducing variability. *Forest Sci.* 16, 433–441.

Williams, R. F. (1946). The physiology of plant growth with special reference to the concept of net assimilation rate. *Ann. Bot.* 10, 41–72.

Zarger, T. G. (1965). Performance of loblolly, shortleaf and eastern white pine superseedlings. *Silvae Genet.* 14, 177–208.

30
Variation in Morphology, Phenology and Nutrient Content among *Picea abies* Clones and Provenances, and its Implications for Tree Improvement

J. KLEINSCHMIT and A. SAUER
Lower Saxony Forest Research Institute, Escherode, West Germany

I.	Introduction	503
II.	Plant Material and Methods	504
III.	Morphological Characters	505
IV.	Phenological Characters	508
V.	Needle Nutrient Contents	509
VI.	Estimated Genetic Gain by Selection	511
VII.	Discussion	512
	A. Plant Material and Methods	512
	B. Tree Improvement	515
VIII.	Summary	516
	Acknowledgement	517
	References	517

I. INTRODUCTION

Much is known about variation at the provenance level, but comparatively little is known about variation among individuals within provenances. This contribution reports differences among 50 clones derived from each of 10 different provenances (500 clones in all). Fifteen characters, which may be particularly important when breeding *Picea abies* (Norway spruce) have been chosen for analysis in this preliminary report. A major aim of the project, in which 34 characters are being measured, is to analyse the structure of natural populations in order to evaluate the likely optimal composition of artificial varieties, and indicate how they may be obtained by selection.

II. PLANT MATERIAL AND METHODS

Clones were propagated from 50 4-year-old trees belonging to each of 10 German provenances of *Picea abies*. Three rooted cuttings of each of the 500 clones were planted in a nursery trial at Escherode in 1971. Plant heights were measured after bud-set in 1972 and 1973; root collar diameters were taken 1 cm above ground level in 1973; at the same time the lengths and angles were measured of 2 branches on the 1972 whorl, and the lengths and widths were taken of 30 1-year-old needles per plant.

Flushing times were recorded every week following the scheme of Volkert and Schelle (1966). Lammas shoot formation was recorded before bud-set in September according to the following classification: 1, no lammas growth; 2, lammas growth only on branches; 3, lammas growth only on leaders; 4, lammas growth on branches and leaders. Bud-set was judged weekly according to the colours of leaders and buds as follows:

Leader	Bud
1 Green	1 green
2 up to 50% brownish	2 top brownish
3 75% brownish	3 total bud light brownish
4 100% brownish	4 total bud brownish
5 leader as brown as the trunk	5 total bud brown

It was supposed that growth had ceased when a clone was scored as 5/5. Thus a morphological indicator was used for physiological dormancy which, we realize, may not be entirely similar (Vegis, 1955; Roberts, 1969; Bhella and Roberts, 1974).

The nutrient contents of needles on branches 15 cm long, removed from the two upper whorls, were determined at the Institute for Soil Science of the Forestry Dept., University of Göttingen. For P, K, Ca and Mg, needles were ashed at 500–600°C, and dissolved in HCl; analyses for K, Ca and Mg were carried out using a Perkin Elmer

Spectrometer; P-analyses were done by the molybdenum-blue method, and N by selenium-sulphuric-acid combustion.

One and two way analyses of variance were done. The latter separated variances between provenances (9 degrees of freedom), between clones (490 d.f.) and within clones (= residual, 1000 d.f.).

III. MORPHOLOGICAL CHARACTERS

The ranges of variation for the different characters, among clone and provenance means, are presented as frequency distributions (e.g. Figs 1 and 2). Highly significant ($P = 0.01$) differences occurred among clones and provenances for all morphological characters.

The mean *heights* of clones ranged from 11 to 45 cm in 1972, and from 20 to 67 cm in 1973 (Fig. 1). These ranges were three times larger than those of provenance means which were 21–30 cm in 1972 and 38–49 cm in 1973. The ranking for heights were similar in both years and were significantly correlated with each other ($r = 0.84**$).† Clearly, the possibilities for genetic gain are considerably greater when selecting clones rather than provenances.

The frequency distributions of *root collar diameters* were similar to those of heights (Fig. 1). Clonal differences accounted for most of the genetic variance (Fig. 2). Extreme clones differed by 7.9 mm in diameter, whereas extreme provenances differed by only 1.9 mm. Root collar diameters were closely correlated with heights, taking single trees ($r = 0.74***$ in 1973), clone means ($r = 0.77***$) and provenance means ($r = 0.87***$). The ratios of height to diameter differed among provenances.

Clonal differences in *branch lengths* accounted for 47% of the total variation, whereas provenance differences accounted for 17% (Figs 1 and 2). Extreme clones had branches differing in length by 11.3 cm, whereas extreme provenances differed by only 3.2 cm. Branch lengths, tree heights and root collar diameters were all significantly correlated with each other ($P > 0.05$), particularly at the clone and provenance levels.

Branch angles were almost an order of magnitude more variable among clones than among provenances (Fig. 1). The maximum difference between provenance means was about 10°, and between clone means about 80°. Narrow branch angles were less frequent

† $*P = 0.05$ $**P = 0.01$ $***P = 0.001$

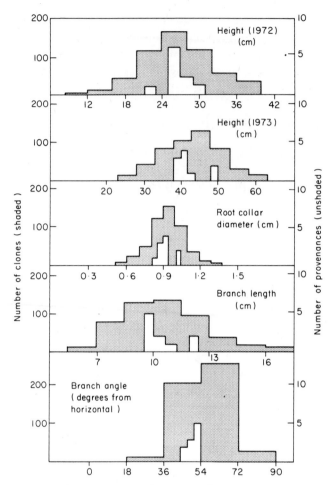

FIG. 1 Frequency distributions of mean heights, root collar diameters, branch lengths and angles for 500 clones (shaded) derived from 10 provenances (unshaded) of 2–3-year-old *Picea abies*. Each clone value is the means of 3 plants; each provenance value the means of 150 plants.

than angles near 90° (Fig. 1). Branch angles were weakly correlated with other characters, although significant correlations occurred at the provenance level (with branch length, $r = -0.69*$; needle length, $r = 0.85**$, and needle width, $r = -0.76*$) indicating that there was a correlated response to natural selection at the provenance level.

Needle lengths varied very little among provenances (1.2 mm, Fig. 3), and provenance variation accounted for only 3.5% of the

30. Variation among *Picea* Clones and Provenances

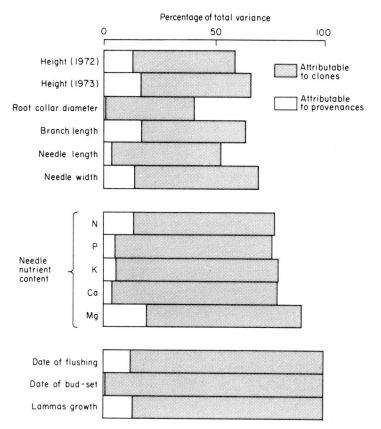

FIG. 2 Percentage of total variance attributable to inherent differences between provenances and clones of *Picea abies* within a nursery trial with 3 replicate trees × 10 provenances × 50 clones (= 1500 vegetatively propagated trees).

total variation (Fig. 2). Clonal variation, on the other hand, accounted for 50% of total variation, and needles on extreme clones differed in length by 8.6 mm. At the provenance level, needle lengths were negatively correlated with needle widths ($r = -0.80^{**}$), whereas at the clone level the correlation was positive ($r = 0.26^{***}$).

Provenances differed more in *needle width* than in *needle length* (Figs 2 and 3), and again, correlations with other characters were closer at the provenance than the clone level. In general, provenances with broad, short needles tended to grow fast and have long branches at narrow angles. However, this was not always so among clones.

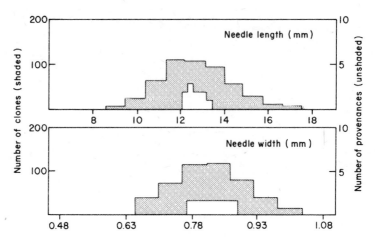

FIG. 3 Frequency distributions of mean needle lengths and widths for 500 clones (shaded) derived from 10 provenances (unshaded) of *Picea abies*. (See legend to Fig. 1.)

IV. PHENOLOGICAL CHARACTERS

Differences in flushing time, bud-set and lammas formation were so small between single plants within clones that they could not be judged by the methods used, and so residual variances were zero (Fig. 2). By far the greatest part of the total variation was due to differences between clones; provenance differences were significant, except for bud-set, but were much smaller (Fig. 4). Thus, clones flushed over a period of 28 days and set buds over 70 days, whereas provenances both flushed and set buds within less than 2 weeks.

The times of flushing and bud-set for all clones within one early, one middle and one late flushing provenance are shown in Fig. 5. Most clones flushed on one of five occasions in April–May in response to high temperatures. The times of bud-set were more variable, and bore no relation to the times of flushing. Thus, there were clones which flushed early and set buds early, and others which flushed early and set buds late, and so on (Fig. 5).

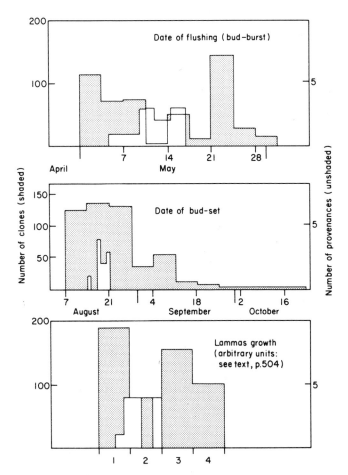

FIG. 4 Frequency distributions of phenological data recorded on 500 clones (shaded) derived from 10 provenances (unshaded) of *Picea abies*. (See legend to Fig. 1.)

V. NEEDLE NUTRIENT CONTENTS

Significant differences occurred among provenances and clones for all five nutrients studied. However, differences between provenance means were small (Fig. 6), and explained between 4% (Ca) and 20% (Mg) of the total variation (Fig. 2), whereas clonal differences

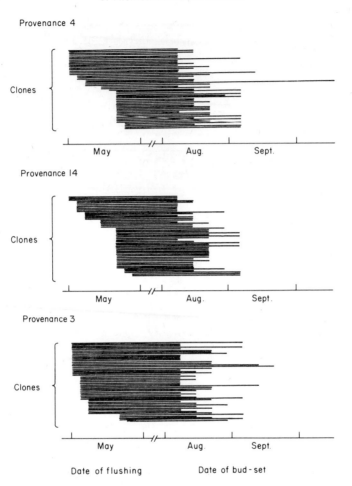

FIG. 5 Dates of flushing (bud-burst) and bud-set on 3-year-old trees of 50 clones taken from three provenances of *Picea abies*. The mean dates of flushing for provenances 3, 4 and 14 were early, intermediate and late, respectively, among 10 provenances studied.

accounted for from 63% (N) to over 70% (Mg). The residual variances were less for these chemical characters than for morphological characters, showing that nutrient contents were under strong genetic control. If clones with the largest concentrations of nutrients grow fastest, this finding suggests that clones could be selected for particular sites.

30. Variation among *Picea* Clones and Provenances

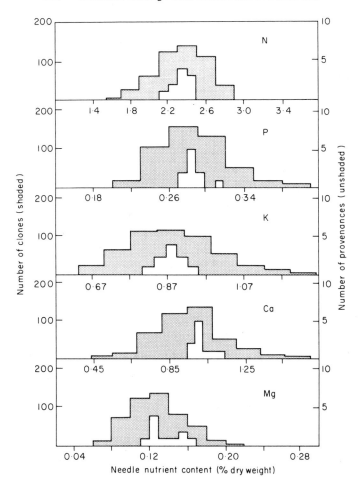

FIG. 6 Frequency distribution of needle nutrient contents determined on 500 clones (shaded) derived from 10 provenances (unshaded) of *Picea abies*. (See legend to Fig. 1.)

VI. ESTIMATED GENETIC GAIN BY SELECTION

Clearly, if clone or provenance differences, which are genetic, represent large proportions of the total variance, the implication is that large genetic gains could be made by selection. The *absolute* values of genetic variances and estimated genetic gains, which can be calculated from this study, are of limited value, as the data were collected from young trees at only one site. However, the *relative* differences are potentially important, because they suggest that selecting the best clones will give greater gains than selecting the best provenances, and also that selection for phenological characters will be more effective than selection for nutrient content, and selection for morphological characters will be least effective (Fig. 2). However, such calculations have limited meaning, because each character is of different economic importance, and certain characters are correlated with each other.

Supposing we were to select the tallest provenance or the 10% tallest clones in 1973. We may then examine the change that would have occurred in the other, unselected characters within the selected populations (Figs 7 and 8). The provenance with the greatest mean height would have the greatest root collar diameter and branch length, as these characters were closely correlated with each other (Fig. 7). More unexpectedly, the selected provenance would be very early flushing, and have high needle Ca and Mg contents. It would also have relatively short, narrow needles with low P and K contents, and be likely to produce lammas growth (Fig. 7).

The consequence of selecting the tallest 10% of all 500 clones, irrespective of provenance, would be quite different (Fig. 8). Using χ^2 to compare original frequency distributions, before selection, with those produced after selection, it was shown that the significant effects of selection would be (a) to increase root collar diameters, branch lengths and needle widths, and (b) to select populations which both flushed early and produced lammas growth. Unlike the tallest provenance, the tallest clones would have relatively long, wide needles, and their nutrient contents would be no different from those of the remaining clones.

VII. DISCUSSION

A. Plant Material and Methods

It is important to realize that only 3-year-old *Picea abies* trees were studied at one site, and each clone was represented by only three replicate plants, which may have been too few to cover the total site

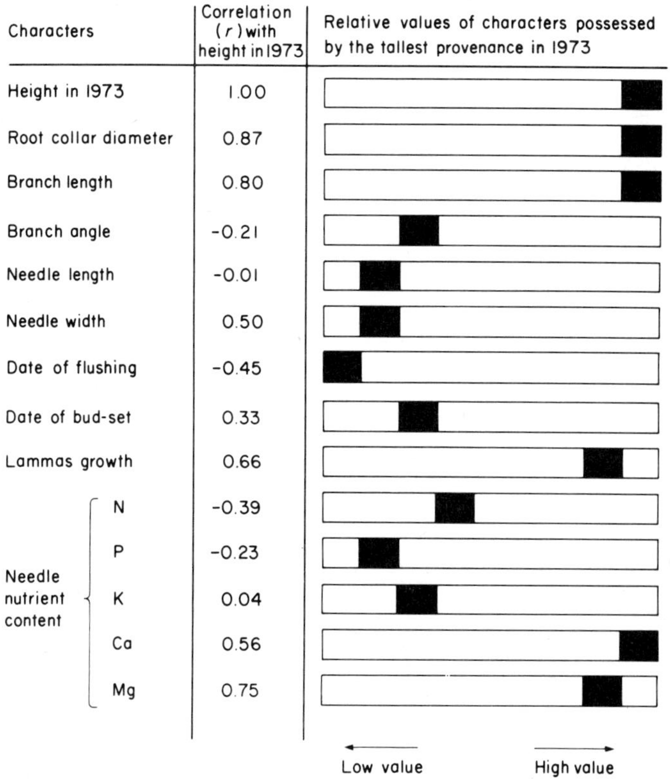

30. Variation among *Picea* Clones and Provenances

Characters		Correlation (r) with height in 1973	Relative values of characters possessed by the tallest provenance in 1973
Height in 1973		1.00	
Root collar diameter		0.87	
Branch length		0.80	
Branch angle		-0.21	
Needle length		-0.01	
Needle width		0.50	
Date of flushing		-0.45	
Date of bud-set		0.33	
Lammas growth		0.66	
Needle nutrient content	N	-0.39	
	P	-0.23	
	K	0.04	
	Ca	0.56	
	Mg	0.75	

Low value ← → High value

FIG. 7 Mean attributes (shaded) of the tallest of 10 provenances of *Picea abies*. Values for unselected provenances are unshaded.

variation. Also, we have limited information on juvenile—mature correlations for the morphological and phenological characters studied and no information on the nutritional characters.

The large variation among clones, as compared to provenances, may have been overestimated for two reasons. First, considerable genotype—site interaction may occur which other field trials suggested would be greater for clones than for provenances, depending on the structure of the populations. Secondly, clone variation was based on the spread of 50 mean values, whereas provenance variation was based on only 10 values. Viewed another way, the genetic components of variance for provenances will have been more exact than for clones because there were so many more values used to estimate provenance means.

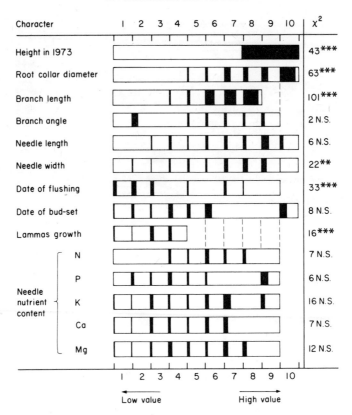

FIG. 8 Frequency distributions with 4–10 equal size classes for 500 clones of *Picea abies* (horizontal bands) showing the classes occupied by the 50 tallest clones (shaded). χ^2 values indicate to what extent the clone frequency distributions for each character were altered by selection.

Clone values may also have been affected by C-effects, that is, effects of the environment in the stock plants. Thus, plant heights in 1972 may have been affected by differences in the nutritional status of stock plants and cuttings, and by differences in rates of rooting. That this was so was suggested by the imperfect correlation between plant heights in 1972 and 1973 ($r = 0.82$). Older provenance and clone trials show that the variation in morphological and phenological characters remains as great as it was in this young material, but there may be changes in rank with time, depending on the site.

B. Tree Improvement

The most important finding was the great variation among clones within provenances, exceeding by far the variation among provenances. This intra-provenance variation enables populations to react in an optimal way to variation in climatic and site conditions, because those genotypes which are best adapted will always grow best. Although clone variances were overestimated, the estimates of genotypic values were better than they would have been without vegetative propagation and replication. It is clear that good prospects exist for genetic gain by clone selection, surpassing those obtained by selecting the best provenances. Clearly, the greatest gain would be achieved by selecting the best clones within the best provenances.

In applied breeding programmes it is important to know how far selection for a certain character influences the frequency distributions of other characters, and hence the population structure; any change could affect the stability or resistance of the breeding population, and have effects on later selection. This study showed that clone selection for height significantly influenced only root collar diameters, branch lengths, needle widths, flushing times and lammas shoot formation (Fig. 8). Except for root collar diameter, the frequency distribution patterns of the characters in the selected population were not changed to such an extent that certain classes were completely excluded. Thus, selecting the tallest clones did not disturb the frequency distributions of other characters as much as might be expected.

Genetic linkage between independent quantitative characters was not expected and did not occur. An increase in root collar diameter with increase in height is as expected. It may be more problematical that the branches on trees in the selected populations were longer than those in the unselected ones.

It was unexpected to find that clone selection changed the frequency distribution for flushing time in favour of early flushing, but not the distribution for bud-set. One might have expected that the larger variation in time of bud-set would have been more closely associated with the variation in height growth than the small variation in flushing time, as has been the case in other studies (Sauer *et al.*, 1973; Ekberg *et al.*, Chapter 11). The climatic conditions in

1972 and 1973 at Escherode, especially autumn temperatures, might account for this odd result. If it is true that early flushing is associated with rapid growth, some growth potential may have to be sacrificed in practical selection programmes when one is forced to select late-flushing genotypes in order to avoid spring frost damage. The alternative is to increase the selection base, and hence the cost, in order to select late-flushing types with increased vigour.

It was important that clone selection did not affect needle nutrient contents, because this can affect resistance to insect attack (Lunderstädt and Claus, 1972; Lunderstädt, 1973).

This study demonstrated the importance of avoiding intensive clone selection early on in a selection programme, in order to maintain genetic variation for later selection and diminish attendant risks (Namkoong, 1974). Clone propagation, combined with a breeding programme, enables the breeding population to be kept separate from the production population, which may be less variable and higher yielding. However, one must stress the need to maintain genetic variation within the production population too, in order to diminish production risks, which may arise for the following reasons.

(1) Selection has to be done before the final harvest (unlike field crops), so that juvenile—mature correlations will be less than unity.

(2) Growing sites vary in time and space, so that genotype x site and genotype x year interactions are greater than zero.

(3) Economic and, consequently, breeding aims will change as time goes on, and this will cause greatest problems when the rotation period is long.

In our view, it would be prudent for breeders to adopt, in part, the strategy of nature; that is, to maintain heterogeneous breeding populations, which are buffered against changes in the environment, excluding only the "negative" parts of populations. It may be less prudent to select only the few best clones. Variable site conditions normally encountered in forestry can only be utilized to the full when there is considerable genetic variation.

VIII. SUMMARY

Three-year-old trees of 50 clones taken from each of 10 provenances of *Picea abies* were studied in a replicated nursery trial. Measurements were taken of six morphological characters (height, root collar diameter, branch length and angle, needle length and width), three phenological characters (times of flushing and bud-set, and lammas growth), and the concentration of five nutrients in needles (N, P, K, Ca and Mg). The range of clone means was several times larger than the range of provenance means for all characters, and the possibilities for genetic gain were in the order phenological traits > needle nutrient contents > morphological characters. Clone selection for height increased root collar diameters, branch lengths, needle widths, and favoured early flushing and lammas formation; other characters were unaffected. The consequences for breeding programmes are discussed, stressing the need to maintain genetic variation.

ACKNOWLEDGEMENT

The project reported here is being supported by funds from the Deutsche Forschungsgemeinschaft.

REFERENCES

Bhella, H. S. and Roberts, A. N. (1974). "Bud and Cambial Activity in Douglas Fir as Related to Stem Cutting Rootability". Department of Horticulture, Oregon State University, Corvallis (Mimeo).

Lunderstädt, J. (1973). Zur Nahrungsqualität von Fichtennadeln für forstliche Schadinsekten. 3. *Z. angew. Ent.* 73, 340–351.

Lunderstädt, J. and Claus, G. (1972). Zur Nahrungsqualität von Fichtennadeln für forstliche Schadinsekten. 2. *Z. angew. Ent.* 70, 386–403.

Namkoong, G. (1974). Breeding for future generations. *Proc. IUFRO Meeting S. 02.04.1–3, Stockholm*, pp. 29–39.

Roberts, A. N. (1969). Timing in cutting propagation as related to developmental physiology. *Proc. Pl. Propag. Soc.* 1969, 1–10.

Sauer, A., Kleinschmit, J. and Lunderstädt, J. (1973). Charakterisierung von Fichten-Klonen (*Picea abies* Karst.) mit Hilfe morphologischer, physiologischer und biochemischer Methoden. *Silvae Genet.* 22, 173–182.

Vegis, A. (1955). Über den Einfluss der Temperatur und der täglichen Licht-Dunkel Periode auf die Bildung der Ruheknospen, zugleich ein Beitrag zur Entstehung des Ruhezustandes. *Symp. Bet. Upsaliensis*, XIV, 1, 1–175.

Volkert, E. and Schnelle, F. (1966). Arboreta Phaenologica, Mitt. Arb. Gemein. Int. Phaenol. Gärten. 1966.

Discussion

Compiled by M. G. R. CANNELL, R. FAULKNER, F. T. LAST,
J. D. MATTHEWS, and discussion group leaders.

During 1959, two somewhat tendentious papers were published in the *Journal of Forestry*. In the first Inman made a number of strident comments about the insular attitudes of practising tree breeders. In replying, Snyder refuted many of Inman's points of detail but commended his underlying theme that "the fundamental principles of plant breeding and genetics are the same for all crops (whether agricultural, horticultural or silvicultural), but the best methods of their application must, of course, be adapted to the biological and cultural characteristics of the species". The same is true for "physiological genetics" as defined, for instance, by Wallace *et al.* (1972), who evaluated the impact that physiological understanding is having on the extensive programmes of field crop improvement. This conference on Physiological Genetics of Forest Tree Yield was called in the belief that such understanding will have a similar beneficial impact on tree yield improvement. The reader is reminded that the aims were (a) to examine physiological and morphological characters which limit wood yield and underlie inherent differences in forest tree yield, and (b) to consider (where possible) the heritability of those characters and the ways in which they might be exploited by breeding.

This chapter collates some of the more important viewpoints expressed during group discussions at the end of the conference.

I. CARBON FIXATION EFFICIENCY

Discussion group: F. T. Ledig (Leader), E. D. Ford, J. C. Gordon, J. Helms, P. G. Jarvis, O. Luukkanen, W. Żelawski.

Net carbon fixation, and net photosynthesis, involve many component processes, which are ultimately linked with other growth processes. It is impossible to decide which processes can or should be manipulated genetically without first defining (a) whether the forest is being grown for pulp or saw-log timber and (b) the crop environment. With these conditions established, physiologists may help (a) to define the balance of attributes that may be desirable, that is the tree "ideotypes", and (b) provide criteria for mass selection.

A. Individual Plant Processes

It is noteworthy that most major genetic gains in field crop yields have resulted from increases simply in the proportion of dry matter that is harvested, often with a decrease in photosynthetic rate per unit of leaf, and that attempts to increase crop yields by selecting for rapid rates of photosynthesis have been singularly unrewarding (e.g. Evans, 1975). In contrast, we believe that there are good prospects that net rates of carbon fixation per unit leaf area can be genetically improved in trees. Trees appear to have been less rigorously selected in nature for maximum rates of carbon fixation. Genetic variants can be identified which are particularly efficient in one or more of the component photochemical, biochemical, gas-conductance, and respiratory processes that make up net photosynthesis. However, most of this work has been done comparing species or widely differing provenances; more work is needed on the pools of variation within breeding populations.

Certain characteristics can be identified as desirable with reasonable confidence, given certain environmental constraints. Three examples merit special note.

(1) The stomata of *Picea sitchensis*, growing in Britain, close in response to small vapour pressure deficits, even in conditions where the supply of soil water is unlikely to be limiting. In such conditions, carbon fixation and growth may be greatly increased if the stomata remained open. Clearly, in drier climates, the advantages of having open stomata would have to be compromised with the need to prevent excessive water loss.

(2) It would be universally advantageous to select trees with photosynthetic apparatus which was able to "adapt" rapidly to changing environments. Thus, it is better to have inductive enzyme systems that allow leaves to adjust their carbon fixation characteristics to their particular conditions rather than systems which respond rigidly irrespective of the current climatic and edaphic conditions. Adaptive systems use resources even more efficiently than systems which confer broad tolerance to different environmental situations.

(3) Another characteristic of general value is an increase in the efficiency with which fixed carbon is converted into structural components; that is, a decrease in constructive respiration. Inherent differences in rates of constructive respiration have been observed, but it is necessary to ensure that they are not merely the result of differences in the proportion of polysaccharides and other components.

B. Crop Processes

Given that the objective is to maximize wood fibre production per unit area, the prime objectives in environments where broad-leaved trees can be grown might be as follows: (1) maximum leaf area duration and dry matter production, obtained by close spacing, early flushing and having leaves which retain high photosynthetic capacities with age; (2) minimum leaf area per individual, and so minimum "competitive influence" among individuals per unit of wood fibre produced; (3) an increase in the "efficiency" with which assimilates are distributed morphologically, anatomically and biochemically. There is evidence that all these processes vary genetically between and within tree populations.

A major area of uncertainty regarding the role of photosynthesis in forest *crop* production is the root structure required to supply minerals and metabolites to the aerial parts, particularly the growing shoot apices. Also, almost no experimental studies have been made comparing net carbon fixation efficiencies of forest canopies with different geometries.

II. SHOOT GROWTH

Discussion group: J. D. Matthews (Leader), M. G. R. Cannell, I. Dormling, I. Ekberg, R. E. Farmer, L. S. Jankiewicz, R. M. Lanner, F. T. Last, K. A. Longman, D. F. W. Pollard, Z. J. Stecki, S. Thompson, P. F. Wareing.

The discussion on shoot growth was focused on (a) increase in height as a widely used criterion of growth rate, (b) the apportionment of growth between new foliage, the cambium and other parts of the tree (the "leaf to total growth ratio" and "harvest index"), and (c) canopy form and structure as they affect volume increment, without necessarily any change in height growth rate.

A. Height Growth as a Selection Criterion

(1) Height increment in one year is not a good measure of the growth rate of individual trees. Increments should be measured over successive years, and related to radial growth and frost or drought damage.

(2) Useful criteria may be obtained for selecting parent trees by examining the component processes responsible for height growth within the available growing season and crossing genotypes which complement each other. Desirable attributes for both broad-leaved and coniferous trees were considered to be: (a)

an ability to generate large numbers of initials (stem units) per year as determined by the rate and duration of apical meristematic activity; and (b) an ability to extend the stem units to the maximum as soon as the environment allowed. Some of the processes underlying these attributes, such as the onset and cessation of apical meristematic activity, and the occurrence of "free" or desirable patterns of "episodic" growth, are heritable adaptive traits which can differ markedly between populations, provenances or species. Such adaptive traits may be expressed equally in forest stands as well as by individuals, and so may be selected for with reasonable confidence, provided juvenile—mature correlations are good.

(3) Height growth components may be inherited independently of height growth itself, and may have higher heritabilities.

B. Distribution of Growth

(1) Differences in periods of leaf initiation and shoot elongation will affect the periods and processes of cambial growth and wood formation, but these important relationships have rarely been examined in a tree improvement context. Differences in the ability of the cambium to respond to stimuli from other meristems may underlie differences in "harvest index" and wood structure.

(2) Many of the variables derived by classical growth analysis do not usefully describe the dynamics of tree growth, for instance, because so much of the dry matter is invested in non-photosynthetic tissue. Nevertheless, growth analysis is a valuable means of studying growth distribution.

(3) At a fundamental level there seems to be some value in continuing research on possible limiting growth regulator levels or balances, and rates of cell division in relation to shoot elongation, primordia initiation, and "form" characters.

C. Canopy Form and Structure

(1) The leaf area durations of several deciduous tree species are related to wood yield and can be increased by selecting or hybridizing provenances adapted to different climatic regimes. The foliation period of locally adapted trees is often undesirably short. Other canopy attributes have to be defined with respect to particular silvicultural systems.

(2) Critical comparative studies are needed to determine the ideal forest canopy structure in particular situations. However, it seems likely that the tall narrow conifer crowns favoured by silviculturists in north temperate regions are

Discussion

desirable. In such forests there can be efficient vertical exchange of heat and CO_2. Non-random groupings of leaves in coniferous forests may be desirable, and further attention should be given to the performance of canopies with different gap areas and degrees of leaf bunching.

(3) Differences in mature crown form can rarely be predicted from juvenile characters, partly because so little is known about ageing processes, and also because crown shape is so much influenced by forest management practices.

III. CAMBIAL PHYSIOLOGY

Discussion group: P. R. Larson (Leader), P. M. Denne, S. M. Hall, C. H. A. Little, M. MacGregor, P. F. Wareing, H. Wellendorf.

Our current knowledge of cambial physiology contains too many gaps and deficiencies to make broad recommendations on likely selection criteria for cambial growth. The main aim here is to focus on those fields of research which are most likely to prove rewarding from a tree improvement point of view.

A. Selection Criteria

For certain species, growing in specific conditions for particular types of wood, physiologists do have some chance of identifying selection criteria to predict cambial growth rate and wood quality. Thus, physiological processes related to bud and needle development appear to be highly correlated with early-wood and late-wood formation in pines, while the balance between leaf production and maturation shows promise of being a predictor of the fibre-vessel ratio in certain diffuse-porous hardwoods.

B. Cambial Development

Because the cambium is a diffuse lateral meristem, any attempt to improve cambial growth of either gymnosperms or angiosperms must consider its intimate ontogenetic relation to other tree parts. Thus, to quantitatively improve cambial growth particular attention must be given to the shoot since the cambium is initially an integral part of the shoot and subsequently a product of it. It should be recognized that the *cessation* of cambial activity may or may not be related to shoot activity, and that products from the root may be involved in regulatory processes.

To quantitatively improve cambial growth, further consideration must be given to the ontogenetic and seasonal development of organs that either directly or indirectly influence the differentiation of cambial derivatives.

C. Recommendations

With the above in mind our recommended research objectives for cambial physiology are as follows:

(1) to determine the interactions among plant parts that directly influence cambial activity during ontogeny and seasonal development;

(2) to pursue developmental studies of the cambium in order to separate the discrete processes of cell production and differentiation;

(3) to investigate the roles of growth hormones and nutritional factors at all stages of tree development; and

(4) to determine whether the above processes are genetically variable and are inherited.

IV. WATER RELATIONS

Discussion group: P. G. Jarvis (Leader), M. P. Coutts, J. Helms, T. T. Kozlowski, O. Luukkanen, M. MacGregor, F. R. Tubbs, M. T. Tyree, W. Żelawski.

Numerous plant characters influence tree water relations, and not all are desirable in situations where the trees are likely to suffer water stress. Three situations were discussed.

A. Ability to Survive Transplanting

Seedlings should be sought which are able to produce large numbers of new roots rapidly after transplanting, in response to either wrenching or undercutting in the nursery, and possibly in response to hormone treatment when grown in containers.

B. Survival of Young Trees in Very Dry Areas

(1) The presence of leaf waxes is of prime importance in restricting water loss through the cuticle and enabling some species to survive drought. It could be a selection criterion for breeders working in seasonally dry climates. Progenies or

provenances should be selected for differences in wax production and tested for survival in field trials.

(2) In drought situations, it should be advantageous to seek plants with "sensitive" stomata that close before severe internal stress develops and soil water reserves become depleted. Further field tests are recommended on differences in the speed at which stomata close in response to increasing water stress.

(3) Root characteristics are known to be important for drought avoidance, particularly those which affect the volume of soil exploited and the ability of a tree to grow in toxic or unstable soils. Such characteristics as deep or spreading root systems may be selected for or may, in special circumstances, be exploited as rootstock characteristics.

(4) Leaf shedding, desiccation tolerance and high base levels of osmotic potential are all considered desirable, but require systematic assessment in progeny trials.

C. Maximizing Forest Growth in Temperate Areas with Summer Drought

(1) It was generally agreed that it would be desirable to select genotypes in which the stomata open and close rapidly in response to changing environmental conditions, especially solar radiation. There was less agreement on whether the stomata should be insensitive to decreases in leaf water potential, down to say -20 bars, although this would enable photosynthesis to continue unabated during normal diurnal periods of water stress.

Certainly, in the long term, breeding for particular patterns of stomatal behaviour may provide one of the most effective means of maintaining wood production in situations where there are periods of moderate water stress. Genotypic differences in stomatal behaviour should be surveyed among species, provenances and progenies, and selections with known stomatal responses should be tested in different environments.

(2) It is also desirable to avoid water stress by selecting for patterns of root growth and morphology which enable the largest possible soil water reserves to be accessed. However, there is very little information on relationships between root development and leaf water potentials.

(3) Experimental selections should be made on the basis of "base levels of osmotic potential" (see Chapter 19) and tested under conditions of water stress. High osmotic potentials maintain turgor pressures and water uptake, and any delay in the loss of turgor pressure with increased water stress should favour growth.

(4) Besides avoiding or tolerating water stress, trees should be selected which have a high "water use efficiency", that is, an ability to photosynthesize rapidly at given stomatal conductances.

V. FROST HARDINESS

Discussion group: C. Glerum (Leader), I. Dormling, O. Juntilla, B. A. Thielges, F. R. Tubbs.

Although in some areas well adapted frost-hardy provenances have already been selected, (and there is little desire to sacrifice genetic diversity by selecting within provenances), in other situations there is a need to predict the frost hardiness of progenies and clones selected within populations. Frost damage is rarely compensated by fast growth, and summer frost hardiness is considered to be virtually unobtainable.

Changes in the frost hardiness of differing tissues should be related to changes in the activity and nutritional status of the shoot and cambial meristems. The assumption that frost-hardy tissues must necessarily have stopped growing was queried.

A. The Ideotype

The frost-hardy "ideotype" flushes late in spring, grows rapidly during the summer, uses and recharges large internal food reserves, and stops elongating and becomes frost hardy early in the autumn. Hardiness criteria are becoming available which may help in selecting such trees either as parents or progenies, certainly to prevent serious errors of judgment. Inherent differences in lethal freezing temperatures of only 2–3 degrees can be crucial.

B. Screening

An immediate practical need exists to decide when seedlings are hardy enough to be planted out on harsh sites. Frost-hardiness screening normally takes about 2 years, but more rapid methods are being developed. Cuttings are rarely hardier than seedlings.

VI. MINERAL NUTRITION

Discussion group: R. Faulkner (Leader), M. P. Coutts, R. E. Goddard, P. J. Goodman, R. Lines, P. A. Mason, P. F. Wareing.

It is well established that inherent differences in nutrient uptake and growth response occur among tree species, provenances, families and clones, particularly to N, P, K and Ca. Nitrogen can increase net assimilation rates as well as leaf production, phosphates enhance root growth, and potash can hasten dormancy onset and increase frost hardiness. The ability to absorb minerals can evidently be enhanced by the presence of mycorrhizas and nitrogen-fixing bacteria. Also, in some conditions it may be hormone limited, particularly when trees appear to be hypersensitive to low threshold levels of nutrients and stop growing without a fall in tissue nutrient concentration.

A. Recommendations

With these facts in mind, the following recommendations were made relating to tree improvement.

(1) Progeny tests should be conducted on sites where major nutrient deficiencies do not exist, or have been corrected.

(2) New clones or varieties should be able to grow well at a wide range of nutrient levels around the "norm" for the species. That is, they should be nutritionally "adaptable" or "variable".

(3) It is preferable to develop varieties which grow well on deficient sites rather than ones which respond well to high nutrient inputs. Desirable genotypes will be those which can extract nutrients effectively from deficient soils, or have small nutrient needs.

(4) Well designed experiments are required to establish nutrient *response curves* for vigorous and slow-growing clones and families, to determine the range of variation at deficient, adequate and high levels of nutrition.

(5) Techniques are being developed to inoculate containerized tree seedlings with effective strains of mycorrhizas. For this, more needs to be known about the compatibility factors, in host and microbe, that control the formation of "effective" mycorrhizal associations. Further work should be done in glasshouse and field conditions, including inheritance studies using methods whereby particular mycorrhizal associates can be identified and the dry-weight gains of the hosts measured.

(6) Further studies are needed to establish whether there are exploitable inherent differences in hormone—mineral balances between genotypes, using trees which respond to bud-applied hormones such as cytokinins by increased nutrient uptake and growth.

VII. EARLY GENETIC SELECTION FOR GROWTH RATE

Discussion group: G. B. Sweet (Leader), P. Baradat, J. Burley, M. Giertych, E. C. Jansen, J. Kleinschmit, T. O. Perry, F. Scholz.

Although most of the increases in forest tree growth rate that can be anticipated will undoubtedly result from improved cultural techniques rather than the initial generations of genetic selection, it is clear that substantial genetic gains in yield are obtainable.

A. Progeny Selection

Most of the genetic gain will be obtained on the basis of evidence from progeny tests, although gains are more certain in pest and pathogen resistance, wood quality and tree form than in yield. It requires sophisticated design and interpretation to derive genetic variances for growth rate, and much of the reported heritability data partly reflect environmental and maternal-genetic factors which influence initial plant size. These factors are most important when the trees are young and make predictions of future performance particularly difficult.

B Recommendations

Because progeny tests are inefficient and difficult to interpret with regard to inherent growth potential it is recommended that: (1) tests always incorporate large numbers of progenies, and growth is adjusted for size differences at planting; and (2) physiologists examine progeny differences in the components of yield, and work with breeders in progeny test designs and assessments.

A further need for physiological understanding exists when selecting for maximum yield per hectare in stands (see Chapter 27), although the importance of "competitive influences" will depend greatly on the species and system of management that is adopted.

VIII. CONCLUSIONS

It was clear from the conference that tree breeders could not immediately expect new and more effective selection criteria to be developed from fundamental studies of growth processes. There was, however, no shortage of ideas for new lines of approach, and "physiological genetics" was foreseen as a productive new area of research on tree and forest yield. Clearly the success of this research depends on the answer to two fundamental questions: (a) can we assume that it is possible greatly to improve the annual dry matter production of forests, and hence increase wood yield, and (b) could breeders exploit knowledge of the "components of yield" even if these were known?.

A. Strategies to Increase Forest Yield

There are two reasons for believing that forest dry matter yields can be increased. First, tree breeders still have to select suitably *adapted* genotypes, that is, genotypes able to exploit the available growing season between damaging frosts or droughts, and able to survive limiting site conditions. There are often no wild ecotypes ideally suited to distant forest sites, nor indeed their local ones, and trees in general appear to have adopted conservative strategies of growth, sacrificing rapid rates of dry matter gain for stress tolerance. This conservatism can be exploited. Secondly, as mentioned above, there is ample reason to believe that trees are inefficient fixers of atmospheric carbon compared with herbaceous plants, and that trees may be one of the few crops where selection for efficient carbon fixation will be rewarding. However, one may ask whether this should be expected as an inevitable outcome of selection for rapid growth, or whether it is worthwhile attempting to cope with the complex of factors involved in measurements of photosynthetic "capacity".

Some conference participants rightly stressed that, at present, the major genetic gains in tree yield are being obtained by eliminating "defects" such as susceptibility to pests, and undesirable silvicultural characters, and that selection for rapid dry matter production will be a future development. One may also add that "defect" traits are being more effectively screened than "vigour" traits which have lower narrow-sense heritabilities. "Vigour" traits will become important when the first round of progeny screening has been completed and the future breeding population determined.

A more arguable point was that more attention and money should be paid to optimize the size of planting stocks in nurseries sited near the forest plantings, since nursery stock size and condition often has a major effect on subsequent absolute rates of growth.

B. The Role of Physiological Genetics in Tree Improvement

There are several ways in which an insight into the physiology of tree yield can be expected to aid tree improvement. The most obvious is to help breeders construct their own "ideotypes", peculiar to their conditions, which may be of value when hybridizing species or ecotypes, or when planning crosses between tested desirable genotypes in breeding programmes. There may be a greater chance of achieving genetic gain, and perhaps heterosis, by crossing particular clones or deviating trees which have superior phenotypes for different known desirable components of yield, rather than crossing "blind" on the basis of yield itself. It is, however, accepted by many breeders that these desirable genotypes (for yield) cannot be obtained by mass selection in wild, outbreeding, heterogeneous stands, particularly if there is competition, unless "competitive indices" can be refined and component characters can be shown to have high narrow-sense heritabilities. It is more likely that most components of yield are no different from yield itself in being under polygenic control, genetically variable in the broad sense, but with a high degree of non-additive genetic variance which can only be manipulated by screening numerous progenies, and may then require special 2-clone orchards in order to exploit specific combinations. In some instances, desirable genotypes can be "frozen" by vegetative propagation, but, more frequently, there are strong arguments for using tree populations with a degree of variability enabling them to accommodate local site differences and changes in weather conditions from year to year.

Many contributors addressed themselves to the urgent need to identify early selection criteria for growth rate and yield which might enable tree breeders to shorten the length of the breeding cycle safely. It normally takes 5—10 years before half-sib families can be evaluated reliably on the basis of height growth alone. There are evidently many *adaptive* traits which show promise, such as the photoperiodic response of the shoots, frost tolerance and mineral nutrient uptake in the presence of specific mycorrhizas. In these instances much of the variation in rates of dry matter accumulation among progenies may centre around a few physiological attributes. However, once the breeding population is well adapted many more factors may contribute to inherent differences in vigour, and it may be more difficult to find single attributes, be they components or sub-components of yield, which can be used reliably as progeny selection criteria. Also, breeders cannot afford to proliferate the number of characters they select for because large numbers of progenies have to be screened in order to find good general or specific combiners for yield. Thus, there are reasons for cautious optimism when pursuing fundamental studies aimed at early selection criteria for yield within populations which are already well adapted to the conditions in which they are grown.

A third way in which physiological understanding may aid tree improvement is in preventing errors of judgment and indicating what is possible. Warnings were noted in Chapter 10 on the implications of the juvenile habit of "free" growth in some species of conifers, and in Chapter 27 on the implications of current studies on competition in monocultures. On the other hand selection for genotypes compatible with specific mycorrhizas or with certain patterns of stomatal behaviour may lift the objectives of tree breeders beyond those formerly thought possible.

In conclusion, we would stress that meaningful advances in physiological genetics can only take place when physiologists and breeders work together. It was clear from this conference that a lot has yet to be done. With one or two notable exceptions, few physiological analyses have yet considered within-species variation in trees. Most physiologists, as a matter of course, consider effects of the physical environment on growth processes, and in some instances have compared species, but can comparisons based on minute samples of the range of variation be valid? If samples were taken from known ecotypes, or, better still, from clones or full-sib families, there would automatically be more information of value to breeders, even if it were impossible to determine the patterns of inheritance. There is, perhaps, a fundamental problem in the short time scale of most physiological studies.

In agriculture, breeders were first concerned with yield in an unsophisticated manner, then sought the help of pathologists and entomologists to locate resistance and tolerance to pathogens and pests, and later searched for a greater understanding of the physiological basis of yield. It is our hope that those conference proceedings will stimulate this development in forestry.

REFERENCES

Wallace, D. H., Ozbun, J. L. and Munger, H. M. (1972). Physiological genetics of crop yield. *Adv. Agron.* **24**, 97–146.

Evans, L. T., ed. (1975). "Crop Physiology, Some Case Histories". Cambridge University Press.

Species Index

Common names are indexed only for tree species which are referred to three or more times. The authorities of Latin names, which have been omitted in the text, are given here.

A

Abies spp.
   free growth, 245
A. balsamea (L.) Mill.
   abscisic acid and xylem growth, 297
   dormancy, 409
   lateral bud formation, 195
   provenances, photosynthesis, 24–25, 27
A. concolor (Gord. & Glend.) Lindl.
   bud development, 176
   competition in mixtures, 67
   hydrostatic gradient, 62
   needle and internode elongation, 191
A. grandis (Dougl.) Lindl.
   gas in, 372
   stomatal response, 313–314
Acacia spp., 16
A. albida Del.
   drought tolerance, 5
A. aneura F. Muell.
   desiccation tolerance, 321
   water relations, 344–345
A. craspedocarpa F. Muell.
   stomatal response, 311
Acer spp.
   shoot and canopy, 129
A. rubrum L.
   shoot and canopy, 129

A. saccharinum L.
   gas in, 372
   stomata, 312
A. saccharum Marsh.
   leaf waxes, 316–317
   rooting and bud break, 257
   shoot and canopy, 129
   stomata, 312, 314–315
   transpiration, 307
   water relations, 337
Adansonia digitata L., 16
Aeschynomene aspera L., 365
Agrostis stolonifera L., 437
Agrostis tenuis Sibth., 437
Aleurites fordii Hemsl.
   CO_2 uptake and nutrition, 30
Alnus in cana (L.) Moench.
   flood tolerance, 367, 394–395
A. rubra Bong.
   flood tolerance, 367
Amanita muscaria (L. ex. Fries) Hooker [fungus], 439, 441–446
American sycamore, *see Platanus occidentalis*
Anabasis spp., 311
Anthocephalus spp., 142
Apple, flood tolerance, 379
   stem anatomy, 283
   see also Malus spp.
Artemisia spp., microphylly, 311
A. cana Pursh., 319

A. *herba-alba* Asso., 344
Ash, *see Fraxinus* spp.
Aspen
   carbon assimilation, 32
   stomatal closure, 67
   see also Populus tremula
Atriplex canescens James, Cat., 320
Avicennia spp., 6
Azadirachta indica A. Juss., 5

B

Balsam fir, *see Abies balsamea*
Barley (*Hordeum sativum* Pers.), 217
Beech, *see Fagus* spp.
Betula spp.
   genetic linkage, phenology, 183
   provenances, photoperiodism, 208
      shoot growth, 129
   stand competition, 464
   see also Birch
B. alleghaniensis Brit.
   (yellow birch) canopy, 57
B. papyrifera Marsh.
   stomata, 312
B. pubescens Ehrh.
   nutrition and mycorrhizas, 439–446
   photoperiodism, 211–212
   rhizosphere, 366
B. verrucosa Ehrh.
   nutrition and mycorrhizas, 439–446
   stomatal response, 314
Birch,
   primary forest, 7
   stomatal closure, 67
   see also Betula spp.
Black poplar, *see Populus nigra*
Black spruce, *see Picea mariana*
Boletus scaber (Bull. ex. Fries) S. F. Gray [fungus], 439
Bulrush millet, *see Pennisetum typhoideum*
Burr oak, *see Quercus macrocarpa*

C

Cabbage (*Brassica oleracea* L.), 31, 406, 411
Calligonum comosum L'Hér., 318
Camellia sinensis L., 3
Canotia halocantha Torr., 318
Cassava (*Manihot esculenta* Crantz), 8
Cecropia spp., 144
Ceiba spp., 142
Cenococcum graniforme (Sow.) Ferd. and Winge [fungus], 439
Ceratonia siliqua L., 344–345
Cercidium floridum Benth.
   bark photosynthesis, 318
Cercis canadensis L.
   leaf waxes, 317
   stomatal response, 314
Chamaecyparis obtusa (S. & Z.) Endl.
   seasonal photosynthesis, 30
Chlorophora excelsa (Welw.) Benth.
   growth habit, 9, 16
Citrus mitis Blanco
   leaf waxes, 317
   stomatal response, 314
Cocoa (*Theobroma cacao* L.), 3
Coffea arabica L., 3, 8, 14
Corylus spp., branch angles, 167
Cotton (*Gossypium* spp.), 12
Cowpea (*Vigna unguiculata* (L.) Walp.), 4, 5
Crataegus spp., stomata, 312
Cryptomeria japonica (L.f.) D. Don
   seasonal photosynthesis, 30
Cupressus spp., waxes and taxonomy, 316
Cupressus arizonica Greene
   photoperiodism and gibberellins, 294–295
Cynosurus cristatus L., 437

D

Dacrydium pectinatum Sol.
   crown type, 144

Species Index

Dactylis glomerata L., 23
Dalea spinosa A. Gray, 318
Douglas fir, *see Pseudotsuga menziesii*

E

Eastern cottonwood, *see Populus deltoides*
Eastern white pine, *see Pinus strobus*
Emmotum fulvum Dev.
 crown type, 144
Encelia farinosa A. Gray ex Torr., 311
Epilobium (Chamaenerion),
 angustifolium L., 319
Eucalyptus spp.
 waxes and taxonomy, 316
 xeromorphic habit, 5, 16, 140
E. gigantea Hook. f.
 leaf waxes, 316
E. globulus Labill.
 crown form, 149
E. gunnii Hoof. f.
 leaf waxes, 316
E. incrassata Sieb.
 root–shoot ratio, 319
E. pauciflora Jieb.
 leaf waxes, 316
E. polyanthemos Schau.
 stomatal response, 313
E. rostrata Schlecht
 stomatal response, 313
E. salicifolia A. Cunn.
 leaf waxes, 316
E. sideroxylon A. Cunn.
 stomatal response, 313
E. socialis F. Muell. ex Miq.
 root–shoot ratio, 319
E. urnigera Hoof. f.
 leaf waxes, 316
European larch, *see Larix decidua*

F

Fagus spp.
 shoot and canopy, 127
Fagus sylvatica L.
 flushing, 127
Festuca ovina L., 437–438

Fouquieria splendens Engelm., 311
Fraxinus spp.
 branch angles, 167
 leaf and cambium development, 272
 provenance study, 129
 stomata, 312
F. americana L.
 leaf waxes, 316–317
 photorespiration, 66
 stomatal response, 314–315
 transpiration, 307
F. pennsylvanica Marsh.
 gas in, 373
 rhizosphere, 366
 stomata, 312

G

Ginkgo biloba L.
 photorespiration, 66
 stomata, 312
 water relations, 342–343
Gleditsia triacanthos L.
 stomata, 312
Gmelina spp., 8

H

Haloxylon spp., 311
Hammada scoparia (Pomel) Ilijn, 340
Helianthus annuus L., 32, 339, 370
Hevea spp., 2, 9

I

Ilex cornuta Lindt., 312–313
Ilex vomitoria Ait., 321

J

Jack pine, *see Pinus banksiana*
Japanese red pine, *see Pinus densiflora*
Juglans spp., allelopathy, 321
J. nigra L.
 provenances, shoots and canopies, 124–125

K

Knobcone pine, see *Pinus attenuata*
Koeberlina spinosa Zucc., 318

L

Laguncularia racenosa Gaerta,
  aerenchyma, 376
Larix spp.
  crowns, 145
  DNA synthesis, 177
  free growth, 245
  growth simulation, 32
  provenances, photoperiodism, 208
L. decidua Mill.
  bud development, 176
  seasonal photosynthesis, 30
  simulation of growth, 32–33
L. laricina (Du Roi) K. Koch
  flood tolerance, 367, 393–395
  frost hardiness, 410
L. leptolepis (Sieb. & Zucc.) Endl.
  seasonal photosynthesis, 30
  simulation of growth, 32–33
L. russica (Endl.) Sabine
  seasonal photosynthesis, 30
  simulation of growth, 32–33
Larrea spp., 321
L. divaricata Cav., 344
Lathyrus odoratus L., 217
Libocedrus decurrens Torrey
  photosynthesis and water relations,
  68–73
Liquidambar styraciflua L.
  flood intolerance, 373
  rhizosphere, 366
  shoot and canopy, 128–129
  winter photosynthesis, 30
Liriodendron tulipifera L.
  flood intolerance, 373
  photoperiodic response, 129
  photorespiration, 66
  rhizosphere, 366
Loblolly pine, see *Pinus taeda*
Lodgepole pine, see *Pinus contorta*
Lolium perenne L., 437

Lombardy poplar, see *Populus nigra
  (italica)*
Lycopersicon esculentum Mill., 312,
  321

M

Maize (*Zea mays* L.), 2, 4, 10, 11, 14,
  27, 59, 121, 370, 389
Malus spp.
  branching, 159–162, 164
  source–sink relationships, 123
  see also Apple
Mangrove, flood tolerance, 365, 378
  salt tolerance, 6
  see also *Avicennia* spp.
Melampsora spp., [fungus], 125
Molinia caerulea (L.) Moench., 364
Monterey pine, see *Pinus radiata*
Musanga cecropoides (parasolier) R.
  Br.
  allometry, 14
  canopy, 7
Myrica gale L., 366, 370–371
M. pennsylvanica Lois. 373

N

Nardus stricta L., 437
Nicotiana tabacum L., 113, 217
Noea spp., 311
Norway spruce, see *Picea abies*
Nyssa aquatica L.
  ethanol in, 367
  rhizosphere, 366
N. sylvatica Marsh.
  flood tolerance, 376
  gas in, 371–373

O

Oak, see *Quercus* spp.
Onion (*Allium cepa* L.), 390

P

Paxillus involutus Fries (Karsten)
  [fungus], 439

Pea (*Pisum sativum* L.), 12, 390
Peach (*Prunus persica* L.), 367, 379
Pear (*Pyrus communis* L.), 367
Pennisetum typhoideum Rich., 10, 344
Phillyrea media Tenmore, 312
Picea spp.
  crowns, 145
  free growth, 245
  photoperiodism, 208
P. abies (L.) Karst.
  seedling apices, 180
  clones, photosynthesis, 115
    variation, Chapter 30
  critical night length for bud set, 214–216
  crown form, 143
  flood tolerance, 370, 374
  growth and gibberellins, 295
  maternal effects, 478
  needle and internode elongation, 191
  primordium size, 182
  provenances, CO_2 compensation point, 116
    photoperiodism, 213
    variation, Chapter 30
  rooting and bud break, 257
  shoot elongation and wood production, 284
  stomatal response, 314
P. glauca (Moench) Voss
  bud development, 176
  DNA content per cell, 180
  frost injury, 414
  photorespiration, 114
  provenances, apical growth, 179–180
  translocation of ^{14}C, 412
P. mariana (Mill.) B.S.P.
  adaptation of CO_2 exchange, 24, 26
  flood tolerance, 367, 393–395
  provenances, free growth, Chapter 13
P. omorika (Pančić) Purkyně
  lateral buds, 196

P. rubens Sarg. (*Picea rubra* Link)
  adaptation of CO_2 exchange, 24, 26–27
P. sitchensis (Bong.) Carr.
  branching, 193, 195, 196
  bud development, 176
  competition, 465
  diallel progenies, 185, 189–190
  flood tolerance, 362, 367, 370, 375, 391–395, 399
  latitude cline, 178
  leaf waxes, 318
  photosynthesis, 63
  pressure bomb used, 339
  provenances, apical activity, 179
    DNA content, 180
    flushing dates, 186
  shoot elongation and leaf size, 286–287
  stomatal reponse, 66
  water relations, 341
Piñon pines, 228–229
Pinus spp.
  gibberellins in, 294
  needle elongation, 192
  provenances, photoperiodism, 208
  shoot development patterns, Chapter 12
  waxes as taxonomic characters, 316
P. armandii Franch.
  bud and shoot development, 229
P. attenuata Lemm.
  free growth, 225
  mycorrhizas, 438
  needle growth, 233
P. ayachahuite Ehrenb.
  bud and shoot development, 229
P. banksiana Lamb.
  branches per whorl, 198
  bud development, 176
  DNA content, 180
  photosynthesis of ^{14}C, 412–413
    and growth, 58
  provenances, DNA content, 180
    photosynthesis, 27–28

P. banksiana Lamb (*Cont.*)
   shoot development, 225, 227,
      229–231, 233
      elongation, 188
P. bungeana Zucc.
   bud and shoot development, 229
P. caribaea Morelet
   bud and shoot development,
      238–239
   foxtailing, 227
   mycorrhizas, 438
   see also *Pinus elliottii* Engelm.
P. cembra L.
   winter buds, 233
P. cembroides Zucc.
   bud and shoot development, 229
P. contorta Dougl.
   apical dome volume doubling, 181
   branching, 193–194, 196–198
   bud development, 176
      morphology, 197–198
      and shoot growth, 229–233,
         237, 239, 240
   flood tolerance, 362, 366–367,
      370–371, 375–378,
      388–396, 399
   latitude cline, 178
   light response, 102
   photoperiodic response, 178
   primordium size, 182
   provenances, apical dome size, 180
      flushing dates, 186
      needle growth, 191–193
      shoot elongation, 187
      stem unit initiation, 180
   shoot elongation and stem units, 188
   stomatal response, 313–314
   water potential, 67
P. coulteri D. Don, 240
P. densiflora S. & Z.
   bud development, 176
   bud and shoot development,
      229–231, 233, 237
   seasonal photosynthesis, 30
P. echinata Mill.
   bud and shoot development, 228

P. edulis Engelm.
   bud and shoot development, 228,
      231–233, 238–239
P. elliottii Engelm. (*P. caribaea*
      Morelet)
   apical activity, 183
   bud and shoot development, 225,
      231, 235–236, 238–239
   fertilizer responses of families,
      Chapter 26
   maternal effects, Chapter 28
   re-assimilation of respiratory CO_2,
      103
P. flexilis James
   winter buds, 233
P. gerardiana Wall.
   bud and shoot development, 229
P. halepensis Mill.
   needle waxes, 318
P. lambertiana Dougl.
   bud and shoot development, 176,
      229–232
P. monophylla T. & F.
   bud and shoot development, 228
P. mugo Turra
   winter buds, 233
P. nelsonii Shaw
   bud and shoot development, 229,
P. nigra Arn.
   bud and shoot development, 229,
      233
   flood tolerance, 370
P. occidentalis Swartz
   mycorrhizas, 438
P. palustris Mill.
   bud and shoot development,
      227–228, 234
P. pinaster Sol.
   bud and shoot development,
      229–233, 240
   flood tolerance, 362
P. pinceana Gord.
   bud and shoot development,
      229
P. pinea L.
   needle waxes, 318

Species Index

P. ponderosa Dougl.
    bud and shoot development, 176, 229, 232
    competition in mixtures, 67
    hydrostatic gradient, 62
    photosynthesis, 61, 68–73
    provenances, needles, 193
    stomatal response, 313–314
    water relations, 67–73

P. pumila (Pall.) Reg.
    5-needle development, 233

P. quadrifolia Parl.
    bud and shoot development, 228

P. radiata D. Don
    abscisic acid, xylem growth, 297
    analysis of clone trials, Chapter 29
    apical activity, 183
    branching, 198
    flood tolerance, 362
    foxtailing, 227, 234, 240
    gibberellin effect on height, 294
    gibberellin and auxin, elongation, 292–293
    mycorrhizas, 438
    needle waxes, 318
    photosynthesis, 61

P. resinosa Ait.
    branching, 193
    bud and shoot development, 176, 229–232, 237
    frost hardiness, 405
    initiation at shoot apices, 177
    needle waxes, 318
    shoot elongation, 188
    stomata,
        response, 315
        size and frequency, 312

P. rigida Mill.
    bud and shoot development, 228
    respiration and growth, 38

P. serotina Michx.
    flood tolerance, 362

P. strobus L.
    bud and shoot development, 176, 229–232, 235, 240
    diallel, maternal effects, 478
    frost hardiness, 407, 409, 412
    needle waxes, 318
    photosynthesis, light response, 102
    proleptic shoots, 158
    provenances, CO_2 uptake, 24
    rooting and bud break, 257
    stomata, 312

P. sylvestris L.
    branching, 198
    bud and shoot development, 229–232, 236–237, 240
    crown types, 168
    flood tolerance, 367, 374, 393–396
    inheritance of photoperiodic response, 213
    manganese uptake, 438
    maternal effects, 478
    needle lengths and shoot elongation, 286
    needle waxes, 318
    photorespiration, 114
    photosynthesis, 27, 30–31, 63, Chapter 5
    provenances,
        needle lengths, 193
        photosynthesis, 24, Chapter 5
        stem units, 236–237
    respiration, 66
    shoot elongation, 158, 188, 284–285
    stomatal response, 314

P. taeda. L.
    apical growth, 183
    bud and shoot development, 228, 233, 240
    drought resistance, Chapter 20
    growth, model predictions, 32
    maternal effects, Chapter 28
    mutual shading of needles, 147
    mycorrhizas, 438
    nutrition, response of families, Chapter 26
    photosynthesis, 24, 30–32, 58
    provenances, drought resistance, Chapter 20
    root–shoot ratios of progenies, 320

P. taeda. L. (*Cont.*)
   shoot elongation and stem units, 188
P. taiwanensis Hayata
   bud and shoot development, 228
P. thunbergii Parl.
   auxins and shoot elongation, 292
Plane tree, *see Platanus occidentalis*
Platanus occidentalis L.
   flood intolerance, 373
   nutrition, 438
   photosynthesis, 31–32, 34
   rhizosphere, 366
   shoot and foliage growth, 129
Ploiarium alternifolium Korthals.
   crown type, 144
Plum (*Prunus domestica* L.), 283
Poliomintha incana A. Gray, 320
Ponderosa pine, *see Pinus ponderosa*
Poplar, *see Populus*
Populus spp.
   aerenchyma, 365
   branching, 159–161, 164, 169
   mixed cropping, 8
   photorespiration, 112–115
   provenances, photoperiodism, 208
   shoot and canopy, 123, 125–126
   stomata, 312
P. balsamifera L.
   hormones and bud break, 257
P. bolleana Lauche
   crown form, 169
P. deltoides Bartr.
   bud break, control and inheritance, Chapter 14
   cambium, origin and development, 262–271
   CO_2 compensation point, 114
   flood intolerance, 373
   hybridization, 126
   transpiration and leaf anatomy, 313
P. x *euramericana* (Dode) Guinier
   photosynthetic and enzymatic criteria, Chapter 4
P. gradidentata Michx x *P. alba* L.
   CO_2 exchange, 27

P. maximowiczii Henry
   CO_2 compensation point, 114
   transpiration and leaf anatomy, 313
P. monolifera (*P. deltoides* Marsh)
   hybridization, 126
P. nigra L. (*P. nigra italica*)
   CO_2 compensation point, 114
   crown form, 169
   hormones in cuttings, 256
   transpiration and leaf anatomy, 313
P. petrowskiana Schneid
   gas in, 371
P. tremula L.
   crown form, 169
   photoperiodism (x *P. tremuloides*), 211
   provenances, heights of crosses, 208–211
   stomatal response, 314
P. tremuloides Michx.
   photoperiodism (x *P. tremula*), 211
   photosynthesis, 23
P. trichocarpa Torr. & Gray
   CO_2 compensation point, 114
   transpiration and leaf anatomy, 313
P. wislizenii Sarg.
   adventitious roots, 320
Prosopis juliflora (Sw.) DC.
   allelopathy, 321
   salt tolerance, 6
Prunus armeniaca L.
   stomatal response, 340
Pseudotsuga menziesii (Mirb.) Franco
   bud break and hormones, 253, 256, 258
   bud development, 176
   bud swelling and enzymes, 272
   flood intolerance, 367
   flushing dates, 186
   frost hardiness, selection for, 416, Chapter 24
   gibberellin metabolism, 295
   lateral bud formation, 195, 196
   mycorrhizas, 439
   photorespiration, 114
   photosynthesis, 57, 61, 63, 68–73

Species Index   541

shoot elongation and auxin, 292
stomatal response, 67, 313–314
water relations, 68–73

Q

Quercus spp. gas in, 373
  shoots and crowns, 127
  stomata, 312
Q. agrifolia Née
  gas in, 370
Q. alba L.
  stomatal response, 314–315
Q. macrocarpa Michx.
  gas in, 372
  leaf waxes, 317
  stomatal response, 314–315
Q. phellos L.
  stem photosynthesis, 34
Q. robur L.
  crown, 127
  stem photosynthesis, 34
Q. rubra L.
  branching, 123
  bud break and rooting, 257
  shoot growth, 41
  stomatal response, 314–315
Q. velutina Lam.
  stomatal response, 314–315

R

Red oak *see Quercus rubra*
Red pine *see Pinus resinosa*
Red spruce *see Picea rubens*
Retana raetam (Forsk.) Webb, 318
Rhagodia baccata (Labill.) Moq. 311
Rhizobium [bacteria], 446
Rhizopogon luteolus Fries [fungus], 439
Rhododendron poukhanensis L., 312–313
Rhus trilobata Nutt.
  adventitious roots, 320
R. typhina L.
  stomata, 312

Rice (*Oryza sativa* L.), 6, 11, 36, 121, 370
Robinia pseudoacacia L.
  frost hardiness, 405
Rubus idaeus L., 161

S

Salix spp.
  flood tolerance, 364, 379, 388
  frost hardiness, 415
  oxygen in, 366
  rooting and bud break, 257
S. alba L.
  gas in, 371
S. atrocinerea (*S. cinerea oleifolia* L.)
  flood tolerance, 370–371
  rhizosphere, 366
S. cinerea L.
  flood tolerance, 367
S. fragilis L.
  flood tolerance, 370–371, 376
  stomata, 312
S. lasiandra Benth.
  water relations, 342–343
S. pentandra L.
  flood tolerance, 376
S. nigricans Sm.
  flood tolerance, 376
S. repens L.
  flood tolerance, 370
S. viminalis L.
  flood tolerance, 365, 376
  hormones and bud break, 256
Salvia mellifera Linn. 321
Scots pine *see Pinus sylvestris*
Sequoia sempervirens (D. Don) Endl.
  gibberellins and apical control, 300
Sequoiadendron spp.
  hydrostatic gradient in, 62
Serbian spruce, *see Picea omorika*
Sitka spruce, *see Picea sitchensis*
Slash pine, *see Pinus elliottii*
Solanum tuberosum L., 312
Sorghum (*Sorghum vulgare* Pers.), 4–5, 14, 59, 217

Soya (*Glycine max* (L.) Merr.), 4–5
Sugar beet (*Beta vulgaris* L.), 3, 7, 10, 59
Sugar maple, *see Acer saccharum*
Sunflower (*Helianthus annuus* L.), 32, 339, 370
Sweetgum, *see Liquidambar styraciflua*

T

Taxodium distichum (L.) Richards
 knee roots, 365, 371
Tilia spp., branch angles, 167
Tobacco (*Nicotiana tabacum* L.), 113, 217
Trema guineensis (Schum. & Thonn.) Ficalho
 allometry, 14
Trifolium repens L., 31
Tsuga canadensis (L.) Carr.
 use of pressure bomb, 339
Tsuga heterophylla (Rafn.) Sarg.
 bud development, 176
Tulip tree, *see Liriodendron tulipifera*

U

Ulmus spp.
 bud burst, 274
 provenance study, 129

V

Vitis vinifera L.
 phloem activation, 272
 stomata, 312

W

Wheat (*Triticum aestivum* L.), 31, 39, 121
White ash, *see Fraxinus americana*
White fir, *see Abies concolor*
White pine, *see Pinus strobus*
White spruce, *see Picea glauca*
Willow, *see Salix*

X

Xerocomus subtomentosus (L. ex. Fries) Quélet [fungus], 439
Xylopia staudii Engl. ex Hutch. & J. M. Dalz.
 pneumatophores, 371

Y

Yucca elata Engelm., 320

Z

Zygophyllum spp., 311
Z. dumosum Boiss., 340

Subject Index

A

ATP, 22, 397
Abortion, of buds, 196–197
   see also Branches, Leaves, Shedding
Abscisic acid,
   assay, 298
   bud dormancy, 272
   shoot elongation, 293, 295
   wilty mutants, 312
   xylem growth, 297–298
Abscission of leaves, 311
   see also Abortion, Branches, Leaves, and Shedding
Absorption,
   capacity of roots, 460
   of ^{32}P, 458–460
Acclimatization,
   needles within crown, 57–58, 100–101
   to temperature, 25–26
   see also Adaptation, Frost hardiness
Acetaldehyde, in anaerobic conditions, 363, 398
Acrotony, terminal bud growth, 161–162
Additive genetic variance,
   nutrient response, 456
   phenological traits, 182
   progeny trial analyses, 491, 498–499
   see also Heritability, Genetic variances, and Variances
Adult stages, see Age

Adventitious roots, desert plants, 320
   flood tolerance, 364–365, 376, 388
   hormones and bud-break, 256–258
   see also Rooting of cuttings
Aerenchyma, flood tolerance, 365–377
Aerial roots, see Pneumatophores
Aerodynamic properties of leaves and canopies, Chapter 8
Agar, culture of *Betula*, 440–441
Age,
   and apical dome size, 180
   and "free" growth, 250
   of leaves and photosynthesis, 12, 24, 57, 100, 84–86
   of leaves and wood formation, 277
   of trees, 167–168
Agri-silviculture, 8, 79
Air, in canopies, 62–63
Alanine, and flood tolerance, 396–398
Alcohol, see Ethanol
Algorithm, for growth models, 34, 40
Alkaloids, and allelopathy, 320
Allele, day-length response, 217
Allelopathy, drought avoidance, 320–321
Allometry,
   and drought avoidance, 320
   predictive growth models, 32, 39–41
   re-investment in leaves, 14
   see also Distribution, Leaf, Root
Alpha-naphthylamine, 366

Altitude, *see* Elevation
Amino acids, and allelopathy, 320
  in flood plants, 397
AMO-1618, gibberellin inhibition, 294–295
Anaerobic conditions, *see* Chapters 21 and 22
Analysis of growth, *see* Growth analysis
Analysis of variance, of young trials, 486–487
Anastomoses, in xylem, 269
Anatomy,
  of cambium, Chapters 15 and 16
  of leaves and growth, 286
  of leaves and photosynthesis, 100, 111–113
  of leaves and transpiration, 313
  of roots, 365, 372–378
  of tracheids and growth, Chapter 16
Angle, of leaves in trees, Chapter 7
  see also Canopy, Crotch angle, Leaf angle
Animals, 63
Annual growth cycle, and photoperiodic response, Chapter 11
  see also Season
Annual rings, and pine shoot growth, 227–228
  see also Cambium, Earlywood, Latewood, Wood, Xylem
Anoxia, in roots, Chapter, 22, 365, 367–369
Antechamber, of stomata, 318
Anthracnose, 125
Apical bud formation, *see* Bud-set
Apical dominance and control,
  and ecological succession, 141
  in deciduous trees, 158–159
  hormone mechanism, 300
  in pines, 158, 194
Apical meristems,
  of baobab shoots, 16
  of conifer shoots,
    primordia initiation, 177–183
    dome activity, 180–181
    lateral bud formation, 196
    gibberellic acid effect on, 296
    gravity effect, 299
    pine shoot growth, 224, 236–239
  of roots and anoxia, 368–391
  see also Cambium, Meristems, and Shoot
Apoplastic water, 338, 344–345
Apparatus, pressure bomb, 335
Arid regions, 308
  see also Desert
Aseptic culture, 439–440
Aspartic acid (aspartate), 395–396, 398
Aspect and environment, 60
Aspiration of tree stems, 378
Assay of hormones in conifers, 298
Assessment of clone and progeny trials, Chapter 29
Assimilates, *see* ^{14}C-labelled assimilates, and Distribution
Assimilation, *see* Photosynthesis
Assimilation chamber, 68
Autolysis, of cells, 276
Autoradiography, 412–413
Autumn, *see* Season
Auxin (indole acetic acid),
  assay, 298
  branch angles, 163
  correlative inhibition, 159–160
  dormancy and bud break, 272–273
  and epinasty, 167
  and photosynthesis, 23
  in poplar cuttings, 256
  and reaction wood, 163
  shoot elongation, 292–296
  tree form, 300–301
  xylem growth, 296–299
Axillary buds, *see* Branches, Buds, and Lateral buds

B

Bark,
  frost hardiness, 405
  gas in, 372–373
  photosynthesis, 34, 103, 318–319
Basitony, basal bud growth, 161–162
Benzoic acid and allelopathy, 320
Bioassay of hormones in conifers, 298
Biochemistry,
  of hormones in conifers, Chapter 17
  mechanism of flood tolerance, 367, Chapter 22
Biomass,
  of C4 plants, 10
  and carbon balance models, 31
  of foliage, 173, 193
  of forests, 121
Biosynthesis, of hormones in conifers, 295, Chapter 17
Bog, 140,
  see also Drainage, Flood tolerance, Waterlogging
Bole, crown : bole ratios, 119, 121, 126
Boundary layer, of leaves, 66, 146
Bowen ratio, 145, 147
Brachyblasts, 311
Branch (branching),
  angle, 36, 504–506
  bud formation on conifers, 194–198
  and canopy structure, 146–150
  of deciduous trees, Chapter 7
  and establishment, 8–9
  hormone effects on, 296, 300
  and morphogenesis in pines, 229, 234
  of mycorrhizas, 444
  of roots and drought avoidance, 351–354
  of roots in flooded soils, 364–365
  shedding, 311
  and tree form, 299, 141–144
  whorls, 16, 195–198

Breeding,
  for drought resistance, Chapter 19, 339–346
  for frost hardiness, 414–417
  for forest yield per hectare, 14–17, Chapter 27
  for height increment, 198–199
  strategy, 516
  for water use efficiency, 308
  for wood volume, 275–279
  see also Genetic gain, Genetic improvement, Selection criteria, Tree improvement
Broad-sense heritability, see Heritability
Broomstick, crown form, 154–155
Bud,
  conifer buds,
    development, 175–183
    elongation, 183–189
    numbers, 193–198
  correlative inhibition, 158–162
  formation, see Bud set
  frost hardiness of, 429–430
  leaf–cambium relation within, 262–271
  and wood production, 278
  see also, Branch, Lateral buds
Bud break, (bud burst),
  control and inheritance, Chapter 14
  on deciduous trees, 124–131
  frost hardiness of, 407, 410
  and gibberellic acid in sap, 294
  and phloem production, 272
  and photoperiod, 207
  and primordia growth, 272
  see also Flushing
Bud scale formation, 248
Bud set, in conifers, 185
  and photoperiod, Chapter 11
  of provenances and clones of *Picea abies*, 504, 508–515
Bulk osmotic pressure, 334–346
Bulk soil water potential, 331–334

C

^{14}C-labelled assimilates, to buds, 161
  in poplars, 82
  to wood, 104, 412–3
C3/C4 photosynthetic pathways, 6, 10, 59, 111–113
Calcium, response, 438
  in needles, 504, 509–511
Callus, on poplar cuttings, 254–255
Cambium (cambial),
  activity and branch angles, 162–167
  anatomy and wood production, 284–288
  gas movement in, 372–373
  genetic improvement, 275–279
  growth and assimilate distribution, 14
  growth and pine shoot elongation, 227–228
  hormones regulating activity, 296–299
  meristem affected by gravity, 297, 301
  origin and development, 262–268
  primary–secondary transition, 268–271
  and reaction wood, 163
  resumed activity in spring, 271–273
  species differences, 273–275
  vascular connections, 160, 262–268
  see also Earlywood, Latewood, Wood and Xylem
Candelabra,
  pine branches, 162
  tree form, 141, 152–154
Canopy, characteristics and environment, Chapter 8
  closure and nutrient cycling, 7–8
  of deciduous trees, Chapter 7
  of poplars, 82, 87–88
  processes, 68–72, 147–151
  structure and carbon fixation, 11–12, 36–37, 41, 57–58
  see also Crown, Leaf angle, and Shade
Capacitance, electrical impedance, 426–427, 430
Capillary forces, 310
Carbohydrates,
  and carbon fixation, 22–23
  equivalents, 37
  import by shoots, 185
  oxidation in flooded plants, 398
  reserves, 13, 113, 123, 161, 397
    see also Reserves
  respiration of, 66
  supply to apices, 180
  see also Photosynthesis and Respiration
Carbon balance, and growth models, 31–33
Carbon dioxide, ambient levels in forests, 60, 62, 64–65
  in C3 and C4 plants, 59
  fixation in dark in roots, 394–396
  in flooded soils, 363
  from flooded roots, 393
  gas in trees, 372
  re-assimilation of respiratory CO_2, 103, 113
  from soil, 62
  starvation method of screening, 113, 116
  transport in stems, 103–104
  uptake, fixation and exchange, see Photosynthesis and Respiration
Carbon fixation, see Photosynthesis
Carboxylase, 38, 41, 111
Cataphylls, 175, 177, 230
Catena, 140
Cell autolysis, 276
Cell contents, and frost hardiness, 405–406, 414
Cell culture,
  of poplars, 90
  fusion of haploid cells, 3

Subject Index

Cell dimensions,
   in needles, 286
   and shoot lengths, 183
Cell disintegration, 365, 404
Cell division and elongation,
   and branch angles, 162
   in cambium, *see* Cambium
   and hormones, 292, 296
   in needles, 189
   in shoots, 174, 183, 185
Cell membranes, and frost damage, 404, 421–426
Cell wall,
   elasticity, 340–343
   synthesis, 38
   water in, 338
Cellulose, 104
Cessation of extension growth,
   and frost hardiness, 409, 415
   and latewood formation, 275
   see also Bud-set
Chilling,
   and frost hardiness, 424
   period and bud-break, 254–256
   requirement of buds, 161, 184–185, 239, 409
   treatment of seedlings, 248
Chi-square, 512, 514
Chlorophyll (chloroplasts),
   in bark, 103
   breakdown and frost hardiness, 412, 422
   photo-bleaching, 4, 57
   and photorespiration, 113
Chromatography, hormone assay, 298
Cinnamic acid, and allelopathy, 320
Citric acid, 396
Clay soil, water potential, 331–333
Cline,
   in leaf waxes on *Eucalyptus*, 316
   in photoperiodic response,
     and needle growth, 191–192
     and shoot growth, 178–179, 208–213, 235
   in temperature response with altitude, 24–25

Clones,
   analysis of clone trials, 485–491
   of poplars analysed, Chapter 4
   see also Species index
Cloud forest, 143
Cold hardiness, Chapter 23 and 24
Colloidal forces, 310
Commensalism, 17, 470
Compensation, *see* Acclimatization
Compensation point, carbon dioxide, Chapter 6, 59, 82–85, 103
   light, 11–12, 57
Competition,
   and forest succession, 145, 148
   and forest yield improvement, Chapter 27, 15–17, 479
   internal, among apices, 158–159, 167–168
   and seed production, 234
   and water relations, 67
   with weeds, 11–12
Components of leaf water potential, 338
Components of shoot growth, Chapter 10
Components of yield,
   analysis of, 2
   selection for, 39, Chapter 7, 464, 469
   in poplars, Chapter 3
Compression wood, 299–300
Cones (strobili), and pine shoot morphogenesis, 196, 229
   in relation to seed size, 477
Conjugation, of hormones, 295
Conservation, 7–8
Contact parastichies, 249
Controlled environments,
   carbon balance studies, 27, 31–33
   cuvette studies, 55, 68–70
   frost hardiness studies, 424–426
   growth of poplars, 83
   photoperiodic response studies, 128, 214–216
Conversion, of hormones, 295, 297

Cork, interxylary, drought adaptation, 319
Correlation,
  CO_2 uptake and dry weight increment, 27–29
  cell dimensions and wood production, Chapter 16
  chilling period and time of bud-break, 255
  foliar characteristics and wood quality, 279
  free growth and height, 246–249
  hormones and growth, Chapter 17
  model prediction and observed, 32–33
  morphological, phenological and nutrient content characters, Chapter 30
  seed weight and height growth, 474–477
  size at planting with subsequent growth, Chapter 29
  stem units and height, 188–189
  see also, Cline, and Juvenile–mature correlations
Correlative inhibition, 158–159, 300
Cortex, aerenchyma, 365
  gas in, 372–373, 376
  width as drought adaptation, 319
Cortical, cell growth and gibberellic acid, 296
  and rib meristems, 194
  sclereids, 178
Cotyledons, 57, 262
Coumarins, and allelopathy, 320
Covariance adjustment, of clone and progeny trial data, 486–493
Criteria of selection, see Selection criteria
Critical night length, for bud-set 214–216
  see also Photoperiod
Crop growth rate, 10
Crop physiology, 2
Cropping, mixed, 8

Crosses, photoperiodic ecotypes, Chapter 11
  specific crosses and nutrient response, 451–452, 457, 461
  of fungal isolates, 445
  see also Diallel, and Families
Crotch angle, 162–167
Crown (form and geometry),
  and branching, 158, 299 see also Branch
  development and seed production, 234
  environment and photosynthesis, 57–58, 100–101
  form and hormones, 299–301
  form and yield of deciduous trees, Chapter 7
  geometry, 36–37, 147–150, Chapter 8
  structure and competition, 467, 470
  types of *Pinus sylvestris*, 168
  see also Leaf angle, Canopy and Shade
Cuticle, resistance to CO_2 transfer, 66
  thickness and waxes, 140, 315–318
  transpiration through, 313
Cyanogenic compounds, in anaerobic conditions, 363
Cyanohydrins, and allelopathy, 320
Cybernetics, 160
Cycles, of pine shoot development, 194, 196
Cycling of nutrients in forests, 6
Cytochrome oxidase, 390
Cytokinins,
  assay, 298
  and bud break, 257, 272
  and bud growth, 160
  effect on photosynthesis, 23
  and xylem growth, 297
Cytoplasm, ice formation in, 404
Cytoplasmic inheritance, 474
Cytoplasmic streaming, gas transport, 378

D

Damage, *see* Animals, Death, Disease, Insects, Frost and Wind
Day-length, *see* Photoperiod
Death,
   of plants in communities, 15, 465–466
   of roots when flooded, 363–364, 389
   of trees with age, 168
Decapitation, and branch development, 158, 163, 167
Defoliation, *see* Abscission, and Leaves
Dehardening, *see* Frost hardiness
Dehydration, and frost, 404
   see Desiccation, Drought and Water stress
Dehydrogenase, and frost survival, 422, 425
Density,
   (plant spacing), 16, 60, 62, 464–467
   (of wood) *see* Specific gravity
Deoxyribonucleic acid, *see* DNA
Derivatives, of cambium, 262–271
Desert plants, 311, 318–322, 340, 344
Desiccation, of tissues, 322
Design, of progeny trials, 483
Diallel,
   of *Picea sitchensis*, 198
   of *Pinus strobus*, 478
   half-diallel, *Picea abies*, 215
   half-diallel, *Pinus sylvestris*, 212–213
Diameter growth,
   and dry matter distribution, 14, 16
   and cambial activity, 275–279
   and hormones, 296–299
   provenance and clone variation, 505–506
   and shoot growth, 159, 125, 128
   see also Cambium, Wood and Xylem

Differentiation,
   of cambium and wood, 263–279
   of xylem in relation to hormones, 296–297
Diffuse porous, 273–279
Diffusion, leaf resistances,
   components, 66
   and leaf age, 86
   and leaf models, 38
   species differences, 70–73
   and transpiration, 330–334
   and waxes, 318
   see also Photosynthesis, Stomata, Transpiration and Water
Diffusion, of oxygen,
   in soils, 363
   in roots, 368–370
Dinitrophenol, 375
Diploid,
   potato, 312
   seed coat, 475
Discriminant function, 427
Disease, 63, 124, 129
Distribution,
   of dry matter within plants, 12–14, 32, 39–42, 59, 119, 126, 299
   see also Allometry, Leaf and Root
   of yield among individuals, 14–17, 463–467
Diurnal, water deficits, 308–309
DNA, 177, 180, 181
Dominance (genetic), nutrient response, 456
Dominant trees, 15, 464–466
Dormancy of buds,
   and abscisic acid, 272, 297
   breaking, Chapter 13
   cambial activity following, 271–273
   correlative inhibition, 160–161
   and frost hardiness, 406, 408–411
   imposed dormancy, 409–410
   morphological indicators of, 504
   and photoperiodism, Chapter 11
   of pine buds, Chapter 12

Drainage, 362–363
Drought,
  and abscisic acid, 297
  adaptations, general, 5, 179
    for avoidance, 310–321
    for tolerance, 140, 143, 146, 321–322, 339–346
    effect on shoot and needle elongation, 192–193
  see also Water stress
Dry weight increment, see Allometry, Photosynthesis, Root and Shoot
Duncan-Stewart model, 36–37
Dune stabilization, 320
Dwarf shoots, see Branches, Buds, and Lateral Buds
Dye, oxidation of, 366, 370

E

Early progeny tests, Chapter 29
Earlywood,
  cambial growth and genetic improvement, 273–279
  tracheid diameter and wood production, 284, 286
  to latewood ratio, 277
Ecological viewpoints,
  net photosynthesis, Chapter 3
  tree form, Chapter 8
Ecosystem, 7–8
Ecotypes, see Species index, provenances
Edaphic factors, 29–30, 438, Chapters 25 and 26
Elastic modulus, of cell walls, 340–343
Electrical viability tests, 423, 425–427
Electrolytic viability test, 423
Electrophoresis, 82
Elevation, provenance adaptation to,
  general, 208
  Abies balsamea, 24–25
  Pinus sylvestris, 102–103
  Quercus, 127

Elongation of shoots,
  general, 183–189
  cessation and photoperiod, Chapter 11
  cessation and wood production, 275, 284–288
  and frost hardiness, 407, 410
  and hormones, 292–296
  see also Shoot
Embolized xylem, 374–375
Embryo, 262, 475
Energy,
  conversion, 122
  cost of branches, 146
  transfer in canopies, 145–151
  of water, 308–310
Enzymes, activity before bud swelling, 272
  frost damage, 421–425
  kinetics, 38
  in poplars, 88–93
  regulation of glycolysis, 398, Chapter 22
  synthesis and hormones, 296
  thermal stability, 39
Epicuticular waxes, 315–318
Epidermis,
  cell widths correlated with shoot elongation, 286
  thickness and transpiration, 313
Epinasty, 167
Ethanol, in anaerobic conditions, 363, 367, 388–398
Ethylene, in anaerobic conditions, 363–364
Eutrophic, marsh plants, 397
Evolution, of pine shoot development, 233–235
Exponential growth, 478
Extinction coefficient, 36
  see also Canopy and Leaf angle

F

F_1 hybrids, see Crosses, Diallel, Families, Hybrids and Species index

Subject Index

F$_2$ segregations, 217
Families, full-sib
   Picea abies, bud-set, 213–217
   Pinus elliottii, nutrient response, Chapter 26
   P. sylvestris, heights, 213
   P. taeda, growth simulation, 24, 31–32
   P. taeda, nutrient response, Chapter 26
   Populus tremula, heights, 208–211
   P. trichocarpa, heights, 211–212
Families, half-sib
   Pinus attenuata, lammas growth, 225, 227
   P. taeda, drought resistance, Chapter 20
   Platanus occidentalis, growth simulation, 32
   Populus tremula, heights, 208–211
   see also Diallel and Hybrids
Fascicles, see Needle
Fastigiate, 169, see also Branch
Fatty acids,
   and allelopathy, 320
   in flooded soils, 364
   and frost hardiness, 405
Feedback mechanisms,
   bud inhibition, 160
   competition, 466
   episodic extension, 41
   models of growth, 40
   photosynthesis, 23, 123
Female (maternal) effects,
   on nutrient response, 456
   on progeny performance, Chapter 28
Fertilization (of ovules), 475
Fertilizers, see Mineral nutrition, Nitrogen, Nutrition, Phosphorous, Potassium, etc.
Fibres,
   living as drought adaptation, 319
   matrix in wood, 268–279
   to vessel ratio, 277

Fixed (predetermined) growth,
   defined, 225
   general, Chapter 10, 175, 245
   hormone regulation, 295
   of pine shoots, Chapter 11
Flavonoids, and allelopathy, 320
Flood tolerance, Chapters 21 and 22
Flowering,
   of *Betula*, 439
   photoperiodic induction, 5, 207
   rootstock effect, 288
   of sorghum, 5
   of tobacco, 217
Flower stalk elongation, 292
Fluorescein di-acetate, viability test, 422
Fluorescence, viability test, 422
Flushing (bud-burst),
   of *Acacia albida*, 5
   of deciduous trees, 124–131
   and frost hardiness, 407, 410, 414–415
   and lateral bud formation, 195
   of northern conifers, 186
   of pines, including recurrent, 227–229, 234–240
   preceded by primordia growth, 272
   variation in *Picea abies*, Chapter 30
   see also Recurrent flushing
Food reserves, see Carbohydrates, and Reserves
Forest,
   communities, 15
   growth of, 14–16, Chapter 27
   leaf surface and water loss, 307
   nutrient cycle, 6
   photosynthesis, Chapter 3
   and tree form, 299
   yield improvement, Chapter 27, 479
   see also Canopy and Succession
Form of trees,
   and environment, Chapter 8
   inherent differences, 168–169
   mechanisms responsible, 157–168
   role of hormones, 300–301

Form of trees, *(Cont.)*
    silvicultural importance, 299
    see also Canopy
Foxtails, 194, 226–227, 234, 240
Free energy, 310
Free growth,
    of conifers, Chapters 11 and 13,
        176, 183–185
    defined, 225
    see also Recurrent flushing
Frequency distributions,
    of characters of *Picea abies*,
        505–511
    of tree size in forests, 465
Frost (hardiness),
    general, Chapter 23
    general phenology
        and avoidance, 124–130
        and flushing date, 186, 516
        and photoperiodic response, 210
        and pine shoot development,
            234, 239
    photosynthesis following, 100
    screening for frost hardiness,
        Chapter 24
    tolerance of maize, 4
    see also Chilling
Full-sib families, *see* Families
Fumaric acid, 396
Fungi, mycorrhizal, 366

G

GA, *see* gibberellic acid
Gametophyte, 475
Gas,
    in pressure bomb, 338–339
    in trees,
        composition, 372
        entry, 371–372
        movement, 372–373, 377–378
        pathways, 374–375
Gas chromatography – mass
        spectrometry, 292, 298
General combining ability,
    shoot growth, 190
    maternal effects, 474

Genes,
    loss of desirable ones, 39
    photoperiodic responses, 217
    single-gene inheritance, 169
    see also Polygenic inheritance
Genetic control,
    of bud break, 254–256
    of nutrient responses, Chapter 25
    of photoperiodic response, 217
    of "plasticity" in forests, 467
    of tree form, 299
Genetic gain,
    by clone selection, 505–516
    estimation in young trials,
        486–500
    see also Breeding, Selection criteria,
        Genetic improvement, and Tree
        improvement
Genetic improvement,
    in shoot growth, 182–183
    in wood yield and quality,
        275–279
    see also Breeding, Selection criteria,
        Genetic gain, and Tree
        improvement
Genetic linkage, 183, 515
Genetic stability of trees, 113
Genetic structure of forests, 464
        467–468
Genetic variances,
    analysis of trials, 491, 498
    characters of *Picea abies*, 511–512
    nutrient responses, 456
    see also Additive genetic variance,
        and Heritability
Genetic variation (within tree species),
    CO_2 compensation point, 114–116
    CO_2 exchange rate, 23–26
        of *Pinus sylvestris*, 101–102
        of *Populus*, 83–88
    characters of *Picea abies*, Chapter 30
    drought resistance, Chapter 20
    early height and diameter growth,
        Chapter 29
    enzyme activity of poplars, 83–88
    frost hardiness, 424

growth in dry weight, 26–28
nutrient responses, Chapters 25 and 26
photoperiodic responses, Chapter 11
shoots and canopies, 129
    deciduous trees, 124–129
    bud break, Chapter 14
    conifers, branch frequency, 198
    free growth, Chapter 13
    needle elongation, 192–193
    shoot apical growth, 178–182
    shoot elongation, 185–189
    tree form, 168–169
Geotropism, 162, 166–167, 301
    see also Gravity
Germination,
    and allelopathy, 321
    rate of, 484
    seedling growth of pines, 227, 234
Gibberellic acid,
    assay, 298
    bud break, 253, 256–258, 272
    photosynthesis, 23
    shoot elongation, 292–296
    tree form, 300–301
    xylem growth, 297–298
Girdling techniques, 373, 405
Glucose, 396, 398
Glutamic acid (glutamate), 395–396, 398
Glycerol, in flooded plants, 398
Glycogen, in flooded plants, 397
Glycolate,
    oxidase activity, 82
    and photorespiration, 111, 113
Glycolysis, regulation in flooded plants, Chapter 22
Glycosides,
    and allelopathy, 320
    from waterlogged bark, 389
Grafting, and growth, 9
    and root–shoot ratio, 320
Gravimetric soil water, 331–332
Gravimorphism, 160–161, 299–301
Growing season, *see* Season

Growth analysis, 9–10, 58–59, 120–124, 130–131, 478–479
Growth cessation, *see* Bud-set
Growth chambers, *see* Controlled environments
Growth cycle, of pine shoots, 224–230
Growth regulators (unspecified),
    and carboxylating enzymes, 41
    feedback inhibition of photosynthesis, 23
    frost-hardiness factor, 407
    and gravimorphism, 160
    hormone-directed transport, 300
    and peroxidase activity, 89
    see also, Abscisic acid, Auxins, Cytokinins and Gibberellic acid
Growth retardant, 294–295
Growth rings, 273–279, 284, 288

H

Half-sib families, *see* Families
Haploid gametophyte, 475
Hardiness, *see* Drought, Frost
Heartwood,
    gas in trees, 372, 375
    water transport, 308
Heat exchange, by leaves and canopies, 11, 61, 72, 146, 149–150
Heat sum, and initiation at apices, 177–179, 183
    and needle elongation, 191
    and shoot elongation, 184
Height-growth,
    analysis of, in conifers, Chapter 10
    cessation, 124–129, Chapter 11
    and "free" growth, 246–247
    and hormone levels in conifers, 292–296
    interpretation shortly after planting, Chapter 29
    and numbers of stem units, 182, 188–189, 236–240
    and nutrition, 453–456

554 Subject Index

Height-growth, *(Cont.)*
  variation in *Picea abies*, 505–506
  see also Bud set, Flushing and Shoot
Heritability,
  bud break, 255–256, 415
  competitive ability, 467
  estimation in young trials, Chapter 29, 486–500
  nutrient response, 455–457
  shoot and canopy characteristics, 124–129
Heterogeneity, of breeding population, 516
Heterosis, *see* Hybrid vigour
Höfler diagram, 336–339, 342
Homeostasis, spatial in forests, 466–467, 469
Horizons, of soil and flooding, 352–353, 363
Hormones, *see* Growth regulators, Abscisic Acid, Auxin, Cytokinins, and Gibberellic acid
Host, tree as fungal host, 439, 441–445
Humidity, stomatal response, 315, 340
  see Stomata and Water
Hybrids, by cell fusion, 3
  between clones of *Picea abies*, 214–217
    plus trees of *Pinus sylvestris*, 212–213
    provenances of *Picea abies*, 213
      Populus tremula, 209–211
      Populus trichocarpa, 211–212
    species of *Picea*, 27, 29
      Pinus, 183
      Populus, 114, 126, 169
      Salix, 379
  see also Families
Hybrid vigour,
  in *Picea sitchensis*, 189
  in *Populus tremula*, 210
  and prolonged apical activity, 183
Hydroactive stomatal closure, 340

Hydrogen acceptors, 398
Hydrogen cyanide, in waterlogged bark, 389
Hydrology, 7, 121, Chapter 19
  see also Water
Hydrostatic tension in trees, 62, 374, 378
Hypocotyl, and auxin, 292

I

IAA, *see* Auxin
Ice formation, inside and outside cells, 404
Ideotype,
  concept, 16
  crown form, 151–154, 168
Impedance, to evaluate frost damage, 423, 425–427, 430–432
Improvement, *see* Genetic Gain, Genetic improvement, Selection criteria and Tree improvement
Inbreeding, 113
Incipient plasmolysis, 337, 340
Indole-3-acetic acid, *see* Auxin
Ingestad solution, 248
Inherent differences, *see* Genetic variation
Inhibitors (unspecified),
  and bud break, 257
  and correlative inhibition, 159
  see also Abscisic acid
Initiation,
  of cambium, 262–268
  of lateral structures on conifers, 193–198, 230–233
    variation, 198–199
  of stem units at conifer apices, 175–183, 225–230
    environment effects on, 177–178
    genetic improvement, 182–183
    periods and rates of, 176–177, 237–240
    variation, 178–182

Inoculation, with mycorrhizal fungi, Chapter 25
Insect damage, to foliage, 63, 123–124
to seeds, 476–477
Intercalary meristems, 191–192
Interception of light, see Canopies
Interconversion, of hormones, 295
International Board of Plant Genetic Resources, 17–18
Internodes,
and cambial development, 263–268
of conifers, see Stem unit (= node plus internode)
elongation and bud burst, 272
elongation and hormones, 293
of poplars, 90–91
and tree form, 141
Interspecific hybrids, see Hybrids
Interxylary cork and periderm, 319
Ionic permeability, and viability, 423, 425
Iron,
in flooded soils, 363, 390
uptake, 366
Isoenzymes, in poplars, 82, 88–91
Isolates, of fungi, 443

J

Johannestriebe, see Lammas
Juglone, and allelopathy, 321
Juvenile growth,
interpretation, Chapter 29
maternal effects, 478
Juvenile–mature correlations, 513, 516
Juvenile–mature transition, 299–300
Juvenile wood, 276

K

Kinetics,
of enzymes, 22
of GA-induced growth, 296
Kinetin, see cytokinins

Knee roots, see Pneumatophores
Krantz syndrome, 6

L

Lactate, 398
Lactones, and allelopathy, 320
Lacunae, 370, 378
Lammas shoots, 176, 183–184, 190, 225
see also Free growth
Lateral buds, of conifers, 158, 193–198, 230–233
see also Branch, and Buds
Lateral leaf traces, 263–268
Latewood, 273–279, 284, 288
Latitude, see Cline, Photoperiod, and Species index, provenances
Leaching of nutrients, 7–8, see also Allelopathy
Leader, "retarded" on *Pinus radiata*, 293
see also Height growth and Shoot
Leaf, see also Needle
Leaf adaptation,
to drought, 311–318, 140–153
to sun and shade, 57, see also Shade
Leaf age,
and cambial development, 264–268
and photosynthesis, 12, 57, 84–86, 100
Leaf anatomy, see Anatomy
Leaf angles, (orientation) and light interception, 36, 88, 121, 124, 146–149
see also Canopy
Leaf area duration, 10, Chapter 7
Leaf area index,
role defined, 10–12
and canopy structure, 148
in growth models, 36, 58
and water use efficiency, 308
Leaf area per hectare, 307
Leaf area ratio, 9, 123, 128

Subject Index

Leaf buoyancy, viability test, 425, 428
Leaf—cambium relation, Chapter 15
Leaf dimensions and wood production, 286
Leaf fall, 124—131
Leafless habit,
   desert adaptation, 318—319
   pea varieties, 12
Leaf plastochron age, 84—93
Leaf plastochron index, 262—271
Leaf posture, see Leaf angle and Canopy
Leaf production and vessel formation, 277
Leaf ratios,
   to roots, 41
   to shoots, 41—42
   to total growth, 13—14
   see also Leaf area ratio, and Leaf area index
Leaf relative growth rate, 122
Leaf resistances, to CO_2 and water transfer, 38—39, 66—67, 70—73, 310—318, 330—345
   see also Diffusion, and Resistance
Leaf respiration, 38
Leaf retention,
   and assimilation, 100
   and fibre formation, 277
Leaf size,
   adaptation, 140—153, 311, 343—344
   and shoot elongation rate, 286
Leaf temperature, 72
Leaf traces, 262—268
Leaf water potential, 62, 330—346
Lenticels, gas entry, 371
Light,
   interception, 36, 87—88, 120—122,
      see also Leaf angle, Canopy
   photosynthetic response, 86—88
   saturation, 27, 60, 63, 70, 85, 101—102
   stomatal response, 314
   see also Shade
Lignification,
   and shoot elongation, 185, 213
   and branch angles, 166

Lignin, 104
Lignotubers, 140
Lipids, and frost hardiness, 404—405
Liquid chromatography, 298
Litter decomposition, 62
Longevity of trees, and forest succession, 9
Long shoots,
   of *Larix*, 245
   of *Pinus*, 158—160
   see also Branch
Lysigenous air spaces, 365

M

Magnesium, content of needles, 504, 509—511
Malate, malic acid and enzyme, in flooded plants, 367, 394—399
Male parent, effect of, 474—475
Manganese,
   uptake by pine, 438
   in waterlogged soils, 364, 390
Mass spectroscopy,
   gas assay, 388
   hormone assay, 292, 298
Maternal effects, Chapter 28
Matric potential, of water, 310
Matrix of fibres in wood, 268—279
Maturation, of leaves and cambial development, 262—268
Mechanical
   damage by frost, 404—405
   pressure on branches, 162—163
   stress and drought injury, 322
Medullary rays, gas in, 373—375
Meiosis, in cell culture, 3
Membranes, see Cell membranes
Mercuric chloride, 375
Meristems,
   cambial, see Cambium
   cortical and rib, 194
   intercalary, 191 see also Needle
   root in waterlogged soils, 390—391
   see also Root

Subject Index

shoot apical and sub-apical, of
    conifers, 175—189
  correlative inhibition, 158
  frost hardiness, 408
  effect of gibberellic acid, 296
  effect of gravity, 299, 301
  production and leaf biomass, 11, 193
  temperature adaptation, 4
  see also Apical meristem
Mesic sites, and crown forms, 140—141, 144—145, 152—153
Mesophyll,
  cell size, 39
  resistance, 66—67
Metabolism,
  adaptations to flooding, 366—367, Chapter 22
  of hormones, 295—296
Metaxylem, 264—271
Methyl-violet dye, 356
Michaelis-Menten relationship, 40
Microphylly, drought adaptation, 141, 153, 311
Mineral nutrition, Chapters 25 and 26
  see also Nitrogen, Nutrition, Phosphorus etc.
Mitochondria, and frost hardiness, 411
Mitosis, 176—177, 180—181, 272
  see also Apical meristem, Cell division and Meristem
Mixed cropping, 8
Models (mathematical),
  of correlative inhibition of buds, 160
  of plant growth, 33—45, 67—68
  for *Pinus sylvestris*, 104
  for *Populus*, 80
Monocultures, competition in, 464—467
Monocyclic, pine buds, 228—229
Morphogenesis,
  of leaf—cambium relation, 262—271
  of pine buds, 224—233

Morphology,
  and competition, 467, 470
  and drought resistance,
    leaves, 311—312, 315—318
    roots, 319—320, 351—353, 355—356
    stem, 318—319
  of leaf—cambium relation, 262—271
  variation in *Picea abies*, 504—508
  see also, Branch, Canopy, Leaf, Stem etc.
Mortality, *see* Death
Multigenic, *see* Polygenic
Multiline mixtures, 469
Multinodal (polycyclic) pine shoots, 228—229
Multivariate statistics, 93—95, 427
Muriate of potash, 452—453
Mutant, wilty tomato, 312
Mutation, 113
Mutual shading, *see* Shade
Mycorrhizas,
  and flood tolerance, 366
  and phosphate cycling, 6
  and tree nutrition, 438—445

N

NAD, $NADH_2$, 398
Needle,
  critical water content, 352
  elongation in pines, 189—193
    and wood quality, 275, 277
  frost hardiness, 429—430
  numbers produced, 176—183, 232—233
  and free growth, Chapter 13
  nutrient content, 504—505, 509—511
  see also Leaf and Canopy
Net assimilation rate, 9—11, 63, 104—105, 119—123
Niches, in tropical forests, 148
Ninhydrin, and frost damage, 423

Nitrate reductase,
  activity, and growth rate, 82, 91–93
  and flood tolerance, 367, 397
Nitrite, accumulation and viability, 422
Nitro-blue tetrazolium, viability test, 422
Nitrogen,
  concentration in needles, 505, 509–511
  gas in pressure bomb, 338
  nutrition, and growth of forest, 60
  and photosynthesis, 30, 101
  and tree growth, 450, 453–457, 460
  reserves, 13
  uptake by flood plants, 397
Nitrogen fixation, 6, 446
Nodal diaphragms, 178
Nucleic acids, and frost hardiness, 404
  see also DNA
Nucleolar bodies, produced by gibberellins, 296
Nucleosides, and allelopathy, 320
Nursery,
  free growth, 246
  management and planting stock size, 478, 484
  root growth, 352–353
Nutrient solution culture, see Water culture
Nutrition,
  adaptation, 5–6, 148, Chapter 25
  cycle of nutrients in forests, 6–8
  fertilizer response, Chapter 26
  and forest growth, 60
  internal competition among apices, 158–160
  leaching, 7–8
  and mycorrhizas, Chapter 25
  and photosynthesis, 30, 301
  and xeromorphy, 140
  see also Nitrogen, Phosphorus, Potassium etc.

O

Occlusion, of stomata, 318
Odour, and frost damage, 423, 431
Ontogeny,
  and carbon balance, 35
  and gibberellins in shoots, 295
  and leaf–cambium relation, 262–271
Organelles, and frost hardiness, 414
Organic acids,
  and allelopathy, 320
  in flooded roots, 367
Orientation, see Branch angle, Leaf angle, and Gravity
Osmotic pressure, 334–346
Osmotic stress, 353–355
Overdominance, see Hybrid vigour
Oxalo-acetate, 398
Oxidation,
  in rhizosphere, 366, 369, 373
  see also Respiration
Oxygen,
  deficiency in flood conditions, Chapters 21 and 22
  balance in roots, 368–370
  debt in trees, 396–399
  transport in trees, 387–388
  depression of photosynthesis, 84–85, 111
  in pressure bomb, 338

P

Pagoda, tree form, 141, 152–154
Palisade,
  cell widths and shoot growth, 286
  parenchyma and transpiration, 313
Parenchyma,
  aeration in wood, 374–375
  and transpiration, 313
Partitioning,
  of assimilates, see Allometry, and Distribution
  of CO_2 exchange, 35–39

Subject Index

Pathways, of gas in trees, 370–378
  see also Carbon dioxide, Diffusion, Resistance, and Water
Pericycle, gas voids in, 370
Periderm, interxylary, drought adaptation, 319
Periodicity, see Bud-set, Flushing, Phenology and Shoot
Permeability, see Cell membranes
Peroxidase activity, and growth, 82, 88–91
Peroxisomes, and photorespiration, 111
Phase angle, and electrical impedance, 426–427, 430
Phellogen, gas in, 372–373
Phenol-amine reactions, tissue death, 422
Phenology,
  of photosynthesis, 57, 100
  of shoot growth, apical activity, 176–182
  conifer shoot elongation, 183–111
  of deciduous trees, Chapter 7
  and frost hardiness, 406–411, 424
  variation in *Picea abies*, 508–509
  see also Bud set, Flushing, Season and Shoot
Phenols, and allelopathy, 320
Phloem, see Cambium and Protophloem
Phloem bridges, 267
Phospholipids, and frost hardiness, 405
Phosphorus (phosphate),
  concentration in needles, 504, 509–511
  cycling in forests, 6
  fertilizer responses, 60, Chapter 26
  leached from Zaire basin, 8
  ^{32}P uptake, 458–459
  response and mycorrhizas, Chapter 25
  and xeromorphy, 140
Photoactive, stomatal opening, 340

Photo-bleaching, of chlorophyll, 4, 57
Photo-oxidation, in leaves, 4, 57
Photoperiod, effect on:
  flowering of sorghum, 5
  frost hardiness, 406, 424
  gibberellic acid levels, 294–295, 300
  needle growth, 192
  peroxidase activity, 90
  shoot growth, apical growth, 178
    bud-break, 184, 254
    bud-set and height growth, 177, 185, 125–126, 128–129, Chapter 11
    free growth, 246–248
Photorespiration, and net photosynthesis, general, 59, 63, 66
  and CO_2 compensation point, Chapter 6
  of *Populus*, 82, 84–85
  of *Pinus sylvestris*, 103
Photosynthate, see ^{14}C-labelled assimilates and Distribution
Photosynthesis,
  and cambial origin, 263–265
  and drought tolerance, 331–334
  environmental factors effecting, 24–26, 59–72, 100–101
  and frost damage, 422, 425, 428, 430–431
  and frost hardiness, 411–414
  genetic variation,
    within species, 23–28, 58, 82–88, 102, 127
    between species, 63, 68–72
  models of, 31–45, 67–68
  plant factors affecting, see Acclimatization, Age, Canopy and Photorespiration
  seasonal pattern, 28–31, 100
  in stem and bark, 34, 103, 318–319
  and yield, 10, 26–28
Phyllodes, 143
Phyllotaxy, 196, 262–267

Physical parameters of water relations, Chapter 19
Phytochrome, 207
Phytotron, see Controlled environments
Pioneers, see Succession
Pith, growth and gibberellins, 296
Pits, of tracheids, 374–375
Plasmolysis, 423
Plasticity, in competition, 467–469
Plastochron, in conifers, 179–181, 199, 237
  see also Leaf plastochron age and index
Plus trees,
  of *Pinus sylvestris*, half-diallel, 212–213
  selected for crown form, 287
    height growth, 182
    wood quality, 287
  value, with regard to forest competition, 464, 469
    maternal effects, Chapter 28
Pneumatophores, 365, 371–372
Podsols, 7, 140
Polar gradient, of vessel differentiation, 269–274
Polarographic methods, 370
Pollen parent, effect of, 474–475
Polycyclic (multinodal), pine buds, 198, 228–229
Polyembryony, 476–477
Polyethyleneglycol, osmotic stress, 353–355
Polygenic inheritance, of crown form, 169
  frost hardiness, 415
  photoperiodic response, 217
  yield, 468–469
Polypeptides, and allometry, 320
Population genetics, 464
Population structure, Chapter 27, 503–508, 512–516
Porometer, 66
Porosity, of roots, 366–370
Potassium napthenate, synthetic auxin, 292

Potassium,
  concentration in needles, 504, 509–511
  nutrition, 450–454
Potential, of water, 309–310, 330–346
Predetermined shoot growth, see Fixed growth
Primary forests, 7, 140
Primary (juvenile) needles, of pines, 100, 227
Primary roots, gas pathways, 376–377
Primary vascular tissues, 262–268
Primordia, see Initiation, Fixed growth, Free growth and Shoot
Pressure,
  bomb apparatus, 334–339, 423, 427
  and branch angles, 162–163
  and gas movement in trees, 377–378
  osmotic, 334–346
  potential of water, 310
  turgor, 334–346
Procambium, 262–271
Progeny, see Families
Progeny tests,
  and fertilizer use, 460–461
  maternal effects, Chapter 28
  statistical analysis, Chapter 29
Proleptic shoots, 158, 169
Propagation effects, 499, 514
Propionic acid, 397
Protein,
  aggregation and drought injury, 322
  and frost hardiness, 404–405
Proton disposal, and anoxia, 392–399
Protophloem, 263–271
Protoplasm,
  augmentation and frost hardiness, 405–406, 408, 414, 417
  hydration, 309, 321–322
Protoxylem, 263–271
Provenance, see Species index
Pruning,
  self-pruning in stands, 299

shoot rejuvenation, 168
Psychrometer, 338–339
Pure lines, 469
Purification of hormone extracts, 298
Purines and allelopathy, 320
Pyruvate, 389–390, 398

Q

Quality,
  of seed, Chapter 28
  of wood, 275–279, 287–288, 461
Quinones and allelopathy, 320

R

Races of mycorrhizal fungi, 439, 441–445
  see also Families, and Species index for provenances
Radial growth, *see* Cambium
Radiation flux, *see* Light
Radio-isotopes,
  ^{14}C-labelled assimilates, 82, 104, 161, 412–413
  ^{14}C fixation in roots, 395–396
  ^{14}C-pyruvate, 389
  (^{3}H) -GA$_4$, 295–296
  ^{32}P uptake, 458–460
Ranking,
  of clones and provenances of *Picea*, 505, 511–514
  of families for nutrient response, 451–453, 457
  of provenances of *Picea*, and free growth, 250
Rays, gas movement in, 373–375
Reaction wood, 163–165
Recessive genes, for photoperiodic response, 217
Recurrent backcrosses in *Picea*, 27
Recurrent flushing, 183, 198, 226–229, 234–235
  effect of hormones, 295
Redox plane, in soils, 364
Regression, and covariance adjustment, 486–500

  see also Correlation
Relative growth rate,
  dry weight, 9
  height, 465, 486, 489–491, 495
Relative water content, 309
Repeatability (heritability) of clonal tests, 488–500
Reserves,
  of carbohydrate,
    and autumn foliage, 123
    attracted to buds, 161
    in desert plants, 308, 319
    and frost hardiness, 405, 413 417
    and plant growth, 13
    oxidation to glycolate, 113
    in roots, 397
    and spring growth, 42
  of nitrogen, 13
  of water in soil, 331–334
Resistances,
  of leaves to diffusion,
    components, 66–67, 38
    in conifer species, 70–73
    in poplars, 84–86
    role of waxes, 318
  of soil–plant system to water, 332–334
  of roots to oxygen diffusion, 368–369
  see also Diffusion, and Leaf resistances
Resistive impedance, frost damage, 426–427, 430
Respiration,
  anaerobic in roots, 395, Chapter 22, esp. 397–399
  component of net assimilation rate, 9–10
  and frost hardiness, 411
  maintenance and constructive, 37–38
  in models of growth, 34–35
  of *Pinus sylvestris*, 103–104
  of *Populus* leaves with age, 84–85, 265

Respiration, (Cont.)
  of reserves and water use efficiency, 308
  see also Photorespiration
Resting buds, see Dormancy
Retardants, 294—295 see also Abscisic acid and Inhibition
Retention of leaves,
  and growth of deciduous trees, 85, Chapter 7
  of pines, 100
  and formation of fibres, 277
Rhizosphere, 6, 366, 369, 373, see also Mycorrhizas
Rhizosphere oxidation, 366, 369, 373
Rhythmic events in cambial origin, 268
Ring porous wood, 273—279
RNA synthesis, 296
Rooting of cuttings, 254—255, 257
  see also Vegetative propagation
Roots,
  adventitious, see Adventitious roots
  exudates, see Allelopathy
  growth, in autumn, 123
    in response to nutrients, 440—446, 458, 460
    in spring before bud break, 253 256—258
  hormones from, and bud break, 253, 256—258, 272
  and photosynthesis, 41
  morphology, adaptations to flooding, 364—366, 369, 376—377
    adaptations to drought, 5, 319—320, 351—354
    and water potential, 333—334
  mortality, 42
  nitrogen reductase activity, 92
  respiration, 35, 38
  shoot : root ratio
    and drought avoidance, 319—320, 354, 357
    and growth, 39—41, 101, 126

Rootstocks, 283, 288, 320
Rosette habit, 209, 211
Rotation, 80, 131

S

Sand culture, 437—438, 458—459
Sap,
  concentration and turgor, 340—341
  contents in flooded plants, 396—399
  gibberellic acid in, 294
Sapwood,
  gas in, 372, 374—375
  water reservoir, 62
  water transport, 308
Saturation
  deficit, 309—310
  light and photosynthesis, 27, 60, 63, 70, 85, 101—102
Scales, 175, 177, 230
Scanning electron microscope, 316—317
Scion, see Rootstock
Schizogenous air spaces, 365
Sclereids, 178
Sclerophyll, 140, 144, 146, 152—153, 319
Screening, see Selection criteria
Season (seasonal),
  adaptation, general, 4—5
  cambial activity and wood production, 272, 276—279
  changes in frost hardiness, 406—411
  needle growth, 191—193
  photosynthesis and dry matter production, 27—31, 42, 58, 100
  shoot development, 178—182, 184—188, 235—240
  transpiration, 67
  see also Phenology
Secondary cambial development, 268—271
Secondary pioneers, 9, 141—142

Subject Index

Seed,
    coat thickness, 476–477
    germination and photoperiod, 207
    production, 127, 233–234, 439
    size and growth, 474–475, 501
    stratification, 476–477
    X-ray treatment, 101
Seed orchards,
    and nutrient responses, 460–461
    and seed size, 477
Selection,
    problems of early assessment, Chapter 29
    maternal factors, Chapter 28
    multiple characters, 511–512
    yield of forest stand, 464, 469
Selection criteria,
    canopy characteristics, Chapter 7
    chemical, 301
    drought avoidance, 345–346, 358
    drought tolerance, 339–345
    dry matter distribution, 14
    enzymatic, 88–93
    flood tolerance, 379, 399
    frost hardiness, 416–417, Chapter 24
    height growth, 182, 208, 250
    leaf area duration, 126
    mycorrhizal formation, 443
    photosynthesis, 23, 31, 39, 83–88
    respiration and photorespiration, 38, 113–116
    for wood volume and quality, 275–279, 287–288
    see also Breeding, Genetic improvement and Tree improvement
Selection efficiency, 488
Selection intensity, 487–489, 491–493
Semi-permeability, *see* Cell membranes
Senescene, *see* Age, and Leaf age
Seral species, 233
Shade,
    mutual shading, and carbon fixation rates, 11–12, 27, 63, 101–102
    within crowns, 57, 63, 87–88, 120–121, 147–150
    sun and shade leaves, 30, 57–58, 86–87, 100–101, 141–142
    tolerance, of coffee, tea and cocoa, 3
    conifer species, 68–71
    Populus clones, 86–88
    Pinus sylvestris, 101–102
Shedding of leaves to avoid drought, 311
    see also Abscission and Retention
Shikimic acid, Shikimate, 367, 398
Shoot,
    analysis of growth in conifers, Chapter 10
    axis development in pines, 224–230
    elongation, characteristics in deciduous trees, 122–129
    episodic, 41
    after flooding, 393
    and frost hardiness, 407, 410
    and hormones, 292–296
    of northern conifers, 183–189
    and wood production, 272, 284–285, 287–288
    and wood quality, 277–279
    see also Height growth
Shoot : root ratio, *see* Roots
Short shoots, of pines, 232, *see also* Needles
Siblings, *see* Families
Silviculture, intensive systems, 79
Simulation, *see* Models
Single-ion-scanning, hormone assay, 298
Sinks,
    number, size and dry matter production, 12–13, 23, 41
    sink–source relationships, 23, 120, 123
Site,
    mesic and xeric, 140–141, 145, 149, 152–153
    and nutrition, Chapter 26, 438, 451, 454, 460

Snow, 148, 150, 163–164, 168
Sodium, uptake by sugar beet, 4
Soil,
   evolution of CO_2, 62
   fertility, Chapters 25 and 26, 438, 449–455, 460–461
   temperature, 4, 253, 256–258
   waterlogged, Chapters 21 and 22
     aeration, 363
     oxygen demand, 368
     pore spaces, 363
     water relations, 330–334
Solar angle (geometry), 36–37, 60, 148
Solutes, in cells, 340–341
   water potential, 310
Spatial,
   arrangement of lateral buds, 195–196
   distribution
     of dominant trees, 15, 466
     of roots in soil, 352–353
Specific crosses, and nutrient response, 451–452, 457, 461
Specific gravity of wood,
   and nutrition, 461
   selection for, 277–279, 288
   in successional species, 141–147
   see also Earlywood and Latewood
Spectrometer, 505
Spring, *see* Season
Spring flush, *see* Flushing
Stability,
   of populations, 515–516
   of yield, 468
   see also Wind
Stands, *see* Forest
Starch, *see* Carbohydrates
Statistics, progeny trial analysis, Chapter 29
Stem,
   elongation, *see* Shoot
   frost hardiness of, 429–430
   photosynthesis, 34, 103, 318–319
   respiration, 38
   see also Form of trees

Stem units,
   defined, 224
   elongation, 183–189, 224–230
   initiation, 175–183, 237–240
   see also Free growth, Needles and Shoot
Stem yield efficiency, 119, 121, 126
Sterile culture, 439–440
Steroids, and allelopathy, 320
Stipules, 311
Stocking density, *see* Density
Stomata,
   characteristics and drought resistance
     closure response, 312–315, 332, 340, 346, 356
     size and frequency, 312, 355–356
     waxes and morphology, 315–318, 355–356
   resistance to CO_2 transfer, 38–39, 66, 82
   see also Diffusion and Resistances
Strains of fungi, 439, 441–445
Stratification of seed, 476–477, 484, 501
Street planting, 288
Stress, *see* Drought, Frost, Mechanical, Nutrition, Temperature and Water
Strobilus, *see* Cone
Sub-apical meristem, 185, *see also* Meristem
Succession, 7–9, 140–147, 233
Succinate, 397
Sucrose (soluble sugars), 177, 276, 396, 404, 407
Sulphides (sulfides),
   and allelopathy, 320
   in flooded soils, 363, 390
Summer, *see* Season
Summer shoots, of pines, 228, 232–240
Sun leaves, *see* Shade
Superphosphate, 452–454

Subject Index

Suppression, forest competition, 464–466, 479
Swamp, 140
Sylleptic shoots, 159
Symbiosis, mycorrhizas, 439, Chapter 25
Symplasm, water in, 336–345
Synthesis, of hormones, 294–295, Chapter 17

T

Tannins, and allelopathy, 320
Taxonomy, use of leaf waxes, 316
Temperature,
    adaptation, general, 3, 24–25
    effect on CO_2 compensation point, 114–115
        net photosynthesis, 24–25, 61, 70–72
        root respiration, 368
        shoot growth, 126, 129, 177, 253, 256–258
        stomatal response, 314–315, 340
    see also Frost hardiness, Heat sum, Season and Winter
Tension wood, *see* Reaction wood
Terpenoids, and allelopathy, 320
Thermocouple psychrometer, 338–339
Thermoperiod, 128
Thinning, 15, 470–471
Tissue culture, 276, 292, 298
Tissues, frost hardiness of, 406, 429–430
Tortuosity,
    leaf resistances to CO_2 and water, 315–318
    root resistances to oxygen, 368
Totipotency, of cambium, 276
Toxins, in anaerobic conditions, 363–364, 389–390
$TPNH_2$, 22
Trace, leaf traces, 262–268
Tracheids,
    ^{14}C incorporation, 412–413

    and branch angles, 166
    gas pathways, 374
    matrix, 274–279
Translocation,
    of frost hardiness factor, 407
    of hormones from roots, 256–258
    of phosphorus, 460
    of photosynthates, and ageing, 167–168
    and leaf dry matter content, 309
    hormone directed, 300
    to new primordia, pathways, 265–267
    resistances in models of growth, 40
    to xylem, 104, 412–413
    see also ^{14}C-labelled assimilates
    of reserves, 13, *see also* Reserves
Transpiration,
    and crown architecture, 145–146
    from forests, 307
    and leaf resistance, 66–72
    of *Pinus taeda* families, 356–357
    see also Chapters 18 and 19, Drought, Resistance, Water
Transport of oxygen in trees, 387–388, Chapter 21
    see also Water
Tree improvement,
    of flood tolerance, 379–380
    of nutrient response, 446, Chapter 26
    strategy and clone selection, 515–516
    of water relations, Chapter 18
    of wood volume and quality, 275–279
    see also Breeding, Genetic gain, Genetic improvement and Selection criteria
Tree line, 143
Tri-phenyl tetrazolium chloride, 422, 425
Tropical,
    deciduous trees, 122

Tropical, (Cont.)
   pine shoot development, Chapter 11
   tree crowns, Chapter 8
Tropisms, geotropism, 162, 166–167, 301
Tube,
   cultures of *Betula*, 440
   root studies, 351
Turgor, 309–310, 315
   negative, 343–345
Tyrosine, 396

U

Umbra (penumbra), 148–149
Understorey plants, 63, 143, 146
Unicyclic (uninodal) pine buds, 198, 228–229
Uridine monophosphate, 396

V

Vapour,
   pressure deficit, 61, 67, 70–73, *see also* Chapters 18 and 19
   transport in soil, 332
Variances,
   analysis of clone and progeny trials, 485–500
   partitioned for clones and provenances, 507
   nutrient responses, 455–456
   see also Additive genetic variances, Genetic variances and Heritability
Variation, *see* Genetic variation (within tree species)
Vascular,
   bundle sheaths in C_4 plants, 111–113
   cell growth and gibberellins, 296
   connections and correlative inhibition, 159–160
   traces, 263–268
   see also Cambium

Vegetative propagation,
   and selection for flood tolerance, 364
   and tree improvement, 515–516
   see also Rooting of cuttings
Vessels,
   size and frequency in apple, 283
   in xylem, general, 268–279
      and gas pathways, 374–376
Viability, tests, 419–427
Volume growth, *see* Cambium and Wood

W

Water,
   conduction, leaf and vessel development, 271
   culture (nutrient culture), 353–355, 391–392
   deficits, *see* Water stress
   loss per acre of forest, 307
   potential, defined, 309–310, *see also* Water stress
   regimes, adaptation to, 5
   relations and growth, general, 61–62
      physical parameters, Chapter 19
      and tree improvement, Chapter 18
   relative water content, 334–346
   reservoir, in sapwood, 62
   resistances to transfer, *see* Resistances
   stress,
      and abscisic acid, 297
      bud development, 177–178
      CO_2 compensation point, 115
      characterization of, 309–310
      development of, 308–309
      drought resistance, 313–314, 353–357
      elongation of shoots and needles, 192–193
      episodic shoot growth, 4
      frost damage, 423, 426–427

Subject Index

photosynthesis, 62, 67, 70, 101
    C_4 pathway, 59
stomatal response, see Stomata
soil water, 143, 330–334
water table, 364–365, 388
water use efficiency, 146, 307–308, 315
Waterlogging, Chapters 21 and 22
Waxes, epicuticular, 315–318
Whorls, see Branch
Wilting, 310, 312
    permanent wilting point, 334
Wind,
    damage, 148, 150
    and evapotranspiration, 145
    stability, 362
    and stomatal response, 315
Winter,
    buds of pines, Chapter 11, see also buds
    dieback of deciduous trees, 128–129
    initiation of stem units, 229
    photosynthesis, 29–30, 102
    see also Frost hardiness, Season and Snow
Wood,
    anatomy, see Chapters 15 and 16
    to bark ratio, 283
    density, see Specific gravity
    production,
        and crown structure, 121, 143
        genetic improvement prospects, 261, 275–279
        link with photosynthesis, 413
        prediction from seedlings, Chapter 16
    quality, 275–279, 287–288, 461
    see also Cambium, Compression wood, Earlywood, Latewood and Reaction wood

X

Xeric sites, 149
Xeromorphy, 311–318, 140, 152–153
X-ray,
    analysis of seeds, 475–476
    autoradiography, 412–413
    treatment of seed, 101
Xylem,
    aerenchyma, 365
    development, 262–271, 412–413
    gas in, 370–378
    growth, see Cambium
    sap hormones, 294
    water, 334–346

Y

Yield,
    analysis of components, 2–3
    of forests as opposed to trees, Chapter 27
    and height growth, 236
    and sink size, 13
    see also Components of yield, and Wood

Z

Zygote, 478